The Problem with Science

The Problem with Science

The Reproducibility Crisis and What to Do About It

R. BARKER BAUSELL

UNIVERSITY PRESS

Oxford University Press is a department of the University of Oxford. It furthers
the University's objective of excellence in research, scholarship, and education
by publishing worldwide. Oxford is a registered trade mark of Oxford University
Press in the UK and certain other countries.

Published in the United States of America by Oxford University Press
198 Madison Avenue, New York, NY 10016, United States of America.

© Oxford University Press 2021

All rights reserved. No part of this publication may be reproduced, stored in
a retrieval system, or transmitted, in any form or by any means, without the
prior permission in writing of Oxford University Press, or as expressly permitted
by law, by license, or under terms agreed with the appropriate reproduction
rights organization. Inquiries concerning reproduction outside the scope of the
above should be sent to the Rights Department, Oxford University Press, at the
address above.

You must not circulate this work in any other form
and you must impose this same condition on any acquirer.

Library of Congress Cataloging-in-Publication Data
Names: Bausell, R. Barker, 1942– author.
Title: The problem with science : the reproducibility crisis and what to do about it /
R. Barker Bausell, Ph.D.
Description: New York, NY : Oxford University Press, [2021] |
Includes bibliographical references and index.
Identifiers: LCCN 2020030312 (print) | LCCN 2020030313 (ebook) |
ISBN 9780197536537 (hardback) | ISBN 9780197536551 (epub) |
ISBN 9780197536568
Subjects: LCSH: Science—Research—Methodology.
Classification: LCC Q126.9 .B38 2021 (print) | LCC Q126.9 (ebook) |
DDC 507.2/1—dc23
LC record available at https://lccn.loc.gov/2020030312
LC ebook record available at https://lccn.loc.gov/2020030313

DOI: 10.1093/oso/9780197536537.001.0001

1 3 5 7 9 8 6 4 2

Printed by Sheridan Books, Inc., United States of America

Contents

A Brief Note	vii
Acknowledgments	ix
Introduction	1

I. BACKGROUND AND FACILITATORS OF THE CRISIS

1. Publication Bias	15
2. False-Positive Results and a Nontechnical Overview of Their Modeling	39
3. Questionable Research Practices (QRPs) and Their Devastating Scientific Effects	56
4. A Few Case Studies of QRP-Driven Irreproducible Results	91
5. The Return of Pathological Science Accompanied by a Pinch of Replication	109

II. APPROACHES FOR IDENTIFYING IRREPRODUCIBLE FINDINGS

6. The Replication Process	133
7. Multiple-Study Replication Initiatives	152
8. Damage Control upon Learning That One's Study Failed to Replicate	173

III. STRATEGIES FOR INCREASING THE REPRODUCIBILITY OF PUBLISHED SCIENTIFIC RESULTS

9. Publishing Issues and Their Impact on Reproducibility   193

10. Preregistration, Data Sharing, and Other Salutary Behaviors   222

11. A (Very) Few Concluding Thoughts   261

Index   271

A Brief Note

This book was written and peer reviewed by Oxford University Press before the news concerning the "problem" in Wuhan broke, hence no mention of COVID-19 appears in the text. Relatedly, since one of my earlier books, *Snake Oil Science: The Truth About Complementary and Alternative Medicine*, had been published by Oxford more than a decade ago, I had seen no need to pursue this line of inquiry further since the bulk of the evidence indicated that alternative medical therapies were little more than cleverly disguised placebos, with their positive scientific results having been facilitated by substandard experimental design, insufficient scientific training, questionable research practices, or worse. So, for this book, I chose to concentrate almost exclusively on a set of problems bedeviling mainstream science and the initiative based thereupon, one that has come to be called "the reproducibility crisis."

However, as everyone is painfully aware, in 2020, all hell broke loose. The internet lit up advocating bogus therapies; the leaders of the two most powerful countries in the world, Xi Jinping and Donald Trump, advocated traditional Chinese herbals and a household cleaner, respectively; and both disparaged or ignored actual scientific results that did not support their agendas. Both world leaders also personally employed (hence served as role models for many of their citizens) unproved, preventive remedies for COVID-19: traditional Chinese herbal compounds by Xi Jinping; hydroxychloroquine (which is accompanied by dangerous side effects) by Donald Trump.

This may actually be more understandable in Xi's case, since, of the two countries, China is undoubtedly the more problematic from the perspective of conducting and publishing its science. As only one example, 20 years ago, Andrew Vickers's systematic review team found that 100% of that country's alternative medical trials (in this case acupuncture) and 99% of its conventional medical counterparts published in China were positive. And unfortunately there is credible evidence that the abysmal methodological quality of Chinese herbal medical research itself (and not coincidentally the almost universally positive results touting their efficacy) has continued to this day.

To be fair, however, science as an institution is far from blameless in democracies such as the United States. Few scientists, including research methodologists such as myself, view attempting to educate our elected officials on scientific issues as part of their civic responsibility.

So while this book was written prior to the COVID-19 pandemic, there is little in it that is not relevant to research addressing future health crises such as this (e.g., the little-appreciated and somewhat counterintuitive [but well documented] fact that early findings in a new area of inquiry often tend to be either incorrect or to report significantly greater effect sizes than follow-up studies). It is therefore my hope that one of the ultimate effects of the reproducibility crisis (which again constitutes the subject matter of this book) will be to increase the societal utility of science as well as the public's trust therein. An aspiration that will not be realized without a substantial reduction in the prevalence of the many questionable research behaviors that permit and facilitate the inane tendency for scientists to *manufacture* (and publish) false-positive results.

Acknowledgments

First, I would like to thank the many conscientious and gifted researchers, methodologists, and statisticians whose insightful work informed this book. I have primarily depended upon their written word and sincerely hope that I have effectively captured the essence of their positions and research. I have not listed individual scientists in this acknowledgment since I have cited or been influenced by so many that I am reluctant to single out individuals for fear of omitting anyone (or including anyone who would prefer not to be so listed).

Finally, I would like to acknowledge my excellent Oxford University Press team who were extremely helpful and competent, and without whom the book would have never seen the light of day. Joan Bossert, Vice President/Editorial Director, for her support, who saw the promise in my original manuscript, selected very helpful peer reviewers, and guided me through the revision process. Phil Velinov, Assistant Editor, who seamlessly and competently coordinated the entire process. Suma George, Editorial Manager, who oversaw production. I would also like to extend my appreciation for my former editors at Oxford: Abby Gross and a special shout-out to the retired Marion Osmun—Editor Extraordinaire.

Introduction

This is a story about science. Not one describing great discoveries or the geniuses who make them, but one that describes the labors of scientists who are in the process of reforming the scientific enterprise itself. The impetus for this initiative involves a long-festering problem that potentially affects the usefulness and credibility of science itself.

The problem, which has come to be known as the *reproducibility crisis*, affects almost all of science, not one or two individual disciplines. Like its name, the problem revolves around the emerging realization that much—perhaps most—of the science being produced cannot be reproduced. And scientific findings that do not replicate are highly suspect if not worthless.

So, three of the most easily accomplished purposes of this book are

1. To present credible evidence, based on the published scientific record, that there exists (and has existed for some time) a serious reproducibility crisis that threatens many, if not most, sciences;
2. To present a menu of strategies and behaviors that, if adopted, have the potential of downgrading the problem from a crisis to a simple irritant; and
3. To serve as a resource to facilitate the teaching and acculturation of students aspiring to become scientists.

The book's potential audience includes

1. Practicing scientists who have not had the time or the opportunity to understand the extent of this crisis or how they can personally avoid producing (and potentially embarrassing) irreproducible results;
2. Aspiring scientists, such as graduate students and postdocs, for the same reasons;
3. Academic and funding administrators who play (whether they realize it or not) a key role in perpetuating the crisis; and

4. Members of the general public interested in scientific issues who are barraged almost daily with media reports of outrageously counterintuitive findings or ones that contradict previously ones.

Some readers may find descriptors such as "crisis" for an institution as sacrosanct as science a bit hyperbolic, but in truth this story has two themes. One involves a plethora of wrongness and one involves a chronicling of the labors of a growing cadre of scientists who have recognized the seriousness of the problem and have accordingly introduced evidence-based strategies for its amelioration.

However, regardless of semantic preferences, this book will present overwhelming evidence that a scientific crisis does indeed exist. In so doing it will not constitute a breathless exposé of disingenuous scientific blunders or bad behavior resulting in worthless research at the public expense. Certainly some such episodes compose an important part of the story, but, in its totality, this book is intended to educate as many readers as possible to a serious but addressable societal problem.

So, in a sense, this is an optimistic story representing the belief (and hope) that the culture of science itself is in the process of being altered to usher in an era in which (a) the social and behavioral sciences (hereafter referred to simply as the *social sciences*) will make more substantive, reproducible contributions to society; and (b) the health sciences will become even more productive than they have been in past decades. Of course the natural and physical sciences have their own set of problems, but only a handful of reproducibility issues from these disciplines have found their way into the present story since their methodologies tend to be quite different from the experimental and correlational approaches employed in the social and health sciences.

For the record, although hardly given to giddy optimism in many things scientific, I consider this astonishing 21st-century reproducibility awakening (or, in some cases, reawakening) to be deservedly labeled as a *paradigmatic shift* in the Kuhnian sense (1962). Not from the perspective of an earth-shattering change in scientific theories or worldviews such as ushered in by Copernicus, Newton, or Einstein, but rather in a dramatic shift (or change) in the manner in which scientific research is *conducted* and *reported*. These are behavioral and procedural changes that may also redirect scientific priorities and goals from a cultural emphasis on *publishing* as many professional articles as humanly possible to one of ensuring that what is published is *correct, reproducible*, and hence has a chance of being at least potentially useful.

However, change (whether paradigmatic or simply behavioral) cannot be fully understood or appreciated without at least a brief mention of what it replaces. So permit me the conceit of a very brief review of an important methodological initiative that occurred in the previous century.

The Age of Internal and External Validity

For the social sciences, our story is perhaps best begun in 1962, when a research methodologist (Donald T. Campbell) and a statistician (Julian C. Stanley) wrote a chapter in a handbook dealing with research on teaching of all things. The chapter garnered considerable attention at the time, and it soon became apparent that its precepts extended far beyond educational research. Accordingly, it was issued as an 84-page paperback monograph entitled *Experimental and Quasi-Experimental Designs for Research* (1966) and was promptly adopted as a supplemental textbook throughout the social sciences.

But while this little book's influence arguably marked the methodological coming of age for the social sciences, it was preceded (and undoubtedly was greatly influenced) by previous methodology textbooks such as Sir Ronald Fisher's *The Design of Experiments* (1935), written for agriculture researchers but influencing myriad other disciplines as well, and Sir Austin Bradford Hill's *Principles of Medical Statistics* (1937), which had an equally profound effect upon medical research.

The hallmark of Campbell and Stanley's remarkable little book involved the naming and explication of two constructs, *internal* and *external validity*, accompanied by a list of the research designs (or architecture) that addressed (or failed to address) the perceived shortcomings of research conducted in that era. Internal validity was defined in terms of whether or not an experimental outcome (generally presumed to be positive) was indeed a function of the intervention rather than extraneous events or procedural confounds. External validity addressed the question of:

> To what populations, settings, treatment variables, and measurement variables can this effect [presumably positive or negative] be generalized? (p. 5)

Of the two constructs, internal validity was the more explicitly described (and certainly the more easily addressed) by a list of 12 "threats" thereto—most of which could be largely avoided by the random assignment of participants to experimental conditions. External validity, relevant only if internal validity was ensured, was so diffuse and expansive that it was basically given only lip service for much of the remainder of the century. Ironically, however, the primary arbiter of external validity (replication) also served as the same bottom line arbiter for the reproducible–irreproducible dichotomy that constitutes the basic subject matter of this book.

Campbell and Stanley's basic precepts, along with Jacob Cohen's (1977, 1988) seminal (but far too often ignored work) on statistical power, were subsequently included and cited in hundreds of subsequent research methods textbooks in just about every social science discipline. And, not coincidentally, these precepts influenced much of the veritable flood of methodological work occurring during the next several decades, not only in the social sciences but in the health sciences as well.

Unfortunately, this emphasis on the avoidance of structural (i.e., experimental design) at the expense of procedural (i.e., behavioral) confounds proved to be insufficient given the tacit assumption that if the architectural design of an experiment was reasonably sound and the data were properly analyzed, then any positive results accruing therefrom could be considered correct 95% of the time (i.e., the complement of the statistical significance criterion of $p \leq 0.05$). And while a vast literature did eventually accumulate around the avoidance of these procedural confounds, less attention was paid to the possibility that a veritable host of investigator-initiated questionable research practices might, purposefully or naïvely, artifactually produce false-positive, hence irreproducible, results.

From a scientific cultural perspective, this mindset was perhaps best characterized by the writings of Robert Merton (1973), a sociologist of science whose description of this culture would be taken as Pollyannaish satire if written today. In his most famous essay ("Science and the Social Order") he laid out "four sets of institutional imperatives—universalism, communism [sharing of information not the political designation], disinterestedness, and organized skepticism—[that] are taken to comprise the ethos of modern science" (p. 270).

Scientific ethos was further described as

> [t]he ethos of science is that affectively toned complex of values and norms which is held to be binding on the man of science. The norms are expressed in the form of prescriptions, proscriptions, preferences, and permissions.

They are legitimatized in terms of institutional values. These imperatives, transmitted by precept and example and reinforced by sanctions, are in varying degrees internalized by the scientist, thus fashioning his scientific conscience or, if one prefers the latter-day phrase, his superego. (1973, p. 269, although the essay itself was first published in 1938)

While I am not fluent in Sociologese, I interpret this particular passage as describing the once popular notion that scientists' primary motivation was to discover truth rather than to produce a publishable p-value ≤ 0.05. Or that most scientists were so firmly enculturated into the "ethos" of their calling that any irreproducible results that might accrue were of little concern given the scientific process's "self-correcting" nature.

To be fair, Merton's essay was actually written in the 1930s and might have been somewhat more characteristic of science then than in the latter part of the 20th and early 21st centuries. But his vision of the cultural aspect of science was prevalent (and actually taught) during the same general period as were internal and external validity. Comforting thoughts certainly, but misconceptions that may explain why early warnings regarding irreproducibility were ignored.

Also in fairness, Merton's view of science was not patently incorrect: it was simply not sufficient. And the same can be said for Campbell and Stanley's focus on internal validity and the sound research designs that they fostered. Theirs might even qualify as an actual methodological paradigm for some disciplines, and it was certainly not incorrect. It was in fact quite useful. It simply was not sufficient to address an as yet unrecognized (or at least unappreciated) problem with the avalanche of scientific results that were in the process of being produced.

So while we owe a professional debt of gratitude to the previous generation of methodologists and their emphasis on the necessity of randomization and the use of appropriate designs capable of negating most experimental confounds, it is now past time to move on. For this approach has proved impotent in assuring the reproducibility of research findings. And although most researchers were aware that philosophers of science from Francis Bacon to Karl Popper had argued that a quintessential prerequisite for a scientific finding to be valid resides in its reproducibility (i.e., the ability of other scientists to replicate it), this crucial tenet was largely ignored in the social sciences (but taken much more seriously by physical scientists—possibly because they weren't required to recruit research participants). Or perhaps it was simply due to their several millennia experiential head start.

In any event, ignoring the reproducibility of a scientific finding is a crucial failing because research that is not reproducible is worthless and, even worse, is detrimental to its parent science by (a) impeding the accumulation of knowledge, (b) squandering increasingly scarce societal resources, and (c) wasting the most precious of other scientists' resources—*their* time and ability to make *their* contributions to science. All failings, incidentally, that the reproducibility initiative is designed to ameliorate.

Another Purpose of This Book

While this book is designed to tell a scientific story, to provide practicing scientists with a menu of strategies to adopt (and behaviors to avoid) for assuring that *their* research can be reproduced by other scientists, or even to serve as a resource for the teaching of reproducibility concepts and strategies to aspiring scientists, these are not the ultimate purposes of the reproducibility initiative—hence not mine either. For while knowledge and altruistic motives may have some traction with those contemplating a career in science or those desiring to change their current practices, such resolutions face powerful competition in the forms of career advancement, families, and the seductive charms of direct deposit.

The ultimate purpose of the myriad dedicated methodologists whose work is described herein involves a far more ambitious task: the introduction of a *cultural* change in science itself to one that demands not only the avoidance of behaviors specifically designed to produce positive results, but also the adoption of a number of strategies that require additional effort and time on the part of already stressed and overworked scientists—a culture dedicated to the production of *correct* inferences to the extent that John Ioannidis (2005) will someday be able to write a rebuttal to his pejorative (but probably accurate) subhead from "Most Research Findings Are False for Most Research Designs and for Most Fields" to "False Positive Results Have Largely Disappeared from the Scientific Literatures." A culture in which the most potent personal motivations are not to produce hundreds of research publications or garner millions of dollars in research funding but to contribute knowledge to their scientific discipline that was previously unknown to anyone, anywhere. And conversely, a culture in which (short of actual fraud) the most embarrassing professional incident that can occur for a scientist is for his or her research to be declared irreproducible due to avoidable questionable research practices when other scientists attempt to replicate it.

The Book's Plan for a Very Small Contribution to This Most Immodest Objective

Almost everyone is aware of what Robert Burns and John Steinbach had to say about the plans of mice and men, but planning is still necessary even if its objective is unattainable or nonexistent. So the book will begin with the past and present conditions that facilitate the troubling prevalence of irreproducible findings in the scientific literature (primarily the odd fact that many disciplines almost exclusively publish positive results in preference to negative ones). Next a *very* brief (and decidedly nontechnical) overview of the role that p-values and statistical power play in reproducibility/irreproducibility along with one of the most iconic modeling exercises in the history of science. The next several chapters delineate the behavioral *causes* (i.e., questionable research practices [QRPs]) of irreproducibility (accompanied by suggested solutions thereto) followed by a few examples of actual scientific pathology which also contribute to the problem (although hopefully not substantially). Only then will the replication process itself (the ultimate arbiter of reproducibility) be discussed in detail along with a growing number of very impressive initiatives dedicated to its widespread implementation. This will be followed by equally almost impressive initiatives for improving the publishing process (which include the enforcement of preregistration and data-sharing requirements that directly impact the reproducibility of what is published). The final chapter is basically a brief addendum positing alternate futures for the reproducibility movement, along with a few thoughts on the role of education in facilitating the production of reproducible results and the avoidance of irreproducible ones.

The Sciences Involved

While the book is designed to be multidisciplinary in nature, as mentioned previously it unavoidably concentrates on the social and health sciences. An inevitable emphasis is placed on psychological research since that discipline's methodologists have unquestionably been leaders in (but by no means the only contributors to) reproducibility thought and the implementation of strategies designed to ameliorate the unsettling preponderance of false-positive results. However, examples from other sciences (and contributions from their practitioners) are provided, including laboratory and preclinical research (on which some of the most crucial human experimentation is often based) with even a nod or two to the physical sciences.

But while some of the book's content may seem irrelevant to practitioners and students of the purely biological and physical sciences, the majority of the key concepts discussed are relevant to almost all empirically based disciplines. Most sciences possess their own problems associated with publishing, the overproduction of positive results, statistical analysis, unrecognized (or hidden) questionable research practices, instrumental insensitivity, inadequate mentoring, and the sad possibility that there are just too many practicing scientists who are inadequately trained to ensure that their work is indeed reproducible.

A Few Unavoidable Irritants

Naturally, some of the content will be presented in more detail than some will prefer or require, but all adult readers have had ample practice in skipping over content they're either conversant with or uninterested in. To facilitate that process, most chapters are relatively self-contained, with cursory warnings of their content posted at the conclusion of their immediately preceding chapter.

Also, while the story being presented almost exclusively involves the published work of others, I cannot in good consciousness avoid inserting my own opinions regarding this work and the issues involved. I have, however, attempted to clearly separate my opinions from those of others.

Otherwise, every topic discussed is supported by credible empirical evidence, and every recommendation tendered is similarly supported by either evidence or reasoned opinions by well-recognized reproducibility thinkers. This strategy has unavoidably necessitated a plethora of citations which only constitute a mere fraction of the literature reviewed. For readability purposes, this winnowing process has admittedly resulted in an unsystematic review of cited sources, although the intent was to represent an overall consensus of those thinkers and researchers who have contributed to this crucial scientific movement.

A Very Little About My Perspective

In my long academic and statistical consulting career, I have personally witnessed examples of pretty much all of the "good," "bad," and "ugly" of scientific practices. Following a brief stint as an educational researcher, my writing and research has largely focused on the methodology of conducting research and the statistical analysis of its results. I even once published an annotated guide to 2,600 published methodological sources encompassing 78 topics, 224 journals, and 125 publishers (Bausell, 1991). The tome was dedicated "to the three generations of research methodologists whose work this book partially represents" (p. viii).

In the three decades that followed, the methodological literature virtually exploded with the publication of more articles, topic areas, and journals (and, of course, blogs) than in the entire history of science prior to that time. And while this work has been extremely beneficial to the scientific enterprise, its main contribution may have been the facilitation of the emergence of a new generation of methodologists studying (and advocating for) the reproducibility of scientific results.

Naturally, as a chronicler, I could hardly avoid recognizing the revolutionary importance of this latter work, not just to research methodology but also to the entire scientific enterprise. My primary motivation for telling this story is to hopefully help promulgate and explain the importance of its message to the potential audiences previously described. And, of course, I dedicate the book "to the present generation of reproducibility methodologists it partially represents."

I must also acknowledge my debt to three virtual mentors who have guided me in interpreting and evaluating scientific evidence over the past two decades, philosophers of science from the recent and distant past whose best-known precepts I have struggled (not always successfully) to apply to the subject matter of this book as well. In chronological order these individuals are

1. William of Occam, the sternest and most fearsome of my mentors, whose most important precept was the *parsimony principle*, which

can be reduced to embracing the least involved explanation for the occurrence of a phenomenon that both fits the supporting data and requires the fewest assumptions (and usually constitutes the simplest explanation);
2. Yogi of Bronx, one of whose more important precepts was the principle of prediction, succinctly stated as "It's tough to make predictions, especially about the future"; and
3. Robert Park, the title of whose most important book pretty much speaks for itself: *Voodoo Science: The Road from Foolishness to Fraud (2000)*. And, as a disclaimer, Bob (the only one of my three mentors I ever actually met) never suggested that mainstream science had arrived at this latter destination—only that far too many scientists have blissfully traveled down that road.

And Finally an Affective Note

Some of the articles, quotations therefrom, and even some of my comments thereupon may appear overly critical. (Although for my part I have made a serious attempt to avoid doing so via the book's many revisions and greatly facilitated by one of its anonymous peer reviewer's very helpful comments.) It is important to remember, however, that what we are dealing with here is a paradigmatic shift (or radical change for those who prefer something less pompous) in the way in which science is conducted and published so *none of us should be overly judgmental*. It takes time for entrenched behaviors to change, and most practicing researchers (including myself) have violated one or more of the movement's precepts in the past, often unthinkingly because we hadn't been aware (or taught) that some of the questionable research behaviors discussed here were actually contraindicated.

After all, it wasn't until sometime around 2011–2012 that the scientific community's consciousness was bombarded with irreproducibility warnings via the work of scientists such as those discussed in this book (although warning shots had been fired earlier by scientists such as Anthony Greenwald in 1975 and John Ioannidis in 2005). However, this doesn't mean that we, as formal or informal peer reviewers, should be all that forgiving going forward regarding obviously contradicted practices such as failures to preregister protocols or flagrantly inflating p-values. We should just be civil in doing so.

So Where to Begin?

Let's begin with the odd phenomenon called *publication bias* with which everyone is familiar although many may not realize either the extent of its astonishing prevalence or its virulence as a facilitator of irreproducibility.

References

Bausell, R. B. (1991). *Advanced research methodology: An annotated guide to sources.* Metuchen, NJ: Scarecrow Press.

Campbell, D. T., & Stanley, J. C. (1966). *Experimental and quasi-experimental designs for research.* Chicago: Rand McNally.

Cohen, J. (1977, 1988). *Statistical power analysis for the behavioral sciences.* Hillsdale, NJ: Lawrence Erlbaum.

Fisher, R. A. (1935). *The design of experiments.* London: Oliver & Boyd.

Greenwald, A. G. (1975). Consequences of prejudice against the null hypothesis. *Psychological Bulletin, 82,* 1–20.

Hill, A. B. (1935). *Principles of medical statistics.* London: Lancet.

Ioannidis, J. P. A. (2005). Why most published research findings are false. *PLoS Medicine, 2,* e124.

Kuhn, T. S. (1962). *The structure of scientific revolutions.* Chicago: University of Chicago Press.

Merton, R. K. (1973). *The sociology of science: Theoretical and empirical investigations.* (N. W. Storer, Ed.). Chicago: University of Chicago Press.

Park, R. (2000). *Voodoo science: the road from foolishness to fraud.* New York: Oxford University Press.

PART I
BACKGROUND AND FACILITATORS OF THE CRISIS

1
Publication Bias

Whether we refer to the subject matter of our story as a crisis or a paradigmatic shift, it must begin with a consideration of a primary facilitator of irreproducibility, one that has come to be called *publication bias*. This phenomenon can be defined most succinctly and nonpejoratively as *a tendency for positive results to be overrepresented in the published literature*—or as "the phenomenon of an experiment's results [here amended to "any study's results" not just experiments] determining its likelihood of publication, often over-representing positive findings" (Korevaar, Hooft, & ter Riet, 2011).

Surprisingly, the artifact's history actually precedes the rise of the journal system as we know it. Kay Dickersin (1991), a methodologist who specialized among other things in biases affecting the conduct of systematic reviews, provides a very brief and fascinating history of publication bias, suggesting that it was first referred to at least as early as the 17th century by Robert Boyle, often referred to as "one of the founders of modern chemistry, and one of the pioneers of modern experimental scientific method" (https://en.wikipedia.org/wiki/Robert_Boyle).

According to Dr. Dickerson, Boyle "was credited [in 1680] with being the first to report the details of his experiments and the precautions necessary for their *replication* [italics added because this process will soon become an integral part of our story]" (p. 1385). And even earlier, long before the rise of scientific journals, "Boyle lamented in 1661 that scientists did not write up single results but felt compelled to refrain from publishing until they had a 'system' worked out that they deemed worthy of formal presentation" (p. 1386). Or, in Boyle's words,

> But the worst inconvenience of all is yet to be mentioned, and that is, that whilst this vanity of thinking men obliged to write either systems or nothing is in request, many excellent notions or experiments are, by sober and modest men, suppressed." (p. 1386)

But, as we all know, over time some things improve while others get even worse. And the latter appears to be the case with publication bias, which has been germinating at least since the mid-20th century and remains in full bloom at the end of the second decades of the 21st century.

By Way of Illustration, a 20th-Century Parable

Suppose a completely fictitious first-year educational research graduate student's introduction to research came in his first methods class taught by a similarly hypothetical senior educational researcher (actually trained as a psychologist) more than a half-century ago. The lecture might have gone something like this:

> Whatever research you do on whatever topic you choose to do it on, always ensure that you can find at least one statistically significant p-value to report. *Otherwise you won't get published,* and you're going to need at least five publications per year if you want to have a halfway decent career. But since your hypothesis may be wrong, always hedge your bet by (a) employing a control group that you know *should* be inferior to your intervention, (b) including several different variables which can be substituted as your outcome if necessary, and (c) gleaning as much information from the student records (e.g., standardized tests, past grades—recalling that he was referring educational research) as possible which can be employed as covariates or blocking variables.

Now it is not known (or remembered) whether this long ago, hypothetical graduate student considered this to be most excellent advice or to reside somewhere on the continuum between absurd and demented. If the latter, he would have soon learned the hard way that the absence of a statistically significant p-value or two did indeed greatly reduce the probability of obtaining a cherished "publish with minor revisions" letter from a journal editor. (The "p" of course denotes probability and an obtained p-value ≤ 0.05 had, for many scientists and their sciences, become synonymous with statistical significance, the correctness of a tested hypothesis, and/or the existence of a true effect.)

In any event much of the remainder of the lecture and the course was given over to avoiding the aforementioned (Campbell & Stanley, 1966) threats to internal validity, with a nod to its external counterpart and an occasional foray into statistical analysis. What was not typically considered (or taught) in those days was the possibility that false-positive results were clogging scientific literatures by unreported (and often unconsidered) strategies perfectly designed to produce publishable positive results—*even when participants were randomly assigned to conditions*. After all, empirical results could always be replicated by other researchers if they were sufficiently interested—which of course very few then were (or even are today).

So, thus trained and acculturated, our long-forgotten (or hypothetical) graduate student conducted many, many educational experiments and obtained statistical significance with some. And those he failed to do so he didn't bother to submit for publication once he had discovered for himself that his instructor had been right all along (i.e., that their rejection rate was several times greater than those blessed with p-values ≤ 0.05 as opposed to those cursed with the dreaded $p > 0.05$). After all, why even bother to write up such "failures" when the time could be more profitably spent conducting more "successful" studies?

Given his ambition and easy access to undergraduate and high school research participants he might have even conducted a series of 22 experiments in a failed effort to produce a study skill capable of producing more learning from a prose passage than simply reading and rereading said passage for the same amount of time—a hypothetical debacle that might have resulting in 22 statistically nonsignificant differences (none of which was ever submitted for publication). And once he might have even had the audacity to mention the quest itself at a brown bag departmental luncheon, which resulted in vigorous criticisms from all sides for wasting participants' time conducting nonsignificant studies.

So the Point of This Obviously Fictitious Parable Is?

Well certainly no one today would conduct such an absurd program of research for no other reason than the creation of an effective learning strategy. So let's just employ it to consider the ramifications of *not* publishing statistically nonsignificant research: some of which include the following.

First, it has come to be realized that it is unethical to ask people to volunteer to participate in research and not make the results available to other scientists. Or, as one group of investigators (Krzyzanowska, Pintilie, &

Tannock, 2003) more succinctly (and bluntly) state, "Nonpublication breaks the contract that investigators make with trial participants, funding agencies, and ethics boards" (p. 496).

Second, regardless of whether or not human participants are employed, publishing negative results

1. Permits other scientists from avoiding dead end paths that may be unproductive.
2. Potentially provides other scientists with an idea for a more effective intervention or a more relevant outcome variable (in the hypothetical scenario, this might have involved using *engagement* in the study activity as the outcome variable, thereby allowing the participants to take as much time as they needed to master the content), and
3. Encourages the creation of useful rather than spurious theories (the latter of which are more likely to be corroborated by positive results in the complete absence of negative ones).

But disregarding hypothetical scenarios or non-hypothetical ethical concerns, the reluctance to publish research associated with p-values > 0.05 has an even more insidious consequence. Decades upon decades of *publication bias* have resulted in a plethora of false-positive results characterizing the literatures of many entire scientific disciplines. And this, in turn, has become a major contributor to the lack of reproducibility in these disciplines since anyone with a sufficiently accomplished "skill set" can support almost any theory or practice.

The Relationship Between Publication and Scientific Irreproducibility

While irreproducibility and publication bias are inextricably linked, their relationship isn't necessarily causal since neither is a necessary nor sufficient condition of the other. For example, there are legitimate reasons why any *given* negative study may not be published, such as its authors' legitimate belief that the study in question was methodologically flawed, underpowered, and/or simply too poorly conducted to merit publication. (Solely for convenience, a "negative study or finding" will be referred to henceforth as one that explicitly or implicitly hypothesizes the occurrence of a statistically significant result, but a nonsignificant one is obtained.) Or, far less commonly, an investigator may hypothesis the *equivalence* between experimental interventions, but these studies require specialized analytic approaches (see Bausell, 2015) and are relatively rare outside of clinical (most commonly pharmaceutical) trials.

Some unacceptable (but perhaps understandable) reasons for investigators' not attempting to publish a negative finding could include

1. Their misremembrance of their introductory statistics professors' cautionary edicts against overinterpreting a negative finding (or their misinterpretation of the precept that it is impossible to "prove" a null hypothesis),
2. Embarrassment resulting from the failure to support a cherished hypothesis that they remain convinced was correct, or, as with our hypothetical student,
3. Their understanding of the difficulties of publishing a negative study leading to a decision to spend their time conducting "positive" research rather than "wasting" time writing up "failed" studies.

When queried via surveys of investigators (e.g., Greenwald, 1975; Cooper, DeNeve, & Charlton, 1997; Weber, Callaham, & Wears, 1998), other reasons listed include (a) lack of time, (b) the belief that a negative study won't be published (as will be discussed shortly, there is definitely an editorial bias against doing so), (c) loss of interest, (d) a realization that the study was too flawed or underpowered (more on this in Chapter 2), and/or (e) the fact that the study was never intended to be published in the first place (presumably because it was a pilot study or was designed to test a single element of a larger study [e.g., the reliability or sensitivity of an outcome variable]). More recently a survey of laboratory animal researchers (ter Riet, Korevaar, Leenaars, et al., 2012) arrived at similar findings regarding reasons for not publishing negative studies and concluded that investigators, their supervisors, peer reviewers, and journal editors all bear a portion of the blame for the practice.

In a sense it is not as important to delineate the contributors to the scarcity of negative studies in the published literature as it is to identify the causes and ameliorate the consequences of the high prevalence of their false-positive counterparts. But before proceeding to that discussion, let's briefly consider a sampling of the evidence supporting the existence and extent of publication bias as well as the number of disciplines affected.

The Prevalence of Positive, Published Results in Science as a Whole

Undoubtedly the most ambitious effort to estimate the extent to which positive results dominate the scientific literature was employed by Daniele Fanelli, who contrasted entire sciences on the acceptance or rejection of their stated hypothesis. In his first paper (2010), 2,434 studies published from 2000 to 2007 were selected from 10,837 journals in order to compare 20 different scientific fields with respect to their rate of positive findings.

The clear "winner" turned out to be psychology-psychiatry, with a 91.5% statistical significance rate although perhaps the most shocking findings emanating from this study were that (a) all 20 of the sciences (which basically constitute the backbone of our species' empirical, inferential scientific effort) reported positive published success rates of greater than 70%, (b) the average positive rate for the 2,434 studies was 84%, and (c) when the 20 sciences were collapsed into three commonly employed categories all obtained positive rates in excess of 80% (i.e., biological sciences = 81%, physical sciences = 84%, and social sciences = 88%).

In his follow-up analysis, Fanelli (2011) added studies from 1990 to 1999 to the 2000 to 2007 sample just discussed in order to determine if these positive rates were constant or if they changed over time. He found that, as a collective, the 20 sciences had witnessed a 22% increase in positive findings over this relatively brief time period. Eight disciplines (clinical medicine, economic and business, geoscience, immunology, molecular biology-genetics, neuroscience-behavior, psychology-psychiatry, and pharmacology-toxicology) actually reported positive results at least 90% of the time by 2007, followed by seven (agriculture, microbiology, materials science, neuroscience-behavior, plants-animals, physics, and the social sciences) enjoying positive rates of from 80% to 90%. (Note that since the author did not report the exact percentages for these disciplines, these values were estimated based on figure 2 of the 2011 report.)

Now, of course, neither of these analyses is completely free of potential flaws (as are none of the other 35 or so studies cited later in this chapter), but they constitute the best evidence we have regarding the prevalence of publication bias. (For example, both studies employed the presence of a key sentence, "test*the hypothes*," in abstracts only, and some disciplines did not rely on p-values for their hypothesis tests.) However, another investigator (Pautasso, 2010) provides a degree of confirmatory evidence for the

Fanelli results by finding similar (but somewhat less dramatic) increases in the overall proportion of positive results *over time* using (a) four different databases, (b) different key search phrases ("no significant difference/s" or "no statistically significant difference/s"), (c) different disciplinary breakdowns, and, for some years, (d) only titles rather than abstracts.

A Brief Disclaimer Regarding Data Mining

Textual data mining meta-research (aka meta-science) is not without some very real epistemological limitations. In the studies cited in this book, a finite number of words or phrases (often only one) is employed which may not capture all of the studies relevant to the investigators' purposes or may be irrelevant to some articles. These choices may inadvertently introduce either overestimates or underestimates of the prevalence of the targeted behavior or phenomena. And, of course, it is a rare meta-research study (which can be most succinctly defined as research on research and which includes a large proportion of the studies discussed in this book) that employs a causal design employing randomization of participants or other entities to groups.

With that said, these studies constitute much of the best evidence we have for the phenomena of interest. And data mining efforts are becoming ever more sophisticated, as witnessed by the Menke, Roelandse, Ozyurt, and colleagues (2020) study discussed in Chapter 11 in which multiple search terms keyed to multiple research guidelines were employed to track changes in reproducibility practices over time.

Other Studies and Approaches to the Documentation of Publication Bias

One of the more commonly employed methods of studying the prevalence of positive findings in published literatures involves examining actual p-values in specific journals rather than the data mining approaches just discussed. One of the earliest of these efforts was conducted by Professor T. D. Sterling (1959) via a survey of four general psychology journals which found that an astonishing 97% of the studies published therein rejected the null hypothesis (i.e., achieved statistical significance at the 0.05 level or below). A little over a decade later, the team of Bozarth and Roberts (1972) found similar results

(94%), as did Sterling and two colleagues (Sterling, Rosenbaum, & Weinkam, 1995) two decades or so later than that. (This time around, Sterling and his co-investigators searched eight psychology journals and found the prevalence of positive results within one or two percentage points [96%] of the previous two efforts.) Their rather anticlimactic conclusion: "These results also indicate that practices leading to publication bias have not changed over a period of 30 years" (p. 108). However, as the previous section illustrated, it has changed for the worst more recently in some disciplines.

While surveys of journal articles published in specific journals constitute the earliest approach to studying publication bias, more recently, examinations of meta-analyses (both with respect to the individual studies comprising them and the meta-analyses themselves) have become more popular vehicles to explore the phenomenon. However, perhaps the most methodologically sound approach involves comparisons of the publication status of positive and negative longitudinal trials based on (a) institutional review board (IRB) and institutional animal care and use committee (IACUC) applications and (b) conference abstracts. Two excellent interdisciplinary reviews of such studies (Song, Parekh-Bhurke, Hooper, et al., 2009; Dwan, Altman, Arnaiz, et al., 2013) found, perhaps unsurprisingly by now, that positive studies were significantly more likely to be published than their negative counterparts.

More entertaining, there are even *experimental* documentations of the phenomenon in which methodologically oriented investigators, with the blessings of the journals involved, send out two almost identical versions of the same bogus article to journal reviewers. "Almost identical" because one version reports a statistically significant result while the other reports no statistical significance. The positive version tended to be significantly more likely to be (a) accepted for publication (Atkinson, Furlong, & Wampold, 1982) and (b) rated more highly on various factors such as methodological soundness (Mahoney, 1977), or (c) both (Emerson, Warme, Wolf, et al., 2010).

A Quick Recap

While psychology may be the scientific leader in reporting positive results, Fanelli (2011) and Pautasso (2010) have demonstrated that the phenomenon is by no means found only in that discipline. In fact the majority of human and animal empirical studies employing p-values as the means for accepting or rejecting their hypotheses seem to be afflicted with this particular bias. In support of this rather pejorative generalization, the remainder of this chapter is given over to the presence of publication bias in a sampling of (a) *subdisciplines* or research *topics within disciplines* and (b) the methodological factors known to be subject to (or associated with) publication bias.

But First, the First of Many Caveats

First, the literature on publication bias is too vast and diverse to be reviewed either exhaustively or systematically here. Second, there is no good reason to do so since Diogenes, even with a state-of-the-art meta-science lantern, would have difficulty finding any methodologically oriented scientist who is not already aware of the existence of publication bias or who does not consider it be a significant scientific problem. And finally, the diversity of strategies designed to document publication bias makes comparisons or overall summaries of point estimates uninterpretable. Some of these strategies (which have been or will be mentioned) include (a) specific journal searches, (b) meta-research studies involving large databases, (c) meta-science examinations of reported p-values or key words associated with statistical significance/nonsignificance, (d) meta-analyses employing funnel plots to identify overabundancies of small studies reporting large effects, and (e) longitudinal follow-ups of studies from conference abstracts and IRB proposals to their ensuing journal publication/nonpublication.

An Incomplete Sampling of the Topic Areas Affected

- Psychology research (e.g., historically: Sterling, 1959; Atkinson et al., 1982; Sterling et al., 1995, and too many others to mention)
- Cancer clinical trials (Berlin, Begg, & Louis, 1989; Krzyzanowska et al., 2003)
- Cancer studies in general (De Bellefeuille, Morrison, & Tannock, 1992)
- Animal studies (Tsilidis, Panagiotou, Sena, et al., 2013)

- Child health (Hartling, Craig, & Russell, 2004)
- Medical randomized clinical trials (RCTs; Dickersin, Chan, & Chalmers, 1987)
- Preclinical stroke (Sena, van der Worp, Bath, et al., 2010)
- Psychotherapy for depression (Cuijpers, Smit, Bohlmeijer, et al., 2010; Flint, Cuijpers, & Horder, 2015)
- Pediatric research (Klassen, Wiebe, Russell, et al., 2002)
- Gastroenterology research (Timmer et al., 2002) and gastroenterological research cancer risk (Shaheen, Crosby, Bozymski, & Sandler, 2000)
- Antidepressant medications (Turner, Matthews, Linardatos, et al., 2008)
- Alternative medicine (Vickers, Goyal, Harland, & Rees, 1998; Pittler, Abbot, Harkness, & Ernst, 2000)
- Obesity research (Allison, Faith, & Gorman, 1996)
- Functional magnetic resonance imaging (fMRI) studies of emotion, personality, and social cognition (Vul, Harris, Winkielman, & Pashler, 2009) plus functional fMRI studies in general (Carp, 2012)
- Empirical sociology (Gerber & Malhotra, 2008)
- Anesthesiology (De Oliveira, Chang, Kendall, et al. 2012)
- Political behavior (Gerber, Malhotra, Dowling, & Doherty, 2010)
- Neuroimaging (Ioannidis, 2011; Jennings & Van Horn, 2012).
- Cancer prognostic markers (Kyzas, Denaxa-Kyza, & Ioannidis, 2007; Macleod, Michie, Roberts, et al., 2014)
- Education (Lipsey & Wilson, 1993; Hattie, 2009)
- Empirical economics (Doucouliagos, 2005)
- Brain volume abnormalities (Ioannidis, 2011)
- Reproductive medicine (Polyzos, Valachis, Patavoukas, et al., 2011)
- Cognitive sciences (Ioannidis, Munafò, Fusar-Poli, et al., 2014)
- Orthodontics (Koletsi, Karagianni, Pandis, et al., 2009)
- Chinese genetic epidemiology (Pan, Trikalinos, Kavvoura, et al., 2005)
- Drug addiction (Vecchi, Belleudi, Amato, et al., 2009)
- Biology (Csada, James, & Espie, 1996)
- Genetic epidemiology (Agema, Jukema, Zwinderman, & van der Wall, 2002)
- Phase III cancer trials published in high-impact journals (Tang, Pond, Welsh, & Chen, 2014)

Plus a Sampling of Additional Factors Associated with Publication Bias

- Multiple publications more so than single publication of the same data (Tramèr, Reynolds, Moore, & McQuay, 1997; Schein & Paladugu, 2001); although Melander, Ahlqvist-Rastad, Meijer, and Beermann (2003) found the opposite relationship for a set of Swedish studies
- The first hypothesis tested in multiple hypothesis studies less than single-hypothesis studies (Fanelli, 2010)
- Higher impact journals more so than low-impact journals (Tang et al., 2014, for cancer studies); but exceptions exist, such as the *Journal of the American Medical Association* (JAMA) and the *New England Journal of Medicine* (NEJM), for clinical trials (Olson, Rennie, Cook, et al., 2002)
- Non-English more so than English-language publications (Vickers et al., 1998; Jüni, Holenstein, Sterne, et al., 2003)
- RCTs with larger sample sizes less so than RCTs with smaller ones (Easterbrook, Berlin, Gopalan, & Matthews, 1991)
- RCTs less often than observational studies (e.g., epidemiological research), laboratory-based experimental studies, and nonrandomized trials (Easterbrook et al., 1991; Tricco, Tetzaff, Pham, et al., 2009)
- Research reported in complementary and alternative medicine journals more so than most other types of journals (Ernst & Pittler, 1997)
- Methodologically sound alternative medicine trials in high impact jounrals less so than their methodologically unsound counterparts in the same journals (Bausell, 2009)
- Meta-analyses more so than subsequent large RCTs on same topic (LeLorier, Gregoire, Benhaddad, et al, 1997).
- Earlier studies more so than later studies on same topic (Jennings & Van Horn, 2012; Ioannidis, 2008)
- Preregistration less often than no registration of trials (Kaplan & Irvin, 2015)
- Physical sciences (81%) less often than biological sciences (84%), less than social sciences (88%) (Fanelli, 2010), although Fanelli and Ioannidis (2013) found that the United States may be a greater culprit in the increased rate of positive findings in the "soft" (e.g., social) sciences than other countries
- Investigators reporting no financial conflict of interest less often than those who do have such a conflict (Bekelman, Li, & Gross, 2003;

Friedman & Richter, 2004; Perlis, Perlis, Wu, et al., 2005; Okike, Kocher, Mehlman, & Bhandari, 2007); all high-impact medical (as do most other medical) journals require a statement by all authors regarding conflict of interest.
- Pulmonary and allergy trials funded by pharmaceutical companies more so than similar trials funded by other sources (Liss, 2006)
- Fewer reports of harm in stroke research in published than non-published studies, as well as publication bias in general (Liebeskind, Kidwell, Sayre, & Saver, 2006)
- Superior results for prevention and criminology intervention trials when evaluated by the program developers versus independent evaluators (Eisner, 2009)
- And, of course, studies conducted by investigators known to have committed fraud or misconduct more so than those not so identified; it is a rare armed robbery that involves donating rather than stealing money.

A Dissenting Voice

Not included in this list is a systematic review (Dubben & Beck-Bornholdt, 2005) whose title, "Systematic Review of Publication Bias in Studies on Publication Bias," might appear to be parodic if it weren't for the fact that there is now a systematic review or meta-analysis available for practically every scientific topic imaginable. And, as should come as no surprise, the vast majority of meta-analytic conclusions are positive since the vast majority of their published literatures suffer from publication bias. (One meta-analytic database, the Cochrane Database of Systematic Reviews, is largely spared from this seemingly slanderous statement when the review involves the efficacy of a specific hypothesis involving a specific outcome; see Tricco et al., 2009.)

The Bottom Line

As a gestalt, the evidence is compelling that publication bias exists, buttressed by supporting (a) studies conducted over a period of decades, (b) the diversity by which the evidence was produced, and (c) the number of disciplines involved. In addition, the Fanelli (2011) and Pautasso (2010) analyses indicate that the phenomenon has been operating for a considerable amount of time and appears to be actually accelerating in most disciplines (although this is close to numerically impossible for some areas whose percentage of positive results approach 100%).

Whether the large and growing preponderance of these positive results in the published scientific literatures is a cause, a symptom, or simply a facilitator of irreproducibility doesn't particularly matter. What is important is that the lopsided availability of positive results (at the expense of negative ones) distorts our understanding of the world we live in as well as retards the accumulation of the type of knowledge that science is designed to provide.

This phenomenon, considering (a) the sheer number of scientific publications now being produced (estimated to be in excess of 2 million per year; National Science Board, 2018) and (b) the inconvenient fact that most of these publications are positive, leads to the following rather disturbing implications:

1. Even if the majority of these positive effects are not the product of questionable research practices specifically designed to produce positive findings, and
2. If an unknown number of *correct* negative effects are not published, then
3. Other investigative teams (unaware of these unpublished studies) will test these hypotheses (which in reality are false) until someone produces a positive result by chance alone—or some other artifact, which
4. Will naturally be published (given that it is positive) even if far more definitive contradictory evidence exists—therefore contributing to an already error-prone scientific literature.

Unfortunately, some very persuasive evidence will be presented in the next few chapters to suggest that the *majority of this welter of positive scientific findings being published today (and published in the past) does not represent even marginally sound scientific findings*, but instead are much more likely to be categorically false. And if this is actually true (which, again, the evidence soon to be presented suggests), then we most definitely have a scientific crisis of epic proportions on our hands.

But What's to Be Done About Publication Bias?

First, given the ubiquitous nature (and long history) of the problem, perhaps some of the strategies that probably won't be particularly helpful to reduce publication bias should be listed. Leaning heavily (but not entirely) on a paper entitled "Utopia: II. Restructuring Incentives and Practices to Promote Truth over Publishability" (Nosek, Spies, & Motyl, 2012), some of these ineffective (or at least mostly ineffective) strategies are

1. *Educational campaigns emphasizing the importance of publishing nonsignificant results as well as statistically significant ones.* While principled education is always a laudable enterprise, writing yet another article extolling the virtues of changing investigator, editorial, or student behaviors is unlikely be especially effective. There have been enough of such articles published over the past half-century or so.
2. *Creating journals devoted to publishing nonsignificant results.* Historically this strategy hasn't been particularly successful in attracting enough submission to become viable. Examples, old and new, include *Negative Results in Biomedicine, Journal of Negative Observations in Genetic Oncology, Journal of Pharmaceutical Negative Results* (this one is definitely doomed), *Journal of Articles in Support of the Null Hypothesis, The All Results Journals, New Negatives in Plant Science, PLoS ONE's Positively Negative Collection, Preclinical Reproducibility and Robustness Gateway,* and probably many others that have come and gone as some of these already have.
3. *Devoting dedicated space to publishing negative results in traditional journals.* Several journals have adopted this strategy and while it isn't a bad idea, it is yet to make any significant impact on the problem.
4. *Putting the burden on peer reviewers to detect false-positive results.* As editor-in-chief of an evaluation journal for 33 years, I can attest to the impossibility of this one. Providing reviewers with a checklist of potential risk factors (e.g., low power, the presence of uncharacteristically large effect sizes, counterintuitive covariates, p-values between 0.045 and 0.0499) might be helpful.
5. *Appealing to the ethical concerns of a new generation of scientists.* Such as, for example, campaigning to designate using human or animal participants in one's research and failing to publish the research

as unethical and a form of actual scientific misconduct (Chalmers, 1990).

Somehow we need to implement a cultural change in the sciences by convincing new and beginning investigators that publishing negative studies is simply a cost of doing business. Certainly it is understandable that conserving time and effort may be a personal priority but *everyone* associated with the publication process bears a responsibility for failing to correct the bias against publishing methodologically sound negative studies. And this list includes

1. Investigators, such as our hypothetical graduate student who slipped his 22 negative studies into Robert Rosenthal's allegorical "file drawer" (1979), never to be translated into an actual research report or accompanied by a sufficiently comprehensive workflow to do so in the future. (See Chapter 9 for more details of the concept of keeping detailed workflows, along with Phillip Bourne's [2010] discussion thereof.) For even though many of us (perhaps even our mythical graduate student) may have good intentions to publish all of our negative studies in the future, in time the intricate details of conducting even a simple experiment fade in the face of constant competition for space in long-term memory. And although we think we will, we never seem to have more time available in the future than we do now in the present.
2. Journal editors who feel pressure (personal and/or corporate) to ensure a competitive citation rate for their journals and firmly believe that publishing negative studies will interfere with this goal as well as reduce their readership. (To my knowledge there is little or no empirical foundation for this belief.) We might attempt to convince editors of major journals to impose a specified percentage annual limit on the publication of positive results, perhaps beginning as high as 85% and gradually decreasing it over time.
3. Peer reviewers with a bias against studies with p-values > 0.05. As mentioned previously, this artifact has been documented experimentally several times by randomly assigning journal reviewers to review one of two identical versions of a fake manuscript, with the exception that one reports statistical significance while the other reports nonsignificance. Perhaps mandatory peer review seminars and/or checklists could be

developed for graduate students and postdocs to reduce this peer reviewer bias in the future.
4. Research funders who would much rather report that they spent their money demonstrating that something works or exists versus that it does not. And since the vast majority of investigators' funding proposals hypothesize (hence promise) positive results, they tend to be in no hurry to rush their negative results into publication—at least until their next grant proposal is approved.
5. The public and the press that it serves are also human, with the same proclivities, although with an added bias toward the "man bites dog" phenomenon.

However, there are steps that all researchers can take to reduce the untoward effects of publication bias by such as

1. Immediately writing and *quickly* submitting negative studies for publication. And, if rejected, quickly resubmitting them to another journal until they are accepted (and they eventually will be given sufficient persistent coupled with the absolute glut of journals in most fields). And, as for any study, (a) transparently reporting any glitches in their conduct (which peer reviewers will appreciate and many will actually reward) and, (b) perhaps especially importantly for negative studies, explaining why the study results are important contributions to science;
2. Presenting negative results at a conference (which does not preclude subsequent journal publication). In the presence of insufficient funds to attend one, perhaps a co-author or colleague could be persuaded to do so at one that he or she plans to attend;
3. Serving as a peer reviewer (a time-consuming, underappreciated duty all scientists must perform) or journal editor (even worse): evaluating studies based on their design, conceptualization, and conduct rather than their results;
4. And, perhaps most promising of all, utilizing preprint archives such as the arXiv, bioRxiv, engrXiv, MedRxiv, MetaArXiv, PeerJ, PsyArXiv, SocArXiv, and SSRN, which do not discriminate against negative results. This process could become a major mechanism for increasing the visibility and availability of nonsignificant results since no peer review is required and there is no page limit, so manuscripts can be as long or as brief as their authors' desire.

An Effective Vaccine

One reason publication bias is so difficult to prevent involves its disparate list of "culprits" (e.g., investigators, journal editors, peer reviewers) and the unassailable fact that negative studies are simply very difficult to get published. One very creative preventive measure was designated as the *Registered Reports process*; it was championed by Chris Chambers and presumably first adopted by the journal *Cortex* and announced via a guest editorial by Brian Nosek and Daniël Lakens in 2014.

This innovative approach to publishing actually *prevents* reviewers and journal editors from discriminating against nonsignificant results and hence greatly incentivizes investigators to submit them for publication in the first place. Perhaps most comprehensively and succinctly described by Nosek, Ebersole, DeHaven, and Mellor (2018), Registered Reports and their accompanying preregistration are potentially one of the most effective strategies yet developed for preventing false-positive results due to publication bias. The process, in the authors' words, occurs as follows:

> With Registered Reports, authors submit their research question and methodology to the journal for peer review before observing the outcomes of the research. If reviewers agree that the question is sufficiently important and the methodology to test it is of sufficiently high quality, then the paper is given in-principle acceptance. The researchers then carry out the study and submit the final report to the journal. At second-stage review, reviewers do not evaluate the perceived importance of the outcomes. Rather, they evaluate the quality of study execution and adherence to the preregistered plan. In addition to the benefits of preregistration, this workflow addresses selective reporting of results and facilitates improving research designs during the peer review process. (p. 2605)

Apparently the Registered Reports initiative is taking hold since, as early as 2018, Hardwicke and Ioannidis found that 91 journals had adopted the procedure across a number of scientific disciplines. A total of 109 Registered Reports had been published in psychology, the majority of which were replication studies (in which case they are referred to as "Registered Replication Reports" which will be discussed in more detail in Chapter 7).

In addition to its potential salutary effects on publication bias, the authors list several other important advantages of the Registered Report process resulting from the a priori peer review process. These include

1. The potential for methodological flaws being identified in time to avoid poorly designed studies being added to the literature;
2. Publishing judgments being based on the merits of the research question and design, rather than on the aesthetic characteristics of the findings;
3. Research being more likely to be comprehensively and transparently reported since authors know they can be held accountable based on the original registered protocol; and
4. Questionable research practices (which will be discussed in detail in Chapter 3) will be greatly reduced since the original protocol can (and hopefully will) be compared with the final research report by journal editors and reviewers.

And as yet another reason for publishing negative results (in the long run they are more likely to be correct than their positive counterparts): Poynard, Munteanu, Ratziu, and colleagues (2002) conducted a completely unique analysis (at least as far I can ascertain) with the possible exception of a book tangentially related thereto (Arbesman, 2012) entitled *The Half-Life of Facts: Why Everything We Know Has an Expiration Date*). The authors accomplished this by collecting cirrhosis and hepatitis articles and meta-analyses conducted between 1945 and 1999 in order to determine which of the original conclusions were still considered "true" by 2000. Their primary results were

> Of 474 [resulting] conclusions (60%) were still considered to be true, 91 (19%) were considered to be obsolete, and 98 (21%) were considered to be false. The half-life of truth was 45 years. The 20-year survival of conclusions derived from meta-analysis was lower (57% ± 10%) than that from nonrandomized studies (87% ± 2%) ($p < 0.001$) or randomized trials (85% ± 3%) ($p < 0.001$). (p. 888)

However, one of the study's subgroup analyses (which should always be interpreted with caution) was of even more interest to us here since it could be interpreted as a comparison between positive and negative findings with respect to their production of false-positive versus false-negative conclusions. Namely that, "in randomized trials, the 50-year survival rate was higher for 52 negative conclusions (68% ± 13%) than for 118 positive conclusions (14% ± 4%, p < 0.001)" (p. 891).

Thus, while based on a single study (and a secondary analysis at that), this finding could constitute a case for the scientific importance of not preferring positive studies to negative studies since the latter may be more likely to be true in the long run.

So What's Next?

Before beginning the exploration of irreproducibility's heart of darkness (and reproducibility's bright potential), it may be helpful to quickly review three introductory statistical constructs (statistical significance, statistical power, and the effect size). These are key adjuncts to reproducibility and touch just about everything empirical in one way or another. However, since everyone who has suffered through an introductory statistics course (which probably encompasses 99% of the readers of this book), a natural question resides in why the repetition is necessary.

The answer, based on my not so limited experience, is that many researchers, graduate students, and postdocs do not have a deep understanding of these concepts and their ubiquitous applications to all things statistical. And even for those who do, the first few pages of Chapter 2 may be instructive since statistical significance and power play a much more important (perhaps *the* most important) role in reproducibility than is commonly realized. The p-value, because it is so easily *gamed* (as illustrated in Chapter 3), is, some would say, set too high to begin with, while statistical power is so often ignored (and consequently set too low).

References

Agema, W. R., Jukema, J. W., Zwinderman, A. H., & van der Wall, E. E. (2002). A meta-analysis of the angiotensin-converting enzyme gene polymorphism and restenosis after percutaneous transluminal coronary revascularization: Evidence for publication bias. *American Heart Journal, 144*, 760–768.

Allison, D. B., Faith, M. S., & Gorman, B. S. (1996). Publication bias in obesity treatment trials? *International Journal of Obesity and Related Metabolic Disorders, 20*, 931–937.

Arbesman, S. (2012). *The half-life of facts: Why everything we know has an expiration date.* New York: Penguin.

Atkinson, D. R., Furlong, M. J., & Wampold, B. E. (1982). Statistical significance reviewer evaluations, and the scientific process: Is there a statistically significant relationship? *Journal of Counseling Psychology, 29,* 189–194.

Bausell, R. B. (2009). Are positive alternative medical therapy trials credible? Evidence from four high-impact medical journals. *Evaluation & the Health Professions, 32,* 349–369.

Bausell, R. B. (2015). *The design and conduct of meaningful experiments involving human participants: 25 scientific principles.* New York: Oxford University Press.

Bekelman, J. E., Li, Y., & Gross, C. P. (2003). Scope and impact of financial conflicts of interest in biomedical research: A systematic review. *Journal of the American Medical Association, 289,* 454–465.

Berlin, J. A., Begg, C. B., & Louis, T. A. (1989). An assessment of publication bias using a sample of published clinical trials. *Journal of the American Statistical Association, 84,* 381–392.

Bourne, P. E. (2010). What do I want from the publisher of the future? *PLoS Computational Biology, 6,* e1000787.

Bozarth, J. D., & Roberts, R. R. (1972). Signifying significant significance. *American Psychologist, 27,* 774–775.

Campbell, D. T., & Stanley, J. C. (1966). *Experimental and quasi-experimental designs for research.* Chicago: Rand McNally.

Carp, J. (2012). The secret lives of experiments: Methods reporting in the fMRI literature. *Neuroimage, 63,* 289–300.

Chalmers, I. (1990). Underreporting research is scientific misconduct. *Journal of the American Medical Association, 263,* 1405–1408.

Cooper, H. M., DeNeve, K. M., & Charlton, K. (1997). Finding the missing science: The fate of studies submitted for review by a human subjects committee. *Psychological Methods, 2,* 447–452.

Csada, R. D., James, P. C., & Espie, R. H. M. (1996). The "file drawer problem" of non-significant results: Does it apply to biological research? *Oikos, 76,* 591–593.

Cuijpers, P., Smit, F., Bohlmeijer, E., et al. (2010). Efficacy of cognitive-behavioural therapy and other psychological treatments for adult depression: Meta-analytic study of publication bias. *British Journal of Psychiatry, 196,* 173–178.

De Bellefeuille, C., Morrison, C. A., & Tannock, I. F. (1992). The fate of abstracts submitted to a cancer meeting: Factors which influence presentation and subsequent publication. *Annals of Oncology, 3,* 187–191.

De Oliveira, G. S., Jr., Chang, R., Kendall, M. C., et al. (2012). Publication bias in the anesthesiology literature. *Anesthesia & Analgesia, 114,* 1042–1048.

Dickersin, K. (1991). The existence of publication bias and risk factors for its occurrence. *Journal of the American Medical Association, 263,* 1385–1389.

Dickersin, K., Chan, S., Chalmers, T. C., et al. (1987). Publication bias and clinical trials. *Controlled Clinical Trials, 8,* 343–353.

Doucouliagos, C. (2005). Publication bias in the economic freedom and economic growth literature. *Journal of Economic Surveys, 19,* 367–387.

Dubben, H-H., & Beck-Bornholdt, H-P. (2005). Systematic review of publication bias in studies on publication bias. *British Medical Journal, 331,* 433–434.

Dwan, K., Altman, D. G., Arnaiz, J. A., et al. (2013). Systematic review of the empirical evidence of study publication bias and outcome reporting bias: An updated review. *PLoS ONE, 8*, e66844.

Easterbrook, P. J., Berlin, J. A., Gopalan, R., & Matthews, D. R. (1991). Publication bias in clinical research. *Lancet, 337*, 867–872.

Eisner, M. (2009). No effects in independent prevention trials: Can we reject the cynical view? *Journal of Experimental Criminology, 5*, 163–183.

Emerson. G. B., Warme, W. J., Wolf, F. M., et al. (2010). Testing for the presence of positive-outcome bias in peer review. *Archives of Internal Medicine, 170*, 1934–1939.

Ernst, E., & Pittler, M. H. (1997). Alternative therapy bias. *Nature, 385*, 480.

Fanelli, D. (2010). "Positive" results increase down the hierarchy of the sciences. *PLoS ONE, 5*, e10068.

Fanelli, D. (2011). Negative results are disappearing from most disciplines and countries. *Scientometrics, 90*, 891–904.

Fanelli, D., & Ioannidis, J. P. (2013). US studies may overestimate effect sizes in softer research. *Proceedings of the National Academy of the Sciences, 110*, 15031–15036.

Flint, J., Cuijpers, P., & Horder, J. (2015). Is there an excess of significant findings in published studies of psychotherapy for depression? *Psychological Medicine, 45*, 439–446.

Friedman, L. S., & Richter, E. D. (2004). Relationship between conflicts of interest and research results. *Journal of General Internal Medicine, 19*, 51–56.

Gerber, A. S., & Malhotra, N. (2008). Publication bias in empirical sociological research: Do arbitrary significance levels distort published results? *Sociological Methods and Research, 37*, 3–30.

Gerber, A. S., Malhotra, N., Dowling, C. M., & Doherty, D. (2010). Publication bias in two political behavior literatures. *American Politics Research, 38*, 591–613.

Greenwald, A. G. (1975). Consequences of prejudice against the null hypothesis. *Psychological Bulletin, 82*, 1–20.

Hardwicke, T., & Ioannidis, J. (2018). Mapping the universe of registered reports. *Nature Human Behaviour, 2*, 10.1038/s41562-018-0444-y.

Hartling, L., Craig, W. R., & Russell, K. (2004). Factors influencing the publication of randomized controlled trials in child health research. *Archives of Adolescent Medicine, 158*, 984–987.

Hattie, J. (2009). *Visible learning: A synthesis of over 800 meta-analyses relating to achievement*. London: Routledge.

Ioannidis, J. P. (2011). Excess significance bias in the literature on brain volume abnormalities. *Archives of General Psychiatry, 68*, 773–780.

Ioannidis, J. P. A. (2008). Why most discovered true associations are inflated. *Epidemiology 19*, 640–648.

Ioannidis, J. P. A., Munafò, M. R., Fusar-Poli, P., et al. (2014). Publication and other reporting biases in cognitive sciences: Detection, prevalence, and prevention. *Trends in Cognitive Science, 19*, 235–241.

Jennings, R. G., & Van Horn, J. D. (2012). Publication bias in neuroimaging research: Implications for meta-analyses. *Neuroinformatics, 10*, 67–80.

Jüni, P., Holenstein, F., Sterne, J., et al. (2003). Direction and impact of language bias in meta-analysis of controlled trials: Empirical study. *International Journal of Epidemiology, 31*, 115–123.

Kaplan, R. M., & Irvin, V. L. (2015). Likelihood of null effects of large NHLBI clinical trials has increased over time. *PLoS ONE, 10,* e0132382.

Klassen, T. P., Wiebe, N., Russell, K., et al. (2002). Abstracts of randomized controlled trials presented at the Society for Pediatric Research Meeting. *Archives of Pediatric and Adolescent Medicine, 156,* 474–479.

Koletsi, D., Karagianni, A., Pandis, N., et al. (2009). Are studies reporting significant results more likely to be published? *American Journal of Orthodontics and Dentofacial Orthopedics, 136,* 632e1–632e5.

Korevaar, D. A., Hooft, L., & ter Riet (2011). Systematic reviews and meta-analyses of preclinical studies: Publication bias in laboratory animal experiments. *Laboratory Animals, 45,* 225–230.

Krzyzanowska, M. K., Pintilie, M., & Tannock, I. F. (2003). Factors associated with failure to publish large randomized trials presented at an oncology meeting. *Journal of the American Medical Association, 290,* 495–501.

Kyzas, P. A., Denaxa-Kyza, D., & Ioannidis, J. P. (2007). Almost all articles on cancer prognostic markers report statistically significant results. *European Journal of Cancer, 43,* 2559–2579.

LeLorier, J., Gregoire, G., Benhaddad, A., et al. (1997). Discrepancies between meta-analyses and subsequent large randomized, controlled trials. *New England Journal of Medicine, 337,* 536–542.

Liebeskind, D. S., Kidwell, C. S., Sayre, J. W., & Saver, J. L. (2006). Evidence of publication bias in reporting acute stroke clinical trials. *Neurology, 67,* 973–979.

Lipsey, M. W., & Wilson, D. B. (1993). Educational and behavioral treatment: Confirmation from meta-analysis. *American Psychologist, 48,* 1181–1209.

Liss, H. (2006). Publication bias in the pulmonary/allergy literature: Effect of pharmaceutical company sponsorship. *Israeli Medical Association Journal, 8,* 451–544.

Macleod, M. R., Michie, S., Roberts, I., et al. (2014). Increasing value and reducing waste in biomedical research regulation and management. *Lancet, 383,* 176–185.

Mahoney, M. J. (1977). Publication prejudices: An experimental study of confirmatory bias in the peer review system. *Cognitive Therapy and Research, 1,* 161–175.

Melander, H., Ahlqvist-Rastad, J., Meijer, G., & Beermann, B. (2003). Evidence b(i)ased medicine-selective reporting from studies sponsored by pharmaceutical industry: Review of studies in new drug applications. *British Medical Journal, 326,* 1171–1173.

Menke, J., Roelandse, M., Ozyurt, B., et al. (2020). Rigor and Transparency Index, a new metric of quality for assessing biological and medical science methods. bioRxiv http://doi.org/dkg6;2020

National Science Board. (2018). *Science and engineering indicators 2018.* NSB-2018-1. Alexandria, VA: National Science Foundation. https://www.nsf.gov/statistics/indicators/.

Nosek, B. A., Ebersole, C. R., DeHaven, A. C., & Mellor, D. T. (2018). The preregistration revolution. *Proceedings of the National Academy of Sciences, 115,* 2600–2606.

Nosek, B. A., & Lakens, D. (2014). Registered reports: A method to increase the credibility of published results. *Social Psychology, 45,* 137–141.

Nosek, B. A., Spies, J. R., & Motyl, M. (2012). Scientific utopia: II. Restructuring incentives and practices to promote truth over publishability. *Perspectives in Psychological Science, 7,* 615–631.

Okike, K., Kocher, M. S., Mehlman, C. T., & Bhandari, M. (2007). Conflict of interest in orthopaedic research: An association between findings and funding in scientific presentations. *Journal of Bone and Joint Surgery, 89,* 608–613.

Olson, C. M., Rennie, D., Cook, D., et al. (2002). Publication bias in editorial decision making. *Journal of the American Medical Association, 287*, 2825–2828.

Pan, Z., Trikalinos, T. A., Kavvoura, F. K., et al. (2005). Local literature bias in genetic epidemiology: An empirical evaluation of the Chinese literature [see comment]. *PLoS Medicine, 2*, e334.

Pautasso, M. (2010). Worsening file-drawer problem in the abstracts of natural, medical and social science databases. *Scientometrics, 85*, 193–202.

Perlis, R. H., Perlis, C. S., Wu, Y., et al. (2005). Industry sponsorship and financial conflict of interest in the reporting of clinical trials in psychiatry. *American Journal of Psychiatry, 162*, 1957–1960.

Pittler, M. H., Abbot, N. C., Harkness, E. F., & Ernst, E. (2000). Location bias in controlled clinical trials of complementary/alternative therapies. *Journal of Clinical Epidemiology, 53*, 485–489.

Polyzos, N. P., Valachis, A., Patavoukas, E., et al. (2011). Publication bias in reproductive medicine: From the European Society of Human Reproduction and Embryology annual meeting to publication. *Human Reproduction, 26*, 1371–1376.

Poynard, T., Munteanu, M., Ratziu, V., et al. (2002). Truth survival in clinical research: An evidence-based requiem? *Annuals of Internal Medicine, 136*, 888–895.

Rosenthal, R. (1979). The file drawer problem and tolerance for null results. *Psychological Bulletin, 86*, 638–641.

Schein, M., & Paladugu, R. (2001). Redundant surgical publications: Tip of the iceberg? *Surgery, 129*, 655–661.

Sena, E. S., van der Worp, H. B., Bath, P. M., et al. (2010). Publication bias in reports of animal stroke studies leads to major overstatement of efficacy. *PLoS Biology, 8*, e1000344.

Shaheen, N. J., Crosby, M. A., Bozymski, E. M., & Sandler, R. S. (2000). Is there publication bias in the reporting of cancer risk in Barrett's esophagus? *Gastroenterology, 119*, 333–338.

Song, F., Parekh-Bhurke, S., Hooper, L., et al. (2009). Extent of publication bias in different categories of research cohorts: A meta-analysis of empirical studies. *BMC Medical Research Methodology, 9*, 79.

Sterling, T. D. (1959). Publication decision and the possible effects on inferences drawn from tests of significance-or vice versa. *Journal of the American Statistical Association, 54*, 30–34.

Sterling, T. D., Rosenbaum, W. L., & Weinkam, J. J. (1995). Publication decisions revisited: The effect of the outcome of statistical tests on the decision to publish and vice versa. *American Statistician, 49*, 108–112.

Tang, P. A., Pond, G. R., Welsh, S., & Chen, E. X. (2014). Factors associated with publication of randomized phase III cancer trials in journals with a high impact factor. *Current Oncology, 21*, e564–572.

ter Riet, G., Korevaar, D. A., Leenaars, M., et al. (2012). Publication bias in laboratory animal research: A survey on magnitude, drivers, consequences and potential solutions. *PLoS ONE, 1*(9), e43404,

Timmer, A., Hilsden, R. J., Cole, J., et al. (2002). Publication bias in gastroenterological research: A retrospective cohort study based on abstracts submitted to a scientific meeting. *BMC Medical Research Methodology, 2*, 7.

Tramèr, M. R., Reynolds, D. J., Moore, R. A., & McQuay, H. J. (1997). Impact of covert duplicate publication on meta-analysis: A case study. *British Medical Journal, 315*, 635–640.

Tricco, A. C., Tetzaff, J., Pham, B., et al. (2009). Non-Cochrane vs. Cochrane reviews were twice as likely to have positive conclusion statements: Cross-sectional study. *Journal of Clinical Epidemiology, 62,* 380–386.

Tsilidis, K. K., Panagiotou, O. A., Sena, E. S., et al. (2013). Evaluation of excess significance bias in animal studies of neurological diseases. *PLoS Biology, 11,* e1001609.

Turner, E. H., Matthews, A. M., Linardatos, E., et al. (2008). Selective publication of antidepressant trials and its influence on apparent efficacy. *New England Journal of Medicine, 358,* 252–260.

Vecchi, S., Belleudi, V., Amato, L., et al. (2009). Does direction of results of abstracts submitted to scientific conferences on drug addiction predict full publication? *BMC Medical Research Methodology, 9,* 23.

Vickers, A., Goyal, N., Harland, R., & Rees, R. (1998). Do certain countries produce only positive results? A systematic review of controlled trials. *Controlled Clinical Trials, 19,* 159–166.

Vul, E., Harris, C., Winkielman, P., & Pashler, H. (2009). Puzzlingly high correlations in fMRI studies of emotion, personality, and social cognition. *Perspectives on Psychological Science, 4,* 274–290.

Weber, E. J., Callaham, M. L., & Wears, R. L. (1998). Unpublished research from a medical specialty meeting: Why investigators fail to publish. *Journal of the American Medical Association, 280,* 257–259.

2
False-Positive Results and a Nontechnical Overview of Their Modeling

Let's begin with a definition of the most worrisome manifestation (and primary constituent) of irreproducible research findings: false-positive results. A *false-positive result* occurs when (a) an impotent intervention results in a statistically significant change in a specific outcome or (b) a relationship between two (or more) *unrelated* variables is found to be statistically significant. Or, more succinctly, it is a positive statistically significant result that cannot be reliably replicated. Or, more pejoratively, it is a positive finding that has resulted from (a) one or more egregious procedural/statistical errors or (b) investigator ignorance, bias, or fraud.

Of course, false-negative results are also problematic, but, given the high prevalence of publication bias in most disciplines, they are considerably more rare and won't be emphasized here. This doesn't mean that false negatives are unimportant since those emanating from clinical trials designed to test the efficacy of actually effective drugs or treatments could lead to tragic events. It is just that scenarios such as this appear to be quite rare in clinical research and almost nonexistent in the social science literatures.

There are three statistical constructs that contribute to the production of false-positive results.

1. *Statistical significance*, defined by the comparison between the probability level generated by a computer following the statistical analysis performed on study results (referred to here as the *p-value*) and the maximum probability level hypothesized by the investigator or based on a disciplinary consensus or tradition (referred to as the *alpha level*). If the obtained p-value is less than or exactly equal to ($\leq$) the hypothesized or disciplinary conventional alpha level (typically 0.05), then statistical significance is declared.
2. *Statistical power* is most succinctly (and rather cavalierly) defined as the probability that a given study will result in statistical significance

(the minimum value of which is most often recommended to be set at 0.80). Statistical power is a function of (a) the targeted alpha level; (b) the study design; (c) the number of participants, animals, or other observations employed; and (d) our third statistical construct, the effect size.

3. The *effect size* is possibly the simplest of the three constructs to conceptualize *but without question* it is the most difficult of the three constructs to predict prior to conducting a study. It is most often predicted based on (a) a small-scale pilot study, (b) a review of the results of similar studies (e.g., meta-analyses), or (c) a disciplinary convention, which, in the social sciences, is often set at 0.50 based on Jacob Cohen's decades-old (1988) recommendation. Its prediction is also the most tenuous of the triad regardless of how it is generated. If the effect size is overestimated, even when a hypothesized effect actually exists, its attendant study will be more difficult to replicate without adjustments such as an increased sample size or the use of questionable research practices (QRPs). If the effect size is underestimated, replication is more likely (even in the absence of QRPs), and, if the true effect size under investigation is sufficiently large, the attendant study will most likely be either trivial or constitute a major scientific finding. Since this latter scenario occurs with extreme rarity and the overestimation of effect sizes is far more common—whether predicted a priori or based on study results—most of what follows will be based on this scenario.

All three constructs are based on a statistical model called the *normal* or *bell-shaped curve*, which is often depicted as shown in Figure 2.1.

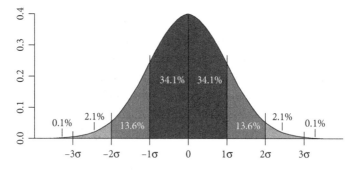

Figure 2.1 The bell-shaped, normal curve.
https://en.wikipedia.org/wiki/File:Standard_deviation_diagram.svg

All three are therefore integrally related to one another. For example, all else being equal:

1. The lower (or more stringently) the alpha level is set (i.e., the manner in which statistical significance is defined), the lower the power will be unless the study design is properly adjusted (e.g., increasing the number of participants or other observations [aka the sample size] to be employed) and/or the larger the hypothesized effect size must be;
2. The higher the desired power, the larger the required sample size and/or the larger the hypothesized effect size must be (if the alpha level is not adjusted); and, obviously,
3. The smaller the hypothesized effect size the less statistical power will be available (unless the sample size is increased and/or the alpha level is not adjusted).

Naturally, since all three of these statistical constructs are based on the normal curve they, too, are all subject to the same rather restrictive governing assumptions plus a few of their own. But the normal curve has proved to be a rather useful model which is surprising robust to minor violations of many of its assumptions.

So let's now begin the modeling of false-positive results based on the statistical models just discussed. As depicted in Table 2.1, this diagram has

Table 2.1 A modeling of the probabilities of the four possible study outcomes in the absence of systematic bias

	What is "actually" true	
	The hypothesis is correct	The hypothesis is incorrect
Possible study outcomes		
The hypothesis is confirmed (*obtained p-value = alpha level = .05*)	[a] Correct Finding (*p of occurrence = statistical power = .80*)	[b] False-Positive Result (*p of occurrence = alpha level = .05*)
The hypothesis is not confirmed (*obtained p-value > .05*)	[c] False-Negative Result] (*p of occurrence = 1-statistical power = .20*]	[d] Correct Finding (*p of occurrence = 1-alpha = .95*)

For present purposes, systematic bias involves questionable research practices (QRPs), whether consciously or unconsciously employed.

confounded students in untold numbers of introductory statistics and research methods books for decades, sometimes with no warning that it represents a most tenuous statistical model that has little or no practical applicability to actual scientific practice.

What could be simpler? Of course, we never know whether the original hypothesis is correct or not, but this is a model, after all, so we must make a few assumptions. And since we know that hypotheses are almost always confirmed in published studies, why even worry about unconfirmed hypotheses? And certainly a false-positive rate of 5% is nothing to worry about and definitely not indicative of a crisis of any sort. So let's just concentrate on the first row of cells.

But, alas, even here some key assumptions are missing governing the use of these concepts (along with the effect size, which does not appear but, as just discussed, is integrally related to statistical significance and power). Assumptions which, if violated, render the false-positive result projection practically useless except as a starting point for additional modeling.

For example, let's start with one of the assumptions related to either the obtained p-value (or alpha level from an a priori perspective).

> A single obtained p-value (or single event alpha level) normally applies to an a priori specified hypothesis involving a *single* result associated with a *single* outcome unless both are adjusted downward to compensate for multiple events or conditions. If no adjustment is made and statistical significance is obtained, the accruing result is more likely to be incorrect (hence a false-positive result) than in the presence of an appropriate adjustment.

To illustrate, let's assume that an investigator conducted a simple two-group experiment designed to ascertain if the two groups differed on a single outcome variable and obtained a p-value of 0.04999. However if *two* outcome variables were employed and the same p-value was obtained for both findings, neither result would actually be statistically significant following an appropriate downward adjustment of either the original alpha level or the obtained p-value. And, not coincidentally, the original power would have been an overestimate because its calculation involved an inflated p-value.

As another example, as mentioned earlier, power is sometimes inadequately defined as the probability of obtaining statistical significance but a more accurate (if unwieldy) definition must take the possibility of systematic error (or bias) into account such as, statistical power is the probability that an experiment [or any type of empirical study for that matter] will result in *statistical significance* if that significance level is appropriate for the design employed [i.e., is properly adjusted], if the study is properly conducted [i.e., in the absence of unavoidable glitches and QRPs], and if its hypothesized effect size is correct. (Bausell & Li, 2002, p. 14)

Many thoughtful reproducibility scholars would probably consider the simplistic model just discussed as too basic to even be considered a false-positive model at all. However, my reason for presenting it in this context is that it provides an opportunity to review some important points regarding the statistical significance and statistical power concepts that are often taken for granted. And, after all, it did provide a barebones model for predicting the prevalence of this most worrisome manifestation (and primary constituent) of irreproducible research findings.

So with these potential precursors to false-positive results at least partially discussed, let's now turn our attention to what is by far the most iconic, influential, and widely read article in the entire scientific reproducibility arena. Authored by John Ioannidis, it has garnered well over 8,000 citations and might just have provided the spark that ignited the reproducibility revolution itself and, not coincidentally, produced a far, far different probability estimate for cell "b" of Table 2.1 (which, it will be remembered, suggested a 5% occurrence of false-positive results in the absence of bias).

Why Most Published Research Findings Are False

John P. A. Ioannidis (2005)

This modeling effort is designed to estimate the rate of false-positive results. It employed the same statistical concepts as did Table 2.1 but added two totally new ones: *the ratio of true to no relationships in a discipline* and the even more important one regarding the likely effect of certain well-known biasing factors (e.g., conducting multiple analyses) known to inflate p-values and hence *produce* false-positive results when these inflated p-values are not corrected.

As an aside, statistical models such as this have important strengths and weaknesses which are bothersome to many scientists. However, they are pervasive in science as a whole, sometimes do not involve actual observations of any sort, are not testable experimentally or observationally (at least when they are first made), and sometimes involve assumptions that may be completely demented but often constitute our only option to predict future events or estimate presently unobservable phenomena. Or, in George Box's much more eloquent and succinct aphorism: "All models are wrong, but some are useful."

But, returning to the model at hand, naturally the use of a *ratio of true to no relationships in a discipline or area of endeavor* immediately raises the question of how anyone could possibly estimate such a value since it is almost impossible to determine when even one research finding is categorically true (short of at least one rigorous replication). But that's the beauty (and utility) of theoretical models: they allow us to *assume* any (or as many different) values as we please for a presently unknowable construct.

The model itself employs a simple algebraic formula (perhaps first proposed by Wacholder et al., 2004) which can be used to ascertain the prevalence of false-positive results in an entire research arena. (Conversely, subtracting this value from 1.0 obviously provides the probability of the average scientific result being correct—hence a *true-positive* result.)

So while correctly estimating the number of true effects that actually exist in an entire scientific field is, to say the least, difficult, Professor Ioannidis wisely chose what appeared to be a realistic example from a field at the border of psychology and genetics—a field with a large literature designed to locate statistically significant correlations between various single-nucleotide polymorphisms (SNPs; of which there are an estimated 10 million) and various psychological constructs and diagnoses, such as general intelligence or schizophrenia (an actual, if unfortunate, empirical example of which will be presented shortly).

SNPs, pronounced "snips," constitute the primary source of genetic variations in humans and basically involve postconception changes in the sequence of a single DNA "building block." The overwhelming majority are benign mutations that have no known untoward or beneficial effects on the individual, but, in the proper location, they have the capacity to affect a gene's functions—one of which is increased susceptibility to a disease.

Ioannidis therefore chose this data mining arena as his example, assuming that 100,000 gene polymorphisms might be a reasonable estimate for the number of possible candidates for such an inquiry (i.e., the denominator of the required ratio), accompanied by a limited but defensible guess at the likely number of SNPs that might actually play a role (the numerator) in the specific psychological attribute of interest. So, given these assumptions, let's suspend judgment and see where this exercise takes us.

For his imputed values, Ioannidis chose 0.05 for the significance criterion (customarily employed in the genomic field at that time but fortunately no longer), 0.60 for the amount of statistical power available for the analysis, and 10 for the number of polymorphisms likely to be associated with the attribute of interest, which Ioannidis hypothetically chose to be schizophrenia. (Dividing the best guess regarding the number of true relationships [10] by the number of analyses [100,000] yields the proportion of "true effects" in this hypothetical domain.)

Plugging these three values into the above-mentioned modeling formula produced an estimated false-positive rate above the 50% level and hence far above the 5% rate of false-positive results posited in Table 2.1. (And this, in turn, indicated that any obtained statistically significant relationship close to a p-value of 0.05 between a gene and the development of schizophrenia would probably be false.)

Ioannidis then went on to add two different scenarios to his model, both of which are known to increase the proportion of published false-positives results.

1. The rate of QRPs (i.e., systematic data analyses and/or investigator procedural practices that have been demonstrated to artifactually enhance the chances of producing statistical significant findings) and
2. A facet of publication bias in which 10 research teams are investigating the same topic but only 1 of the 10 finds statistically significant results (which of course means that this single positive finding would be considerably more likely to be published—*or even submitted for publication*—than would the other nine non-statistically significant results).

Not surprisingly the results of these additions to the model produced estimated false-positive percentages even higher than the original less restrictive model. So Ioannidis, not one to mince words, summarized his conclusions under the previously mentioned, very explicit subhead: "Most Research Findings Are False for Most Research Designs and for Most Fields"—which very nicely mirrors the article's equally pejorative title.

It is hard to say whether this subheading and the article's title are accurate or not, but certainly both are absolute "generalization whoppers." And, as is true for most models' assumptions, anyone can quibble with their real-world validity, including those employed in this study or even the almost universally accepted constructs of p-values, statistical power, and the normal curve.

However, since we *do* design and analyze our research based on these particular assumptions, it makes sense to use them in our efforts to estimate false-positive rates. So regardless of whether or not we agree with Professor Ioannidis's point estimates, we all owe him a debt for his modeling efforts and the following six corollaries he presents related to the genesis of false-positive results.

Hopefully, he will not object to these "corollaries" being repeating verbatim here or my attempts at succinct explanations for their mechanisms of action.

1. *The smaller the studies conducted in a scientific field, the less likely the research findings are to be true.* Studies with small sample sizes are generally associated with less statistical power, but when such studies happen to generate positive results they are considerably more likely to be incorrect than their high-powered counterparts. However, since low-powered studies are more common than high-powered ones in most disciplines, this adds to these disciplines' false-positive rates. (As mentioned, low statistical power can also be a leading cause of false-negative results, but this is less problematic because so few negative studies are published.)

2. *The smaller the effect sizes in a scientific field, the less likely the research findings are to be true.* As fields mature and the low hanging fruit has already been harvested, their effects become smaller and thus require increasingly larger samples to maintain acceptable statistical power. If these sample size compensations are not made accordingly, then power decreases and the rate of false positives increases. (And when effects move toward the limits of our instruments' capacity to reliably detect them, erroneous findings increase accordingly.)

3. *The greater the number and the lesser the selection of tested relationships in a scientific field, the less likely the research findings are to be true.* This corollary might have been stated more clearly, but what Ioannidis apparently means is that fields with lower pre-study

probabilities of being true (e.g., genetic association studies in which thousands upon thousands of relationships are tested and only a few true-positive effects exist) have a greater prevalence of false-positive results in comparison to fields in which fewer hypotheses are tested since said hypotheses must be informed by more and better preliminary, supportive data (e.g., large medical randomized controlled trials [RCTs], which are quite expensive to mount and must have preliminary data supporting their hypotheses before they are funded). In addition, clinical RCTs (perhaps with the exception of psychological and psychiatric trials) tend to be more methodologically sophisticated (e.g., via the use of double-blinded placebo designs) and regulated (e.g., the requirement that detailed protocols be preregistered, thereby decreasing the prevalence of a posteriori hypothesis changes). These conditions also reduce the prevalence of false-positive results.

4. *The greater the flexibility in designs, definitions, outcomes, and analytical modes in a scientific field, the less likely the research findings are to be true.* Here, the difference between publishing practices in high-impact, hypothesis-testing medical journals versus social science outlets is even greater than for the previous corollary. Efficacy studies such as those published in the *Journal of the American Medical Association* or the *New Journal of Medicine* customarily involve randomization of patients; a detailed diagram of patient recruitment, including dropouts; double blinding; veridical control groups (e.g., a placebo or an effective alternative treatment); a recognized health outcome (as opposed to idiosyncratic self-reported ones constructed by the investigators), pre-registration of the study protocol, including data analytic procedures; intent-to-treat analyses; and the other strategies listed in the Consolidated Standards of Reporting Trials (CONSORT) Statement of clinical medical trials (Schulz, Altman, & Moher for the Consort Group, 2010). Publication standards in the social sciences are far more "flexible" in these regards, most notably perhaps in the sheer number of self-reported idiosyncratic outcome variables that are quite distal from any recognized veridical social or behavioral outcome. More importantly the "greater flexibility" mentioned in this corollary also entails a greater prevalence of QRPs in the design and conduct of studies (see Chapter 3).

> 5. *The greater the financial and other interests and prejudices in a scientific field, the less likely the research findings are to be true.* Rather self-explanatory and encompasses self-interest, bias, fraud, and misconduct—all of which will be discussed in later chapters. An example of the biasing effects due to financial interests will be discussed with respect to pharmaceutical research in Chapter 10.
> 6. *The hotter a scientific field (with more scientific teams involved), the less likely the research findings are to be true.* If nothing else, "hotter" fields encourage publication bias such as via the author's genetic scenario in which the first team to achieve statistical significance is more likely to publish its results than the first team that finds no statistically significant effect.

And while Dr. Ioannidis' assumptions regarding genetic association studies (definitely a hot field) may appear unrealistic, an extremely impressive review of such studies involving a single genetic location and susceptibility to a specific disease suggests otherwise (Hirschhorn, Lohmueller, Byrne, & Hirschhorn, 2002). This latter team initially found 603 statistically significant findings associated with 268 individual genes that were associated with 603 statistically significant findings, 166 of which had been studied three or more times, and only six of these proved to be consistently replicable. So for those not keeping track, this makes Ioannidis's assertion that "most positive findings are false" appear rather modest.

A Second Modeling Exercise

Let's now consider a second modeling exercise that may have more applicability for social and behavioral experimentation with actual human participants. This model also requires an estimate regarding the prior probability of true effects and basically employs the same formula used by Ioannidis and proposed by Wacholder et al. However, since this one targets an entire discipline's experimentation rather than multiple analyses on the same dataset, its prior probability estimate may be somewhat of a greater stretch (but perhaps more applicable to experimental research).

> **Is the Replicability Crisis Overblown? Three Arguments Examined**
>
> Harold Pashler and Christine R. Harris (2012)
>
> Targeting psychological experimentation as a whole, Pashler and Harris define the proportion of true-positive effects in their discipline as the percentage of effects that "researchers look for actually exist." They posit 10% as the most reasonable estimate (which is obviously a far cry from Ioannidis's 10^{-4} choice for genetic data monitoring). However, it may actually be an overestimate for disciplines such as educational research, which appears to lean heavily on trivial and repetitive hypotheses (Bausell, 2017) and therefore would presumably be excluded from Pashler and Harris's 10% estimate.
>
> Inserting this 10% value into the formula along with the average power for psychological research (≈ 0.50) for detecting a titular alpha level of 0.05 (assuming an average effect size and a sample size of 20 participants per group in a two-group study) yielded a *discipline false-positive* estimate of 56% within the experimental psychology literature. Which, coincidentally (or not), is quite consonant with Ioannidis's "inflammatory" 2005 conclusion that "most research findings are false for most research designs and for most fields"—or at least for the field of experimental psychology.
>
> Again, the primary weakness of this model resides in the choice of the imputed value for the prior probability of true effects (that "researchers look for actually exist"), which, of course, is subject to change over time. But since Harold Pashler and Christine Harris *are* well regarded psychological researchers, 10% is probably as reasonable an estimate as any.

However, as previously mentioned, the primary advantage of modeling resides in the ability to input as many different determinants of the endpoint of interest as the modeler desires. So, in this case, I have taken the liberty of expanding Pashler and Harris's illustrative results by adding a few additional values to the three input constructs in Table 2.2. Namely:

1. The prevalence of "true" disciplinary effects (.05 and .25 in addition to .10),
2. statistical power (.80 currently most often recommended and .50 to .35), and

Table 2.2 Estimation of false-positive results model

A: Proportion of discipline-wide studies assumed to have true effect	B: Power	C: Alpha (actual)	D: Proportion of false-positive results
.050	.800	.050	.54
.050	.500	.050	.66
.050	.350	.050	.73
.100	.800	.050	.36
.100	.500	.050	.47
.100	.350	.050	.56
.250	.800	.050	.16
.250	.500	.050	.23
.250	.350	.050	.30
.050	.800	.025	.37
.050	.500	.025	.49
.050	.350	.025	.58
.100	.800	.025	.22
.100	.500	.025	.31
.100	.350	.025	.39
.250	.800	.025	.09
.250	.500	.025	.13
.250	.350	.025	.18
.050	.800	.010	.19
.050	.500	.010	.28
.050	.350	.010	.35
.100	.800	.010	.10
.100	.500	.010	.15
.100	.350	.010	.20
.250	.800	.010	.04
.250	.500	.010	.06
.250	.350	.010	.08

Table 2.2 Continued

A: Proportion of discipline-wide studies assumed to have true effect	B: Power	C: Alpha (actual)	D: Proportion of false-positive results
.050	.800	.005	.11
.050	.500	.005	.16
.050	.350	.005	.21
.100	.800	.005	.05
.100	.500	.005	.08
.100	.350	.005	.11
.250	.800	.005	.02
.250	.500	.005	.03
.250	.350	.005	.04

3. in addition to the alpha level of .05 (.025, .01, and .005)—the latter, incidentally, being recommended by a consensus panel (Benjamin et al., 2017) for improving the reproducibility of new discoveries.

The operative component of this table is the final column (the proportion of positive results that are false), the values of which range from .73 to .02 (i.e., 73% to 2% of false positives in the psychological literature—or whatever that literature happens to be to which the inputted constructs might apply). That's obviously a huge discrepancy so let's examine the different assumptive inputs that have gone into this estimate (and it definitely is an estimate).

When the hypothesized proportion of true effects that scientists happen to be looking for ranges between 5% and 25% (the first column), the proportion of false-positive results (Column D) are powerfully affected by these hypothesized values. Thus if the discovery potential (Column A) is as low as .05, which might occur (among other possibilities) when scientists in the discipline are operating under completely fallacious paradigms, the average resulting proportion of false-positive results in the literature is .39 and ranges from .11 to .73. When true effects of .10 and .25 are assumed, the estimated published positive effects that are false drop to averages of .25 and .11, respectively.

Similarly as the average statistical power in a discipline increases from .35 to .80 the rate of published false-positive effects (irrespective of the alpha level and the estimated rate of false-positive effects in the literature) drops from .31 to .19. However it is the alpha level which is the most powerful independent determinant of false-positive results in this model. When the alpha is set at .05, the average rate of false-positive results in the literature averaged across the three levels of power and the four modeled assumed level of true effects is .46 or almost half of the published positive results in many scientific literatures. (And positive results, it will be recalled, comprise from .92 to .96 of psychology's published literature.)

An alpha level of .01 or .005 yields much more acceptable false-positive results (.16 and .09, respectively, averaged across the other two inputs). An alpha of .005, in fact, produces possibly acceptable false-positive results (i.e., < .20 or an average of .075) for all three projected rates of true effects and power levels of .80 and .50. Not coincidentally, a recent paper (Benjamin et al., 2017) in *Nature Human Behavior* (co-authored by a veritable Who's Who host of reproducibility experts) recommended that studies reporting *new discoveries* employ a p-value of .005 rather than .05. Ironically, a similar modeling conclusion was reached more than two decades ago in the classic article entitled "Effect Sizes and p Values: What Should Be Reported and What Should Be Replicated" (Greenwald, Gonzalez, Harris, & Guthrie, 1996).

For those interested in history, it could be argued that the first author, Anthony Greenwald, one of the 73 authors of the *Nature Human Behavior* paper just cited, foresaw the existence of the reproducibility crisis almost half a century ago in a classic paper "Consequences of Prejudice Against the Null Hypothesis" (1975). This paper also detailed what surely must have been the first (and, if not the first, surely the most creative) modeling demonstration of false-positive results so, just for the fun of it, let's briefly review that truly classic article.

Consequences of Prejudice Against the Null Hypothesis

Anthony G. Greenwald (1975)

To begin with, Greenwald sent a brief questionnaire to 48 authors and 47 reviewers of the *Journal of Personality and Social Psychology* querying them about practices regarding the alpha level, statistical power, and null results. (A quite acceptable 78% response rate was obtained.)

The results (recall that this survey was conducted more than four decades ago) of most interest to us today were as follows:

1. The mean probability level deemed most appropriate for rejecting a null hypothesis was .046 (quite close to the conventional .05 level).
2. The available statistical power deemed satisfactory for accepting the null hypothesis was .726 (also quite close to the standard .80 power recommendation for design purposes). Interestingly, however, only half of the sample responded to this latter query, and, based on other questions, the author concluded that only about 17% of the sample typically considered statistical power or the possibility of producing false-*negative* results prior to conducting their research. (In those days the primary effects of low power on reproducibility were not widely appreciated, and low power was considered problematic primarily in the *absence* of statistical significance.)
3. After conducting an initial full-scale test of the primary hypothesis and not achieving statistical significance, only 6% of the researchers said that they would submit the study without further data collection. (Hence publication bias is at least half a century old, as is possibly this QRP as well [i.e., presumably collecting additional data to achieve statistical significance without adjusting the alpha level].) A total of 56% said they would conduct a "modified" replication before deciding whether to submit, and 28% said they would give up on the problem. Only 10% said that they would conduct an exact replication.

Greenwald then used these and other questionnaire responses to model what he concluded to be a dysfunctional research publication system in which "there may be relatively few publications on problems for which the null hypothesis is (at least to a reasonable approximation) true, and of these, a high proportion will erroneously reject the null hypothesis" (p. 1). It is worth repeating that this prescient statement regarding reproducibility was issued almost a half-century ago and was, of course, largely ignored.

Greenwald therefore went on to conclude that by the time the entire process is completed the actual alpha level may have been raised from .05 to as high as .30 and the researcher "because of his investment in confirming his theory with a rejection of the null hypothesis, has overlooked the possibility that the observed x-y relationship may be dependent on a specific manipulation, measure, experimenter, setting, or some combination of them" (p. 13). Then, under the heading "Some Epidemics of Type I Error" (aka false-positive results), he buttresses his case involving the inflation of alpha levels via several well-received past studies that found their way into textbooks but were later discarded because they couldn't be replicated.

So not only did Greenwald recognize the existence of a reproducibility crisis long before the announcement of the present one, he also (a) warned his profession about the problems associated with publication bias, (b) advocated the practice of replication at a time when it was even less common that it is today, and (c) provided guidelines for both the avoidance of publication bias and false-positive results. And while he is recognized to some extent for these accomplishments, he deserves an honored place in the pantheon of science itself (if there is such a thing).

So What Should We Make of These Modeling Efforts?

Are these models correct, incorrect, evidence of the existence of a true scientific crisis, or overblown empirical warnings that the sky is falling? No definitive answers exist for these questions because the questions themselves are inappropriate.

What these models provide, coupled with the incontrovertible existence of publication bias, is a clear warning that the published scientific literature may be characterized by far more false-positive results than either the scientific community or the public realize. But we've only started this story. We haven't even considered the calamitous effects of the primary villains of this story: a veritable host of QRPs and institutional impediments that foster the production of false-positive results. So let's consider some of these along with what appears to be an oxymoron: a downright amusing model illustrating the effects of QRPs on irreproducibility.

References

Bausell, R. B. (2017). *The science of the obvious: Education's repetitive search for what's already known.* Lanham, MD: Rowman & Littlefield.

Bausell, R. B., & Li, Y. F. (2002). *Power analysis for experimental research: A practical guide for the biological, medical, and social sciences.* Cambridge: Cambridge University Press.

Benjamin, D. J., Berger, J. O., Johannesson, M., et al. (2017). Redefine statistical significance. *Nature Human Behavior, 2,* 6–10.

Cohen, J. (1988). *Statistical power analysis for the behavioral sciences* (2nd ed.). Hillsdale, NJ: Lawrence Erlbaum.

Greenwald, A. G. (1975). Consequences of prejudice against the null hypothesis. *Psychological Bulletin, 82,* 1–20.

Greenwald, A. G., Gonzalez, R., Harris, R. J., & Guthrie, D. (1996). Effect sizes and p values: What should be reported and what should be replicated? *Psychophysiology, 33,* 175–183.

Hirschhorn, J. N., Lohmueller, K., Byrne, E., & Hirschhorn, K. (2002). A comprehensive review of genetic association studies. *Genetics in Medicine, 4,* 45–61.

Ioannidis, J. P. A. (2005). Why most published research findings are false. *PLoS Medicine, 2,* e124.

Pashler, H., & Harris, C. R. (2012). Is the replicability crisis overblown? Three arguments examined. *Perspectives on Psychological Science, 7,* 531–526.

Schulz, K. F., Altman, D. G., & Moher, D. for the Consort Group. (2010). CONSORT 2010 Statement: Updated guidelines for reporting parallel group randomized trials. *British Medical Journal, 340,* c332.

Wacholder, S., Chanock, S., Garcia-Closas M., et al. (2004). Assessing the probability that a positive report is false: An approach for molecular epidemiology studies. *Journal of the National Cancer Institute, 96,* 434–442.

3

Questionable Research Practices (QRPs) and Their Devastating Scientific Effects

Publication bias is the most commonly ascribed cause and/or facilitator of the presumed high prevalence of false-positive results in the scientific literatures. However, in this case and facilitators are co-dependent with even more distal antecedents such as investigator ambitions, the need to provide for one's families, ignorance (caused by inadequate mentoring or day-dreaming in class), and undoubtedly several others that don't need to be identified.

So suffice it to say that while publication bias appears to be a facilitator of false-positive results, it is neither the only culprit nor probably the most important one. What are more impactful are an impressive set of investigative behaviors that conspire to produce both publication bias and false-positive results. And that dubious honor is reserved for questionable research practices (QRPs) that are almost entirely behavioral in nature.

In a sense the sheer number, diversity, and astonishingly high disciplinary prevalence of these miscreants constitute the primary linchpin of our story. What makes them even more problematic is the fact that the vast majority are never reported in the studies employing them. So our task in this chapter is to list and explicate these culprits. But first, a natural question presents itself.

Just How Prevalent Are These So-Called Contraindicated Practices?

The short answer is that we don't know, and there is really no completely accurate way of finding out. However there have been a number of surveys estimating the prevalence of both QRPs and outright fraudulent behaviors, and, like just about everything else scientific, these efforts have been the subject of at least one meta-analysis.

We owe this latter gift to Daniele Fanelli (2009), already introduced as an important and frequent contributor to the reproducibility literature, who has graced us with a meta-analysis involving 18 such surveys. Unfortunately the response rates of most of them were unimpressive (some would say inadequate), and, since they unavoidably involved self- or observed reports of antisocial behaviors, the results produced were undoubtedly underestimates of the prevalence of QRPs and/or fraud.

While the survey questions and sampling procedures in the studies reviewed by Dr. Fanelli were quite varied, the meta-analysis' general conclusions were as follows:

1. The percentage of self-reported actual *data fabrication* (the behavior universally deemed to constitute the worst example of fraudulent scientific practice) was low (approximately 2%),
2. Inevitably the respondents suspected (or had observed) higher percentages of misconduct among other scientists than they themselves had committed, and
3. "Once methodological differences were controlled for, cross study comparisons indicated that samples drawn exclusively from medical (including clinical and pharmacological) research reported misconduct more frequently than respondents in other fields or in mixed samples." (p. 10)

However, with respect to this third point, a more recent and unusually large survey (John, Loewenstein, & Prelec, 2012) casts some doubt thereupon and may have moved psychology to the head of the QRP class.

In this huge survey of almost 6,000 academic psychologists (of which 2,155 responded), questionnaires were emailed soliciting self-reported performance of 10 contraindicated practices known to bias research results. An intervention was also embedded within the survey designed to increase the validity of responses and permit modeling regarding the prevalence of the 10 targeted QRPs although, for present purposes, only the raw self-admission rates of these 10 QRPs listed in Table 3.1 will be discussed.

A cursory glance at these results indicates an unusually high prevalence of many of these practices, especially since they are based on self-reports. While a great deal of additional information was collected (e.g., respondents' opinions of the justifiability of these practices and whether their prevalence

Table 3.1 Self-admitted questionable research practices (QRPs) (for at least one episode) among academic psychologists

Questionable research practices	Prevalence
1. Failing to report all outcome variables	66.5%
2. Deciding whether to collect more data	58.0%
3. Selectively reporting studies that "worked"	50.0%
4. Deciding whether to exclude data following an interim analysis	43.4%
5. Reporting an unexpected finding as a predicted one	35.0%
6. Failing to report all study conditions	27.4%
7. Rounding off p-values (e.g., .054 to .05)	23.3%
8. Stopping study after desired results are obtained	22.5%
9. Falsely claiming that results are unaffected by demographic variables	4.5%
10. Falsifying data	1.7%

was greater in universities other than the respondents' [they were]), four of the author's conclusions stand out.

1. "Cases of clear scientific misconduct have received significant media attention recently, but less flagrantly *questionable research practices may be more prevalent and, ultimately, more damaging to the academic enterprise* [emphasis added]" (p. 524).
2. "Respondents considered these behaviors to be defensible when they engaged in them . . . but considered them indefensible overall" (p. 530) [scientific methodology evolves over time, which is one reason tolerance was previously recommended for some past sins—even the one attributed to our hypothetical graduate student].
3. "All three prevalence measures [which also included modeled QRP rates plus a follow-up survey of respondents] point to the same conclusion: a surprisingly high percentage of psychologists admit to having engaged in QRPs" (p. 530).
4. And most poignantly: "QRPs can waste researchers' time and stall scientific progress, as researchers fruitlessly pursue extensions of effects that are not real and hence cannot be replicated. More generally, the prevalence of QRPs raises questions about the credibility of research findings and threatens research integrity by producing unrealistically elegant results that may be difficult to match without engaging in such practices oneself. This can lead to a 'race to the bottom,' with questionable research begetting even more questionable research" (p. 531).

However, disciplinary differences in the prevalence of QRP practice (and especially when confounded by self-reports vs. the observance of others) are difficult to assess and not particularly important. For example, in the year between the Fanelli and the John et al. publications, Bedeian, Taylor, and Miller (2010) conducted a smaller survey of graduate business school faculty's *observance* of colleagues' committing 11 QRPs during the year previous to their taking the survey. This likewise produced a very high prevalence of several extremely serious QRPs, with the fabrication of data being higher in this survey (26.8%) than any I have yet encountered. Recall, however, that this and the following behaviors are reports of others' (not the respondents') behaviors:

1. A posteriori hypothesizing (92%),
2. Not reporting some methodological details or results (79%),
3. Selectively reporting data that supported the investigators' hypotheses (78%), and
4. Using ideas without permission or giving due credit (70%).

Now, of course, surveys are near the bottom of most triangles of evidence if they even make the list in the first place, and we shouldn't put too much stock in their point estimates given response rates and other biases inherent in self-reports. In addition, making comparisons between surveys is also challenging since (a) even questionnaires addressing the same topics usually contain slightly different items or item wordings, (b) the response instructions often differ (e.g., time period covered or self- vs. observed behaviors), and (c) the classification of behaviors differs (e.g., scientific fraud vs. QRPs vs. misconduct).

In way of illustration, while they would probably agree with some of the conclusions of the large John et al. survey, Drs. Fiedler and Schwarz (2015) argue (buttressed by a survey of their own) that the former overestimated the prevalence of QRPs due to ambiguities in wording and the fact that "prevalence" of a behavior cannot be calculated or inferred from the proportions of people who engaged in these behaviors only once. As an example, they pose the following explanatory question: "What does the proportion of people who ever told a lie in their life reveal about the prevalence of lying?"

For our purposes, however, the actual prevalence of individual untoward research behaviors is not as important as the fact that so many scientists do appear to engage in at least some of them and/or have observed others doing so. Also, how we classify these behaviors is not particularly important, although one way to draw the line of demarcation is between ignorance and willfulness. *Everyone*, for example, knows that fabricating data is fraudulent, but not bothering to report all of ones variables might simply be due to substandard training and mentoring. Or it might be cultural based on the science in question.

However, while ignorance may have served as a passable excuse for engaging in some detrimental practices in the past, it does nothing to mitigate their deleterious effects on science and should no longer be tolerated given the amount of warnings promulgated in the past decade or so. For it is worth repeating that the ultimate effect of QRPs is the potential *invalidation* of the majority of some entire empirical literatures in an unknown number of scientific disciplines. And that, going forward, is simply not acceptable.

Modeling the Effects of Four Common Questionable Research Practices

Fortunately, just as a meta-analysis can be (and has been) conducted on just about every scientific topic imaginable, so can just about anything be modeled. And that latter truism (or unsubstantiated exaggeration) leads to one of the most iconoclastic (and definitely one of my favorite) articles in psychology—and most certainly possessive of one of that discipline's most pejorative titles.

> **False-Positive Psychology: Undisclosed Flexibility in Data Collection and Analysis Allows Presenting Anything as Significant**
>
> Joseph Simmons, Leif Nelson, and Uri Simonsohn (2011)
>
> This seminal article presents us with two genres of simulations quite different from the ones discussed in the previous chapter, the second of which is actually quite entertaining in a masochistic sort of way. But, regardless of the effect it has upon us, it is unlike anything that I have personally encountered in a peer reviewed journal.

Both models address the previous contention that the practice of a few inflationary QRPs allows savvy investigators (who must value reaping the rewards provided by their discipline over preserving its integrity) to increase their chances of producing statistically significant results to the point that a p-value < 0.05 is *more* likely to occur than a p-value > 0.05.

Simulation 1: How to subvert the already generous alpha level of .05 and continue the decade- old process of constructing a trivial (but entertaining) science.

This simulation is the more conventional of the two. Here 15,000 random samples were drawn from a normal distribution to assess the impact of the following four QRPs: (a) choosing which of two correlated outcome variables to report (plus an average of the two), (b) not specifying sample sizes a priori but beginning with 20 observations per cell and adding 10 more observations if statistical significance is not yet obtained, (c) using three experimental conditions and choosing whether to drop one (which produced four alternate analytic approaches), and (d) employing a dichotomous variable and its interaction with the four combinations of analytic alternatives detailed in (c).

Letting the computer do the heavy lifting, the p-values actually obtained for the 15,000 samples were computed and contrasted to three titular alpha levels (< 0.1, <0.05, and $< .01$). For the most commonly used level of 0.05, the percentage of false-positive results resulting from the four chosen scenarios were (recalling that a 5% rate would be expected to occur by chance alone) as follows:

(a) Use of two moderately correlated outcome variables (9.5%),
(b) Addition of 10 more Ss (7.7%) when statistical significance in not found,
(c) Dropping a treatment group or using all three (12.6%), and
(d) Adding a dichotomous variable plus its interaction with the treatments as a covariate (11.7%).

Now, as bad as this seems, anyone disingenuous (or untrained) enough to use any one of these strategies is also quite likely to use more than one (or even some additional QRPs) so the authors also included the effects of some of the combinations of the four which produced false-positive estimates ranging from 14.4% to a whopping 60.7%.

It is worth repeating that the just listed percentages assumed an alpha level of .05, but the authors also provided the same information for alphas of .10 and .01 (see table 1 in the original article). As would be expected, the QRP effects are much higher for an alpha of .10 (which is comparable to a one-tailed alpha of .05) and considerably lower for .01. (For example, the 14.4% and 60.7% range estimated for multiple QRP sins reduces to 3.3% and 21.5%, respectively, for an alpha of .01). And, as suggested by Benjamin, Berger, Johannesson, et al. (2017); Greenwald, Gonzalez, Harris, and Guthrie (1996); and the simulations in Table 2.1, the deleterious effects of the individual QRPs and their combinations would be greatly reduced if the titular alpha level were to be decreased to .005. However, since psychology and the vast majority of other scientific disciplines (at least the social sciences) aren't likely to adopt an alpha of .005 anytime soon, the criterion of 0.05 was employed in Simmons, Nelson, and Simonsohn's other astonishing simulation.

Simulation 2: Also, how to subvert the already generous titular alpha level of .05 and continue the process of constructing an irreproducible (but entertaining) science.

This one must surely be one of the first modeling strategies of its kind. Two experiments were reported, the first apparently legitimate (if trivial since it was based on Daryl Bem's infamous study "proving" that future events can influence past events) using a soft, single item and 30 undergraduates (also a typical sample size for psychology experiments) who were randomized to listen to one of two songs: "Hot Potato" (a children's tune that the undergraduates would most likely remember from childhood as the experimental condition) versus a rather blah instrumental control tune ("Kalimba"). Note that the experimental children's song was perfectly and purposefully selected to create an immediate reactive response by asking the undergraduates if they felt older immediately after listening to it. And, sure enough, employing the age of the participants' father as a covariate (which basically made no sense and was not justified) the experimental group listening to "Hot Potato" reported that they had felt significantly older ($p = .033$) on a 5-point scale (the study outcome) than the group hearing the nonreactive control song.

For their second study the authors performed a "conceptual" replication of the one just described, but this time embedding all four of the computer-modeled QRPs from Simulation 1. However, the authors first reported the study's design, analytic procedure, and results *without* mentioning any of these QRPs, which made it read like an abstract of a typically "successful" psychology publication:

> Using the same method as in Study 1, we asked 20 University of Pennsylvania undergraduates to listen to either "When I'm Sixty-Four" by The Beatles or "Kalimba." Then, in an ostensibly unrelated task, they indicated their birth date (mm/dd/yyyy) and their father's age. We used father's age to control for variation in baseline age across participants. An ANCOVA revealed the predicted effect: According to their birth dates, people were nearly a year-and-a-half younger after listening to "When I'm Sixty-Four" (adjusted M = 20.1 years) rather than to "Kalimba." (Adjusted M = 21.5 years), p = .040). (p. 1360)

Then, they confessed their sins:

1. A second intervention (listening to the Beatles' "When I'm Sixty-Four") was employed in lieu of "Hot Potato" ("Hot Potato" did not reach statistical significance this time around as compared to the "Kalimba" control and hence was dropped from the analysis and not mentioned in the simulated report);
2. The participants' (a) father's age, (b) gender, (c) and the gender's interaction with the experimental conditions were employed as covariates (a plethora of other variables were also included and apparently auditioned as covariates, such as participants' mother's age, their political persuasion, an item about Canadian quarterbacks, and so forth);
3. Following the analytic process a new outcome variable (the participants' *adjusted* ages) was employed since the first study's self-reported variable (i.e., whether the respondents felt older) also did not reach statistical significance this time; and
4. The analytic process itself included post hoc interim analyses conducted after additional participants were run until statistical significance was achieved—at which time the exercise was terminated.

The Authors' (Simmons, Nelson, and Simonsohn) Suggestions for Improvement

These suggestions were presented via two headings ("Requirements for Authors" and "Guidelines for Reviewers"). Some of these are obvious given the second simulation and some, unfortunately, may not go far enough. First the author requirements (these are in italics and numbered as they appear on pages 1362–1363 of the original article):

1. *Authors must decide the rule for terminating data collection before data collection begins and report this rule in the article.* Any serious institutional review board (IRB) requires such a statement in proposals submitted to them along with a rationale for prematurely terminating a study or adding more participants to the original sample size justification if applicable. Submission of these documents should probably be required by journals prior to publication if the same information is not preregistered.

2. *Authors must collect at least 20 observations per cell or else provide a compelling cost-of-data-collection justification.* This one is unclear since an N of 68 per group is necessary to produce adequate statistical power for a typical social science effect size of 0.50. Ironically, a number of authors (e.g., Bakker, van Dijk, & Wicherts, 2012) have lamented the fact that the typical power available for psychological experiments can be as low as 0.35, and (also ironically) 20 participants per cell doesn't even quite meet this low criterion for a two-group study.

3. *Authors must list all variables collected in a study.* This one is quite important and of course should be an integral part of the preregistration process for relatively simple experiments such as the ones described here. Large clinical random controlled trials (RCTs) (as well as databases used for correlational studies) often collect a large number of demographic, background, health, and even cost data, a simple list of which might run several pages in length. Thus perhaps this suggestion could be loosened a bit for some types of research. However, those used as covariates, blocking variables, subgroup analyses, and, of course, primary outcomes must be prespecified accordingly.

4. *Authors must report all experimental conditions, including failed manipulations.* The failure to include an extra intervention or comparison group should be considered censorable misconduct.
5. *If observations are eliminated, authors must also report what the statistical results are if those observations are included.* And, of course, a rationale for the inclusion-exclusion criteria and the treatment of outliers (with definitions) should be provided a priori.
6. *If an analysis includes a covariate, authors must report the statistical results of the analysis without the covariate.* This is an excellent point and is seldom adhered to. Additionally, the actual covariate–outcome correlation should be reported (which is almost never done). Covariates always adjust the meanings of outcomes to a certain extent, so it is extremely important to ensure that the adjusted outcome conceptually remains a variable of interest. It is not immediately apparent, for example, exactly what adjusting "feeling older" in the first experiment or adjusting "participants' ages" based upon "fathers' ages" in the second experiment winds up producing.

The second list of guidelines is presented for peer reviewers. These will be supplemented in Chapter 9, which discusses publishing concerns.

1. *Reviewers should ensure that authors follow the requirements* [presumably set by the journal or professional guidelines]. This is especially important for preregistration of key elements of the experimental process, as illustrated in the authors' second simulation. And as the authors note: "If reviewers require authors to follow these requirements, they will" (p. 1363).
2. *Reviewers should be more tolerant of imperfections in results.* "Underpowered studies with perfect results are the ones that should invite extra scrutiny" (p. 1363).
3. *Reviewers should require authors to demonstrate that their results do not hinge on arbitrary analytic decisions.* It might even be suggested that arbitrary analytic decisions shouldn't be made in the first place.
4. *If justifications of data collection or analysis are not compelling, reviewers should require the authors to conduct an exact replication.* With apologies to Drs. Simmons, Nelson, and Simonsohn, one might wonder if it makes sense to perform an "exact" self-replication of a study with design flaws or non-compelling analytic procedures.

> While the original effect did not replicate in this simulation, might not some flaws that produced a false-positive result in an original study (such as an obvious demand characteristic coupled with a lack of blinding) also produce a false-positive result in an exact replication? Which leads to an even more bizarre question.

Would the Two Studies Conducted by the Simmons Team Replicate?

Now, as previously mentioned, I am quite fond of these studies but we all like research results that reinforce our biases and predilections. So what about the reproducibility of these two studies? Well, their authors tell us that the first one didn't replicate when employing identical procedures and analyses. But what about the second study?

There's no way of knowing short of replicating it, but that's the wrong question anyway. A better one might be

> Certainly the contrived "Hot Potato" experiment is a tour de force illustration of how skillful manipulations of QRBs are *capable* of creating non-replicable, false-positive results, but do these miscreants also produce comparable false-positive result in the *published* literature?

Said another way, the Simmons et al. study demonstrated the effects that QRPs *could* have on the artifactual achievement of statistically significant results. So while the study may not have demonstrated (a) the actual occurrence of artifactually significant results or (b) that their four QRPs actually *do* result in artifactual statistical significance in the published literature, surely it demonstrated their *potential* for doing so.

Of course, the authors' first simulation involving the likely effects of their four key QRPs provides strong evidence that such practices also have the potential to dramatically inflate the obtained p-value and hence produce false-positive results—evidence buttressed by the preceding survey results demonstrating the high prevalence of these and other QRPs in the actual conduct of scientific research.

But even more convincingly, a group of management investigators fortuitously picked up where the Simmons et al. study left off and demonstrated the *actual* effects of QRPs on the production of statistically significant findings. In my personal opinion this particular study provides one of the best *empirical* documentations of the untoward effects of QRPs and their implicit relationship to both publication bias and false-positive results.

The investigators accomplished this impressive feat by longitudinally following a group of studies from their authors' dissertation to their subsequent publication in peer reviewed journals. And, as if this wasn't sufficiently impressive, the title of their article rivals the iconic entries of both Ioannidis's ("Why Most Published Research Findings Are False") and Simmons et al.'s ("False-Positive Psychology: Undisclosed Flexibility in Data Collection and Analysis Allows Presenting Anything as Significant").

The Chrysalis Effect: How Ugly Initial Results Metamorphosize into Beautiful Articles

Ernest O'Boyle, Jr., George Banks, and
Erik Gonzalez-Mule, E. (2014)

Without going into excruciating detail on the study methods, basically, the authors identified management-related dissertations registered between 2000 and 2012 that possessed a formal hypothesis and were subsequently published. The task wasn't as simple as it sounds since the published articles often had different titles and/or failed to mention that they were based on dissertation research (the latter being mandated by published ethics codes of the Academy of Management and the American Psychological Association).

Thankfully the investigators persevered and were able to identify 142 dissertations where there was "overwhelming" evidence that the studies had been subsequently published in a refereed journal. (The average time to publication was 3.29 years.) Altogether (i.e., in both dissertations and journal articles), there were 2,311 hypotheses, 1,978 of which were tested in the dissertations and 978 in the paired articles.

Overall differences between dissertation and journal article results showed that of the 1,978 hypotheses contained in the dissertations, 889 (44.9%) were statistically significant while, of the 978 hypotheses tested in the publications, 645 (65.9%) achieved that status. Or, in the authors' conceptualization of their results, "Our primary finding is that from dissertation to journal article, the ratio of supported to unsupported hypotheses more than doubled (0.82 to 1.00 versus 1.94 to 1.00)" (p. 376).

Another way to view these results is to consider only the 645 hypothesis tests which were common to both dissertations and journal articles. Here, 56 of the 242 (20.6%) negative hypothesis tests in the dissertations somehow changed into positive findings in the published articles, while only 17 of the 323 (4.6%) positive dissertation findings were changed to negative ones. That in turn reflects a greater than four-fold negative to positive change as compared to a positive to negative one.

As for results due to QRPs, perhaps the most creative aspect of this seminal study involved estimating the effects of individual QRPs on the bottom-line inferential changes occurring over time for the 142 paired study versions. (Perhaps not coincidentally the five QRPs in this study basically overlap the four modeled in the Simmons et al. study, which, in all but one case, overlapped the preceding John et al. survey.)

QRP 1: Deletion or addition of data after hypothesis tests. Across the 142 projects, 14 (9.9%) added subjects (as evidenced by increases in sample size from dissertation to journal) and 29 (20.4%) dropped subjects. Overall both adding and deleting participants resulted in increased statistical significance over time (24.5% vs. 10.2%, respectively).

When broken down by adding versus deleting participants, 19% of the effects changed from negative to positive when the sample size was *increased*, while 8.9% (a two-fold reduction) changed in the opposite direction. (Note that this contrast was not statistically significant because of the relatively few studies that increased their sample size over time.) Among the studies that dropped subjects, there was a 2.5-fold difference favoring changes from non-significance to statistical significance as compared to positive to negative changes (28.1% vs. 11.1%, respectively).

QRP 2: Altering the data after hypothesis testing. This potential QRP was assessed in 77 studies in which the sample size did not change over time. (There were 22 cases in which it was not possible to determine whether data were altered.) The authors rationale was that "those studies that added or deleted data have a logical (*but not necessarily appropriate* [emphasis added]) reason why descriptive statistics would change from dissertations to their matched journal publications" (p. 386). Of these 77 studies, 25 (32.5%) showed changes in the means, standard deviations, or interrelations of the included variables, which represented 47 nonsignificant hypothesis tests and 63 statistically significant ones. Following publication, 16 (34%) of the negative studies became positive and 0% changed from positive to negative.

QRP 3: Selective deletion or addition of variables. Deleting the same 22 studies as in QRP 2 (i.e., for which data alteration couldn't be ascertained) left 120 pairs (i.e., 142 – 22). Of these, 90 included instances "where not all of the variables included in the dissertation appeared in the publication and 63 (52.5%) instances where not all of the variables found in the article were reported in the dissertation." (There were 59 studies which both added and dropped variables.) In the dissertations, there were 84 negative tests and 136 positive ones. Adding variables to the published studies resulted in a change from negative to positive of 29.8% as compared to an 8.1% change in the negative direction—a three-fold migration favoring negative to positive change.

QRP 4: Reversing the direction or reframing hypotheses to support data. The authors note that this artifact doesn't necessarily include changing a hypothesis from "the intervention will be efficacious" to "it won't work." Instead it might involve adding a covariate (recall the iconoclastic Simmons et al. simulations) or slightly changing a predicted three-way interaction effect that might actually be significant as hypothesized but not in the expected direction. Here, the good news is that only eight studies representing 22 hypothesis tests were guilty of substantively reframing the original hypothesis. But of course "bad" news usually follows good news in this arena so, in this case, the bad news is that none (0%) of these 22 dissertation hypotheses was originally statistically significant while 17 (77.3%) of the p-values "somehow" changed from $p > 0.05$ to $p < 0.05$ when published.

QRP 5: Post hoc dropping or adding of hypotheses. Of the 142 paired studies, 126 (87.7%) either dropped or added a hypothesis, with 80 doing both. This translates to (a) 1,333 dropped hypotheses of which 516 (38.7%) were statistically significant as opposed to (b) 333 added hypotheses of which 233 (70.0%) were statistically significant. In other words, the new hypotheses were almost twice as likely to be positive as the ones that were dropped from the dissertation.

Qualifiers: The authors quite transparently describe a number of alternative explanations to some of their findings. For example, they note that

> between the dissertation defense and journal publication, it is possible, even likely, that mistakes were identified and corrected, outliers removed, new

> analytic techniques employed, and so on that would be classified as questionable by our criteria [i.e., labeled as QRPs in their study] but were nevertheless wholly appropriate to that particular project. That being said, these changes consistently coincided with increases in statistical significance and increases in the ratio of supported to unsupported hypotheses, and on this basis, we conclude that the preponderance of QRPs are engaged in for nonideal [a nice euphuism] reasons. (p. 392)
>
> Other possible weaknesses identified by the authors in their data included the possibilities that some of the discrepancies noted between dissertation and journal publications might (a) have been mandated by journal editors or peer reviewers based on space limitations or (b) unique to the academic doctoral process (e.g., "forced" on the doctoral students by one or more committee members.)

However, it should be remembered that both the O'Boyle et al. and the Simmons et al. designs do not necessarily lead to definitive *causal* conclusions. With that said, I personally consider these studies to be extremely creative and both their data (actual or simulated) and the conclusions based on them quite persuasive, especially when considered in the context of the other previously presented observational and modeling studies coupled with actual replication results that will be presented shortly.

Perhaps, then, the primary contribution of this and the previous chapter's simulations resides in the facts that

1. The disciplinary-accepted or prespecified alpha level for any given study is probably almost *never* the actual alpha level that winds up being tested at study's end—unless, of course, a study is properly designed, conducted, and analyzed;
2. Many, many investigators—while complying with *some* important and obvious bias-reducing strategies—still conduct studies that are improperly designed, conducted, and analyzed (hence biased in other ways).

So although we may never be capable of ascertaining the true false-positive rate of any discipline (or even the actual correct p-value emanating from any imperfectly designed, conducted, and/or analyzed study), we do know that the percentage of false-positive results for most studies employing an alpha level of 0.05 will be considerably above the 5% figure suggested in cell b of Table 2.1. And assuming the validity of the modeled false-positive rate for the presence of one or more of the four contraindicated practices employed in the Simmons et al. 2011 paper, the false-positive rate would be expected to mutate from the theoretical 5% level to between 7.7% and (a downright alarming) 60.7%.

But while this minimal estimate of an alpha increase of 2.7% (7.7% −5%) may not appear to be a particularly alarming figure, it is important to remember that this translates into tens of thousands of false-positive results. (However, it should also be noted that correlations among outcome variables greater than the 0.50 value employed in this simulation would result in a greater increase in the veridical alpha level.) And, even more discouraging, the projections for the untoward effects of both the four unitary QRPs and their combinations may well be underestimates. And to make matters worse still, the total menu of QRPs soon to be enumerated ensures many more options for producing fallacious statistically significant findings than those modeled in the Simmons et al. paper and validated in the later O'Boyle et al. study.

So perhaps this is a good time to introduce of few more QRPs. But perhaps we should first differentiate between investigator-driven QRPs and inane institutional policies (IIPs) which, like questionable investigator behavior, are also capable of contributing to the prevalence of irreproducible results.

A Partial List of Inane Institutional Scientific Policies and Questionable Research Practices

Inane institutional scientific policies (IISPs) are not directly under the personal control of individual investigators, but this does not imply that they cannot be changed over time by individual scientists through their own advocacy, professional behaviors, or collectively via group pressures. QRPs, on the other hand, are almost exclusively under individual investigators' personal control, although adversely influenced by IISPs and inadequate scientific mentorship.

So first consider the following list of the more common IISP culprits:

1. *Publication bias,* which has already been discussed in detail and is partially due to institutional behaviors involving journal editors, funders,

peer reviewers, publishers, the press, and the public, in addition to individual investigator behaviors. (So this one, like some that follow, constitutes combination QRP-IISP issues.) Researchers often bemoan the fact that journal editors and peer reviewers make negative studies so difficult to publish, but who, after all, are these nefarious and short-sighted miscreants? Obviously the vast majority are researchers themselves since they typically serve in these publishing and funding capacities. It is therefore incumbent upon these individuals to not discriminate against well-conducted nonsignificant studies and to so lobby the institutions for which they work.

2. A *flawed peer review system* that encourages publication bias, ignores certain QRPs, does not always enforce journal guidelines, and sometimes engages in cronyism. But again, who are these peer reviewers? The answer is that almost everyone reading this is (or will be) a peer reviewer at some point in their career. And some, heaven forbid, may even become journal editors which will provide them with an even greater opportunity to influence attitudes and practices among both their peer reviewers and their publishers. (Suggestions for reforming the peer review process will be discussed in some detail in Chapter 9.)

3. *Insufficient scientific mentoring and acculturation of new or prospective investigators.* This one is tricky because senior mentors have important experiential advantages but some are completely "set in their ways," resistant to change, and may not even be aware of many of the issues discussed in this book. However, one doesn't have to be long of tooth to adopt an informal mentoring role and guide new or prospective researchers toward the conduct of methodologically sound research.

4. *A concomitant lack of substantive disciplinary and methodological knowledge on the part of many of these insufficiently mentored investigators.* Some of the onus here lies with these individuals to supplement their own education via the many online or print sources available. However, institutions also bear a very real responsibility for providing ongoing educational opportunities for their new faculty researchers—as well as inculcating the need for self-education, which is freely and conveniently available online.

5. *Institutional fiscal priorities* resulting in untoward pressure to publish and attract external research funds. These issues are largely outside any single individual's personal control but someone must at least attempt

to educate the perpetrators thereof to change their policies—perhaps via organized group efforts.
6. Related to this is the academic administration's seeming adoption of the corporate model of perpetual expansion (physical and financial) resulting in evaluating department heads based on the amount of external grant funds their faculties manage to garner. Such pressures can force senior scientists to spend too much of their time pursuing funding opportunities at the expense of actually engaging in scientific activities or adequately supervising their staff who do. And many of the latter in turn understand that *they* will be evaluated on the number of publications their work generates, which leads us back to publication bias and the multiple geneses of the reproducibility crisis.
7. *The institutionalization of too many disciplines possessing no useful or truly unique knowledge base* and thereby ensuring the conduct of repetitive, obvious, and trivial research. (In the spirit of transparency, this is one of my personally held, idiosyncratic opinions.) One option for individuals stuck in such pseudoscientific professions is to seek opportunities to work with research teams in other more propitious arenas. Alternately (or additionally) they can try mightily to discover what might be a useful and/or parsimonious theory to guide research in their field, which in turn might eventually lead to a scientifically and societally useful knowledge base.
8. And related to Number 7 is the institutional practice of never abandoning (and seldom downsizing) one of these disciplines while forever creating additional ones. Or recognizing when even some mature, previously productive disciplines have exhausted their supply of "low hanging fruit" and hence may be incapable of advancing beyond it. This, of course, encourages trivial, repetitive, and obvious studies over actual discoveries as well as possibly increasing the pressure to produce exciting, counterintuitive (hence often false positive) findings. The only cure for this state of affairs is for investigators to spend less time conducting tautological studies and spend more time searching for more propitious avenues of inquiry. It is worth noting, however, that publishing obviously trivial studies whose positive effect are already known *should* result in statistical significance and not contribute to a discipline's false positive rate.

74 BACKGROUND AND FACILITATORS OF THE CRISIS

9. And related to *both* of the previous IISPs is the reluctance of funding agencies to grant awards to speculative or risky proposals. There was even once an adage at the National Institutes of Health (NIH) to the effect that the agency seldom funds a study to which the result isn't already known. But of course the NIH is not a stand-alone organism nor do its employees unilaterally decide what will be funded. Scientists are the ones who know what is already known, and they have the most input in judging what is innovative, what is not, what should be funded, and what shouldn't be.

10. *Using publication and citation counts as institutional requirements for promotion, tenure, or salary increases.* Both practices fuel the compulsion to publish as much, as often, and keyed to what investigators believe will result in the most citations as humanly possible. We all employ numeric goals in our personal life such as exercise, weight loss, and wealth (or the lack thereof), but excessive publication rates may actually decrease the probability of making a meaningful scientific contribution and almost certainly increases publication bias. One study (Ioannidis, Klavans, & Boyack, 2018) reported that, between 2000 and 2016, 9,000 individuals published one paper every 5 days. True, the majority of these were published in high-energy and particle physics (86%) where the number of co-authors sometimes exceeded a thousand, but papers in other disciplines with 100 co-authors were not uncommon. (Ironically the lead author of this paper [John Ioannidis, who I obviously admire] has published more than a thousand papers himself in which he was either first or last author. But let's give him a pass here.)

And Now a Partial List of Individual Questionable Research Practices

It should be noted that some of these QRPs are discipline- or genre-specific but the majority are applicable (perhaps with a bit of translation) to most types of empirical research. There is also some unavoidable interdependence among the list's entries (i.e., some share superordinate methodological components) as well as possessing similar redundancies with a number of the QRPs previously mentioned. But with these caveats and disclaimers dutifully disclosed, hopefully the following annotated list represents a reasonably comprehensive directory of the most relevant QRPs to scientific irreproducibility:

1. *The use of soft, reactive, imprecise, self-reported, and easily manipulated outcome variables that are often chosen, created, or honed by investigators to differentially fit their interventions.* This one is especially endemic to those social science investigators who have the luxury of choosing or constructing their own outcomes and tailoring them to better match (or be more reactive to) one experimental group than the other. From a social science perspective, however, if an outcome variable has no social or scientific significance (such as how old participants feel), then the experiment itself will most likely also have no significance. But it will most likely add to the growing reservoir of false-positive results.
2. *Failure to control for potential experimenter and participant expectancy effects.* In some disciplines this may be the most virulent QRP of all. Naturally, double-blinded randomized designs involving sensible control/comparison groups are crucial in psychology and medicine, given demand and placebo effects, respectively, but, as will be demonstrated in Chapter 5, they are equally important in the physical sciences as well. As is the necessity of blinding investigators and research assistants to group membership in animal studies or in any research that employs variables scored by humans or that require human interpretation. (The randomization of genetically identical rodents and then blinding research assistants to group membership is a bit more complicated and labor intensive in practice than it may appear. Also untoward effects can be quite subtle and even counterintuitive, such as male laboratory rats responding differentially to male research assistants.)
3. *Failure to report study glitches and weaknesses.* It is a rare study in which no glitches occur during its commission. It is true, as Simmons et al. suggest, that some reviewers punish investigators for revealing imperfections in an experiment, but it has been my experience (as both a journal editor-in-chief and investigator) that many reviewers appreciate (and perhaps reward) transparency in a research report. But more importantly, hiding a serious glitch in the conduct of a study may result in a false-positive finding that of course can't be replicated.
4. *Selective reporting of results.* This has probably been illustrated and discussed in sufficient detail in the Simmons et al. paper in which

selected outcomes, covariates, and experimental conditions were deleted and not reported in either the procedure or result sections. Another facet of this QRP involves "the misreporting of true effect sizes in published studies . . . that occurs when researchers try out several statistical analyses and/or data eligibility specifications and then selectively report those that produce significant results" (Head, Holman, Lanfear, et al., 2015, p. 1). Suffice it to say that all of these practices are substantive contributors to the prevalence of false-positive results.

5. *Failure to adjust p-values based on the use of multiple outcomes, subgroup analyses, secondary analyses, multiple "looks" at the data prior to analysis, or similar practices resulting in artifactual statistical significance.* This is an especially egregious problem in studies involving large longitudinal databases containing huge numbers of variables which can easily yield thousands of potential associations there among. (Using nutritional epidemiology as an example, the European Prospective Investigation into Cancer and the Nutrition, Nurses' Health Study has resulted in more than 1,000 articles each [Ioannidis, 2018].) I personally have no idea what a reasonable adjusted p-value should be in such instances, although obtaining one close to 0.05 will obviously be completely irrelevant. (Perhaps somewhere in the neighborhood [but a bit more liberal] to the titular alpha levels adopted by the genomic and particle physics fields, which will be discussed later.)

6. *Sloppy statistical analyses and erroneous results in the reporting of p-values.* Errors such as these proliferate across the sciences, as illustrated by David Vaux's (2012) article in *Nature* (pejoratively titled "Know When Your Numbers Are Significant"), in which he takes biological journals and investigators to task for simple errors and sometimes absurd practices such as employing complex statistical procedures involving *N*s of 1 or 2. As another very basic example, Michal Krawczyk (2008), using a dataset of more than 135,000 p-values found (among other things) that 8% of them appeared to be inconsistent with the statistics upon which they were based (e.g., t or F) and, à propos of the next QRP, that authors appear "to round the p- values down more eagerly than up." Perhaps more disturbing, Bakker and Wicherts (2011), in an examination of 281 articles, found "that around 18% of statistical results in the psychological literature are incorrectly reported . . . and around 15% of the articles contained at least one statistical conclusion

that proved, upon recalculation, to be incorrect; that is, recalculation rendered the previously significant result insignificant, or vice versa" (p. 666). And it should come as no surprise by now that said errors were most often in line with researchers' expectations, hence an example of confirmation bias (Nickerson, 1998).

7. *Procedurally lowering p-values that are close to, but not quite ≤ 0.05.* This might include suspicious (and almost always unreported) machinations such as (a) searching for covariates or alternate statistical procedures or (b) deleting a participant or two who appears to be an outlier in order to whittle down a p of, say, 0.07 a couple of notches. Several investigators in several disciplines (e.g., Masicampo & Lalande, 2012; Gerber, Malhotra, Dowling, & Doherty, 2010; Ridley, Kolm, Freckelton, & Gage, 2007) have noted a large discrepancy between the proportions of p-values found just below 0.05 (e.g., 0.025 to 0.049) as opposed to those just above it (vs. 0.051 to 0.075).

8. *Insufficient attention to statistical power issues* (e.g., conducting experiments with too few participants). While most past methodology textbooks have emphasized the deleterious effects of low power on the production of negative studies, as previously discussed, insufficient power has equally (or greater) unfortunate effects on the production of false-positive findings. The primary mechanism of action of low power involves the increased likelihood of producing unusually large effect sizes by chance, which in turn are more likely to be published than studies producing more realistic results. And, as always, this QRP is magnified when coupled with others such as repeatedly analyzing results with the goal of stopping them as soon as an effect size large enough emerges to produce statistical significance. (Adhering to the prespecified sample size as determined by an appropriate power analysis would completely avoid this artifact.)

9. *Gaming the power analysis process*, such as by hypothesizing an unrealistically large effect size or not upwardly adjusting the required sample size for specialized designs such as those involving hierarchical or nested components.

10. *Artifactually sculpting (or selecting) experimental or control procedures to produce statistical significance* during the design process which might include:

    a. *Selecting tautological controls*, thereby guaranteeing positive results if the experiment is conducted properly. The best examples of this involve comparisons between interventions which have a recognized mechanism of action (e.g., sufficient instruction delivered at an appropriate developmental level in education or a medical intervention known to elicit a placebo effect) versus "instruction or treatment as usual." (It could be argued that this is not a QRP if the resulting positive effect is not interpreted as evidence of efficacy, but, in the present author's experience, it almost always is [as an example, see Bausell & O'Connell, 2009]).

b. *Increasing the fit between the intervention group and the outcome variable or, conversely, decreasing the control–outcome match.* (The first, semi-legitimate experiment described in the Simmons et al. paper provides a good example of this although the study was also fatally underpowered from a reproducibility perspective.) While speculative on the present author's part, one wonders if psychology undergraduates or Amazon Mechanical Turk participants familiar with computer-administered experiments couldn't surmise that they had been assigned the experimental group when a song ("Hot Potato" which they had heard in their childhood) was interrupted by asking them how old they felt. Or if their randomly assigned counterparts couldn't guess that they were in the control group when a blah instrumental song was similarly interrupted. (Relatedly, Chandler, Mueller, and Paolacci [2014] found [a] that investigators tend to underestimate the degree to which Amazon Turk workers participate across multiple related experiments and [b] that they "overzealously" exclude research participants based on the quality of their work. Thirty-three percent of investigators employing crowdsourcing participants appear to adopt this latter approach, thereby potentially committing another QRP [i.e., see Number 11].)

11. *Post hoc deletion of participants or animals for subjective reasons.* As an extreme example, in one of my previous positions I once witnessed an alternative medicine researcher proudly explain his criterion for deciding which observations were legitimate and which were not in his animal studies. (The latter's lab specialized in reputably demonstrating the pain-relieving efficacy of acupuncture resulting from tiny needles being inserted into tiny rat legs and comparing the results to a placebo.) His criterion for deciding which animals to delete was proudly explained as "sacrificing the non-acupuncture responding

animals" with an accompanying smile while drawing his forefinger across his neck. (No, I am not making this up nor do I drink while writing. At least not at this moment.)

12. *Improper handling and reporting of missing data.* Brief-duration experiments normally do not have problematically high dropout rates, but interventions whose effects must be studied over time can suffer significant attrition. Preferred options for compensating for missing data vary from discipline to discipline and include regression-based imputation of missing values (available in most widely used statistical packages) and intent-to-treat. (The latter tending to be more conservative than the various kinds of imputation and certainly the analysis of complete data only.) Naturally, the preregistration of protocols for such studies should describe the specific procedures planned, and the final analyses should comply with the original protocol, preferably presenting the results for both the compensatory and unvarnished data. Most funding and regulatory agencies for clinical trials require the prespecification of one or more of these options in their grant proposals, as do some IRBs for their submissions. Of course it should come as no surprise that one set of investigators (Melander, Ahlqvist-Rastad, Meijer, & Beermann, 2003) found that 24% of standalone studies neglected to include their preregistered intent-to-treat analysis in the final analysis of a cohort of antidepressant drug efficacy experiments—presumably because such analyses produce more conservative (i.e., less positive) results than their counterparts. (Because of the high financial stakes involved, positive published pharmaceutical research results tend to be greatly facilitated by the judicious use of QRPs [see Turner, Matthews, Linardatos, et al., 2008, for an especially egregious example]).

13. *Adding participants to a pilot study in the presence of a promising trend, thereby making the pilot data part of the final study.* Hopefully self-explanatory, although this is another facet of performing interim analyses until a desired p-value is obtained.

14. *Abandoning a study prior to completion based on the realization that statistical significance is highly unlikely to occur (or perhaps even that the comparison group is outperforming the intervention).* The mechanism by which this behavior inflates the obtained p-value may not be immediately apparent, but abandoning an ongoing experiment (i.e., not a pilot study) before its completion based on (a) the perceived

impotence of the intervention, (b) the insensitivity of the outcome variable, or (c) a control group that might be performing above expectations allows an investigator to conserve resources and immediately initiate another study until one is found that is sufficiently promising. Continually conducting such studies until statistical significance is achieved ultimately increases the prevalence of false, non-replicable positive results in a scientific literature while possibly encouraging triviality at the same time.

15. *Changing hypotheses based on the results obtained.* Several facets of this QRP have already been discussed, such as switching primary outcomes and deleting experimental conditions, but there are many other permutations such as (a) obtaining an unexpected result and presenting it as the original hypothesis or (b) reporting a secondary finding as a planned discovery. (All of which fit under the concepts of *HARKing* [for Hypothesizing After the Results are Known, Kerr, 1991] or *p-hacking* [Head et al., 2015], which basically encompasses a menu of strategies in which "researchers collect or select data or statistical analyses until nonsignificant results become significant.") As an additional example, sometimes a plethora of information is collected from participants for multiple reasons, and occasionally one unexpectedly turns out to be influenced by the intervention or related to another variable. Reporting such a result (or writing another article based on it) without explicitly stating that said finding resulted from an exploratory analysis constitutes a QRP in its own right. Not to mention contributing to publication bias and the prevalence of false-positive results.

16. *An overly slavish adherence to a theory or worldview. Confirmation bias*, a tendency to search for evidence in support of one's hypothesis and ignore or rationalize anything that opposes, it is subsumable under this QRP. A more extreme manifestation is the previously mentioned animal lab investigator's literal termination of rodents when they failed to respond to his acupuncture intervention. But also, perhaps more commonly, individuals who are completely intellectually committed to a specific theory are sometimes capable of actually seeing phenomenon that isn't there (or failing to see disconfirmatory evidence that is present). The history of science is replete with examples such as craniology (Gould, 1981), cold fusion (Taubes, 1993), and a number of other pathologies which will be discussed in

Chapter 5. Adequate controls and effective blinding procedures are both simple and absolutely necessary strategies for preventing this very troublesome (and irritating) QRP.

17. *Failure to adhere to professional association research standards and journal publishing "requirements,"* such as the preregistration of statistical approaches, primary hypotheses, and primary endpoints before conducting studies. Again, as Simmons and colleagues state: "If reviewers require authors to follow these requirements, they will" (p. 1363).

18. *Failure to provide adequate supervision of research staff.* Most experimental procedures (even something as straightforward as the randomization of participants to conditions or the strict adherence to a standardized script) can easily be subverted by less than conscientious (or eager to please) research staff, so a certain amount of supervision (e.g., via irregular spot checks) of research staff is required.

19. *Outright fraud,* of which there are myriad, well-publicized, and infamous examples, with perhaps the most egregious genre being data fabrication such as (a) painting patches on mice with permanent markers to mimic skin grafts (Hixson, 1976), (b) Cyril Burt making a splendid career out of pretending to administer IQ tests to phantom twins separated at birth to "prove" the dominance of "nature over nurture" (Wade, 1976), or (c) Yoshitaka Fujii's epic publication of 172 fraudulent articles (Stroebe, Postmes, & Spears, 2012). While data fabrications such as these are often dismissed as a significant cause of scientific irreproducibility because of their approximately 2% self-reported incidence (Fanelli, 2009), even this probable underestimate is problematic when one considers the millions of entries in published scientific databases.

20. *Fishing, data dredging, data torturing* (Mills, 1993), *and data mining.* All of which are used to describe practices designed to reduce a p-value below the 0.05 (aka p-hacking) threshold by analyzing large numbers of variables in search of statistically significant relationships to report—but somehow forgetting to mention the process by which these findings were obtained.

21. *The combined effects of multiple QRPs,* which greatly compounds the likelihood of false-positive effects since (a) a number of these practices are independent of one another and therefore their effects on false-positive results are cumulative and (b) individuals who knowingly

commit one of these practices will undoubtedly be inclined to combine it with others when expedient.

22. *Failing to preregistering studies and ensure their accessible to readers.* It is difficult to overemphasize the importance of preregistering study protocols since this simple strategy would avoid many of the QRPs listed here *if* preregistrations are routinely compared to the final research reports during the peer review process. Or, barring that, they are routinely compared by bloggers or via other social media outlets.

23. *Failure to adhere to the genre of established experimental design standards discussed in classic research methods books and the myriad sets of research guidelines discussed later.* Common sense examples include (a) randomization of participants (which should entail following a strict, computer-generated procedure accompanied by steps to blind experimenters, participants, and principal investigators); (b) the avoidance of experimental confounds, the assurance of reliability, and the validity of measuring instruments, taking Herculean steps (if necessary) to avoid attrition; and (c) a plethora of others, all of which should be common knowledge to anyone who has taken a research methods course. However, far and away the most important of these (with the possible exception of random assignment) is the blinding of experimenters (including animal and preclinical researchers), research assistants, and participants (including everyone who comes in contact with them) with respect to group membership and study hypotheses/purposes. Of course this is not possible in some genres of research, as when having research assistants count handwashing episodes in public lavatories (Munger & Harris, 1989), employing confederates in obedience studies (Milgram, 1963), or comparing the effects of actual knee surgery to placebo surgery (Moseley, O'Malley, Petersen, et al., 2002), but in most research scenarios blinding can and must be successfully instituted.

It may be that imperfect blinding of participants and research staff (at least those who come in contact with participants) may be among the most virulent QRPs in experiments in which investigators tend to *sculpt experimental or control* procedures to produce statistical significance (QRPs Numbers 2 and 10) and/or are themselves metaphorically blinded, given the degree to which they are wedded to their theory or word view (QRP Number 16).

If this is true, it follows that investigators should (and should be required to) employ blinding *checks* to ascertain if the procedures they put (or failed to put) into place were effective. Unfortunately a considerable amount of evidence exists that this seemingly obvious procedure is seldom employed. Much of this evidence comes from individual trials, such as the classic embarrassment in which 311 NIH employees were randomly assigned to take either a placebo or ascorbic acid capsule three times a day for 9 months to ascertain the effectiveness of vitamin C for the treatment and prevention of the common cold. Unfortunately the investigators failed to construct a placebo that matched the acidic taste of the intervention, and a blinding check revealed that many of the NIH participants broke said blind attempt by tasting the capsules (Karlowski, Chalmers, Frenkel, et al., 1975).

Unfortunately a number of methodologists have uncovered substandard blinding efforts (most notably failures to evaluate their effectiveness) in a number of different types of studies. Examples include

1. Fergusson, Glass, Waring, and Shapiro (2004) found that only 8% of 191 general medical and psychiatric trials reported the success of blinding;
2. A larger study (Baethge, Assall, & Baldessarini, 2013) found even worse results, with only 2.5% of 2,467 schizophrenia and affective disorders RCTs reported assessing participant, rater, or clinician blinding; and, not to be outdone,
3. Hróbjartsson, Forfang, Haahr, and colleagues (2007) found an ever lower rate (2%) of blinding assessments for a sample of 1,599 interdisciplinary blinded RCTs.

However, every scientist residing outside of a cave knows that participants, research assistants, and clinicians *should* be blinded (at least when feasible, since studies involving surgery or acupuncture cannot blind the individuals administering the treatments). Unfortunately compliance with this knowledge is a bit short of perfect.

Colagiuri and Benedetti (2010), for example, quite succinctly sum up the importance of universal blinding checks in a carefully crafted criticism of the otherwise excellent CONSORT 2010 Statement's updated guidelines for randomized trials which inexplicably deleted a provision to check blinding and downgraded it to a recommendation. The Colagiuri and Benedetti team explained the rationale for their criticism (via a *British Medical Journal* Rapid Response article):

Testing for blinding is the *only* [emphasis added] valid way to determine whether a trial is blind. Trialists conducting RCTs should, therefore, report on the success of blinding. In situations where blinding is successful, trialists and readers can be confident that guesses about treatment allocation have not biased the trial's outcome. In situations where blinding fails, trialists and readers will have to evaluate whether or not bias may have influenced the trial's outcomes. Importantly, however, in the second situation, while trialists are unable to label their trial as blind, the failure of blinding should not be taken as definitive evidence that bias occurred. Instead, trialists should provide a rationale as to why the test of blinding was unsuccessful and a statement on whether or not they consider the differences between treatment arms to be valid. (2010, p. 340)

While these authors' comments were directed at medical researchers, one wonders just how successful double-blinding is in the average low-powered social science experiment—especially given the high prevalence of positive results in these latter literatures. Why not, therefore, take the time to administer a one-item blinding check to participants after their completion of the experimental task by simply asking to which treatment group they believed they had been assigned? This is especially important in psychology since Amazon Mechanical Turk participants and psychology students are probably more experienced and savvy in gleaning the purpose of a study than investigators realize.

Comparing participants' guesses regarding assignment to their actual assignment would be an excellent mechanism for evaluating the success of whatever blinding strategy was implemented. And perhaps the very act of evaluating this strategy might induce investigators to be more careful about the design of their studies and their crafting of experimental conditions. (Especially if the blinding check was included in the preregistration.) So let's add the following, perhaps idiosyncratic, QRP to our burgeoning (but still incomplete) list:

24. *Failing to check and report experimental participants' knowledge (or guesses) regarding the treatments they received.* Again, this could be done by administering a single force-choice item (e.g., "To what condition [experimental or control] do you believe you were assigned?") at the end of an experiment and then correlating said answer not only to actual group assignment but also to ascertain if there was an

interaction between said guesses and actual assignment with respect to outcome scores. The answers would not necessarily be causally definitive one way or another, but a large portion of the participants correctly guessing their treatment assignment by study's end would be rather troublesome. And it would be even more troublesome if individuals in the control group who suspected they were in the intervention scored differentially higher or lower on the outcome variable than their control counterparts who correctly guessed their group assignment.

A Few "Not Quite QRPs" but Definitely Irritating Reporting Practices

These do not necessarily impact either publication bias or the prevalence of false-positive results, but they are, at the very least, disingenuous, irritating, and possibly becoming more prevalent.

1. *Downplaying or failing to mention study limitations* (Ioannidis, 2007);
2. *Hyping results* via such increasingly common descriptors as "robust, novel, innovative, and unprecedented" (Vinkers, Tijdink, & Otte, 2015); and
3. *"Spinning" study results*; Boutron, Dutton, Ravaud, and Altman (2010), for example, found that "a majority of 72 trials reporting non-statistically significant results had included *spin* type descriptors in both the abstract and the conclusion sections [which the authors found particularly problematic since many clinicians read only those two sections]. A number of these even attributed beneficial effects for the treatment being evaluated despite statistically *nonsignificant results*.

As a counter-example to these hyping and spinning practices, consider the announcement of James Watson and Francis Crick's (the former who definitely did not having a propensity for minimizing his accomplishments) introductory statement in the paper announcing the most heralded biological discovery of the twentieth century: "We wish to suggest a structure for the salt of deoxyribose nucleic acid (D.N.A.). This structure has novel features which are of considerable biological interest" (1953, p. 737).

QRPs and Animal Studies

While there are many different types of laboratory experiments, those employing in vivo animals undoubtedly have the most in common with human experimentation and hence are most susceptible to the types of QRPs of concern to us here. There is, in fact, a surprising similarity in requisite behaviors and procedures required for producing valid, reproducible results in the two genres of research—at least with a bit of translation.

Certainly some investigative behaviors are unique to humans, such as the necessity of ensuring that participants can't guess their group membership or querying them thereupon (QRP Number 24). But, as with experiments employing human participants, a disquieting amount of literature exists chronicling the shortcomings of published in vivo animal studies. In fact it is probably safe to say that, historically, preclinical investigators have been among the leaders in failing to report their methods thoroughly and their adherence to recommended experimental practices.

This is especially problematic in animal studies that precede and inform randomized clinical human trials designed to test the efficacy of pharmaceutical therapies. Unfortunately, the track record for how well animal studies actually do inform human trials borders on the abysmal. It has been estimated, for example, that only about 11% of therapeutic agents that have *proved promising* in animal research and have been tested clinically are ultimately licensed (Kola & Landis, 2004). In addition, Contopoulos-Ioannidis, Ntzani, and Ioannidis (2003) found that even fewer (5%) of "high impact" basic science discoveries claiming clinical relevance are *ever* successfully translated into approved therapies within a decade.

Naturally, every preclinical study isn't expected to result in a positive clinical result, but these statistics relate to *positive* animal studies, not those that initially "failed." It is therefore highly probable that a significant number of the former reflect false-positive results due to many of the same QRPs that bedevil human experimentation.

This supposition is buttressed by Kilkenny, Parsons, Kadyszewski, and colleagues (2009) who, surveying a large sample of published animal studies, found that only 13% reported randomizing animals to treatments and only 14% apparently engaged in blinded data collection. This particular study, incidentally, apparently led to the Animal Research: Reporting of in Vivo Experiments (ARRIVE) guidelines (Kilkenny, Browne, Cuthill, et al., 2010) which is closely modeled on the CONSORT statement—yet another commonality between the two genres of research. And, like its predecessor, ARRIVE also has its own checklist (http://www.nc3rs.org.uk/ARRIVEchecklist/) and has likewise been endorsed by an impressive number of journals.

So, as would be expected, the majority of the ARRIVE procedural reporting guidelines (e.g., involving how randomization or blinding was performed if instituted) are similar to their CONSORT counterparts although others are obviously unique to animal research. (For a somewhat more extensive list of methodological suggestions for this important genre of research, see Henderson, Kimmelman, Fergusson, et al., 2013.)

Unfortunately, while the ARRIVE initiative has been welcomed by animal researchers and a wide swath of preclinical journals, enforcement has been a recurring disappointment, as demonstrated in the following study title "Two Years Later: Journals Are Not Yet Enforcing the ARRIVE Guidelines on Reporting Standards for Pre-Clinical Animal Studies" (Baker, Lidster, Sottomayor, & Amor, 2014). The investigators, in examining a large number of such studies published in the journals *PLoS* and *Nature* (both of which officially endorsed the ARRIVE reporting guidelines) found that

1. The reporting of blinding "was similar to that in past surveys (20% in *PLoS* journals and 21% in *Nature* journals)," and
2. "Fewer than 10% of the relevant studies in either *Nature* or *PLoS* journals reported randomisation (10% in *PLoS* journals and 0% in *Nature* journals), and even fewer mentioned any power/sample size analysis (0% in *PLoS* journals and 7% in *Nature* journals)" (p. 3).

From one perspective, perhaps 2 years is not a great deal of time for comprehensive guidelines such as these to be implemented. But from a scientific perspective, this glass is neither half full nor half empty because behaviors such as blinding, randomization, and power analyses should not require guidelines in the 21st century. Rather their commission should be ironclad prerequisites for publication.

Whether the primary etiology of this disappointing state of affairs resides in journal policies, mentoring, or knowledge deficiencies is not known. However, since hopefully 99% of practicing scientists know that these three methodological procedures are absolute requirements for the production of valid experimental findings, the major onus probably involves journal involvement, such as the suggestion by Simmons, Nelson, and Simonsohn (2012) (based on their iconic 2011 article) that investigators affix a 21-word statement to their published experiments (i.e., "We report how we determined our sample size, all data exclusions (if any), all manipulations, and all measures in the study"). A simple innovation which inspired both the PsychDisclosure initiative (LeBel, Borsboom, Giner-Sorolla, et al., 2013) and the decision of the editor of the most prestigious journal in the field (*Psychological Science*) to *require* authors' disclosure via a brief checklist. (Checklists, incidentally have been shown to be important peer review aids and are actually associated with improved compliance with methodological guidelines [Han, Olonisakin, Pribis, et al., 2017]).

Next Up

The next chapter features a discussion of some especially egregious case studies graphically illustrating the causal link between QRPs and irreproducibility, the purpose of which is not to pile demented examples upon one another but rather to suggest that the QRP → irreproducibility chain will undoubtedly be more difficult to sever than we would all like.

References

Baethge, C., Assall, O. P., & Baldessarini, R. J. (2013). Systematic review of blinding assessment in randomized controlled trials in schizophrenia and affective disorders 2000–2010. *Psychotherapy and Psychosomatics, 82*, 152–160.

Baker, D., Lidster, K., Sottomayor, A., & Amor, S. (2014). Two years later: Journals are not yet enforcing the ARRIVE guidelines on reporting standards for pre-clinical animal studies. *PLoS Biology, 11*, e1001756.

Bakker, M., van Dijk, A., & Wicherts, J. M. (2012). The rules of the game called psychological science. *Perspectives on Psychological Science, 7*, 543–554.

Bakker, M., & Wicherts, J. M. (2011). The (mis)reporting of statistical results in psychology journals. *Behavior Research, 43*, 666–678.

Bausell, R. B., & O'Connell, N. E. (2009). Acupuncture research: Placebos by many other names. *Archives of Internal Medicine, 169*, 1812–1813.

Bedeian, A. G., Taylor, S. G., & Miller, A. N. (2010). Management science on the credibility bubble: Cardinal sins and various misdemeanors. *Academy of Management Learning & Education, 9*, 715–725.

Benjamin, D. J., Berger, J. O., Johannesson, M., et al. (2017). Redefine statistical significance. *Nature Human Behavior, 2*, 6–10.

Boutron, I., Dutton, S., Ravaud, P., & Altman, D. G. (2010). Reporting and interpretation of randomized controlled trials with statistically nonsignificant results for primary outcomes. *Journal of the American Medical Association, 303*, 2058–2064.

Chandler, J., Mueller, P., & Paolacci, G. (2014). Nonnaivete among Amazon Mechanical Turk workers: Consequences and solutions for behavioral researchers. *Behavioral Research Methods, 46*, 112–130.

Colagiuri, B., & Benedetti, F. (2010). Testing for blinding is the only way to determine whether a trial is blind. *British Medical Journal, 340,* c332. https://www.bmj.com/rapid-response/2011/11/02/testing-blinding-only-way-determine-whether-trial-blind

Contopoulos-Ioannidis, D. G., Ntzani, E., & Ioannidis, J. P. (2003). Translation of highly promising basic science research into clinical applications. *American Journal of Medicine, 114,* 477–484.

Fanelli, D. (2009). How many scientists fabricate and falsify research? A systematic review and meta-analysis of survey data. *PLoS ONE, 4,* e5738.

Fergusson, D., Glass, K. C., Waring, D., & Shapiro, S. (2004). Turning a blind eye: The success of blinding reported in a random sample of randomized, placebo controlled trials. *British Medical Journal, 328,* 432.

Fiedler, K., & Schwarz N. (2015). Questionable research practices revisited. *Social Psychological and Personality Science 7,* 45–52.

Gerber, A S., Malhotra, N., Dowling, C. M, & Doherty, D. (2010). Publication bias in two political behavior literatures. *American Politics Research, 38,* 591–613.

Gould, S. J. (1981). *The mismeasure of man.* New York: Norton.

Greenwald, A. G., Gonzalez, R., Harris, R. J., & Guthrie, D. (1996). Effect sizes and p values: what should be reported and what should be replicated? *Psychophysiology, 33,* 175–183.

Han, S., Olonisakin, T. F., Pribis, J. P., et al. (2017). A checklist is associated with increased quality of reporting preclinical biomedical research: A systematic review. *PLoS ONE, 12,* e0183591.

Head, M. L., Holman, L., Lanfear, R., et al. (2015). The extent and consequences of p-hacking in science. *PLoS Biology, 13,* e1002106.

Henderson, V. C., Kimmelman, J., Fergusson, D., et al. (2013). Threats to validity in the design and conduct of preclinical efficacy studies: A systematic review of guidelines for in vivo animal experiments. *PLoS Medicine, 10,* e1001489.

Hixson, J. R. (1976). *The patchwork mouse.* Boston: Anchor Press.

Hróbjartsson, A., Forfang, E., Haahr, M. T., et al. (2007). Blinded trials taken to the test: An analysis of randomized clinical trials that report tests for the success of blinding. *International Journal of Epidemiology, 36,* 654–663.

Ioannidis, J. P. A. (2007). Limitations are not properly acknowledged in the scientific literature. *Journal of Clinical Epidemiology, 60,* 324–329.

Ioannidis, J. P. A., Klavans, R., & Boyack, K. W. (2018). Thousands of scientists publish a paper every five days, papers and trying to understand what the authors have done. *Nature, 561,* 167–169.

Ioannidis, J. P. A. (2018). The challenge of reforming nutritional epidemiologic research. *JAMA, 320,* 969–970.

John, L. K., Loewenstein, G., & Prelec, D. (2012). Measuring the prevalence of questionable research practices with incentives for truth-telling. *Psychological Science, 23,* 524–532.

Karlowski, T. R., Chalmers, T. C., Frenkel, L. D., et al. (1975). Ascorbic acid for the common cold: A prophylactic and therapeutic trial. *Journal of the American Medical Association, 231,* 1038–1042.

Kerr, N. L. (1991). HARKing: Hypothesizing after the results are known. *Personality and Social Psychology Review, 2,* 196–217.

Kilkenny, C., Browne, W. J., Cuthill, I. C., et al. (2010). Improving bioscience research reporting: The ARRIVE Guidelines for reporting animal research. *PLoS Biology, 8,* e1000412.

Kilkenny, C., Parsons, N., Kadyszewski, E., et al. (2009). Survey of the quality of experimental design, statistical analysis and reporting of research using animals. *PLoS ONE, 4,* e7824.

Kola, I., & Landis, J. (2004). Can the pharmaceutical industry reduce attrition rates? *Nature Reviews Drug Discovery, 3*, 711–715.

Krawczyk, M. (2008). *Lies, Damned lies and statistics: The adverse incentive effects of the publication bias.* Working paper, University of Amsterdam. http://dare.uva.nl/record/302534

LeBel, E. P., Borsboom, D., Giner-Sorolla, R., et al. (2013). PsychDisclosure.org: Grassroots support for reforming reporting standards in psychology. *Perspectives on Psychological Science, 8*, 424–432.

Masicampo E. J., & Lalande D. R. (2012). A peculiar prevalence of p values just below .05. *Quarterly Journal of Experimental Psychology, 65*, 2271–2279.

Melander, H., Ahlqvist-Rastad, J., Meijer, G., & Beermann, B. (2003). Evidence b(i)ased medicine-selective reporting from studies sponsored by pharmaceutical industry: Review of studies in new drug applications. *British Medical Journal, 326*, 1171–1173.

Milgram, S. (1963). Behavioral study of obedience. *Journal of Abnormal and Social Psychology, 67*, 371–378.

Mills, J. L. (1993). Data torturing. *New England Journal of Medicine, 329*, 1196–1199.

Moseley, J. B., O'Malley, K., Petersen, N. J., et al. (2002). A controlled trial of arthroscopic surgery for osteoarthritis of the knee. *New England Journal of Medicine, 347*, 82–89.

Munger, K., & Harris, S. J. (1989). Effects of an observer on hand washing in public restroom. *Perceptual and Motor Skills, 69*, 733–735.

Nickerson, R. S. (1998). Confirmation bias: A ubiquitous phenomenon in many guises. *Review of General Psychology, 2*, 175–220.

O'Boyle, Jr., E. H., Banks, G. C., & Gonzalez-Mule, E. (2014). The Chrysalis effect: How ugly initial results metamorphosize into beautiful articles. *Journal of Management, 43*, 376–399.

Ridley, J., Kolm, N., Freckelton, R. P., & Gage, M. J. G. (2007). An unexpected influence of widely used significance thresholds on the distribution of reported P-values. *Journal of Evolutionary Biology, 20*, 1082–1089.

Simmons, J. P., Nelson, L. D., & Simonsohn, U. (2011). False-positive psychology: undisclosed flexibility in data collection and analysis allows presenting anything as significant. *Psychological Science, 22*, 1359–1366.

Simmons, J. P., Nelson, L. D., & Simonsohn, U. (2012). A 21 word solution. https://ssrn.com/abstract=2160588

Stroebe, W., Postmes, T., & Spears, R. (2012). Scientific misconduct and the myth of self-correction in science. *Perspectives on Psychological Science, 7*, 670–688.

Taubes, G. (1993). *Bad science: The short life and weird times of cold fusion.* New York: Random House.

Turner, E. H., Matthews, A. M., Linardatos, E., et al. (2008) Selective publication of antidepressant trials and its influence on apparent efficacy. *New England Journal of Medicine, 358*, 252–260.

Vaux D. (2012). Know when your numbers are significant. *Nature, 492*, 180–181.

Vinkers, C. H., Tijdink, J. K., & Otte, W. M. (2015). Use of positive and negative words in scientific PubMed abstracts between 1974 and 2014: Retrospective analysis. *British Medical Journal, 351*, h6467.

Wade, N. (1976). IQ and heredity: Suspicion of fraud beclouds classic experiment. *Science, 194*, 916–919.

Watson, J. D., & Crick, F. (1953). A structure for deoxyribose nucleic acid. *Nature, 171*, 737–738.

4
A Few Case Studies of QRP-Driven Irreproducible Results

The title of our first entry unambiguously describes the article's bottom-line conclusions as well as several of the other case studies presented in this chapter.

> **The Statistical Crisis in Science: Data-Dependent Analysis—A "Garden of Forking Paths"—Explains Why Many Statistically Significant Comparisons Don't Hold Up**
>
> Andrew Gelman and Eric Loken (2014)
>
> The authors illustrate their point by briefly mentioning several psychological studies that either failed to replicate (e.g., women being more likely to wear pink or red at peak fertility [Beall & Tracy, 2013]) or that report unrealistically large effect sizes (e.g., women changing their voting preferences based on their ovulatory cycles [Durante, Rae, & Griskevicius, 2013]). While these and other examples are linked to specific statistical questionable research practices (QRPs) listed in the previous chapter, Gelman and Loken metaphorically articulate the problem in terms of taking a scientific journey in which there are many forks in the road, thereby necessitating the many decisions that must be made before arriving at the final destination.
> Or, in their far more eloquent words:
>
> > In this garden of forking paths, whatever route you take seems predetermined, but that's because the choices are done *implicitly* [emphasis added]. The researchers are not trying multiple tests to see which has the best p-value; rather, they are using their scientific common sense

> to formulate their hypotheses in a reasonable way, given the data they have. The mistake is in thinking that, if the particular path that was chosen yields statistical significance, this is strong evidence in favor of the hypothesis. (p. 464)
>
> The authors go on to rather compassionately suggest how such practices can occur without the conscious awareness of their perpetrators.
>
> Working scientists are also keenly aware of the risks of data dredging, and they use confidence intervals and p-values as a tool to avoid getting fooled by noise. Unfortunately, a by-product of all this struggle and care is that when a statistically significant pattern does show up, it is natural to get excited and believe it. The very fact that scientists generally don't cheat, generally don't go fishing for statistical significance, makes them vulnerable to drawing strong conclusions when they encounter a pattern that is robust enough to cross the $p < 0.05$ threshold. (p. 464)
>
> As for solutions, the authors suggest that, in the absence of pre-registration, almost all conclusions will be data-driven rather than hypothesis-driven. And for observational disciplines employing large *unique* databases (hence impossible to replicate using different data), they suggest that *all of the relevant comparisons* in such databases be more fully analyzed while not concentrating only on statistically significant results.

But now let's begin our more detailed case studies by considering a hybrid psychological-genetic area of study involving one of the former's founding constructs (general intelligence or the much reviled [and reified] *g*). We won't dwell on whether this construct even exists or not, but for those doubters who want to consider the issue further, Jay Gould's *Mismeasure of Man* (1981) or Howard Gardner's (1983) *Frames of Mind: The Theory of Multiple Intelligences* are definitely recommended.

What makes the following study so relevant to reproducibility are (a) its demonstration of how a single QRP can give rise to an entire field of false-positive results while at the same time (b) demonstrating the power of the *replication process* to identify these fallacious results.

Most Reported Genetic Associations with General Intelligence Are Probably False Positives

Christopher Chabris, Benjamin Herbert,
Daniel Benjamin, et al. (2012)

In their introduction, these authors justify their study by arguing that "General cognitive ability, or g, is one of the most heritable behavioral traits" (p. 1315). (Apparently a large literature has indeed found statistically significant correlations between various single-nucleotide polymorphisms [SNPs; of which there are estimated to be ten million or so] and g.)

The present example represented a *replication* of 13 of the SNPs previously found to be related to g in an exhaustive review by Antony Payton (2009) spanning the years from 1995 to 2009; these SNPs happened to be located near 10 potentially propitious genes. The authors' replications employed three large, "well-characterized" longitudinal databases containing yoked information on at least 10 of these 13 SNPs: (a) the Wisconsin high school students and a randomly selected sibling ($N = 5,571$), (b) the initial and offspring cohorts of the Framingham Heart Study ($N = 1,759$), and (c) a sample of recently genotyped Swedish twins born between 1936 and 1958 ($N = 2,441$).

In all, this effort resulted in 32 regression analyses (controlling for variables such as age, gender, and cohort) performed on the relationship between the appropriate data on each SNP and IQ. Only one of these analyses reached statistical significance at the 0.04 level (a very low bar), and it was in the *opposite direction to that occurring in one of the three original studies*. Given the available statistical power of these replications and the number of tests computed, the authors estimate that at least 10 of these 32 associations should have been significant at the .05 level by chance alone.

Their conclusions, while addressed specifically to genetic social science researchers, is unfortunately relevant to a much broader scientific audience.

> Associations of candidate genes with psychological traits and other traits studied in the social sciences should be viewed as tentative until they

> have been replicated in multiple large samples. Failing to exercise such caution may hamper scientific progress by allowing for the proliferation of potentially false results, which may then influence the research agendas of scientists who do not realize that the associations they take as a starting point for their efforts may not be real. And the dissemination of false results to the public may lead to incorrect perceptions about the state of knowledge in the field, especially knowledge concerning genetic variants that have been described as "genes for" traits on the basis of unintentionally inflated estimates of effect size and statistical significance. (p. 1321)

A Follow-Up

Partially due to replication failures such as this one (but also due to technology and lowering costs of a superior alternative), candidate gene analyses involving alpha levels of .05 have largely disappeared from the current scientific literature. Now genome-wide association studies are in vogue, and the genetics research community has reduced the significance criterion by several orders of magnitude (i.e., $p \leq .0000005$). Obviously, as illustrated in the simulations discussed in Chapter 2, appropriate reductions in titular alpha levels based on sensible criteria will greatly reduce the prevalence of false-positive results, and perhaps, just perhaps, psychology will eventually follow suit and reduce its recommended alpha level for new "discoveries" to a more sensible level such as 0.005.

Unfortunately, psychological experiments would be considerably more difficult to conduct if the rules were changed in this way. And much of this science (and alas others as well) appears to be a game, as suggested by the Bakker, van Dijk, and Wicherts (2012) title ("The Rules of the Game Called Psychological Science") along with many of the previously discussed articles. But let's turn our attention to yet another hybrid psychological-physiological foray somewhere past the boundaries of either irreproducibility or inanity, this time with a somewhat less pejorative title.

Puzzlingly High Correlations in fMRI Studies of Emotion, Personality, and Social Cognition

Edward Vul, Christine Harris, Piotr Winkielman, and Harold Pashler (2009)

We've all been regaled by functional magnetic resonance imaging (fMRI) studies breathlessly reporting that one region or another of the brain is associated with this or that attribute, emotion, or whatever. Thousands have been conducted, and 41 meta-analyses were located by the intrepid and tireless John Ioannidis who definitively documented an excess of statistically significant findings therein (2011).

However, the Vul et al. study went several steps further than is customary in research such as this. First, the authors examined not only statistical significance but the effect sizes produced, noting that many of the Pearson correlations coefficients (which serve as effect sizes in correlational research) between fMRI-measured brain activity and myriad other psychosocial constructs (e.g., emotion, personality, and "social cognition") were in excess of 0.80. (Correlation coefficients range from −1.00 to +1.00, with +/−1.00 representing a perfect correspondence between two variables and zero indicating no relationship whatever.) Then, given the "puzzling" (some would unkindly say "highly suspicious") size of these correlations, the team delved into the etiology of these astonishingly large (and decidedly suspect) values.

Suspect because veridical correlations (as opposed to those observed by chance, data analysis errors, or fraud) of this size are basically impossible in the social sciences because the reliability (i.e., stability or reliability) of these disciplines' measures typically fall *below* 0.80 (and the reliability of neuroimaging measures, regardless of the disciplines involved, usually falls a bit short of 0.70). Reliability, as the authors of this study note, places an algebraic upper limit on how high even a perfect correlation (i.e., 1.00) between two variables can be achieved via the following very simple formula:

Formula 4.1: *The maximum correlation possible between two variables given their reliabilities*

Corrected Perfect Correlation

$$= \sqrt{\text{Measure 1}\left(\text{Reliability of Psychological Measures}\right) \times \text{Reliability of fMRI Mesurements}}$$

$$= \sqrt{.80\left(\text{Psychological Measure}\right) \times .70\left(\text{FMRI Measures}\right)}$$

$$= \sqrt{.56} = 0.75$$

So, to paraphrase Shakespeare, perhaps something is rotten in brain imaging research? Alas, let's all hope that it's only brain imaging research's olfactory output that is so affected.

To get to the bottom of what the authors charitably described as "puzzling," a literature search was conducted to identify fMRI studies involving correlations between the amount of deoxygenated hemoglobin in the blood (called the BOLD signal, which is basically a measure of blood flow) and psychosocial constructs as measured by self-reported questionnaires. The search "resulted in 55 articles, with *274 significant correlations* [emphasis added] between BOLD signal and a trait measure" (p. 276), which if nothing else is a preconfirmation of Ioannidis's previously mentioned 2011 analysis concerning the extent of publication bias in fMRI research in general.

The next step in teasing out the etiology of this phenomenon involved contacting the authors of the 55 studies to obtain more information on how they performed their correlational analyses. (Details on the fMRI data points were not provided in the journal publications.) Incredibly, at least some information was received from 53 of the 55 articles, close to a record response for such requests and is probably indicative of the authors' confidence regarding the validity of their results.

Now to understand the etiology of these "puzzling" correlation coefficients, a bit of background is needed on the social science stampede into brain imaging as well as the analytics of how increased blood flow is measured in fMRI studies in general.

First, Vul and colleagues marveled at the eagerness with which the social sciences (primarily psychology) had jumped into the neuroimaging fad only a few years prior to the present paper, as witnessed by

1. The creation of at least two new journals (*Social Neuroscience* and *Social Cognitive and Affective Neuroscience*),
2. The announcement of a major funding initiative in the area by the National Institute of Mental Health in 2007, and (and seemingly most impressive to Vul and his co-authors),
3. "The number of papers from this area that have appeared in such prominent journals as *Science*, *Nature*, and *Nature Neuroscience*" (p. 274). (The first two of these journals have been accused of historically exhibiting a penchant for publishing sensationalist, "man bites dog" studies while ignoring the likelihood of their being reproducible.)

Second, three such studies were briefly described by the authors (two published in *Science* and one in a journal called *NeuroImage*) that associated brain activity in various areas of the brain with self-reported psychological scales while

1. Playing a game that induced social rejection (Eisenberger, Lieberman, & Williams, 2003),
2. Completing an empathy-related manipulation (Singer, Seymour, O'Doherty, et al., 2004), or
3. Listening to angry versus neutral speech (Sander, Grandjean, Pourtois, et al., 2005).

Incredibly, the average correlation for the three studies was 0.77. But, as explained earlier, this is statistically impossible given the amount of error (i.e., 1– reliability) present in both the psychological predictors and the fMRI scans. So even if the true (i.e., error-free) relationship between blood flow in the studied regions and the psychological scales was *perfect* (i.e., 1.00), the maximum numerically possible correlation among these variables in our less than perfect scientific world, would be *less* than 0.77. Or, said another way, the results obtained were basically statistically impossible since practically no social science scales are measured with sufficient precision to support such high correlations.

The authors also kindly and succinctly provided a scientific explanation (as opposed to the psychometric one just tendered) for why perfect correlations in studies such as these are almost impossible.

First, it is far-fetched to suppose that only one brain area influences any behavioral trait. Second, even if the neural underpinnings of a trait were confined to one particular region, it would seem to require an *extraordinarily favorable set of coincidences* [emphasis added] for the BOLD signal (basically a blood flow measure) assessed in one particular stimulus or task contrast to capture all functions relevant to the behavioral trait, which, after all, reflects the organization of complex neural circuitry residing in that brain area. (p. 276)

And finally, to complete this brief tour of fMRI research, the authors provide a very clear description of how brain imagining "works," which I will attempt to abstract without botching it too badly.

A functional scanning image is comprised of multiple blood flow/oxygenation signals from roughly cube-shaped regions of the brain called voxels (*volumetric pixels*, which may be as small as 1 mm^3 or as large as 125 mm^3). The number of voxels in any given image typically ranges from 40,000 to 500,000 of these tiny three-dimensional pieces of the brain, and the blood flow within each of these can be correlated with *any* other data collected on the individual (in this case, psychosocial questionnaires involving self-reports).

Each voxel can then be analyzed separately with any variable of interest available on (or administered to) the individuals scanned—normally 20 or fewer participants are employed (sometimes considerably fewer) given the expense and machine time required. As mentioned in the present case, these dependent variables are psychosocial measures of perhaps questionable validity but which can reflect *anything*. The intervention can also encompass a wide range of manipulations, such as contrasting behavioral scenarios or gaming exercises.

Thus we have a situation in which there are potentially hundreds of thousands of correlations that *could* be run between each voxel "score" and a single independent variable (e.g., a digital game structured to elicit an emotional response of some sort or even listening to "Hot Potato" vs. "Kalimba," although unfortunately, to my knowledge, that particular study has yet to be conducted via the use of fMRI). Naturally, reporting thousands of correlation coefficients would take up a good deal of journal space so groups of voxels are selected in almost any manner investigators choose in order to "simplify" matters. And therein resides the solution to our investigators' "puzzlement" because the following mind-bending strategy was employed by a *majority* of the investigators of the 53 responding authors:

> First, the investigator computes a separate correlation of the behavioral measure of interest with each of the voxels (fig. 4 in the original article). Then, he or she selects those voxels that exhibited a sufficiently high correlation (by passing a statistical threshold; fig. 4b). Finally, an ostensible measure of the "true" correlation is aggregated from the [subset of] voxels that showed high correlations (e.g., by taking the mean of the voxels over the threshold). With enough voxels, such a biased analysis is *guaranteed* [emphasis added] to produce high correlations even if none are truly present [i.e., by chance alone]. Moreover, this analysis will produce visually pleasing scatter grams (e.g., fig. 4c) that will provide (quite meaningless) reassurance to the viewer that s/he is looking at a result that is solid, is "not driven by outliers," and so on. . . . This approach amounts to selecting one or more voxels based on a functional analysis and then reporting the results of the same analysis and functional data from just the selected voxels. This analysis distorts the results by selecting noise that exhibits the effect being searched for, and any measures obtained from a non-independent analysis are biased and untrustworthy. (p. 279)

As problematic as this strategy is of reporting only the mean correlation involved in a subset of voxels highly correlated with the psychosocial intervention, 38% of the respondents actually "reported the correlation of the *peak* [emphasis added] voxel (the voxel with the highest observed correlation)" (p. 281).

The authors conclude their truly astonishing paper by suggesting alternative strategies for reducing the analytic biases resulting from this genre of research, including (a) ensuring that whoever chooses the voxels of interest be *blinded* to the voxel–behavioral measure correlations and (b) not to "peek" at the behavioral results while analyzing the fMRI output.

Another Perspective on the fMRI Studies

Undoubtedly the title of another article ("Voodoo Correlations Are Everywhere: Not Only in Neuroscience") represents a nod to Robert Park's well-known book titled *Voodoo Science: The Road from Foolishness to Fraud* (2000). The article (Fiedler, 2011) is mentioned here because it also makes mention of the Vul et al. article but places it within the broader framework regarding the effects of idiosyncratic sampling decisions and their influence on the reproducibility of a much wider swath of scientific enquiry.

While Professor Fiedler acknowledges that "sampling" procedures such as the ones just discussed involving brain scanning studies are unusually egregious, he argues that psychological research as a whole is plagued with selective sampling strategies specifically designed to produce not only p-values ≤ .05, but what Roger Giner-Sorolla (2012) terms *aesthetically pleasing* results to go along with those propitious p-values. But let's allow Professor Fiedler to speak for himself in the study abstract by referring to the "voodoo correlations" just discussed.

> Closer inspection reveals that this problem [the voxel correlations] is only a special symptom of a broader methodological problem that characterizes *all paradigmatic research* [emphasis added] not just neuroscience. *Researchers not only select voxels to inflate effect size, they also select stimuli, task settings, favorable boundary conditions, dependent variables and independent variables, treatment levels, moderators, mediators, and multiple parameter settings in such a way that empirical phenomena become maximally visible and stable* [emphasis added again because this long sentence encapsulates such an important point]. In general, paradigms can be understood as conventional setups for producing idealized, inflated effects. Although the feasibility of representative designs is restricted, a viable remedy lies in a reorientation of paradigmatic research from the visibility of strong effect sizes to genuine validity and scientific scrutiny. (p. 163)

Fiedler's methodological language can be a bit idiosyncratic at times, such as his use of the term "metacognitive myopia" to characterize "a tendency in sophisticated researchers, who only see the data but overlook the sampling filters behind, [that] may be symptomatic of an industrious period of empirical progress, accompanied by a lack of interest in methodology and logic of science" (p. 167). However, he minces no words in getting his basic message across via statements such as this:

> Every step of experimental design and scholarly publication is biased toward strong and impressive findings, starting with the selection of a

research question; the planning of a design; the selection of stimuli, variables and tasks; the decision to stop and write up an article; the success to publish it; its revision before publication; and the community's inclination to read, cite and adopt the results. (p. 167)

And while this statement may seem uncomfortably radical and unnecessarily pejorative, a good bet would be that the vast majority of the reproducibility experts cited in this book would agree with it. As they might also with the statement that "as authors or reviewers, we have all witnessed studies not published because [the experimental effect] was too small, but hardly any manuscript was rejected because the treatment needed for a given outcome was too strong" (p. 166).

But is it really fair to characterize the results of fMRI studies such as these as "voodoo"? Robert Park certainly would, but it isn't important how we classify QRP-laden research such as this. What is important is that we recognize the virulence of QRPs to subvert the entire scientific process and somehow find a way to agree upon a means to reduce the flood of false-positive research being produced daily. For it is all types of empirical studies and many entire scientific literatures, not just psychology or experimental research, that have ignited the spark that has grown into a full-blown initiative designed to illuminate and ultimately deflate what is a full-blown crisis. But since it may appear that psychology is being unduly picked on here, let's visit yet another discipline that is at least its equal in the production of false-positive results.

Another Irreproducibility Poster Child: Epidemiology

Major components of this science involve (a) case control studies, in which individuals with a disease are compared to those without the disease, and (b) secondary analyses of large databases to tease out risk factors and causes of diseases. (Excluded here are the roots of the discipline's name, tracking potential epidemics and devising strategies to prevent or slow their progress—obviously a vital societal activity upon which we all depend.)

Also, while case control studies have their shortcomings, it the secondary analysis wing of the discipline that concerns us here. The fodder for these analyses is mostly comprised of surveys, longitudinal studies (e.g., the Framingham Heart Study), and other large databases (often constructed for other purposes but fortuitously tending to be composed of large numbers of variables with the potential of providing hundreds of secondary analyses and hence publications).

The most serious problems with such analyses include already discussed QRPs such as (a) data involving multiple risk factors and multiple conditions which permit huge fishing expeditions with no adjustments to the alpha level, (b) multiple confounding variables which, when (or if) identifiable can only be partially controlled by statistical machinations, and (c) reliance on self-reported data, which in turn relies on faulty memories and under- or overreporting biases.

These and other problems are discussed in a very readable article titled "Epidemiology Faces Its Limits," written more than two decades ago (Taubes, 1995) but which still has important scientific reproducibility implications today.

Taubes begins his article by referencing conflicting evidence emanating from analyses designed to identify cancer risk factors—beginning with those garnering significant press coverage during the year previous to the publication of his article (i.e., 1994).

1. Residential radon exposure caused lung cancer, and yet another study that found it did not.
2. DDT exposure was not associated with breast cancer, which conflicted with the findings of previous, smaller, positive studies.
3. Electromagnetic fields caused brain cancer, which conflicted with a previous study.

These examples are then followed by a plethora of others, a sample of which is included in the following sentence:

Over the years, such studies have come up with a "mind-numbing array of potential disease-causing agents, from hair dyes (lymphomas, myelomas, and leukemia) to coffee (pancreatic cancer and heart disease) to oral contraceptives and other hormone treatments (virtually every disorder known to woman). (p. 164)

Of course we now know the etiology and partial "cures" for preventing (or at least lowering the incidence and assuaging the impact of) false-positive results (key among them in the case of epidemiology, and in specific analyses of large databases in particular, being lowering the titular alpha level for such analyses to at *least* 0.005). However, while none of the plethora of epidemiologists Taubes interviewed in 1995 considered this option, several did recognize the related solution of considering only large relative risks (epidemiology's effect size of choice) including a charmingly nontechnical summarization of the field articulated by Michael Thun (the then-director of analytic epidemiology for the American Cancer Society): "With epidemiology you can tell a little [i.e., effect size] from a big thing. What's very hard to do is to tell a little thing from nothing" (p. 164).

And, to be fair, among the welter of false-positive results, we do owe the discipline for a few crucially important and exhaustively replicated relationships such as smoking and lung cancer, overexposure to sunlight and skin cancer, and the ill effects of obesity. (Coincidentally or not, all were discovered considerably before 1995. And definitely not coincidentally, all qualified as "big" effect sizes rather than "little" ones.)

Unfortunately while this article was written more than two decades ago, things don't appear to have improved all that much. And fortunately, one of my virtual mentors rescued me from the necessity of predicting what (if any) changes are likely to occur over the next 20 years.

Still, Taubes suggests (under the heading "What to Believe?") that the best answer to this question is to believe only strong correlations between diseases and risk factors which possess "a highly pausible biological mechanism."

Other steps are suggested by Stanley Young and Alan Karr (2011), including preregistration of analysis plans and (for large-scale observational studies) the use of a data cleaning team separate from the data analyst. These two epidemiologists also gift us with a liturgy of counterintuitive results that have failed to replicate, such as:

- Coffee causes pancreatic cancer. Type A personality causes heart attacks. Trans-fat is a killer. Women who eat breakfast cereal give birth to more boys. [Note that women do not contribute the necessary Y chromosome for producing male babies.] (p. 116)

Their next volley involved the presentation of a set of 12 observational studies involving 52 significant relationships that were subsequently subjected to randomized clinical trials. None of the 52 associations replicated, although some resulted in statistical results in the *opposite* direction.

Next, an observational study conducted by the US Centers for Disease Control (CDC) was described which correlated assayed urine samples for

275 chemicals with 32 medical outcomes and found, among other things, that bisphenol A was associated with cardiovascular disease, diabetes, and a marker for liver problems. However, there was a slight problem with this finding, but let's allow the authors to speak for themselves:

> There are 275 × 32 = 8800 potential endpoints for analysis. Using simple linear regression for covariate adjustment, there are approximately 1000 potential models, including or not including each demographic variable [there were 10]. Altogether the search space is about 9 million models and endpoints. The authors remain convinced that their claim is valid. (p. 120)

And finally a special epidemiological favorite of mine, involving as it does the secondary analysis of the data emanating from (of all things) one of the largest and most important randomized experiments in the history of education evaluating the learning effects of small versus large class sizes (Word, Johnston, Bain, et al., 1990; Mosteller, 1999).

The Effect of Small Class Sizes on Mortality Through Age 29 Years: Evidence from a Multicenter Randomized Controlled Trial

Peter Muennig, Gretchen Johnson, and Elizabeth Wilde (2011)

These authors, in excellent epidemiological tradition, obtained the original data emanating from what is often referred to as the Tennessee Class-Size study and linked it to the National Death Index records to identify the former's relationship to the latter (i.e., the relationship between students who had been randomly assigned to large vs. small classrooms and their subsequent deaths between 1985 and the end of 2007). The authors' rationale for the study was succinctly stated as:

> A large number of nonexperimental studies have demonstrated that improved cognition and educational attainment are associated with large health benefits in adulthood [seven citations were provided]. However, the short-term health effects of different schooling policies are largely

unknown, and long-term effects have never been evaluated using a randomized trial. (p. 1468)

And there is a good reason why a randomized trial has never investigated the long-term health implications of an instructional strategy. Why should anyone bother?

The authors thus imply that correlating an extraneous, non-hypothesized variable occurring subsequent to a randomized study constitutes a causal relationship between the intervention and the dependent variable (death in this case)—which of course is absurd, but absurdity in some disciplines seldom precludes publication.

What the authors found was that, through age 29, the students randomized to the small class size group experienced statistically significantly higher mortality rates than those randomized to regular size classes. (The original experiment also consisted of class sizes with and without an assigned teacher's aide, but fortunately the presence of an aide didn't turn out to be lethal.) Therefore in the authors' words:

> Between 1985 and 2007, there were 42 deaths among the 3,024 Project STAR participants who attended small classes, 45 deaths among the 4,508 participants who attended regular classes, and 59 deaths among the 4,249 participants who attended regular classes with an aide. (p. 1468)

Interestingly, the authors never discuss the possibility that this might be a chance finding or that the relationship might be non-causal in nature or that the entire study was an obvious fishing expedition. Instead they came up with the following mechanism of action:

> It is tempting to speculate that the additional attention children received in their smaller classes—and possibly in the classes with teacher's aide—helped them to become more outgoing and affirmed their intellectual curiosity. . . . However, this will occasionally have negative outcomes. Poisonings, drugs, drinking and driving, and firearms account for a greater degree of exploration (e.g., poisonings in childhood) and extroversion (e.g., social drug use in adolescence). However, this hypothesis remains highly speculative [*Do you think?*]. (p. 1473)

> It is also tempting to speculate that this finding might have run counter to the investigators' original expectations, given their above-quoted rationale for the study and the fact that the class size study in question resulted in significant learning gains (i.e., "improved cognition and educational attainment are associated with large health benefits in adulthood)" (p. 1468)

Epistemologically, the results of this study can probably be explained in terms of the existence of a number of hidden QRPs, such as not reporting multiple analyses involving multiple "outcome" variables or *harking* (a previously mentioned acronym [but soon to be discussed in more detail] invented by Norbert Kerr [1991] for *h*ypothesizing *a*fter the *r*esults are *k*nown).

Suffice it to say that much of epidemiology involves locating and often (as in this case) combining datasets, bleeding them of all the potential bivariate relationship possible, and then throwing in multiple covariates until something (anything) surfaces with a p-value ≤ 0.05. And persistent investigators will find such a p-value, no matter how many analyses or different datasets are required because, sooner or later, a statistically significant relationship *will* occur. And with a little creativity and a lot of chutzpah, said finding can be accompanied by some sort of mechanism of action. (In this case there is absolutely nothing in the entire educational literature [or within the much more rarified confines of common sense] to suggest that an innocuous behavioral intervention such as smaller class sizes, which have been proved to increase learning, are also lethal.)

Unfortunately, epidemiological practice doesn't appear to have changed a great deal between Taubes's 1995 article and this one published more than a decade later—or as of this writing more than two decades later. Just recently we have found that alcohol consumption in moderation is protective against heart disease, or it isn't, or that red wine is but white wine isn't, or that all alcohol is bad for you regardless of amount. Fish oil prevents heart attacks or it doesn't; this vitamin or supplement prevents this or that disease or it doesn't; antioxidants are the best drugs since antibiotics, or they aren't; vigorous exercise is beneficial or it isn't if someone sits at their desk too long and doesn't walk enough; ad nauseam.

So what are the solutions to this tragedy of inanities? The same as for psychology and most other similarly afflicted disciplines, some which have been discussed, some which will be discussed, and some with no plausible mechanism of action and are so methodologically sloppy that they can be perfunctorily dismissed as not falling under the rubric of legitimate science.

This latter category happens to constitute the subject matter of our next chapter, which deals with a topic sometimes referred to as "pathological" science. It is certainly several steps down the scientific ladder from anything we've discussed so far, but it must be considered since it is actually a key component of our story.

References

Bakker, M., van Dijk, A., & Wicherts, J. M. (2012). The rules of the game called psychological science. *Perspectives on Psychological Science, 7*, 543–554.

Beall, A. T., & Tracy, J. L. (2013). Women are more likely to wear red or pink at peak fertility. *Psychological Science, 24*, 1837–1841.

Chabris, C., Herbert, B., Benjamin, D., et al. (2012). Most reported genetic associations with general intelligence are probably false positives. *Psychological Science, 23*, 1314–1323.

Durante, K., Rae, A., & Griskevicius, V. (2013). The fluctuating female vote: Politics, religion, and the ovulatory cycle. *Psychological Science, 24*, 1007–1016.

Eisenberger, N. I., Lieberman, M. D., & Williams, K. D. (2003). Does rejection hurt? An FMRI study of social exclusion. *Science, 302*, 290–292.

Fiedler, K. (2011). Voodoo correlations are everywhere—not only in neuroscience. *Perspectives on Psychological Science, 6*, 163–171.

Gardner, H. (1983). *Frames of mind: The theory of multiple intelligences*. New York: Basic Books.

Gelman, A., & Loken, E. (2014). The statistical crisis in science: Data-dependent analysis—a "garden of forking paths"—explains why many statistically significant comparisons don't hold up. *American Scientist, 102*, 460–465.

Giner-Sorolla, R. (2012). Science or art? How aesthetic standards grease the way through the publication bottleneck but undermine science. *Perspectives on Psychological Science, 7*, 562–571.

Gould, S. J. (1981). *The mismeasure of man*. New York: Norton.

Ioannidis, J. P. (2011). Excess significance bias in the literature on brain volume abnormalities. *Archives of General Psychiatry, 68*, 773–780.

Kerr, N. L. (1991). HARKing: Hypothesizing after the results are known. *Personality and Social Psychology Review, 2*, 196–217.

Mosteller, F. (1995). The Tennessee study of class size in the early school grades. *The Future of Children, 5*, 113–127.

Muennig, P., Johnson, G., & Wilde, E. T. (2011). The effect of small class sizes on mortality through age 29 years: Evidence from a multicenter randomized controlled trial. *American Journal of Epidemiology, 173*, 1468–1474.

Park, R. (2000). *Voodoo science: The road from foolishness to fraud*. New York: Oxford University Press.

Payton, A. (2009). The impact of genetic research on our understanding of normal cognitive ageing: 1995 to 2009. *Neuropsychology Review, 19*, 451–477.

Sander, D., Grandjean, D., Pourtois, G., et al. (2005). Emotion and attention interactions in social cognition: Brain regions involved in processing anger prosody. *NeuroImage, 28*, 848–858.

Singer, T., Seymour, B., O'Doherty, J., et al. (2004). Empathy for pain involves the affective but not sensory components of pain. *Science, 303*, 1157–1162.

Taubes, G. (1995). Epidemiology facts its limits. *Science, 269*, 164–169.

Vul, E., Harris, C., Winkielman, P., & Pashler, H. (2009). Puzzlingly high correlations in fMRI studies of emotion, personality, and social cognition. *Perspectives on Psychological Science, 4*, 274–290.

Word, E., Johnston, J., Bain, H. P., et al. (1990). *Student/teacher achievement ratio (STAR): Tennessee's K–3 class size study: Final summary report 1985–1990.* Nashville: Tennessee Department of Education.

Young, S. S., & Karr, A. (2011). Deming, data and observational studies: A process out of control and needing fixing. *Significance, 9*, 122–126.

5
The Return of Pathological Science Accompanied by a Pinch of Replication

In 1953, Irving Langmuir (a Nobel laureate in chemistry) gave what must have been one of the most entertaining colloquia in history (*Colloquim on Pathological Science*). Fortunately, it was recorded and later transcribed for posterity by R. N. Hall (http://galileo.phys.virginia.edu/~rjh2j/misc/Langmuir.pdf) for our edification.

In the talk, Professor Langmuir discussed a number of blind alleys taken over the years by scientists who failed to understand the importance of employing appropriate experimental *controls*, thereby allowing them to see effects "that just weren't there." Today we lump these failures under the methodological category of *blinding*—the failure of which, as mentioned previously, is one of the most pernicious and common of the questionable research practices (QRPs).

Several examples of counterintuitive physical science discoveries ascribable to this failing were presented by Professor Langmuir, such as n-rays in 1903 and mitogenic rays a couple of decades later. Both findings elicited huge excitement in the scientific community because they were unexplainable by any known physical theory of their times (i.e., had no plausible mechanism of action) and consequently generated hundreds (today this would be thousands) of scientific publications. Both ray genres gradually fell out of favor following the realization that their existence could only be observed by zealots and could not be *replicated* or *reproduced* when experimenter expectations were effectively eliminated via blinding controls.

However, lest we look back at these miscues too condescendingly, many of us can probably vaguely recall the cold fusion debacle of a couple of decades ago in Utah, which suffered a similar natural history: huge excitement, considerable skepticism, a few early positive replications, followed by many more failures to replicate, and capped by everyone in science except the lunatic fringe soon moving on. (Note that here the term "failure to replicate" does not refer to a replication of an original study not being undertaken but instead to a divergent result and conclusion being reached from a performed replication.)

While the cold fusion episode will be discussed in a bit more detail later, an additional psychological example proffered by Langmuir is probably even more relevant to our story here, given the chemist's par excellence personal investigation into research purporting to prove the existence of extrasensory perception (ESP) conducted by the Duke psychologist Joseph Banks Rhine (1934).

At the time (1934), Langmuir was attending a meeting of the American Chemical Society at Duke University and requested a meeting with Rhine, who quite enthusiastically agreed due to Langmuir's scientific eminence. Interestingly, the visitor transparently revealed his agenda at the beginning of the meeting by explaining his opinion about "the characteristics of those things that aren't so" and that he believed these applied to Rhine's findings.

Rhine laughed and said "I wish you'd publish that. I'd love to have you publish it. That would stir up an awful lot of interest. I'd have more graduate students. We ought to have more graduate students. This thing is so important that we should have more people realize its importance. This should be one of the biggest departments in the university."

Fortunately for Duke, the athletic department was eventually awarded that status but let's return to our story. Basically, Rhine revealed that he was investigating both *clairvoyance*, in which an experimental participant was asked to guess the identity of facedown cards, and *telepathy*, in which the participant was required to read the mind of someone behind a screen who knew the identity of each card. Both reside in the extrasensory perception realm, as did Daryl Bem's precognition studies that had such a large impact on the reproducibility initiative.

As designed, these experiments were quite easy to conduct and the results should have been quite straightforward since chance occurrence was exactly five (20%) correct guesses with the 25-card deck that Rhine employed. His experimental procedures also appeared fine as well (at least for that era), so, short of fraud or some completely unanticipated artifact (such as occurred with the horse Clever Hans who exhibited remarkable arithmetic talents [https://en.wikipedia.org/wiki/Clever_Hans]), there was no obvious way the results could have been biased. (*Spoiler alert*: all scientific results *can* be biased, purposefully or accidentally.)

After conducting thousands of trials (Langmuir estimated hundreds of thousands, and he was probably good with numbers), Rhine found that his participants correctly guessed the identity of the hidden cards 28% of the time. On the surface this may not sound earth-shattering but given the number of experiments conducted, the probability associated with these thousands upon thousands of trials would undoubtedly be equivalent to a randomly chosen individual winning the Mega Millions lottery twice in succession.

Needless to say this result met with a bit of skepticism on Langmuir's part, and it wasn't greatly assuaged when Rhine mentioned that he had (a) filled several filing cabinets with the results of experiments that had produced only chance results or lower and (b) taken the precaution of sealing each file and placing a code number on the outside because he "Didn't trust anybody to know that code. Nobody!"

When Langmuir impoliticly (after all he was a guest even if an enthusiastically welcomed one) expressed some incredulity at Rhine's ignoring such a mountain of negative evidence locked away on the theory that his distractors had deliberately guessed incorrectly just to "spite" him, Rhine was not in the least nonplussed. After a bit more probing on Langmuir's part, Rhine did amend his reason for not at least mentioning these negative results in his book (1934) on the topic to the fact that he hadn't had time to digest their significance and, furthermore, didn't want to mislead the public.

Naturally, Rhine's work has been replicated, but his results have not (Hines, 2003). As an interesting aside, in preparation for his meeting at Duke, Langmuir even "commissioned" his own replication by convincing his nephew, an employee of the Atomic Energy Commission at the time, to recruit some of his friends to spend several of their evenings attempting to replicate Rhine's experiments. At first the group became quite excited because their results (28% or 7 correct guesses out of 25) almost perfectly reflected those of Rhine's, but soon thereafter the results regressed down to chance (i.e., 5 correct guesses out of 25 cards).

Langmuir concluded his fascinating talk as follows (remember this is a transcription of a poorly recorded, informal lecture):

> The characteristics of [the examples he discussed], they have things in common. These are cases where there is no dishonesty involved but where people are tricked into false results by a lack of understanding about what human beings can do to themselves in the way of being led astray by subjective effects, wishful thinking or threshold interactions. These are examples of *pathological science* [emphasis added]. These are things that attracted a

great deal of attention. Usually hundreds of papers have been published upon them. Sometimes they have lasted for fifteen or twenty years and then they gradually die away. (p. 13 of Hall's transcript of Langmuir's *Colloquim on Pathological Science* (1953).

Langmuir may have been a bit too kind in his conclusion of no dishonesty being involved since Rhine reported in his second book that the coded file cabinets originally designated as containing his enemies' purposely incorrect guesses had somehow morphed into an average of seven (28%) correct guesses—thereby exactly replicating his initially reported (and unfiled) results.

But in the final analysis it doesn't matter whether we label substandard scientific procedures as fraud or ignorance or stupidity or QRPs. What is important is that we must avoid allowing them to produce a crisis of confidence in the scientific process itself, not to mention impeding scientific progress in an era in which society increasingly depends on it.

So let's transition now to one of the parents of the modern reproducibility awakening. "Transition," because there is a centuries-old tradition of this genre of experimentation dating back at least to Benjamin Franklin's dismissal of Franz Mesmer's (another early parapsychologist) discovery of animal magnetism at the behest of King of France (Kaptchuk, 1999).

The Odd Case of Daryl Bem

In 2011, an apparently well-respected psychologist and *psi* devotee (*psi* is a "branch" of ESP involving precognition) named Daryl Bem published in the prestigious *Journal of Personality and Social Psychology* a series of nine experiments conducted over a 10-year period and entitled "Feeling the Future: Experimental Evidence for Anomalous Retroactive Influences on Cognition and Affect." The author's premise was that some individuals (or at least some Cornell undergraduates) could not only predict future events a la Joseph Banks Rhine work but, going one step farther, that the direction of causation is not limited to past events influencing future ones, but can travel from the future to the past.

Rather than describing all nine experiments, let's allow Bem to describe the basic methodology of his last two presented experiments, which were very similar in nature and, of course, positive. (Authors of multiexperiment studies often save what they consider to be the most definitive studies for last and sometimes even present a negative study first to emphasize later, more positive findings). His abbreviated description of the eighth experiment's objective follows:

> Inspired by the White Queen's (a character in Lewis Carroll's *Through the Looking Glass—And What Alice Found There*) claim, the current experiment tested the hypothesis that memory can "work both ways" by testing whether rehearsing a set of words makes them easier to recall— even if the rehearsal takes place after the recall test is given. Participants were first shown a set of words and given a free recall test of those words. They were then given a set of practice exercises on a randomly selected subset of those words. The psi hypothesis was that the practice exercises would retroactively facilitate the recall of those words, and, hence, participants would recall more of the to-be-practiced words than the unpracticed words. (p. 419)

And who could argue with such a venerable theoretician as the White Queen? So to shorten the story a bit, naturally the hypothesis was supported:

> The results show that *practicing a set of words* after the recall test *does, in fact, reach back in time* [emphasis added] to facilitate the recall of those words. (p. 419)

After all, what else could it have been but "reaching back into time?" Perhaps William of Occam (unquestionably my most demanding virtual mentor) might have come up with some variant of his hair-brained parsimony principle, but the good man's work is *really* outdated so let's not go there.

To be fair, Professor Bem does provide a few other theoretical mechanisms of action related to quantum mechanics emanating from "conversations" taking place at an "interdisciplinary conference of physicists and psi researchers sponsored by the American Association for the Advancement of Science" (Bem, 2011, p. 423). (Perhaps some of which were centered around one such application advanced by homeopathy advocates to explain how water's *memory* of a substance—the latter dissolved therein but then completely removed—can still be there and be palliative even though the original substance elicited anti-palliative symptoms.)

Perhaps due to the unusual nature of the experiments, or perhaps due to Bem's repeated quoting of Carroll's White Queen (e.g., "memory works both ways" or "It's a poor sort of memory that only works backwards"), some psychologists initially thought the article might be a parody. But most didn't since the study of psi has a venerable history within psychological research, and one survey (Wagner & Monnet, 1979) found that almost two-thirds of academic psychologists believed that psi was at least possible. (Although in the discipline's defense, this belief was actually lower for psychologists than for other college professors.)

However, it soon became apparent that the article was presented in all seriousness and consequently garnered considerable attention both in the professional and public press—perhaps because (a) Bem was an academically respectable psychological researcher (how "academic respectability" is bestowed is not clear) housed in a respectable Ivy League university (Cornell) and (b) the experiments' methodology seemed to adequately adhere to permissible psychological research practice at the time. (Permissible at least in the pre-reproducibility crisis era but not so much now, since, as mentioned by another Nobel Laureate, "the times they are [or may be] a changing.")

And therein lay the difficulties of simply ignoring the article since the methodological quality of the study mirrored that of many other published studies and few psychologists believed that Professor Bem would have been untruthful about his experimental methods, much less have fabricated his data. So, playing by the rules of the game at the time and presenting considerable substantiating data, it could be argued that Bem's methods were marginally adequate. The problem was that by 2011 (the publication date of the article in question), (a) the rules of the game were actually beginning to change and (b) everyone beyond the lunatic fringe of the discipline knew that the positive results Bem reported were somehow simply *wrong*.

Unfortunately for Bem, for while scientific journals (especially those in the social science) have been historically reluctant to publish replications unless the original research was sufficiently controversial, interesting, or counterintuitive, this, too, was beginning to change. And his series of experiments definitely qualified on two (and for some all three) accounts anyway.

Two teams of researchers (Galak, LeBoeuf, Nelson, & Simmons, 2012; Ritchie, Wiseman, & French, 2012) both quickly performed multiple replications of Bem's most impressive studies (the Galak team choosing Bem's eighth and ninth and the Ritchie Wiseman, and French teams his eighth study) and promptly submitted their papers for publication. The Galak group submitted to the same journal that had published Bem's original series (i.e., *The Journal of Personality and Social Psychology*), and, incredibly to some (but not so much to others), the editor of that journal promptly rejected the paper on the basis that it was his journal's policy not to publish replications. According to Ed Yong (2012), the Ritchie team encountered the same resistance in *Science* and *Psychological Science,* which both said that they did not publish "straight replications." A submission to the *British Journal of Psychology* did result in the paper being sent out for peer review but it was rejected (although Bem having been selected as one of the peer reviewers surely couldn't have influenced that decision) before *PLoS ONE* finally published the paper.

Now, as previously mentioned, it is not known whether this episode marked the birth of the reproducibility movement or was simply one of its several inaugural episodes. And whether it will have an effect on the course of scientific progress (or simply serve as another publishing opportunity for academics), my Bronx mentor will not permit me to guess. But surely the following article is one of the most impactful replications in this very unusual and promising movement (with kudos also to the less cited, but equally impressive, Ritchie et al. replication).

Correcting the Past: Failures to Replicate Psi

Jeff Galak, Robyn A. LeBoeuf, Leif D. Nelson, and Joseph P. Simmons (2012)

Since it took Bem a decade to conduct his nine studies, it is perhaps not surprising that these authors chose to replicate only two of them. Both dealt with the "retroactive facilitation of recall," and the replicating authors reported choosing them because they were (a) the "most impressive" (both were statistically significant and the ninth experiment [as is customary] reported the largest effect size), and (b) the findings in the other seven studies employed affective responses that were more difficult to reproduce and might have been, by then, time-specific (no pun intended).

In all, seven replications were conducted, four for Bem's eighth experiment and three for the ninth. Undergraduates were used in three studies and online participants in the other four. Additional methodological advantages of the replications included:

1. They all used predetermined sample sizes and all employed considerably more participants than Bem's two studies. As previously

discussed, the crucial importance of (a) employing sufficiently large sample sizes for ensuring adequate statistical power and (b) the a priori decision of deciding how many participants to employ avoid two of the most virulent QRPs.

2. The replications used both identical and different words (categories of words) to Bem's. This was apparently done to ensure both that (a) the studies were direct replications of the originals and (b) there wasn't something unusual about Bem's choice of words (a replicator's version of both "having one's cake and eating it, too").

3. There was less contact between research staff and participants in the replications, which is important because the former can unconsciously (or in some cases quite consciously) cue responses from the latter.

4. Post-experimental debriefing included the following question for online samples: "Did you, at any point during this study, do something else (e.g., check e-mail)?" Participants were assured that their answer would not influence their payments for participating. This was done to help ensure that respondents were taking their task seriously and following the protocol. If they were not, then obviously the original findings wouldn't replicate.

5. At least two of the four relevant QRPs that were responsible for producing false-positive results modeled by Simmons, Nelson, and Simonsohn (2011) and *may* have characterized Bem's experiments were completely avoided in the seven replications. One involved choosing when to stop running participants and when to add more (the replicators did not look at their data until the end of the experiments, whereas Bem apparently did and adjusted his procedures accordingly as he went along based on how things were going). (This obvious QRP was reported by a former research assistant in an excellent article on the subject in *Slate Magazine* [Engber, 2017].) The other avoided QRP involved not choosing which dependent variables to report on. (Bem reputably conducted multiple analyses but emphasized only the statistically significant ones).

Of course Bem's original results did not replicate, but our protagonists (Galak, LeBoeuf, Nelson, and Simmons) weren't quite through. For while they had apparently (a) repeated Bem's analyses as closely as possible,

> (b) used more participants than he, and (c) ensured that certain QRPs did *not* occur in their replications, who is to say that they themselves might be wrong and Bem himself might be correct?
>
> So, the replicating authors went one step farther. Through an exhaustive search of the published literature they located 10 independent replications of Bem's most impressive two experiments (i.e., numbers 8 and 9) other than their own (of which, it will be recalled, there were seven). Numerically, five of these were in a positive *direction* (i.e., produced a differential between recalled words that were reinforced after the recall test vs. words that were not reinforced), and five favored the words not reinforced versus those there were. (In other words, the expected result from a coin flipping exercise.)
>
> Next our heroes combined all 19 studies (i.e., the 2 by Bem, the 7 by the authors themselves, and the 10 conducted by other researchers) via a standard meta-analytic technique. Somewhat surprisingly, the results showed that, as a gestalt, there was no significant evidence favoring reverse causation even while including Bem's glowingly positive results. When the analysis was repeated using only *replications* of Bem's work (i.e., without including his work), the evidence was even more compelling against psi—reminiscent of Langmuir's description of those "things that aren't so."

So, was the issue settled? Temporarily perhaps, but this and other nonsense resurfaces every few years and surveys of young people who are deluged with (and enjoy watching) screen adaptations of Marvel and DC comic books, devotees of conspiracy theories on YouTube, and adults who continue to frequent alternative medical therapists in order to access the placebo effect will probably continue to believe in the paranormal until the planet suffers a major meteor strike.

And, as a footnote, obviously Bem—like Rhine before him—remained convinced that his original findings were correct. He even

1. Conducted his own meta-analysis (Bem, Tressoldi, Rabeyron, & Duggan, 2016) which, of course, unlike Galak et al.'s was positive;
2. Preregistered the protocol for a self-replication in a *parapsychology registry* (which I had no idea existed) but which theoretically

prevented some of the (probable) original QRPs (http://www.koestler-parapsychology.psy.ed.ac.uk/Documents/KPU_registry_1016.pdf), which it will be recalled involved (a) deep-sixing negative findings, (b) cherry-picking the most propitious outcome variables, (c) tweaking the experimental conditions as he went along, as Bem (in addition to his former research assistant) admitted doing in the original studies, and (d) abandoning "false starts" that interim analyses indicated were trending in the "wrong" direction (which Bem also admitted doing although he could not recall the number of times this occurred); and, finally,

3. Actually conducted said replication with Marilyn Schlitz and Arnaud Delorme which, according to his preregistration, failed to replicate his original finding. However, when presenting his results at a parapsychological conference, Engber reports a typically happy, pathological scientific ending for our intrepid investigator.

They presented their results last summer, at the most recent [2016] annual meeting of the Parapsychological Association. According to their preregistered analysis, there was no evidence at all for ESP, nor was there any correlation between the attitudes of the experimenters—whether they were believers or skeptics when it came to psi—and the outcomes of the study. In summary, their large-scale, multisite, pre-registered replication ended in a failure. [That's the bad news, but there's always good news in pathological science.] In their conference abstract, though, Bem and his co-authors found a way to wring some droplets of confirmation from the data. After adding in a set of new statistical tests, *ex post facto*, they concluded that the evidence for ESP was indeed "highly significant."

Of course, pathological science isn't limited to psychology. In fact, Professor Langmuir's famous colloquium (given his professional interests) was more heavily weighted toward the physical than the social sciences. So let's go back a few decades and revisit a much more famous pathological example generated by QRPs involving the physical science, the desire for fame and fortune, and the willingness to see and engage in "the science of things that aren't so" (which, incidentally, would have constituted a better label than "irreproducible science").

The Possibly Even Odder Case of the Discovery of Cold Fusion

While the cold fusion debacle of some three decades ago may have begun to fade from memory (or was never afforded any space therein by some), it still holds some useful lessons for us today. Ultimately, following a few initial bumps in the road, the scientific response to the event should go down as a success story for the replication process in the physical sciences just as the psi replications did for psychology. But first a bit of background.

Nuclear fusion occurs when the nuclei of two atoms are forced into close enough proximity to one another to form a completely different nucleus. The most dramatic example of the phenomenon occurs in stars (and our sun, of course) in the presence of astronomically (excuse the pun) high temperatures and pressures.

Ironically, this celestial process turns out to be the only mechanism by which the alchemists' dreams of changing lighter (or less rare) elements into gold can be realized since their crude laboratory apparatuses couldn't possibly supply the necessary energy to duplicate the fusion process that occurs at the center of stars or the explosion of a nuclear fusion (hydrogen) bomb. But as Irving Langmuir (pathological science), Robert Park (voodoo science), and even Barker Bausell (snake oil science) have illustrated, all of the laws of science can be subverted by a sufficient amount of ambition, ignorance, and disingenuousness.

So it came to pass on March 23, 1989, that the University of Utah held a press conference in which it was breathlessly announced that two chemists, Stanley Pom and Martin Fleischmann, had invented a method that gave promise to fulfilling the dream of eventually producing unlimited, nonpolluting, and cheap energy using a simple tabletop device that would have charmed any alchemist of centuries past. The apparatus (Figure 5.1) itself generating this earthshaking discovery was comprised of

1. An unimposing, tabletop-sized, insulated container designed to maintain the temperature of the contents therein independently of the outside temperature for relatively long periods of time;
2. Heavy water (an isotope of hydrogen containing a proton and a neutron as opposed to only a single proton), plus an electrolyte to facilitate the flow of electricity through it; and
3. A cathode made of palladium (a metallic element similar to platinum).

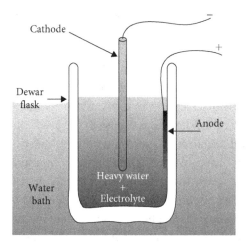

Figure 5.1 The remarkably simple and cheap device that ostensibly produced a cold fusion reaction.

https://en.wikipedia.org/wiki/Cold_fusion#/media/File:Cold_fusion_electrolysis.svg

The press conference occurred before any of these results had been submitted to a peer reviewed journal, apparently to preempt another cold fusion researcher (Stephen Jones) at another Utah university (Brigham Young). Jones's version of inducing nuclear fusion actually had a recognized scientific mechanism of action but was considered completely impractical due to the inconvenient fact (which he apparently recognized) that the tiny amount of energy apparently emanating from his procedures was far exceeded by the amount of energy required to produce it.

In any event, a double-helix style race commenced (although perhaps even more vituperative between Pom and Fleischmann vs. Jones), with Pom suspecting Jones of trying to steal his work (although there is no evidence of this). Unlike the race to characterize the structure of DNA, however, the financial stakes here were so high (potentially involving trillions of dollars) that the University of Utah's press conference was characterized by exaggerated claims about their researchers' actual progress.

As a result, both the press and a large swath of the scientific community appeared to lose their respective minds, with Pom and Fleischmann immediately receiving rock star status accompanied by dozens of laboratories all over the world beginning the process of attempting to *replicate* their results. Very shortly, aided by the simplicity of the intervention, "confirmations" of the experiment were issued by researchers at Georgia

Tech and Texas A&M, but, before long, laboratories at MIT, Cal Tech, Harwell, and others reported failures to do so. To greatly oversimplify the entire drama, along with the back-and-forth accusations and counter claims, Georgia Tech and A&M retracted their confirmations upon re-examining their results.

However, exact replications can be a bit difficult without knowledge of the actual methods employed in the original work, and Stanley Pom (who basically became the spokesperson for the entire fiasco) wasn't about to share *anything* with *anybody* including high-impact peer reviewed journals or scientific competitors. So although there were dozens of attempted replications accompanied by an ever decreasing number of positive results, Pom assured the pathologically gullible press (*The Wall Street Journal* being the primary advocate of cold fusion research, given the trillion-dollar industry it would potentially spawn) that the many failures to replicate could be explained by the simple fact that the labs producing them had not used the exact and proper Pom-Fleischmann procedures (which, of course, was probably true since Pom refused to share those details with them).

But soon, more and more conventional physicist and chemists became increasingly critical (and indeed incredulous) that neither the original Utah experiments nor their "successful" replications bothered to run controls involving, say, plain water instead of its heavy counterpart. For experimental controls are not only *absolute* prerequisites for findings to be considered credible in *any* discipline (social, biological, or physical), but also their absence constitutes an egregious QRP in and of itself. And almost equally important, for some reason, Pom and Fleischmann failed to secure sufficiently sensitive (and available) instrumentation to filter out background environmental contaminants or other sources of laboratory noise. A few sticklers were even concerned that the etiology of the effect violated the first law of thermodynamics (i.e., while energy can be transformed from one form to another, it cannot be created or destroyed, even in a closed system such as our protagonists' tabletop device). But as the old saying goes, "laws are made to be broken."

But counter-arguments such as these were impotent compared to those of a consummate snake oil salesman such as Stanley Pom who dismissed anything with negative implications as unimportant or part of a witch hunt by jealous East Coast physicists, especially those affiliated with MIT and Yale. In fact, Pom was reported by Gary Taubes (1993) in his splendid 473-page history of the incident (entitled *Bad Science: The Short Life and Weird Times of Cold Fusion*) as saying, when confronted by the ever-growing number of laboratories' failure to find *any* effect, "I'm not interested in negative effects!" (Perhaps he should have been a journal editor.)

And so it went, press conference after press conference, professional conference after professional conference. Positive results were highlighted and negative results were summarily dismissed. Even the advocates who occasionally produced small degrees of excess heat (a supposed indicator of the fusion process but obviously of many other more mundane processes as well) could only do so occasionally. But this was enough to keep the process alive for a while and seemed to add to its mysteriousness. For others, it was obvious that something was amiss, and it wasn't particularly difficult to guess what that was.

What was obvious to Pom, however, was what was really needed: more research, far more funding, partnerships with corporations such as General Electric, and, of course, *faith*. And, every so often, but with increasing rarity, one of the process's proponents would report an extraordinary claim, such as a report from the Texas A&M laboratory that their reaction had produced tritium, a ubiquitous byproduct of fusion. And while this was undoubtedly the most promising finding emanating from the entire episode, like everything else, it only occurred once in a few devices in a single lab which eventually led to a general consensus that the tritium had been spiked by a single individual.

So, gradually, as the failures to replicate continued to pile up and Dr. Pom kept repeating the same polemics, the press moved on to more newsworthy issues such as a man biting a dog somewhere in the heartland. And even the least sensible scientists appeared to have developed a modicum of herd immunity to the extravagant claims and disingenuous pronouncements by the terminally infected. So, in only a very few years, the fad had run its course, although during one of those heady years cold fusion articles became the most frequently published topic area in all of the physical sciences.

Today, research on "hot" (i.e., conventional) fusion continues, the beneficiary of billions in funding, but cold fusion investigations have all but disappeared in the mainstream scientific literature. Most legitimate peer reviewed journals in fact now refuse to even have a cold fusion study reviewed, much less publish one. However, a few unrepentant investigators, like their paranormal and conspiracy theory compatriots, still doggedly pursue the dream and most likely will continue to do so until they die or become too infirm to continue the good fight.

RETURN OF PATHOLOGICAL SCIENCE 123

Lessons Learned

Unfortunately, this subheading is an oxymoron when applied to pathological, voodoo, or snake oil science researchers, but it is unlikely that any of these practitioners will ever read a book such as this (if any such exist). The rest of us should always keep Irving Langmuir's lecture on pathological science in mind, especially his six criteria for what makes a scientific discovery pathological, be it cold fusion, N-rays, mitogenetic rays, or photographic evidence proving the existence of flying saucers—the latter of which loosely qualifies as a physical science phenomenon since Langmuir, after examining what investigators considered the "best evidence," concluded that "most of them [extraterrestrial vehicles] were Venus seen in the evening through a murky atmosphere."

So perhaps, with a little effort, present-day scientists can translate at least some of Langmuir's criteria to their own disciplines. Let's use cold fusion as an example.

1. "The maximum effect that is observed is produced by a causative agent of barely detectable intensity, and the magnitude of the effect is substantially independent of the intensity of the cause." This one certainly applies to cold fusion since astronomically large amounts of heat are required to generate nuclear fusion while cold fusion required only a simple, low-voltage electrical current flowing through a liquid medium at room temperature.
2. "The effect is of a magnitude that remains close to the limit of detectability, or many measurements are necessary because of the very low statistical significance of the results." Over and over again responsible physicists unsuccessfully argued that the small amount of heat generated in the tabletop device was millions of times less than that generated by the smallest known fusion reactions. The same was true for the tiny number of emitted neutrons (a byproduct of the process) occasionally claimed to have been measured.
3. "There are claims of great accuracy." As only two examples, the occasional reports of tiny amounts of increased heat as measured by our heroes apparently involved a substandard calorimeter and did not take into account the facts that (a) different solutions result in different degrees of conductivity (hence influencing measurable heat) or even that (b) the amount of commercial electric current (that helped produce

said heat) is not constant and varies according to various conditions (e.g., the performance of air conditioners on hot summer days).
4. "Fantastic theories contrary to experience are suggested." This one is obvious.
5. "Criticisms are met by ad hoc excuses." As well as downright lies, a paranoid belief that the individuals failing to replicate positive findings had hidden or nefarious agendas and/or were not able to employ the original procedures (since, in the case of cold fusion, these happened to be closely guarded secrets). Furthermore, even the most avid supporters of the process admitted that their positive results occurred only sporadically (hence were not reproducible by any scientific definition of the term). Various excuses were advanced to explain this latter inconvenient truth, although only one excuse (admitted ignorance of what was going on) was not disingenuous.
6. "The ratio of supporters to critics rises and then falls gradually to oblivion." As mentioned previously, the number of supporters quickly approached almost epidemic levels, perhaps to a greater extent than for any other pathologically irreproducible finding up to that point. However, facilitated by a discipline-wide replication initiative, the epidemic subsided relatively quickly. But a good guess is that Pom continues to believe in cold fusion as fervently as Bem still believes in psi.

But since Dr. Langmuir was not privy to the cold fusion epidemic or Daryl Bem's landmark discoveries, perhaps he wouldn't object too strenuously if three additional principles were added: one advanced by a philosopher of science centuries ago who would have also been a Nobel laureate if the award existed, one attributable to a number of scientific luminaries, and one to completely unknown pundit:

7. "What is done with fewer assumptions is done in vain with more," said William of Occam, who counseled scientists choosing between alternative theories or explanations to prefer the one that required the fewest unproved assumptions.
8. "Extraordinary claims require extraordinary evidence," which, according to Wikipedia was only popularized by Carl Sagan but originally was proposed in one form or another by David Hume, Pierre-Simon Laplace, and perhaps some others as well. A more extraordinary claim is difficult to concoct than the contention that a humble tabletop apparatus could subvert the laws of physics and fuel the world's energy

needs for millennia to come. Or that future events can effect past ones for that matter.
9. When a scientific finding sounds too good (or too unbelievable) to be true, it most likely isn't (pundit unknown).

However, Gary Taubes (1993) probably deserves the final word on the lessons that the cold fusion fiasco has for science in general as well as its direct applicability to the reproducibility crisis.

> Of all the arguments spun forth in defense of cold fusion, the most often heard was *there must be something to it*, otherwise the mainstream scientific community would not have responded so vehemently to the announcement of its discovery. What the champions of cold fusion never seemed to realize, however, or were incapable of acknowledging, was that the vehemence was aimed not at the science of cold fusion, but at the method [i.e., experimental methodology]. Positive results in cold fusion were inevitably characterized by sloppy and amateurish experimental techniques [we could substitute QRPs here]. If these experiments, all hopelessly flawed, were given the credibility for which the proponents of cold fusion argued, the science itself would become an empty and meaningless endeavor. (p. 426)

But Was It Fraud?

As mentioned previously, such judgments are beyond my paygrade. However, one of my three very accomplished virtual mentors (the physicist Robert Park, head of the Washington office of the American Physical Society at the time) was eminently qualified to render such a judgment. So, according to Gary Taubes (to whom I apologize for quoting so often but his book truly is am exemplary exposition of the entire fiasco and should be read in its entirety):

> When the cold fusion announcement was made, Robert Bazell, the science reporter for NBC News, interviewed Robert Park.... He asked Park, off the record, whether he thought cold fusion was fraud, and Park said, "No but give it two months and it will be." (p. 314)

And again according to Taubes, Bob was quite proud of this piece of prognostication, even though he admitted that his timeline was off by about 6 weeks.

But how about Daryl Bem and the ever-increasing plethora of positive, often counterintuitive, findings published in today's literature? Is that fraud? Unfortunately, Professor Park is quite ill and not able to grace us with his opinion. So let's just label psi a QRP-generated phenomena and let it go at that. In science, being wrong for the wrong reasons is bad enough.

But Should Outrageous Findings Be Replicated?

"Should" is probably the wrong word here. And anyway, some of us simply don't have the self-control to avoid debunking high-profile, ridiculous, aesthetically offensive (to some of us at least) nonsense. So the higher the profile, the more likely findings are to be replicated.

Of course some topics are politically immune to evidence (think climate change), some are outside the purview of scientific investigation (e.g., religion), and some scientists and non-scientists are so tightly wrapped into their theories or political views that no amount of evidence can change their opinions. Scientists are also humans, after all, so they can be equally susceptible to foibles such as shamelessly hyping the importance of their own work while debunking any conflicting evidence. Whether such behaviors qualify as QRPs is a matter of opinion.

But pathological findings aside, should scientists conduct replications as part of their repertoire? Here, "should" *is* the correct word. First, all scientists should replicate their own work whenever possible, but not as a means of convincing others of their finding's veracity. Scientists in general tend to be skeptical, so many are likely to consider successful self-replications to be too governed by self-interest to be taken at face value. Instead, self-replications should be used as a means of ensuring that one's own work is valid to avoid continuing down a dead end street and wasting precious time and resources. And regardless of whether scientists replicate their findings or not, they should also (a) check and recheck their research findings with an eye toward identifying any QRPs that somehow might have crept into their work during its conduct and (b) cooperate fully with colleagues who wish to independently replicate their work (which, in his defense, Daryl Bem apparently did).

Of course, given the ever increasing glut of new studies being published daily, everything obviously can't be replicated, but when an investigative team plans to conduct a study based on one of these new findings, replication is a sensible strategy. For while a replication is time- and resource-consuming, performing one that is directly relevant to a future project may actually turn out to be a cost- *and* time-saving device. Especially if the modeling results involving the prevalence of false-positive results (e.g., Ioannidis, 2005; Pashler & Harris, 2012) previously discussed have any validity.

However, some studies *must* be replicated if they are paradigmatically relevant enough to potentially challenge the conventional knowledge characterizing an entire field of study. So all scientists in all serious disciplines *should* add the methodologies involved in performing replications to their repertoires.

In the past, "hard" sciences such as physics and chemistry have had a far better record for performing replications of potentially important findings quickly and thoroughly than the "softer" social sciences, but that difference is beginning to fade. As one example, ironically, in the same year (2011) that Daryl Bem published his paradigm-shifting finding, physics experienced a potentially qualitatively similar problem.

In what a social scientist might categorize as a post hoc analysis, the Oscillation Project Emulsion-t Racking Apparatus (OPERA) recorded neutrinos apparently traveling faster than the speed of light (Adam, Agafonova, Aleksandrov, et al., 2011). If true (and not a false-positive observation), this finding would have negated a "cornerstone of modern physics" by questioning a key tenet of Einstein's theory of general relativity.

Needless to say, the physics community was more than a little skeptical of the finding (as, in their defense, were most psychologists regarding the existence of psi), especially given the fact that only seven neutrinos were observed traveling at this breakneck speed. In response, several labs promptly conducted exact replications of the finding within a few months (similar to the speed at which the Galak and Ritchie et al. teams replicated Bem's work). Also not unlike these tangentially analogous social science replications, all failed to produce any hint of faster-than-light travel while concomitantly the original OPERA team discovered the probable cause of the discrepancy in the form of a faulty clock and a loose cable connection. The ensuing embarrassment, partially due to the lab's possibly premature announcement of the original finding, reputably resulted in several leaders of the project to submit their resignations (Grossman, 2012).

So Are the "Hard" Sciences All That Different From Their "Softer" Counterparts?

At present there is little question that the physical sciences are at least more prestigious, successful, and gifted with a somewhat lower degree of publication bias (and thus perhaps more likely to have a lower prevalence of false-positive findings) than the social sciences. The former's cumulative success in generating knowledge, theoretical and useful, is also far ahead of the social sciences although at least some of that success may be due to the former's head start of a few thousand years.

Daniele Fanelli (2010), a leading meta-scientist who has studied many of the differences between the hard and soft sciences, has argued that, to the extent that the two scientific genres perform methodologically comparable research, any differences between them in other respects (e.g., subjectivity) is primarily only "a matter of degree." If by "methodological comparability" Dr. Fanelli means (a) the avoidance of pathological science and (b) the exclusion of the extreme differences in the sensitivity of the measurement instrumentation available to the two genres, then he is undoubtedly correct. However, there seems to be a huge historical affective and behavior gap between their approaches to the replication process which favors the "hard" sciences.

As one example of this differential disciplinary embrace of the replication process, a *Nature* online poll of 1,575 (primarily) physical and life scientists (Baker, 2016) found that 70% of the respondents *had tried and failed* to replicate someone else's study, and, incredibly, almost as many had tried and failed to replicate one of their own personal finding. Among this group, 24% reported having published a successful replication while 13% had published a failure to replicate one, both an interesting twist on the publication bias phenomenon as well as indirect evidence that the replication process may not be as rare as previously believed—especially outside the social sciences.

If surveys such as this are representative of the life sciences, their social counterparts have a long way to go despite the Herculean replication efforts about to be described here. However, it may be that if the social sciences continue to make progress in replicating their findings and begin to value being correct over being published, then perhaps the "hierarchy of science" and terms such as "hard" and "soft" sciences will eventually become archaic.

The Moral Being (Besides Raising the Strawman of Hard Versus Soft Science)

As mentioned previously (e.g., the classic Watson and Crick example), when announcing an honest-to-goodness new discovery, employing circumspect (or at least cautious) language to announce it is a good practice. Even more importantly, an actual new discovery probably shouldn't even be announced unaccompanied by a sufficiently stringent a priori registered alpha level (perhaps 0.005 in the social sciences) and/or before performing a direct replication of said finding.

So with all this in mind, it is now time to finally begin discussing the replication process in a bit more detail since it remains the most definitive strategy available for ferreting out irreproducible scientific findings. In so doing a somewhat heavier emphasis will be placed on the social sciences since the performance of replications appears to have been a neglected component of their repertoires—at least until the second decade of this century.

References

Adam, T., Agafonova, A., Aleksandrov, A., et al. (2011). Measurement of the neutrino velocity with the OPERA detector in the CNGS beam. *airXiv: 1109.4897v1*.

Baker, M. (2016). Is there a reproducibility crisis? *Nature, 533*, 452–454.

Bem D. J. (2011). Feeling the future: Experimental evidence for anomalous retroactive influences on cognition and affect. *Journal of Personality and Social Psychology, 100*, 407–425.

Bem, D., Tressoldi, P., Rabeyron, T., & Duggan, J. (2016). Feeling the future: A meta-analysis of 90 experiments on the anomalous anticipation of random future events. *F1000Research, 4*, 1188.

Engber, D. (2017). Daryl Bem proved ESP is real: Which means science is broken. *Slate Magazine*. https://slate.com/health-and-science/2017/06/daryl-bem-proved-esp-is-real-showed-science-is-broken.html

Fanelli, D. (2010). "Positive" results increase down the hierarchy of the sciences. *PLoS ONE, 5*, e10068.

Galak, J., LeBoeuf, R. A., Nelson, L. D., & Simmons, J. P. (2012). Correcting the past: Failures to replicate psi. *Journal of Personality and Social Psychology, 103*, 933–948.

Grossman, L. (2012). Leaders of controversial neutrino experiment step down. *New Scientist*. www.newscientist.com/article/dn21656-leaders-of-controversial-neutrino-experiment-step-down/

Hines, T. (1983). *Pseudoscience and the paranormal: A critical examination of the evidence*. Buffalo, NY: Prometheus Books.

Ioannidis, J. P. A. (2005). Why most published research findings are false. *PLoS Medicine, 2*, e124.

Kaptchuk, T. (1999). Intentional ignorance: A history of blind assessment and placebo controls in medicine. *Bulletin of the History of Medicine, 72*, 389–433.

Pashler, H., & Harris, C. R. (2012). Is the replicability crisis overblown? Three arguments examined. *Perspectives on Psychological Science, 7*, 531–526.

Rhine, J. B. (1934). *Extra-sensory perception*. Boston: Bruce Humphries.

Ritchie, S., Wiseman, R., & French, C. (2012). Failing the future: Three unsuccessful attempts to replicate Bem's retroactive facilitation of recall effect. *PLoS ONE, 7*, e33423.

Simmons, J. P., Nelson, L. D., & Simonsohn, U. (2011). False-positive psychology: Undisclosed flexibility in data collection and analysis allows presenting anything as significant. *Psychological Science, 22*, 1359–1366.

Taubes, G. (1993). *Bad science: The short life and weird times of cold fusion*. New York: Random House.

Wagner, M. W., & Monnet, M. (1979). Attitudes of college professors toward extra-sensory perception. *Zetetic Scholar, 5*, 7–17.

Yong, E. (2012). Bad copy. *Nature, 485*, 298–300.

PART II
APPROACHES FOR IDENTIFYING IRREPRODUCIBLE FINDINGS

6
The Replication Process

Previous chapters have hinted at the key role that the replication process plays in enhancing scientific progress and ameliorating the reproducibility crisis. However, there is little about the scientific process that is either easy or perfect—replication included—so two points should probably be reviewed, one affective and one epistemological.

First, *the affective perspective*: those who replicate a study and fail to confirm the original finding should not expect to be embraced with open arms by the original investigators(s). No scientist wishes to be declared incorrect to the rest of the scientific world and most will undoubtedly continue to believe (or at least defend) the validity of their results. So anyone performing a replication should be compulsively careful in its design and conduct by avoiding any potential glitches or repeating the mistakes of the original. And from both a personal and a scientific perspective, all modern replications should

1. *Be preregistered on a publically accessible website prior to data collection.* And as important as this dictum is for original research, it may be even more important for a replication given the amount of blowback likely to be generated by an offended original investigator whose results failed to replicate—or by her or his passionate defenders—as often or not via social media.
2. *Be designed with considerably more statistical power than the original (preferably 0.90).* For, as mentioned in Chapter 2, if a replication employs the same amount of power as a typical original study (0.50 for psychological experiments), then it will have only a 50% chance of obtaining statistical significance even if a true effect exists and the original finding was correct.
3. *Follow the original design as closely as possible (with the exception of repeating any questionable research practices [QRPs] therein)* since even the slightest deviation constitutes fodder for a counterattack. Exceptions exist here, but any design or procedural changes should be justified (and justifiable) and preferably informed by pilot work.

4. *Attempt to engage the original investigators in the replication process as much as possible.* There are at least two reasons for this. First, it will help assure the original investigators that the replication is not meant to be an attack on their integrity, hence some of the virulence of their objections to a disconfirming replication can be deflected by requesting any feedback they might have regarding the design of the proposed replication. (This is not only a professional courtesy but good scientific practice since it may result in an offer to share scientific materials and other key information seldom available in a typical journal publication.)
5. *And, perhaps most importantly of all, to take the time to examine the Open Science Framework website* (https://osf.io) *and its abundance of highly recommended instructions, examples, and information available for both the preregistration and replication processes.*

And second, *from an epistemological perspective*, it is important to remember that failures to replicate can occur for a number of reasons, including

1. The original study's methods were flawed and its results were incorrect,
2. The replicating study's approach was flawed and its results were false,
3. Both studies were incorrect, or
4. The methods or participants used in the second study were substantively different from those used in the first study (hence the replication does not match the original in terms of key conditions).

But to make things a bit murkier in social and behavioral research, it is also always possible that the original study finding could have been correct *at the time* but no longer reproducible because (to quote Roland, the Stephen King character) "the world has moved on." This possibility is especially problematic in the scenario-type, often culture-related studies of which psychology, political science, and economics are so fond and in which undergraduates and/or Amazon's Mechanical Turk participants are almost universally employed.

Said another way, it may be that constantly evolving cultural changes can influence responses to interventions. Or, given the number of brief interventional studies being conducted, participants may become more experienced, sophisticated, and therefore more difficult to blind to group membership. Or become more susceptible to demand characteristics purposefully or accidentally presented in the experimental instructions provided to them. Or, if the original (or a very similar) study has achieved a degree of notoriety by finding its way into the press or disciplinary textbooks, participants may have been exposed to it, recognize the replication's intent, and consequently respond accordingly.

But setting such quibbles aside, replications remain the best indicators we have for judging the validity of social scientific results. So let's begin by examining the types or genres of replications available to investigators.

Exact Replications

While preferable, exact replications are not typically possible in most social sciences. They can be feasible in disciplines such as chemistry, physics, or (theoretically) animal research in which genetically identical rodents reared under standardized conditions are employed. (The concept itself was first proposed and named [at least to my knowledge] by the previously mentioned classic modeling article by Greenwald, Gonzalez, Harris, and Guthrie [1996] who basically dismissed the possibility of its implementation in research involving human participants.)

Research involving human (and probably the majority of animal) experiments aside, exact replications are possible (and recommended) for modeling studies and analyses involving existing databases designed to tease out relationships among variables. The latter (aka a*nalytic replications*) can (and should) be performed to ascertain if the same results can be reproduced using the original data and code *if* the latter is available with adequate documentation, which, unfortunately, as will be discussed in Chapter 10, appears to be surprisingly rare.

So let's now concentrate on the replication of experimental findings, which leads us to the most recommended genre by just about everyone interested in reproducibility, although all forms of replication have their special charms.

Direct (aka Close) Replications

This form of replication involves employing the same procedures used in the original study of interest including (a) instructions to participants and experimenters, (b) types of participants, (c) interventions, (d) measures, and (e) statistical analyses. Possible exceptions involve different subject pools, such as the use of Amazon's Mechanical Turk employees in lieu of undergraduates and definitely should involve (a) increased sample sizes if the original study is underpowered (which in the social sciences is more likely than not), (b) the avoidance of QRPs present in the original (and, of course, avoiding new ones in the replication), and (c) more appropriate data analytic techniques if the original's are patently unacceptable.

As mentioned, direct replications normally require information from the original investigators since journal articles do not communicate anything approaching the required level of detail needed to actually replicate the procedures described therein. Experimental materials should also be solicited from the original investigators, as well as feedback regarding the proposed protocol before it is preregistered. (There is currently a robust movement to persuade all original investigators to preregister their materials, data, code, and detailed experimental protocols, which, if successful, will greatly facilitate the replication process.)

From a strict reproducibility perspective, this genre of replication is preferable to those that follow because the closer the replication matches the original study, the more confidence can be had in the bottom-line inference (i.e., the original results replicated or they did not). A change in a replication's procedures (other than the correction of an obvious methodological or statistical flaw) will almost certainly be cited by the original author(s) as the reason for a failure to replicate if one occurs.

Several original investigators, for example, have blamed a failure to replicate (occasionally vituperatively, sometimes reasoned) on the computerized administration of interventions (often using Amazon Turk participants) as a substitute for laboratory presentations employing college students. Since this change in the *presentation* of stimuli sometimes results in subtle changes to the interventions themselves, it is reasonable to question whether such studies can still be classified as *direct* replications. However, it may be more reasonable to question whether an effect has any scientific importance if it is so tenuous and fragile that it can't be replicated if respondents read their instructions rather than listening to a research assistant recite them.

Many if not most of the 151 replications of psychological studies conducted by the Open Science Collaboration and the three "Many Labs" efforts discussed in the next chapter were conducted using approaches and participants similar to those used in the original studies, hence any such discrepancies were not likely to have contributed to the disappointing failure-to-replicate rate in these initiatives—especially since the results from the first "Many Labs" study found that undergraduate and Amazon Mechanical Turk employees responded similarly to one another in its replications.

In addition, all 151 replications were highly powered and at least as methodologically sound as the studies they replicated. The replications also employed dozens of sites and investigators, thereby reducing the possibility of systematic biases due to settings or individual researchers. Of course it is always possible that subtle changes in the presentation or timing of an intervention (which are often necessary in the translation from laboratory to computer) might affect an outcome, but again, if a finding is this fragile, how likely is it to be relevant to human behavior in the noisy milieu of everyday life?

A registered replication report (more on that later) published in *Perspective on Psychological Science* performed by the team of Alogna, Attaya, Aucoin, and colleagues (2014) provides an interesting perspective on both of these issues (i.e., fragility and "minor" alternations in study procedures) in a replication of a somewhat counterintuitive study on a concept (or theory) referred to as "verbal overshadowing." The original study (Schooler & Engstler-Schooler, 1990) involved a scenario in which all participants watched a video of a simulated bank robbery. One group then verbally described the robber while the other performed an irrelevant task listing US states and capitals.

Attempting to commit something to memory normally facilitates later recall but in this case the participants who verbally described the appearance of the culprit were significantly less successful in identifying said culprit from a mock lineup than the comparison group who performed an irrelevant task instead.

The replication study initially failed to support the original finding, but its first author (Jonathan Schooler) objected to a timing change between the two events employed in the replication. Accordingly, the replication team repeated that aspect of the study and the original effect reached statistical significance, although, as usual (Ioannidis, 2008), the initial effect size was larger than the replicated one. (Hence, if nothing else this represents a positive case study involving cooperation between replicators and original investigators.)

As for computer-based versus laboratory-based differences between replications and original studies, the jury remains out. For example, in a response to Hagger, Chatzisarantis, Alberts, and colleagues (2016) failure to replicate something called the "ego depletion effect," the original investigators (Baumeister & Vohs, 2016) argued that "the admirable ideal that all meaningful psychological phenomena can be operationalized as typing on computer keyboards should perhaps be up for debate" (p. 575)—an argument that Daryl Bem and others have also used to suggest a reason for their results' failures to replicate. (Recall that a "failure to replicate" here is meant to represent a study that was replicated but failed to reproduce the original study's bottom-line result.)

Of course a truly unrepentant curmudgeon might suggest that the idea than any societally meaningful real-world phenomena that can be discovered in an academic psychological laboratory employing undergraduate psychology students "should perhaps be up for debate" as well. By way of example, returning to the successful cooperative "verbal overshadowing" example following the tweaking of the time interval separating the video from the pictorial lineup, one wonders how likely this finding would translate to a real-life "operationalization" of the construct? Say, to someone (a) actually witnessing a real armed bank robbery in person (possibly accompanied by fear of being shot), followed by (b) questions regarding the appearance of the robbers anywhere from a few minutes to a few hours later by police arriving on the scene, and then (c) followed by a live police lineup several days or weeks later?

Conceptual (aka Differentiated, Systematic) Replications

This genre of replication is more common (at least in the social sciences) than direct replications. Different authors have slightly different definitions and names for this genre, but basically *conceptual replications* usually involve purposefully changing the intervention or the outcome measure employed in the original study in order to extend the concept or theory guiding that study. (The experimental procedures may also be changed as well, but the underlying purpose of this type of study is normally not to validate the original finding since it is tacitly assumed to be correct.)

Of course different investigators have different objectives in mind for conducting a conceptual replication such as

1. Supporting the original study by extending the conditions and circumstances under which its effect occur (similar to the "external validity" concept),
2. Determining whether a concept or construct replicates from one theoretical arena to another using the same or a similar paradigmatic approach, or (less commonly), and
3. Ascertaining if the original effect is an artifact of the specialized manner in which the study was designed or conducted.

Differing objectives such as these (coupled with investigator expectations and attendant [possibly unconscious] design decisions) tend to dilute the primary advantage of their direct counterparts: namely, to determine whether or not original inferential results replicated. In addition, some reproducibility experts (e.g., Pashler & Harris, 2012) consider conceptual replications to actually increase the incidence of false-positive results while others (e.g., Lindsay & Ehrenberg, 1993) argue that direct replications are primarily "important early in a research program to establish quickly and relatively easily and cheaply whether a new result can be repeated at all." After which, assuming a positive replication, conceptual (aka "differentiated") replications are indicated "to extend the range of conductions under which the [original] result... still holds" (p. 221).

Nosek and Lakens (2014) list the following three advantages of direct replications along with a quibbling aside or two on my part which hopefully the authors will forgive:

1. "First, direct replications add data to increase precision of the effect size estimate via meta-analysis" (p. 137).
2. "Second, direct replication can establish generalizability of effects. There is no such thing as an exact replication. [Presumably the authors are referring to psychological experiments involving human participants here.] Any replication will differ in innumerable ways from the original.... Successful replication bolsters evidence that all of the sample, setting, and procedural differences presumed to be irrelevant are, in fact, irrelevant." It isn't clear why a "successful," methodologically sound conceptual replication wouldn't also "establish generalizability," but I'll defer here.
3. "Third, direct replications that produce negative results facilitate the identification of boundary conditions for real effects. If existing theory

anticipates the same result should occur and, with a high-powered test, it does not, then something in the presumed irrelevant differences between original and replication could be the basis for identifying constraints on the effect" (p. 137). William of Occam might have countered by asking "Wouldn't an at least equally parsimonious conclusion be that the theory was *wrong*?"

However, another case *against* conceptual replications (or perhaps even calling them "replications" in the first place) is made quite succinctly by Chris Chambers and Brian Nosek in the previously cited and very informative article by Ed Yong (2012):

From Chambers: "You can't replicate a concept.... It's so subjective. It's anybody's guess as to how similar something needs to be to count as a conceptual replication." [He goes on to illustrate via a priming example how the practice also produces a "logical double-standard" via its ability to verify but not falsify ... thereby allowing weak results to support one another."] And Brian Nosek adds in the same article that conceptual replications are the "scientific embodiment of confirmation bias.... Psychology would suffer if it [the conceptual replication process] wasn't practiced but it doesn't replace direct replication. *To show that "A" is true, you don't do "B." You do "A" again* [emphasis added because this is one of the most iconic quotes in the replication literature]." (p. 300)

While I was initially unmoved by these opinions, after some thought buttressed by the iconic simulation by Simmons, Nelson, and Simonsohn (2011), I have come to believe that the etiology of many conceptual replications involves investigators being seduced by the potential of an exciting, highly cited article and then fiddling with the applicable QRPs to ascertain if a desirable p-value can be obtained with a sufficient plot twist to avoid the study being considered a direct replication. If "successful," the resulting underpowered conceptual replication is published; if not it is deep-sixed. (To go one step farther, it may be that many researchers consider the ability to produce a statistically significant study to be the primary indicator of scientific skill rather than discovering something that can stand the test of time or actually be useful—but this is definitely an unsupported supposition.)

Replication Extensions

Since direct replications have historically been difficult to publish and are perceived by some investigators as less interesting than "original research," one strategy to overcome this prejudice may be to conduct a hybrid extension or two accompanying a direct replication. This is especially feasible in disciplines such as psychology in which multiple experiments are typically included in the same publication, with the first study often involving a direct self-replication of a previously published study. In that case, the remaining experiments could involve conceptual replications thereof in which certain facets of the original intervention, experimental procedures, or outcome variables are changed.

However, Ulrich Schimmack (2012) makes a compelling case against psychological science's love affair with multiple studies in the same publication (think of Daryl Bem's nine-study article in which all but one supported his underlying hypothesis). He demonstrates, among other things, that a surplus of low-powered, statistically significant studies in the same publication can be shown to be

1. Highly improbable (via his "Incredibility-Index"),
2. Less efficient than employing these combined sample sizes into the effort's focal study, and
3. Most likely non-replicable due to their probable reliance on QRPs.

As an example, using a typical power level of 0.50 in psychology (recall that 0.80 is the minimum recommended level), the probability of all five experiments in a five-experiment article reporting statistical significance at the 0.05 level would itself be less than 0.05 even if all five results were in fact "true" positive effects. This is comparable to the probability of obtaining five "heads" on five consecutive coin flips when "heads" was prespecified. Alternatively, the probability of a perfect set of 10 statistically significant studies occurring with a power of 0.50 would be extremely improbable ($p < .001$). And, of course, if some one or more of these experiments reported a p-value substantively less than 0.05, then these two probability levels (i.e., $< .05$ or $< .001$) would be even lower. So perhaps an astute reader (or peer reviewer) should assume that something untoward might be operating here in these multiexperiment scenarios? Perhaps suspecting the presence of a QRP or two? For a more thorough and slightly more technical explication of these issues, see Ulrich Schimmack's 2012 article aptly entitled "The Ironic Effect of Significant Results on the Credibility of Multiple-Study Articles" or the Uri Simonsohn, Leif Nelson, and Joseph Simmons (2014) prescient article, "P-Curve: A Key to the File-Drawer."

Partial Replications

Of course some studies are impossible, too difficult, or too expensive to replicate in their entirety. An example resides in what is probably the best known series of psychological experiments yet conducted (i.e., Stanley Milgram's 1963 obedience studies). In the most famous of these experiments, a confederate encouraged participants to administer electric shocks in increasing doses to another confederate in a separate room who pretended to fail a learning task and was accordingly shocked into unconsciousness.

More than four decades later, and using the original equipment, Jerry Burger (2009) replicated this fifth and most famous experiment in this series up to the point at which the second participant's first verbal complaint occurred. This early stoppage in the replication was necessitated by the likelihood that no modern institutional review board (IRB) would approve the original protocol, in which the participants were subjected to an untoward amount of stress based on the "pain" they were administering to the ostensibly slow "learners." (Burger in turn justified this termination point based on the fact that 65% of the participants in the original experiment continued to administer electric shocks all the way to the end of the fake generator's range, at which point the second confederate was ostensibly unconscious.)

His conclusions, following appropriate caveats, were that

> Although changes in societal attitudes can affect behavior, my findings indicate that the same situational factors that affected obedience in Milgram's participants still operate today. (p. 9)

However it is hard to imagine that many educated people by the turn of this century had not heard of Milgram's obedience study, thereby creating a powerful demand effect. (The only relevant screening procedure employed to ascertain this problem was the exclusion of participants who had taken *more than two* psychology courses. Surely those with one or two psychology courses would have heard of these experiments.)

However, regardless of opinions surrounding the validity of either the original study or its replication, Burger's study is a good example of how a little creativity can provide at least a partial replication when a direct one is infeasible—whether for ethical or practical reasons.

Hypothetical Examples of Direct, Conceptual, Extension, Independent, and Self-Replications

Returning to our hypothetical, long-ago graduate student, let's pretend that he and his advisor were wont to replicate their more exciting (at least to them) findings to ensure their reproducibility (although that word wasn't used in that context in those days). Their motivations would have been lost to time and memory by now if this wasn't a hypothetical scenario, but since it is, let's ascribe their motives to a combination of (a) insecurity that the original findings might not have been "real" (after all, our protagonist was a graduate student) and (b) an untoward degree of skepticism, an occasionally useful scientific trait (but always a socially irritating and not a particularly endearing scientific one).

Let's pretend that our duo's first foray into the replication process involved a relatively rare conceptual replication of a pair of negative studies conducted by a famous educational researcher at the time (James Popham, 1971) who was subsequently served as president of the American Educational Research Association. Jim's original purpose was to find a way to measure teaching proficiency on the basis of student achievement but his studies also demonstrated that trained, experienced teachers were no more effective than mechanics, electricians, and housewives in eliciting student learning. (Some may recall the hubbub surrounding the failure of value-added teacher evaluations [Bausell, 2010, 2017] based on student test scores of a decade or so ago which later proved to be a chimera.)

Naturally any educational graduate student (then or now) would have considered such a finding to be completely counterintuitive (if not demented) since everyone knew (and knows) that schools of education train teachers to produce superior student learning. However, since all doctoral students must find a dissertation topic, let's pretend that this hypothetical one convinced his advisor that they should demonstrate some of the *perceived* weaknesses of Popham's experimental procedures. Accordingly, a conceptual replication (also a term not yet introduced into the professional lexicon) was performed comparing trained experienced classroom teachers with entry-level undergraduate elementary education students. (The latter participants were selected based on their self-reported lack of teaching experience and were known to have no teacher training.)

Alas, to what would have been to the chagrin of any educational researchers, the replication produced the same inference as Popham's original experiments (i.e., no statistically significant learning differences being obtained between the experienced and trained classroom teachers and their inexperienced, untrained undergraduate counterparts). However, being young, foolish, and exceedingly stubborn, our hero might have decided to replicate his and his advisor's own negative finding using a different operationalization of teacher experience and training involving a controlled comparison of tutoring versus classroom instruction—*an extremely rare self-replication of a replication of a negative finding.*

So, in order to replicate their conceptual replication of their negative teacher experience and training study, they might have simply added that comparison as a third factor in a three-way design producing a 2 (tutoring vs. classroom instruction) by 2 (undergraduate elementary education majors who had completed no mathematics instructional courses or teaching experience vs. those who had received both) by 3 (high, medium, and low levels of student ability based on previous standardized math scores).

Naturally, like all hypothetical studies this one was completed without a hitch (well, to be more realistic, let's pretend that it did possess a minor glitch involving a failure to counterbalance the order of instruction in one of the several schools involved). Let's also pretend that it produced the following results for the replication factor (of course, everyone knows [and had known for a couple of millennia] that tutoring was more effective than classroom instruction):

1. No difference (or even a trend) surfaced between the learning produced by the trained and experienced student teachers and their beginning level undergraduate counterparts, and
2. No interaction (i.e., differential learning results produced) occurred between the two types of teachers within the tutoring versus classroom interventions.

Now, even the most skeptical of graduate students (real or imagined) would probably have been convinced at this point that their original negative findings regarding teacher training were probably valid—at least within the particular experimental contexts employed. Accordingly, this particular graduate student and his advisor might have abandoned the lack of teacher training/experience as a viable variable in their program of research and instead conducted a single-factor study designed to both replicate the originally positive effect for tutoring and extend it to differently sized instructional groups. Diagrammatically, this latter design might have been depicted as in Table 6.1.

The first two cells (tutoring vs. classroom instruction) involved a near perfect *direct* replication of the factorial study's tutoring versus classroom instruction while the final two cells constituted an *extension* (or, in today's language, a *conceptual* replication) thereof. (The construct underlying both replications was conceptualized as *class size*.) The "near perfect" disclaimer was due to the hypothetical addition to the study outcome of two instructional objectives accompanied by two items each based on said objectives in the hope that they would increase the sensitivity of the outcome measure.

For the results of the single-factor study to have been perfect (a) the direct replication of tutoring versus classroom instruction would have reproduced the investigators' original finding (i.e., the tutored students would have learned more than their classroom counterparts), and (b) the conceptual replications would have proved statistically significant in an incrementally ascending direction as well. To make this myth a bit more realistic (but still heartening), let's pretend that

1. Tutoring was significantly superior to classroom instruction (defined as one teacher to 23 students or 1:23) as well as to the 1:2 and 1:5 small group sizes,
2. Both the 1:2 and 1:5 groups learned significantly more than students taught in a 1:23 classroom setting, but
3. There was no statistically significant difference (other than an aesthetically pleasing numerical trend in the desired direction) between 1:2 and 1:5.

Table 6.1 Direct replication combined with an extension (aka conceptual) replication

Classroom Instruction	Tutoring	2-Student Small Group Instruction	5-Student Small Group Instruction

But would these latter findings replicate? And herein resides a nuanced issue and one of the reasons this allegory was presented. The finding probably would replicate if the same instructional objectives, students with similar instructional backgrounds (i.e., who had not been exposed to the experimental curriculum), and the same amounts of instructional time were all employed.

However, if a completely different instructional unit were substituted (e.g., some facet of reading instruction or even a different mathematical topic), the effect might not have replicated since the original unit was specifically chosen (and honed based on pilot studies) to be capable of registering learning within a classroom setting within the brief instructional time employed. It was not honed to produce superiority between tutoring, small group, and classroom instruction, which arguably would have constituted a QRP (a term which also hadn't been yet coined and which the hypothetical graduate student would have probably simply called "stupid"), but rather to ensure that the subject matter and the test could detect learning gains within the 1:23 comparison group sans any significant "ceiling" or "basement" effects.

Alternately, if the same experimental conditions had been employed in a replication involving 60 minutes of instruction, the effect might not have replicated because too many student in all four conditions might have performed close to the maximum score possible (and, of course, substantively less instructional time would have produced the opposite reason for a failure to replicate). Or if the instructional period had been extended to several weeks and regular classroom teachers had been employed in lieu of supervised undergraduates, the original effect might not have replicated given the reality of teacher noncompliance, which has long bedeviled educational research studies.

So the purpose of these hypothetical examples was to simply illustrate some of the complexities in performing and interpreting different types of replications.

A note on independent versus self-replication: As previously mentioned there is no question that replications performed by independent investigators are far more credible than those performed by the original investigators—if nothing else because of the possibility of fraud, self-interest, self-delusion, and/or the high prevalence of unreported or unrecognized QRPs. Makel, Plucker, and Hegarty (2012), for example, in a survey of more than a century of psychological research found that replications by the same team resulted in a 27% higher rate of confirmatory results than replications performed by an independent team. And incredibly, "when at least one author was on both the original and replicating articles, only three (out of 167) replications [< 2%] failed to replicate *any* [emphasis added] of the initial findings" (p. 539). Also, using a considerably smaller sample (67 replications) and a different discipline (second-language research), Marsden, Morgan-Short, Thompson, and Abugaber (2018) reported a similar finding for self-replications (only 10% failed to provide any confirmation of the original study results).

So the point regarding self-replications is? Self-replications are most useful for allowing researchers to test the validity of their *own* work in order to avoid (a) wasting their time in pursuing a unprofitable line or inquiry and (b) the embarrassment of being the subject of a negative replication conducted by someone else. But, unfortunately, the practice is too fraught with past abuses (not necessarily fraudulent practices but possibly unrecognized QRPs on the part of the original investigators) to provide much confidence among peer reviewers or skeptical readers. And perhaps self-replications (or any replications for that matter) should not even be submitted for publication in the absence of a preregistered protocol accompanied by an adequate sample size.

How Can Investigators Be Convinced to Replicate the Work of Others?

After all, myriad calls for more replications have historically been to no avail (e.g., Greenwald, 1975; Rosenthal, 1991; Schmidt, 2009). In fact the publication rates in some literatures comprise 2% or less (e.g., Evanschitzky, Baumgarth, Hubbard, & Armstrong, 2007, in marketing research; Makel, Plucker, & Hegarty, 2012, in psychology; Makel & Plucker, 2014, in education)—the latter study being the standard-bearer for non-replication at 0.13% for the top 100 educational journals.

True, there have been a number of replications overturning highly cited (even classical) studies, examples of which have been described decades ago by Greenwald (1975) and more recently in Richard Harris's excellent book apocalyptically titled *Rigor Mortis: How Science Creates Worthless Cures, Crushes Hope, and Wastes Billions* (2017). But despite this attention, replications have remained more difficult to publish than original work—hence less attractive to conduct.

However, to paraphrase our Nobel Laureate one final time, things do appear to be changing, as witnessed by a recently unprecedented amount of activity designed to actually conduct replications rather than bemoan the lack thereof. The impetus for this movement appears to be a multidisciplinary scientific anxiety regarding the validity of published scientific results—one aspect of which involves the modeling efforts discussed previously. This fear (or belief) that much of the scientific literature is false has surfaced before (e.g., in the 1970s in psychology), but this time a number of forward-looking, methodologically competent, and very energetic individuals have made the decision to do something about the situation as chronicled by this and previous chapters' sampling of some very impressive individual initiatives involving a sampling of high-profile, negative replications of highly questionable constructs such as psi and priming, poorly conceived genetic studies, and high-tech, extremely expensive fMRI studies.

All of which have "gifted" our scientific literatures with thousands (perhaps tens of thousands) of false-positive results. But a propos the question of how scientists can be convinced to conduct more replications given the difficulties of publishing their findings, let's move on to a discussion of some of the solutions.

Registered Replication Reports

The registered report process (as mentioned in Chapter 1) was primarily designed to decrease publication bias by moving the peer review process from the end of a study to its beginning, thus procedurally avoiding the difficulties of publishing negative results. This basic concept has been extended to the replication process in the form of a *registered replication report* (RRR) in which publication–nonpublication decisions were made based almost exclusively on the replication protocol.

A number of journals have now adopted some form of this innovation; one of the first being *Perspectives on Psychological Science*. So let's briefly examine that journal's enlightened version as described by Simons, Holcombe, and Spellman (2014). Basically, the process involves the following steps:

1. A query is submitted to the journal making "a case for the 'replication value' of the original finding. Has the effect been highly influential? Is it methodologically sound? Is the size of the effect uncertain due

to controversy in the published literature or a lack of published direct replications?" (p. 552).
2. If accepted, "the proposing researchers complete a form detailing the methodological and analysis details of the original study, suggesting how those details will be implemented in the replication, and identifying any discrepancies or missing information" (p. 553). Therein follows a back-and-forth discussion between the replicating proposers, the original author(s) of the research to be replicated, and the editors, which, if promising, results in a formal proposal. How this actually plays out is unclear but most likely some objections are raised by the original investigator(s) who hopefully won't have the last word in how the replication study will be designed and conducted.
3. Since the RRR normally requires the participation of multiple laboratories due to the necessity of recruiting large numbers of participants quickly (as well as reducing the threat of replicator bias), the journal facilitates the process by putting out a call for interested participants. This step is not necessarily adopted by all journals or replicators who may prefer either to select their own collaborators or perform the replication themselves using a single site.
4. But, back to the *Perspectives on Psychological Science* approach, the selected laboratories "document their implementation plan for the study on OpenScienceFramework.org. The editor then verifies that their plan meets all of the specified requirements, and the lab then creates a registered version of their plan. The individual labs must conduct the study by following the preregistered plan; their results are included in the RRR regardless of the outcome" (p. 553).
5. All sites (if multiple ones are employed) employ identical methods, and results accruing therefrom are analyzed via a meta-analytic approach which combines the data in order to obtain an overall p-value. Each participating site's data is registered on the Open Science Framework repository and freely available for other researchers to analyze for their own purposes.

While there are significant advantages to a multiple-lab approach, single-site replications should also be considered if they adhere to the designated journal's specified replication criteria. Obviously what makes this RRR initiative appealing to replicating investigators resides in the commitment made by journal editors that, if the replication is carried off as planned, the resulting report will be published by their journals regardless of whether the original study results are replicated or not. However, as promising as this innovation is, another one exists that is even more impressive...

What's Next?

Now that the basics of the replication process has been discussed, it is now time to examine the most impressive step yet taken in the drive for increasing the reproducibility of published research. For as impressive as the methodological contributions discussed to this point have been in informing us of the existence, extent, etiology, and amelioration of the crisis facing science, it is now time to consider an even more ambitious undertaking. Given that the replication process is the ultimate arbiter of scientific reproducibility, a group of forward-thinking and energetic scientists have spearheaded the replication of large clusters of original findings. And it is these initiatives (one of which involved replicating 100 different experiments involving tens of thousands of participants) that constitute the primary subject of Chapter 7.

References

Alogna, V. K., Attaya, M. K. Aucoin, P., et al. (2014). Registered replication report: Schooler and Engstler-Schooler (1990). *Perspectives on Psychological Science, 9*, 556–578.

Baumeister, R. F., & Vohs, K. D. (2016). Misguided effort with elusive implications. *Perspectives on Psychological Science, 11*, 574–575.

Bausell, R. B. (2010). *Too simple to fail: A case for educational change.* New York: Oxford University Press.

Bausell, R. B. (2017). *The science of the obvious: Education's repetitive search for what's already known.* Lanham, MD: Rowman & Littlefield.

Burger, J. M. (2009). Replicating Milgram: Would people still obey today? *American Psychologist, 64*, 1–11.

Evanschitzky, H., C., Baumgarth, Hubbard, R., & Armstrong, J. S. (2007). Replication research's disturbing trend. *Journal of Business, 60*, 411–414.

Greenwald, A. G. (1975). Consequences of prejudice against the null hypothesis. *Psychological Bulletin, 82*, 1–20.

Greenwald, A. G., Gonzalez, R., Harris, R. J., & Guthrie, D. (1996). Effect sizes and p values: what should be reported and what should be replicated? *Psychophysiology, 33*, 175–183.

Hagger, M. S., Chatzisarantis, N. L. D., Alberts, H., et al. (2016). A multilab preregistered replication of the ego-depletion effect. *Perspectives on Psychological Science, 11*, 546–573.

Harris, R. (2017). *Rigor mortis: How sloppy science creates worthless cures, crushes hope and wastes billions*. New York: Basic Books.

Ioannidis, J. P. A. (2008). Why most discovered true associations are inflated. *Epidemiology, 19*, 640–648.

Lindsay, R. M., & Ehrenberg, A. S. C. (1993). The design of replicated studies. *American Statistician, 47*, 217–228.

Makel, M. C., Plucker, H. A., & Hegarty, B. (2012). Replications in psychology research: How often do they really occur? *Perspectives in Psychological Science, 7*, 537–542.

Makel, M. C., & Plucker, J. A. (2014). Facts are more important than novelty: Replication in the education sciences. *Educational Researcher, 43*, 304–316.

Marsden, E., Morgan-Short, K., Thompson, S., & Abugaber, D. (2018). Replication in second language research: Narrative and systematic reviews and recommendations for the field. *Language Learning, 68*, 321–391.

Milgram, S. (1963). Behavioral study of obedience. *Journal of Abnormal and Social Psychology, 67*, 371–378.

Nosek, B. A., & Lakens, D. (2014). Registered reports: A method to increase the credibility of published results. *Social Psychology, 45*, 137–141.

Pashler, H., & Harris, C. R. (2012). Is the replicability crisis overblown? Three arguments examined. *Perspectives on Psychological Science, 7*, 531–526.

Popham, W. J. (1971). Performance tests of teaching proficiency: Rationale, development, and validation. *American Educational Research Journal, 8*, 105–117.

Rosenthal, R. (1991). Replication in behavioral research. In J. W. Neuliep (Ed.), *Replication research in the social sciences* (pp. 1–39). Newbury Park, CA: Sage.

Schmidt, S. (2009). Shall we really do it again? The powerful concept of replication is neglected in the social sciences. *Review of General Psychology, 13*, 90–100.

Schimmack, U. (2012). The ironic effect of significant results on the credibility of multiple-study articles. *Psychological Methods, 17*, 551–566.

Schooler, J. W., & Engstler-Schooler, T. Y. (1990). Verbal overshadowing of visual memories: Some things are better left unsaid. *Cognitive Psychology, 22*, 36–71.

Simmons, J. P., Nelson, L. D., & Simonsohn, U. (2011). False-positive psychology: Undisclosed flexibility in data collection and analysis allows presenting anything as significant. *Psychological Science, 22*, 1359–1366.

Simonsohn, U., Nelson, L. D., & Simmons, J. P. (2014). P-curve: A key to the file drawer. *Journal of Experimental Psychology: General, 143*, 534–547.

Simons, D. J., Holcombe, A. O., & Spellman, B. A. (2014). An introduction to registered replication reports. *Perspectives on Psychological Science, 9*, 552–555.

Yong, E. (2012). Bad copy. *Nature, 485*, 298–300.

7
Multiple-Study Replication Initiatives

As crucial as individual replications are to the scientific process, future scientists may look back on the first two decades of this century and conclude that one of the greatest impacts made by the reproducibility initiative involved the replication of sets of studies—hundreds of them, in fact, often conducted by teams of researchers conducted in multiple cooperating laboratories.

This chapter reviews some of these initiatives, along with the lessons they provide. Lessons related not only to the conduct of replications, but for safeguarding the integrity of science itself. So let's begin with, perhaps surprisingly, two large-scale corporate replication efforts.

Preclinical Medical Research

Many scientists may have been unaware of the fact that biotech companies have been replicating promising published preclinical results for some time, especially those published by university laboratories. Anecdotally, an "unspoken rule" in the early venture capital industry is reported to be that at least 50% of preclinical studies, even those published in high-impact academic journals, "can't be repeated with the same conclusions by an industrial lab" (Osherovich, 2011, quoting Bruce Booth, a venture capitalist).

The reproducibility of such studies is especially important for the pharmaceutical industry because preclinical work with cells, tissues, and/or animals form the basis for the development of new clinical drugs. And while this preliminary research is costly, it pales in comparison to the costs of drug development and the large controlled efficacy trials required by the US Food and Drug Administration (FDA) prior to clinical use and marketing.

So if these positive preclinical results are wrong, the drugs on which they are based will almost surely not work and the companies that develop and test them will lose a formidable amount of money. And we all know that pharmaceutical companies are very much more interested in making money than losing it.

The Amgen Initiative

And thus here enters Glenn Begley into our drama, stage left (Begley & Ellis, 2012). As a senior researcher in Amgen, a major biotech firm, Dr. Begley, like his counterparts in many such companies, constantly monitored the published literature for preclinical results that might have important clinical implications. And whenever an especially promising one was found, Dr. Begley would either have the published results replicated in his own company's hematology and oncology department labs to see if they were valid (which often they were not) or file the study away for future consideration.

What makes Dr. Begley so central to this part of our story is that, prior to leaving Amgen for an academic position, he decided to "clean up his file cabinet" of 53 promising effects to see if any of them turned out to be as promising as their published findings suggested. (The papers themselves "were deliberately selected that described something completely new, such as fresh approaches to targeting cancers or alternative clinical uses for existing therapeutics" [p. 532].)

Of the 53 studies replicated, the results from only 6 (or 11%) were reproducible. This is obviously a finding with very disturbing implications for this genre of research, medical treatment, and science in general as succinctly explained in the authors' words:

> Some non-reproducible preclinical papers [i.e., of the 53 studies replicated] had spawned an entire field, with hundreds of secondary publications [i.e., conceptual replications] that expanded on elements of the original observation, but did not actually seek to confirm or falsify its fundamental basis. More troubling, some of the research has triggered a series of clinical studies—suggesting that many patients had subjected themselves to a trial of a regimen or agent that probably wouldn't work. (p. 532)

Following Begley's exit from the company, Amgen apparently took steps to continue his initiative. In an interview (Kaiser, 2016) with Bruce Alberts, former editor-in-chief of *Science Magazine*, Amgen announced the creation of a new online journal (*The Preclinical Reproducibility and Robustness Gateway*) designed to publish failed efforts to replicate other investigators' findings in biomedicine and "is seeding the publication with reports on its own futile attempts to replicate three studies in diabetes and neurodegenerative disease in the hope that other companies will follow suit." The purpose of this initiative, according to Sasha Kamb (Amgen's senior vice president for research) is to reduce wasted time and resources following-up on flawed findings as well as "to help improve the self-correcting nature of science to benefit society as a whole, including those of us trying to create new medicines."

The Bayer Initiative

In another of these "mass" replication efforts, the huge pharmaceutical company Bayer Heath Care performed a replication of 67 published projects as part of their "target identification and validation" program (Prinz, Schlange, & Asadullah, 2011). Comparing their results to the published data (with the cooperation of the 23 heads of the laboratories involved in producing the original studies), the following results were obtained:

> This analysis revealed that only in ~20–25% of the projects were the relevant published data completely in line with our in-house findings. In almost two-thirds of the projects, there were inconsistencies between published data and in-house data that either considerably prolonged the duration of the target validation process or, in most cases, resulted in termination of the projects because the evidence that was generated for the therapeutic hypothesis was insufficient to justify further investments into these projects. (p. 713)

While this is a seminal reproducibility initiative in its own right, it would have been preferable if the authors were to have provided a more precise definition and estimate of reproducibility–irreproducibility than approximately 20–25% of the replicated data not being completely in line with the original results or that in, "*almost* two-thirds" of the cases the two data sources were inconsistent enough "*in most cases*" to result in termination. However, point estimates in "mass" replication studies such as this are not as important as their bottom line, which is that "a majority of these potentially important studies failed to replicate."

The Cancer Biology Initiative

Perhaps in response to these rather disheartening studies, another ambitious initiative spearheaded by an amazing organization that will be described shortly (the Center for Open Science, headed by a whirling dervish of a psychologist named Brian Nosek) secured funding to replicate 50 high-impact cancer biology studies. Interestingly, Glenn Begley is reported (Harris, 2017) to have resigned from this particular project because he argues that repeating their abysmally poor design will produce meaningless results even if they do replicate.

Unfortunately, this project has run into a number of other setbacks as of this writing (January 2019). First, the costs were more than anticipated (originally budgeted at $25,000 per study, the actual costs rose to more than $60,000 [Kaiser, 2018]). This, among other problems (e.g., the difficulties of reproducing some of the laboratory materials and the unexpected amount of time necessary to troubleshoot or optimize experiments to get meaningful results), has resulted in reducing the originally planned 60 replications to 37 in 2015 and then further down to 29 as of 2017.

The online open-access journal *elife* has been keeping a running tab of the results, which are somewhat muddled because clear replicated versus non-replicated results aren't as definitive as they would be for a single social science finding involving one intervention and one outcome variable (or a physics experiment measuring the speed of neutrinos). Again, as of early 2019, the following 12 results were reported (https://elifesciences.org/collections/9b1e83d1/reproducibility-project-cancer-biology):

1. Results were replicated: 4
2. Results were not replicated: 2
3. Some results were replicated and some were not: 4
4. Results could not be interpreted: 2

In the spirit of transparency, after reading each study, I could not definitively categorize results 3 and 4. Hence, a 33% failure-to-replicate in some aspect based upon these four categories are reported in Table 7.1, which summarizes the 11 replication initiatives discussed in this chapter.

Table 7.1 Results of 11 major replication initiatives

Study	# Replications	# Replication failures	% Failures
Amgen	53	47	89%
Bayer	67	44[a]	66%
Preclinical Cancer Biology	6	2	33%
Preclinical Materials	238	109[b]	46%
Psychology (Self-Reported)[c]	257	130	51%
Open Science Collaboration	100	64	64%
Many Labs I	13	3	23%
Many Labs II (*In Press*)	28	13	46%
Many Labs III[d]	10	5	50%
Experimental Economics	18	7	39%
Social Science Research	21	8	38%
Total	811	432	53.3%

[a] As mentioned in the text, this is an estimate since the authors reported that in only 20–25% of the cases were the results of the replication identical, and, in *most* of "*almost two-thirds*" of the cases, the results were not sufficiently close to merit further work. Many reports citing this study report a failure rate of 75–80%, but this seems to ignore the "almost two-thirds" estimate, whatever that means.

[b] Results were not reported in terms of number of studies but instead used the number of resources employed across all 238 studies. Therefore this figure (46%) was applied to the number of studies which would add a (probably relatively) small amount of imprecision to this number.

[c] Hartshorne and Schachner (2012), based on a survey of self-reported replications.

[d] This estimate was based on the nine direct and one conceptual replication and not the added effects or interactions.

Another Unique Approach to Preclinical Replication

One of the lessons learned from the cancer biology project is that the replication of preclinical findings is more involved than in social science research due to the complexity of the experimental materials—exacerbated by laboratories not keeping detailed workflows and the effects of passing time on the realities of fading memories and personnel changes.

This is perhaps best illustrated by Vasilevsky, Brush, Paddock, and colleagues (2013) who examined this very problematic source of error via a rather unique approach to determining reproducibility in laboratory research. While this study does not employ actual replications, its authors (as well as Freedman, Cockburn, & Simcoe [2015] before them) argue that without sufficient (or locatable) and specific laboratory materials (e.g., antibodies, cell lines, knockout reagents) a study cannot be definitively replicated (hence is *effectively irreproducible*). In fact, Freedman and colleagues advance an interesting definition of irreproducibility that is probably applicable to all research genres. Namely, that irreproducibility:

> Encompasses the existence and propagation of one or more errors, flaws, inadequacies, or omissions (collectively referred to as errors) that prevent replication of results. (p. 2)

So, incorporating this definition, and based on 238 life science studies (e.g., biology, immunology, neuroscience), Vasilevsky et al.'s calculation of the number of unique (or specific) experimental materials that could *not* be identified were 56% of antibodies, 57% of cell lines, 75% of constructs such as DNA synthesized for a single RNA strand, 17% of knockout regents, and 23% of the organisms employed.

The Costs of All of These Irreproducible Results

Well, fortuitously, someone has provided us with what appears to be a reasonable estimate thereof via the following unique analysis of the costs of irreproducibility for one meta-discipline.

The Economics of Reproducibility in Preclinical Research

Leonard Freedman, Iain Cockburn, and Timothy Simcoe (2015)

This is the only study documenting the costs of irreproducibility of which I am aware. Extrapolating from 2012 data, the authors estimate that $56.4 billion per year is spent on preclinical research, of which a little over two-thirds is funded by the government. Assuming that 50% of this research is irreproducible (which is certainly not an unreasonable estimate based on Table 7.1), then $28 billion is spent on irreproducible preclinical research. When this is broken down by category, the authors offer the following estimates of the causes of preclinical reproducibility:

1. Biological reagents and reference materials (36.1%)
2. Study design (27.6%)

3. Data analysis and reporting (25.5%)
4. Laboratory protocols (10.8%)

Using one example from the first category (experimental materials, which you may recall constituted the sole target of the Vasilevsky et. al. [2013] study with an overall failure rate of 46%), the authors use a single component of *that* category (cell lines) to illustrate the severe problems inherent in conducting reproducible preclinical cancer studies:

> An illustrative example [i.e., of the problems involving cell lines in general] is the use and misuse of cancer cell lines. The history of cell lines used in biomedical research is riddled with misidentification and cross-contamination events [Lorsch, Collins, & Lippincott-Schwartz, 2014, is cited here, who incidentally report that more than 400 widely used cell lines worldwide have been shown to have been misidentified] which have been estimated to range from 15% to 36% (Hughes, Marshall, Reid, et al., 2007). Yet despite the availability of the short tandem repeat (STR) analysis as an accepted standard to authenticate cell lines, and its relatively low cost (approximately $200 per assay), only one-third of labs typically test their cell lines for identity. (p. 5)

The authors go on to list a number of potential solutions (some of which are generic to research itself and have been [or will be] discussed here, whereas some are primarily targeted at preclinical researchers and won't be presented). However, their concluding statement is worthy of being chiseled on a stone tablet somewhere and (with a disciplinary word change or two) considered by everyone interested in the integrity and reproducibility of science:

> Real solutions, such as addressing errors in study design and using high quality biological reagents and reference materials, will require time, resources, and collaboration between diverse stakeholders that will be a key precursor to change. Millions of patients are waiting for therapies and cures that must first survive preclinical challenges. Although any effort to improve reproducibility levels will require a measured investment in capital and time, the long term benefits to society that are derived from increased scientific fidelity will greatly exceed the upfront costs. (p. 7)

Experimental Psychology: The Seminal "Reproducibility Project"

The psychological version of the Amgen and Bayer initiatives (but comprised of even more replications reported in much more detail) was first announced in 2012 (Brian Nosek, corresponding author, representing the Open Science Collaboration). The results followed 3 years later in a report targeting the scientific community as a whole and published in *Science Magazine* (again with Brian Nosek as the corresponding author along with a Who's Who of psychological investigators with a special interest in scientific reproducibility).

The following abstract represents an attempt to summarize the scope of this seminal scientific effort and describe some of its more salient results. Naturally, the article itself should be read in its entirety (and probably already has been by the majority of the readers of this book, for whom the abstraction will be a review).

> **Estimating the Reproducibility of Psychological Science**
>
> The Open Science Collaboration (2015)
>
> Somehow, 270 investigators were recruited and convinced to participate in the replication of 100 psychological experiments published in three high-impact psychology journals (*Psychological Science, Journal of Personality and Social Psychology,* and *Journal of Experimental Psychology: Learning, Memory, and Cognition*). The logistics of the effort itself were historically unprecedented in the social sciences, and their successful consummation represented (at least to me) an almost incomprehensible feat.
>
> Unfortunately, it wasn't possible for the sampling of articles to be random since it was necessary to match the studies themselves with the expertise and interests of the replicating teams. In addition, some studies were judged too difficult or expensive to replicate (e.g., those employing such difficult-to-recruit participants as autistic children or that required the availability of specialized and expensive tests such as magnetic resonance imaging [MRI]). While these practical constraints may not have provided a precise disciplinary reproducibility rate (which of course neither did the preclinical initiatives just discussed), the project may

have resulted in a reasonable estimate thereof for brief psychological interventions.

But, these caveats aside, the design and methodology of the replications were nevertheless both exemplary and remarkable. Equally impressive, of the 153 eligible studies available, 111 articles were selected for replication, and 100 of these were actually completed by the prespecified deadline. Fidelity of the replications was facilitated by using the actual study materials supplied by the original investigators. These investigators were also consulted with respect to their opinions regarding any divergences from the original design that might interfere with replicability. The average statistical power available for the 100 replications was in excess of 0.90 based on the originally obtained effect sizes. And, of course, all replications were preregistered.

Overall, the replication effect sizes were approximately 50% less than those of the original studies, and their published statistical significance decreased from 97% for the 100 original studies to 36% in the replications. (A statistically significant reduction of less than 10% would have been expected by chance alone.) In addition, the large number of studies replicated provided the opportunity to identify correlates of replication successes and failures, which greatly expands our knowledge regarding the replication process itself. Examples include the following:

1. The larger the original effect size, the more likely the finding was to replicate [recall that the larger the effect size, the lower the obtained p-value when the sample size is held constant a la the simulations discussed previously], so this finding also extends to the generalization that, everything else being equal, *the lower the obtained p-value the more likely a finding is to replicate.*
2. The more scientifically "surprising" the original finding was (in the a priori opinion of the replication investigators, who were themselves quite conversant with the psychological literature), the *less* likely the finding was to replicate.
3. The more difficult the study procedures were to implement, the less likely the finding was to replicate.
4. Studies involving cognitive psychology topics were more likely to replicate than those involving social psychology.

In their discussion, the authors make a number of important points as well, most notably

1. "It is too easy to conclude that successful replication means that the theoretical understanding of the original finding is correct. Direct replication mainly provides evidence for the *reliability of a result. If there are alternative explanations for the original finding, those alternatives could likewise account for the replication* [emphasis added]. Understanding is achieved through multiple, diverse investigations that provide converging support for a theoretical interpretation and rule out alternative explanations" (p. aac4716-6).
2. "It is also too easy to conclude that a failure to replicate a result means that the original evidence was a false positive. Replications can fail if the replication methodology differs from the original in ways that interfere with observing the effect" (p. aac4716-6). [As, of course, does random error and the presence of one or more questionable research practices (QRPs) performed by the replicators themselves.]
3. "How can we maximize the rate of research progress? Innovation points out paths that are possible; replication points out paths that are likely; progress relies on both" (p. aac4716-7).

The authors clearly identify publication bias (investigator and reviewer preferences for positive results) along with insufficient statistical power as primary villains in the failure of 36% of the original studies to replicate. Inextricably coupled with this, but certainly to a lesser extent, is regression to the mean, given the fact that 97% of the studies reviewed were published as positive (recall how close this was to the earlier estimates [96%] of positive psychology results conducted more than four decades ago).

Many Labs 1

This is another extremely impressive cooperative psychology replication project initiated just a bit later than the Open Science Project (Klein, Ratliff, Vianello, et al., 2014). It involved 36 samples totaling 6,344 participants in the replication of 13 "classic and contemporary" studies. Of the 13 effects, 11 (85%) replicated, although this may partly be a function of the authors' declaration that "some" of the 13 elected effects had already been replicated and were "known to be highly replicable."

This and the other two "Many Labs" projects employed methodologies similar to the Open Science replications, although in addition to their replications they possessed other agenda which were methodological in nature and involved exploring the possibility that the failure to replicate an effect might be due to factors unique to the process itself. Thus, in this first project, differences in replicating sites and/or type of participants (e.g., online respondents vs. students) were explored to ascertain their relationship (if any) to replication/non-replication. Perhaps not surprisingly, the interventions themselves accounted for substantively more of the between-study variation than the sites or samples employed.

Many Labs 2

This study (Klein, Vianello, Hasselman, et al., 2018) was basically an expansion of the Many Labs 1 effort exploring the effects of variations in replicability across samples and settings. This time, 28 published findings were selected for replication of which 15 (54%) were declared replicable, although the obtained effect sizes were considerably smaller than in the original studies, which is commonly the case in replication research.

To determine variations across samples and settings, "each protocol was administered to approximately half of 125 samples and 15,305 total participants from 36 countries and territories." As in the first study, the variability attributable to the different samples and types of participants was relatively small.

Many Labs 3

In keeping with the Many Labs approach to coupling methodological concerns with experimental replications, the purpose of this study (Ebersole. Athertomb, Belangeret, et al., 2016) was to ascertain if the point in the academic semester at which student participants engaged in an experiment was related to reproducibility. More than 3,000 participants were employed involving 20 sites in addition to online participants. Counting the conceptual replication, 5 of the 10 studies apparently replicated the originals, yielding a 50% success rate. In general, time during the academic semester in which participation occurred did not surface as a substantive moderating variable.

A Summary of the Methodologies Employed

Both the Open Science and the Many Labs replications (the latter borrowing much of its infrastructure from the former) appear to have been designed and reported as definitively, transparently, and fairly as humanly possible. In addition, these initiatives' procedural strategies are relevant for any future replications of experiments and should be followed as closely as possible: a sampling follows:

1. First, the original investigators were contacted in order to (a) obtain study materials if available and necessary, (b) apprise said investigators of the replication protocol, and (c) obtain any feedback these investigators might have regarding the project. (The latter is both a courtesy to the original investigators and in some cases a necessary condition for conducting a direct replication if the original materials are not otherwise accessible.) The replicating team was not required to accept any suggested changes to the planned protocol, but, generally speaking, the Open Science investigators appeared to accept almost all such feedback when proffered since (again) not everything occurring in a study tends to be reported in a journal article.
2. Next, and most importantly, the design of the replication and the planned analysis were preregistered and any necessary divergences from this plan were detailed in the final report. Both of these steps should occur whenever a replication (or any study for that matter) is published. Not to do so constitutes a QRP (at least for replications occurring after 2016—the publication date for this Open Science report—or perhaps more fairly after 2018 as suggested in Chapter 10).
3. The Open Science replications employed considerably larger sample sizes than the original studies in order to ensure statistical power levels that were at least 0.80 (usually ≥ 0.90). Sufficient power is essential for all replications since inadequate power levels greatly reduce the credibility of any research. (Recall that since the typical statistical power for

a psychological experiment is 0.35 for an effect size of 0.50; hence a replication employing exactly the same sample size would have only a 35% chance of replicating the original study even if the original positive result was valid.)

4. It is a rare psychological research publication that reports only one study (76% of those replicated by the Open Science Collaboration reported two or more) and an even rarer one that reports only a single p-value. To overcome this potential problem the Open Science replicators typically chose the final study along with the p-value reported therein which they considered to be associated with the most important result in that study. (If the original author disagreed and requested that a different effect be selected instead, the replicating investigators typically complied with the request.)

5. The replication results were reanalyzed by an Open Science-appointed statistician to ensure accuracy. This is not a bad idea if the analysis of a replication is performed by a non-statistician and should probably be universally copied—as should the other strategies just listed for that matter.

A brief note on the use of multiple sites in the replication of single studies: Medical research has employed multisite clinical trials for decades due to the difficulty of recruiting sufficient numbers of patients with certain rare diagnoses. Psychological studies are not commonly affected by this problem but they do often require more participants than a single site can supply, hence the discipline has invented its own terms for the strategy such as "crowdsourcing science" or "horizontal versus vertical approaches to science."

However, regardless of terminology, multicenter trials unquestionably have a number of advantages over single-site research, as well as unique organizing and coordinating challenges of their own—some of which are specific to the discipline. The ultimate impact of this approach on the future of psychology is, of course, unknown, but if nothing else high-powered studies (both original and replicates) are considerably more likely to be valid than their underpowered counterparts. And while the use of multiple sites introduces additional sources of variance, these can be handled statistically. (As well, this variance, systematic or erroneous, is normally overwhelmed by the increased sample sizes that "crowdsourcing" makes possible.)

Advice for assembling, designing, and conducting all aspects of such studies from a psychological perspective is clearly detailed in an article entitled "Crowdsourcing Science: Scientific Utopia III" (Uhlmann, Ebersole, & Chartier, 2019) and need not be delineated here. (Utopias I and II will be discussed shortly.)

A Survey Approach to Tracking Replications

Joshua Hartshorne and Adena Schachner (2012) report an interesting approach to tracking the frequency of replication attempts by psychological researchers. Employing a sample of 100 "colleagues of the authors," 49 responded, reporting a total of 257 replications in answer to the following question:

> Approximately how many times have you attempted to replicate a published study? Please count only completed attempts—that is, those with at least as many subjects as the original study. (Hartshorne & Schachner, 2012, Appendix)

Of the 257 replications performed, only 127 (49%) studies fully validated the original results. And if these self-reports are applicable, this suggests that replications are considerably more common in psychology than generally supposed.

Given the sampling procedure employed, no projections to a larger population are possible (e.g., 14 of the respondents were graduate students), but that caveat applies in one degree or another to all of the multiple replication initiatives presented in this chapter. With that said, some variant of this survey approach could constitute a promising method for tracking replications if the identity of the original studies was to be obtained and some information regarding the replication attempt was available (preferably with sharable data).

Experimental Economics

Colin Camerer, Anna Dreber, Eskil Forsell, et al. (2016) replicated 18 studies published in two prominent economic journals between 2011 and 2014. All replications were powered at the 0.90 level or above, and 11 of the 18 studies replicated, yielding a 61% success rate. (As in the Open Science Collaboration and the Many Labs, study procedures were preregistered.) Unlike the Open Science initiative, which also employed expert predictions regarding replication results using an economic tool called *prediction markets* (Dreber, Pfeiffer, Almenberg, et al., 2015), the expert predictions in the Camerer et al. effort were not particularly accurate.

Social Science Studies in General

Science and *Nature* are among the highest profile, most often cited, and most prestigious journals in science. They are also the most coveted publishing outlets as perhaps illustrated by the average bounty (i.e., in excess of $40,000) paid by China (Quan, Chen, & Shu, 2017) to any home-grown scientists who are fortunate enough to garner an acceptance email therefrom, although that bounty has apparently been terminated recently.

As such, these journals have the pick of many litters on what to publish, and their tendency appears (perhaps more than any other extremely high-impact journals) to favor innovative, potentially popular studies. However, their exclusiveness should also enable them to publish methodologically higher quality research than the average journal, so these disparate characteristics make their studies an interesting choice for a replication initiative.

Accordingly, in 2018, Colin Camerer, Anna Dreber, Felix Holzmeister, and a team of 21 other investigators (several of whom were involved in the previously discussed replication of 18 economics studies) performed replications of 21 experimental social science studies published in *Science* and *Nature* between 2010 and 2015. The selected experiments were required to (a) report a p-value associated with at least one hypothesis and (b) be replicable with easily accessible participants (e.g., students or Amazon Mechanical Turk employees). The replicating team followed the original studies' procedures as closely as possible, secured the cooperation of all but one of the original authors, and ensured adequate statistical power via the following rather interesting two-stage process:

> In stage 1, we had 90% power to detect 75% of the original effect size at the 5% significance level in a two-sided test. If the original result replicated in stage 1 (a two-sided $P < 0.05$ and an effect in the same direction as in the original study), no further data collection was carried out. If the original result did not replicate in stage 1, we carried out a second data collection

in stage 2 to have 90% power to detect 50% of the original effect size for the first and second data collections pooled. (p. 2)

Of the 21 findings employed, 13 (or 61.9%) replicated the original results employing p-values. (A number of other analyses and replication criteria were also used, including the omnipresent reduction in effect sizes, this time being slightly greater than 50%.) One rather creative analysis involved meta-analyses of the original and replicated studies, which resulted in an important (although generally known) conclusion: namely, that "true-positive findings will overestimate effect sizes on average—even if the study replicates."

A Summary of the 11 Replication Initiatives

First a tabular summarization of the 11 projects just discussed (Table 7.1). (Note that programmatic replication of *existing datasets* are not included here, such as Gertler, Baliani, and Romero's 2018 failure to find "both raw data and usable code that ran" in 84% of 203 published economic studies or the International Initiative for Impact Evaluation's 2018 considerably more successful program for validating their investigators' datasets https://www.3ieimpact.org/evidence-hub/publications/replication-papers/savings-revisited-replication-study-savings. Note also that the Chabris, Herbert, Benjamin, and colleagues' (2012) failure to replicate individual gene–intelligence associations in Chapter 4 were not included because (a) none of the reported associates replicated (hence would skew the Table 7.1 results) and (b) these studies represent an approach that is not used in the discipline following its migration to genome-wide associations.)

Also not included in the preceding calculations are a "mass" (aka crowdsourced) replication initiative (Schweinsberg, Madana, Vianello, et al., 2016) involving the replication of a set of 10 unpublished psychological studies conducted by a single investigator (Eric Uhlmann and colleagues) centered on a single theoretical topic (moral judgment).

In one sense this replication effort is quite interesting because the investigator of the 10 studies reports that two of the key QRPs believed to be responsible for producing irreproducible results were not present in his original studies: (a) repeated analyses during the course of the study and (b) dropping participants for any reason. The authors consider these methodological steps to be major factors in the production of false-positive results and avoiding them presumably should increase the replicability of the 10 studies.

In another sense, however, the original investigator's choosing of the replicators and the studies to be replicated is somewhat problematic in the sense that it positions the process somewhere between self- and independent replication efforts. The authors of the study, on the other hand, consider this to be a major strength in the sense that it helped (a) duplicate the original contexts of the 10 studies and (b) ensure the experience of the replicating labs in conducting such studies. In any event, the resulting replications produced positive evidence for the reproducibility of 8 of the 10 originally positive studies (1 of the 2 originally negative studies proved to be statistically significant when replicated whereas the other did not).

All in all it is difficult to classify the positive replications in this effort. They appeared to be well-conducted, highly powered, preregistered, and methodologically sound (i.e., by employing a variant of the "Many Labs" approach). So it may be unfair to exclude them from the Table 7.1 results simply because they were selected in part because they were expected to produce positive replications and the replicating laboratories were personally selected by the original investigator. So, for the hopefully silent majority who wish to take issue with this decision, adding these eight out of eight positive replications of positive studies to Table 7.2 produces the overall results shown in Table 7.2.

Close but still an apples versus oranges comparison and weak support for the Ioannidis and Pashler and Harris modeling results.

And, as always, there are probably additional multiple replication initiatives that escaped my search, hence the list presented here is undoubtedly incomplete. In addition, an impressive replication of 17 structural brain–behavior correlations (Boekel, Wagenmakers, Belay, et al., 2015) was not included because it relied on a Bayesian approach which employed different replication/non-replication criteria from the previous 12 efforts. This study's finding is as follows:

Table 7.2 Revision: Results of 12 major replication initiatives

Study	# Replications	# Replication failures	% Failures
Total	811 + 8	432 + 0	52.7%

For all but one of the 17 findings under scrutiny, confirmatory Bayesian hypothesis tests indicated evidence in favor of the null hypothesis [i.e., were negative] ranging from anecdotal (Bayes factor < 3) to strong (Bayes factor > 10). (p. 115)

Five Concluding Thoughts Regarding These Multiple Replications

First of all, the Amgen and Bayer preclinical initiatives basically provided no procedural details regarding how the replications were conducted or how replication versus non-replication was operationalized (other than they didn't merit follow-up work). However, these two efforts were of significant importance because they were early multiple replication attempts and their topic area was arguably of greater scientific, economic, and societal import than any of the other disciplinary initiatives except perhaps the troubled biology cancer initiative.

Second, while all 11 of these initiatives were impressive and important contributions to the scientific reproducibility knowledge base, none provides a generalizable estimate of the prevalence of false-positive results in their literatures due to their understandably unsystematic and non-random selection criterion. Recognizing the very real dangers here of combining apples and oranges, the 11 initiatives as a whole involved a total of 811 studies, of which 432 failed to replicate. *This yielded a 53.3% failure-to-replicate rate.* Not a particularly heartening finding but surprisingly compatible with both Ioannidis's and Pashler and Harris's Chapter 2 modeling estimates.

Third, from an overall scientific perspective, the importance of these initiatives is that if 25% were to be considered an acceptable level for irreproducible results, only 1 (the first "many labs" project) of the 11 initiatives reached this level. (And recall that an unspecified number of the "Many Labs I" studies were reported to have been selected *because* they had already been successfully replicated.)

Fourth, although I have reported what amounts to hearsay regarding the details of Glenn Begley's resignation from the Open Science cancer biology initiative designed to replicate 50 high-impact cancer biology studies, I do agree with Professor Begley's reported objection. Namely, that if a study's design and conduct are sufficiently deficient, a high-fidelity replication thereof employing the same QRPs (sans perhaps low statistical power) is uninformative. And while psychological versus preclinical experiments may differ with respect to the types and prevalence of these artifacts, the end result of such failings will be the same in any discipline: a deck carefully and successfully stacked to increase the prevalence of false-positive results in both original research *and* its replication.

And finally, all of these initiatives are basically exploratory demonstration projects conducted for the betterment of science and for the benefit of future scientists. Furthermore, none of the authors of these papers made any pretense that their truly impressive approaches would solve the reproducibility crisis or even that their results were representative of their areas of endeavor. They have simply taken the time and the effort to do what they could to alert the scientific community to a serious problem for the betterment of *their* individual sciences.

And an apology: there are undoubtedly more multiple-study replications under way than have been discussed here. Engineering, for example, which has here been given short shrift, has apparently employed a form of replication for some time to ensure compatibility of electronic and other parts in order to market them to different companies and applications. Loosely based on this model, the Biological Technologies Office of the US Defense Advanced Research Projects Agency (DARPA) has actually initiated a randomized trial to evaluate the effects of requiring (as a condition of funding) the primary awardees to cooperate and facilitate (sometimes via in person visits or video presentations) independent shadow teams of scientists in the replication and validation of their study results (Raphael, Sheehan, & Vora, 2020). The results of this initiative or its evaluation are not yet available as of this writing, but it is intriguing that the costs that this replication add on typically range between 3% and 8% of the original study's overall budget.

Next

The next chapter looks at two affective views of the replication process based on (a) the reactions of scientists whose original studies have been declared irreproducible and (b) professional (and even public) opinions regarding the career effects thereupon (along with a few hints regarding the best way to respond thereto).

References

Begley, C. G., & Ellis, L. M. (2012). Drug development: raise standards for preclinical cancer research. *Nature, 483*, 531–533.

Boekel, W., Wagenmakers, E.-J., Belay, L., et al. (2015). A purely confirmatory replication study of structural brain-behavior correlations. *Cortex, 66*, 115–133.

Camerer, C., Dreber, A., Forsell, E., et al. (2016). Evaluating replicability of laboratory experiments in economics. *Science, 351*, 1433–1436.

Camerer, C. F., Dreber, A., Holzmeister, F., et al. (2018). Evaluating the replicability of social science experiments in *Nature* and *Science* between 2010 and 2015. *Nature Human Behaviour, 2*, 637–644.

Chabris, C., Herbert, B., Benjamin, D., et al. (2012). Most reported genetic associations with general intelligence are probably false positives. *Psychological Science, 23*, 1314–1323.

Dreber, A., Pfeiffer, T., Almenberg, J., et al. (2015). Using prediction markets to estimate the reproducibility of scientific research. *Proceedings of the National Academy of Sciences, 112*, 15343–15347.

Ebersole, C. R., Athertonb, A. E., Belangeret, A. L., et al. (2016). Many Labs 3: Evaluating participant pool quality across the academic semester via replication. *Journal of Experimental and Social Psychology, 67*, 68–82.

Freedman, L. P., Cockburn, I. M., & Simcoe, T. S. (2015). The economics of reproducibility in preclinical research. *PLoS Biology, 13*, e1002165.

Gertler, P., Baliani, S., & Romero, M. (2018). How to make replication the norm: The publishing system builds in resistance to replication. *Nature, 554*, 417–419.

Harris, R. (2017). *Rigor mortis: How sloppy science creates worthless cures, crushes hope and wastes billions*. New York: Basic Books.

Hartshorne, J. K., & Schachner, A. (2012). Tracking replicability as a method of postpublication open evaluation. *Frontiers in Computational Neuroscience, 6*, 8.

Hughes, P., Marshall, D., Reid, Y., et al. (2007), The costs of using unauthenticated, overpassaged cell lines: How much more data do we need? *Biotechniques, 43*, 575–582.

International Initiative for Impact Evaluation. (2018). http://www.3ieimpact.org/about-us

Kaiser, J. (2016). If you fail to reproduce another scientist's results, this journal wants to know. https://www.sciencemag.org/news/2016/02/if-you-fail-reproduce-another-scientist-s-results-journal-wants-know

Kaiser, J. (2018). Plan to replicate 50 high-impact cancer papers shrinks to just 18. *Science*. https://www.sciencemag.org/news/2018/07/plan-replicate-50-high-impact-cancer-papers-shrinks-just-18

Klein, R., Ratliff, K. A., Vianello, M., et al. (2014). Investigating variation in replicability: A "many labs" replication project. *Social Psychology, 45*, 142–152.

Klein, R. A., Vianello, M., Hasselman, F., et al. (2018). Many labs 2: Investigating variation in replicability across sample and setting. *Advances in Methods and Practices in Psychological Science, 1*, 443–490.

Lorsch, J. R., Collins, F. S., & Lippincott-Schwartz, J. (2014). Cell biology: Fixing problems with cell lines. *Science, 346*, 1452–1453.

Open Science Collaboration. (2012). An open, large-scale, collaborative effort to estimate the reproducibility of psychological science. *Perspectives in Psychological Science, 7*, 657–660.

Open Science Collaboration. (2015). Estimating the reproducibility of psychological science. *Science, 349,* aac4716-1-7.

Osherovich, L. (2011). Hedging against academic risk. *SciBX,* 4.

Prinz, F., Schlange, T., & Asadullah, K. (2011). Believe it or not: How much can we rely on published data on potential drug targets? *Nature Reviews Drug Discovery, 10,* 712-713.

Quan, W., Chen, B., & Shu, F. (2017). Publish or impoverish: An investigation of the monetary reward system of science in China (1999-2016). *Aslib Journal of Information Management, 69,* 1-18.

Raphael, M. P., Sheehan, P. E., & Vora, G. J. (2020). A controlled trial for reproducibility. *Nature, 579,* 190-192.

Reproducibility Project: Cancer Biology. (2012). Brian Nosek (correspondence author representing the Open Science Collaboration). https://elifesciences.org/collections/9b1e83d1/reproducibility-project-cancer-biology

Schweinsberg, M., Madana, N., Vianello, M., et al. (2016). The pipeline project: Pre-publication independent replications of a single laboratory's research pipeline. *Journal of Experimental Social Psychology, 66,* 55-67.

Uhlmann, E. L., Ebersole, C. R., & Chartier, C. R. (2019). Scientific utopia III: Crowdsourcing science. *Perspectives on Psychological Science, 14,* 711-733.

Vasilevsky, N. A., Brush, M. H., Paddock, H., et al. (2013). On the reproducibility of science: Unique identification of research resources in the biomedical literature. *PeerJ, 1,* e148.

8
Damage Control upon Learning That One's Study Failed to Replicate

So far we've established that replication is the best (if imperfect) means we have for determining a study's reproducibility. But what happens if one's study, perhaps conducted years ago, fails to replicate and its carefully crafted conclusions are declared to be wrong?

A flippant answer might be that individuals should feel flattered that their study was considered important enough to replicate. Most scientists have not, nor probably ever will be, awarded such a distinction.

And why should colleagues of someone whose study has failed to replicate be surprised since they should know by now that most studies aren't reproducible? But of course we're all human and emotion usually trumps knowledge. And for scientists, their work constitutes a major component of their self-worth.

So the purpose of this brief chapter is to explore the implications for scientists who find themselves on the wrong side of the replication process. And to begin, let's consider two high-profile case studies of investigators who found themselves in this particular situation, the negative aspects of which were exacerbated by the ever increasing dominance of social media—scientific and otherwise.

Case Study 1: John Bargh's Defense of Behavioral Priming

Without getting bogged down in definitions of different types of theories, the more successful a science is, the fewer theories it requires to explain its primary phenomena of interest. Hopefully psychology is an exception here, because a large proportion of its investigators appear to aspire to eventually developing a theory of their very own. And for those not quite ready for this feat, many psychological journal editors encourage investigators to yoke their experiments to the testing of an existing one.

One of the more popular of these theories is attributed to John Bargh and designated as "behavioral priming"—which might be defined in terms of exposure to one stimulus influencing response to another in the absence of other competing stimuli. Literally hundreds of studies (naturally most of them positive) have been conducted supporting the phenomenon, probably the most famous of which was conducted by John Bargh, Mark Chen, and Lara Burrows (1996), in which participants were asked to create sentences from carefully constructed lists of scrambled words.

The paper itself reported three separate experiments (all positive confirmations of the effect), but the second garnered the most interest (and certainly helped solidify the theory's credence). It consisted of two studies—one a replication of the other—with both asking 30 participants to reconstruct 30 four-word sentences from 30 sets of five words. The participants were randomly assigned to one of two different conditions, one consisting of sets embedded with elderly stereotypic words, the other with neutral words. Following the task they were told to leave the laboratory by taking the elevator at the end of the hall during which a research assistant unobtrusively recorded their walking speed via a stopwatch.

The results of both studies (the original and the self-replication) showed a statistically significant difference between the two conditions. Namely, that the students who had been exposed to the stereotypic aging words walked more slowly than the students who had not—a scientific slam dunk cited more than 5,000 times. Or was it a chimera?

Apparently some major concerns had surfaced regarding the entire concept over the ensuing years, which may have motivated a team of researchers to conduct a replication of the famous study during the magical 2011–2012 scientific window—accompanied, of course by one of the era's seemingly mandatory flashy titles.

Behavioral Priming: It's All in the Mind, But Whose Mind?

Stéphane Doyen, Olivier Klein, Cora-Lise Pichon, and Axel Cleeremans (2012)

These investigators attempted to replicate the Bargh team's influential study via the following two experiments. The first was designed as a direct replication of the 1996 study and employed the same intervention and control tasks (constructing brief sentences from scrambled word sets) followed by the respondents' timed walking speeds.

However, the replication instituted three methodological improvements:

1. The experimenters were *blinded* to group membership (always a requirement in any serious scientific experiment). The original experiment reported that the "experimenter kept himself blind to condition by prepackaging the various scrambled-sentence tasks and picking packets randomly when the participant arrived at the laboratory waiting area" (p. 236).
2. An automated timing mechanism (infrared sensor) was instituted to measure participants' walking speed as they exited the laboratory since the experimenters' use of a stopwatch could theoretically result in biased or inaccurate measurements. And, finally, since Bargh's studies were *severely* underpowered (even for psychology experiments).
3. The replication employed four times as many participants to mitigate that very serious weakness (as explained in Chapter 2).

The bottom line was that despite considerably greater statistical power, the priming effect failed to replicate. But now our investigators got *really* creative and designed their second experiment to ascertain the most plausible reason (inadequate blinding) for Bargh's originally positive findings (as well as their failure to replicate it).

The Blinding Follow-Up Experiment

Fifty undergraduates were randomly assigned to either the priming or non-priming conditions as in the first experiment. Five priming and five control participants were then assigned to each of 10 *experimenters* (i.e., research assistants), five of whom were led to believe that their participants would walk more slowly at the end of the study while the other five experimenters were assured that their participants would walk faster. These "indoctrinations" were an hour long, so obviously a good deal of effort was spent on shaping their expectations in order to ascertain if experimenter beliefs or expectations might have influenced the discrepancy between the original study and its replication. In addition to the electronic timing device, experimenters were asked to also employ a stopwatch as in the first experiment via the bogus explanation that the investigators had found the infrared sensors to be inaccurate.

The results obtained by the research assistants using stopwatches (which the authors' termed "subjective timing") were as follows:

1. This time around a variation of the Bargh et al. priming effect was replicated in the sense that the primed participants of the five research assistants who had been led to believe that the priming intervention would induce participants to walk more slowly did indeed register significantly slower walking times than their non-primed counterparts.
2. Interestingly, however, the five research assistants who had been led to believe that their primed participants would walk faster actually did record faster walking times than their five control counterparts who had not been fictitiously "sensitized."

And, as the investigators hypothesized, differing results were obtained for the infrared sensors (which it will be recalled involved measuring exactly the same walking behavior for all 50 participants).

1. For the five slow-walking indoctrinated research assistants, their primed participants did walk significantly more slowly than their non-primed participants as in the Bargh et al. study.
2. There was no difference in walking speed between priming and non-priming participants for the fast walking condition.

The investigators therefore concluded, in addition to the fact that despite using a much larger sample they were unable to replicate the Bargh et al. "automatic effect of priming on walking speed," that

> in Experiment 2 we were indeed able to obtain the priming effect on walking speed for both subjective and objective timings. Crucially however, this was *only possible* [emphasis added] by manipulating experimenters' expectations in such a way that they would expect primed participants to walk slower. (p. 6)

John Bargh's Reaction

Although perhaps understandable since much of Bargh's reputation was built on the priming construct (he quoted himself a dozen times in the 1996 study alone), the good professor went absolutely ballistic at the replication results and especially the scientific media's reaction to it. According to Ed Yong (2012), who has written on reproducibility a number of times for *Science* and writes a blog for *Discover Magazine* http://blogs.discovermagazine.com/notrocketscience/failed-replication-bargh-psychology-study-doyen/), Bargh called the authors of the replication "incompetent or ill-informed," defamed the journal *PLoS One* as not receiving "the usual high scientific journal standards of peer review scrutiny" (a damning criticism for which there is absolutely no supporting evidence), and attacked Yong himself for "superficial online science journalism"—all via *his* online blog. (Reactions, incidentally, that Joseph Simmons suggested constituted a textbook case in how *not* to respond to a disconfirming replication.)

The Professional Reaction

Since the Doyen team's failure to replicate Bargh's famous study, the priming construct has understandably lost a considerable amount of its former luster. For example, in the first "Many Labs" replication project (Klein, Ratliff, Vianello, et al., 2014, discussed in the previous chapter), only two replications (of 13) provided no support for their original studies' findings and both of these involved this once popular construct (i.e., flag priming [influencing conservatism] and currency priming [influencing system justification]). And, according to an article in *Nature,* Chivers (2019) maintained that dozens of priming replications did not confirm the effect—perhaps leading Brian Nosek to state concerning the effect that " I don't know a replicable finding. It's not that there isn't one, but I can't name it" (p. 200).

Of course a meta-analysis (Weingarten, Chen, McAdams, et al., 2016) on the topic assessing the word priming effect found a small but statistically significant effect size of 0.33 for 352 effects (not a typo), but that is customary in meta-analyses. However, the average number of participants employed was only 25 per condition, which implied that the average statistical power thereof was 0.20 (which it will be recalled translates to a 20% chance of obtaining statistical significance if a real effect exists—which in the case of priming an increasing percentage of scientists no longer believe exists).

Case Study 2: The Amy Cuddy Morality Play

What started out as another typically underpowered social psychology experiment involving a short-term, extremely artificial, media-worthy intervention morphed into something that may ultimately have an impact on the discipline comparable to Daryl Bem's psi studies. Perhaps this event served notice that counterintuitive, headline-catching studies could constitute a double-edged sword for their authors by encouraging methodologically oriented critics to mount a social media counterattack—especially if the original investigators appeared to be profiting (financially, publicly, or professionally) by conducting questionable research practice (QRP)-laced, irreproducible science.

What this episode may also herald is an unneeded illustration of the growing use of the internet to espouse opinions and worldviews concerning findings published in traditional peer reviewed journals, a phenomenon that might also reflect a growing tendency for an already broken peer review/editorial propensity to allow political and social biases to influence not just publication decisions, but the hyperbolic language with which those publications are described. (The latter of which is apparently on an upswing; see Vinkers, Tijdink, & Otte, 2015.)

But let's begin by considering the study itself (Carney, Cuddy, & Yap, 2010) whose unimposing title ("Power Posing: Brief Nonverbal Displays Affect Neuroendocrine Levels and Risk Tolerance") hardly suggested the controversy that would follow. And neither did its abstract portend anything controversial, or at least not until its final sentence.

> Humans and other animals express power through open, expansive postures, and they express powerlessness through closed, contractive postures. But can these postures actually cause power? The results of this study confirmed our prediction that posing in high-power nonverbal displays (as opposed to low-power nonverbal displays) would cause neuroendocrine and behavioral changes for both male and female participants: High-power posers experienced elevations in testosterone,

decreases in cortisol, and increased feelings of power and tolerance for risk; low-power posers exhibited the opposite pattern. In short, posing in displays of power caused advantaged and adaptive psychological, physiological, and behavioral changes, and these findings suggest that embodiment extends beyond mere thinking and feeling, to physiology and subsequent behavioral choices. *That a person can, by assuming two simple 1-min poses, embody power and instantly become more powerful has real-world, actionable implications* [emphasis added]. (p. 1363)

For those unfamiliar with scientific abstracts, this concluding sentence reflects a completely unacceptable departure from conventional (or at least past) scientific publishing conventions. Had a single word such as "might" been inserted (which a competent journal editor should have insisted upon—even prior 2011–2012 [note the study's publication date of 2010, which is a year before Professor Bem's magnum opus was published]), it is quite possible that the vituperative reactions to the study *might* have been avoided. That and Dr. Cuddy's popularization of her finding.

But alas no such addition (or disclaimer) was added, probably because its authors appeared to belong the school of belief that just about anything to which a $p < 0.05$ can be affixed does translate to "real-world" behavior and scientific meaningfulness. Hence our little morality play, which is eloquently presented in a *New York Times Magazine* piece written by Susan Dominus and aptly titled "When the Revolution Came for Amy Cuddy" (2017, Oct. 18).

Apparently (at least from her critics' perspectives), Dr. Cuddy parlayed her often cited 2010 study (note that she was the second author on the study in question) into two extremely popular YouTube presentations (one of which constituted Ted Talks' second most popular offering with 43 million views and counting), a best-selling book (*Presence*), and myriad paid speaking engagements.

While the study in question was neither her first nor last on the topic, the 2010 article—coupled with her new-found fame—engendered a replication (Ranehill, Dreber, Johannesson, et al., 2015) employing a sample five times as large as the original study's. Unfortunately the original study's positive results for cortisol, testosterone, or risk-taking did not replicate, although a quarter of a point difference was observed on a 4-point rating scale soliciting the degree to which the participants *felt* powerful following the brief set of "power" poses—an effect that was also significant in the original article but which could be interpreted as primarily an intervention "manipulation check" (i.e., evidence that participants could perceive a difference between the "high-power nonverbal poses" and the "low-power nonverbal poses" but little else).

Naturally Cuddy defended her study as almost any investigator would and continued her belief in the revolutionary societal effects of empowering the un-empowered (e.g., young women, female children, and even black men) via her 1-minute poses. As part of this effort and in response to their critics, she and her original team (note again that Dana Carney was the first author on this paper as well) even accumulated a group of 33 studies involving "the embodied effects of expansive (vs. contractive) nonverbal displays" (Carney, Cuddy, & Yap, 2015), none of which, other than *theirs*, found an effect for testosterone or cortisol. As a group, however, the cumulative results were overwhelmingly positive, which of course is typical of a science which deals in publishing only positive results involving soft, self-reported, reactive outcomes.

Also, in adddtion to appealing to the authority of William James, the authors also presented a list of rebuttals to the Ranehill et al. failure to replicate—one of which was that, for some reason, the latter announced to their participants that their study was designed to test the effects of physical position upon hormones and behavior. (It is unclear why the replicating team did this, although its effect [if any] could have just as easily increased any such difference due to its seeming potential to elicit a demand effect.)

So far, all of this is rather typical of the replication and rebuttal process since no researcher likes to hear or believe that his or her findings (or interpretations thereof) are incorrect. (Recall that even Daryl Bem produced a breathlessly positive meta-analysis of 90 experiments on the *anomalous anticipation of random future events* in response to Galak, LeBoeuf, Nelson, and Simmons's failure to replicate psi.)

But while no comparison between Amy Cuddy and Daryl Bem is intended, Drs. Galak, Nelson, and Simmons (along with Uri Simonsohn) soon became key actors in our drama as well. This is perhaps due to the fact that, prior to the publication of the 33-experiment rejoinder, Dana Carney (again the original first author of both the review and the original power posing study) sent the manuscript along with her version of a p-curve analysis (Simonsohn, Nelson, & Simmons, 2014) performed on these 33 studies to Leif Nelson (who promptly forwarded it on Simmons and Simonsohn).

COPING WITH A PERSONAL FAILURE TO REPLICATE 181

The *p-curve* is a statistical model designed to ascertain if a related series of *positive* studies' p-values fit an expected distribution, the latter being skewed to the right. Or, in the words of its originators,

> Because only true effects are expected to generate right-skewed *p*-curves—containing more low (.01s) than high (.04s) significant *p* values—only right-skewed *p*-curves are diagnostic of evidential value. (p. 534).

It also doesn't hurt to remember that the p-curve is a statistical model (or diagnostic test) whose utility has not been firmly established. And while it probably is useful for the purposes for which it was designed, its results (like those of all models) do not reflect absolute mathematical certainty. Or, as Stephan Bruns and John Ioannidis (2016) remind us, the exact etiology of aberrant effects (skewness in the case of p-curves) remains "unknown and uncertain."

In any event, at this juncture our story begins to get a bit muddled to the point that no one comes across as completely righteous, heroic, *or* victimized. According to Susan Dominus, Simmons responded that he and Simonsohn had conducted their own p-curve and came to a completely different conclusion from Carney's, whose version they considered to be incorrect, and they suggested that "conceptual points raised before that section [i.e., the "incorrect" p-curve analysis] are useful and contribute to the debate," but, according to Dominus they advised Carney to delete her p-curve and then "everybody wins in that case."

Carney and Cuddy complied by deleting their p-curve but they (especially Amy Cuddy) apparently weren't among the universal winners. Simonsohn and Simmons, after giving the original authors a chance to reply online, then published a decidedly negative blog on their influential *Data Coda* site entitled "Reassessing the Evidence Behind the Most Popular TED Talk" (http://datacolada.org/37), accompanied by a picture of the 1970s television version of *Wonder Woman*. (The latter being a stark reminder of the difference between internet blogs and peer reviewed scientific communication.)

The rest, as the saying goes, is history. Cuddy temporarily became, even more so perhaps than Daryl Bem and John Bargh, the poster child of the reproducibility crisis even though her research was conducted prior to the 2011–2012 enlightenment, as were both Bem's and Bargh's.

Dominus's *New York Times Magazine* article sympathetically detailed the emotional toll inflicted upon Dr. Cuddy, who was portrayed as a brain-injured survivor who had overcome great obstacles to become a respected social psychologist, even to the point of surviving Andrew Gelman's "dismissive" (Dominus' descriptor) blogs (http://andrewgelman.com/) along with his 2016 *Slate Magazine* article (with Kaiser Fung) critical of her work and the press' role in reporting it. But perhaps the unkindest cut of all came when her friend and first author (Dana Carney) completely disavowed the power pose studies and even recommended that researchers abandon studying power poses in the future.

In response to her colleague's listing of methodological weaknesses buttressing the conclusion that the study was not reproducible, Dr. Cuddy complained that she had not been apprised by Dr. Carney of said problems—which leads one to wonder why Dr. Cuddy had not learned about statistical power and demand characteristics in her graduate training.

So what are we to make of all of this? And why is this tempest in a teapot even worth considering? Perhaps the primary lesson here is that scientists should approach with caution their Facebook, Twitter, and the myriad other platforms that encourage brief, spur-of-the-moment posts. Sitting alone in front of a computer makes it very easy to overstep one's scientific training when responding to something perceived as ridiculous or methodologically offensive.

However. online scientific commentaries and posts are not likely to go away anytime soon and, in the long run, may even turn out to be a powerful disincentive for conducting QRP-laden research. But with that said, perhaps it would be a good idea to sleep on one's more pejorative entries (or persuade a friend or two to serve as one's private peer reviewer). A bit of time tends to moderate our immediate virulent reactions to something we disagree with, which offends our sensibilities, or serves as the subject matter for an overdue blog.

As for Amy Cuddy's situation, it is extremely difficult for some social scientists to avoid formulating their hypotheses and interpreting their results independently of their politico-social orientations—perhaps as difficult as the proverbial camel traversing the eye of a needle. And it may be equally difficult to serve in the dual capacity of scientist *and* entrepreneur while remaining completely unbiased in the interpretation of one's research. Or, even independent of both scenarios, not to feel a decided sense of personal and professional indignation when one's work is subjected to a failed replication and subsequently labeled as not reproducible.

As for articles in the public press dealing with scientific issues, including Susan Dominus's apparently factual entry, it is important for readers to understand that these writers often do not possess a deep understanding of the methodological issues or cultural mores underlying the science they are writing about. And as a result, they may be more prone to allow their personal biases to surface occasionally.

While I have no idea whether any of this applies to Susan Dominus, she did appear to be unusually sympathetic to Amy Cuddy's "plight," to appreciate Joseph Simmons's and Uri Simonsohn's apparent mea culpa that they could have handled their role somewhat differently (which probably would have had no effect on the ultimate outcome), and to not extend any such appreciation to Andrew Gelman, whom she appeared to consider too strident and "dismissive" of Dr. Cuddy.

So while treating Dr. Cuddy's *work* "dismissively" is understandable, it might also be charitable to always include a brief reminder that studies such as this were conducted prior to 2011–2012—not as an excuse but as a potentially mitigating circumstance. If nothing else, such a disclaimer might constitute a subliminal advertisement for the many available strategies for decreasing scientific irreproducibility.

But Is a Failure to Replicate Really All That Serious to a Scientist's Reputation?

If scientists' memories (or attention spans) possess anything resembling those of the general public, the best guess is "not very since it will soon be forgotten." But while no researcher likes to have a cherished finding disparaged, the failure of one's results to replicate is hardly a professional death sentence for anyone's career. But even though opinions and self-reported affects don't amount to a great deal in some sciences, they may have greater importance in the social sciences, hence the following two surveys.

The Ebersole, Axt, and Nosek (2016) Survey

This is the larger survey of the two (4,786 US adults and 313 researchers) with also the more iconic title: "Scientists' Reputations Are Based on

Getting It Right, Not Being Right." Employing brief scenarios in two separate surveys which were then combined, the more germane results of this effort were:

1. In general, investigators who produced replicable but mundane results were preferred to their more exciting counterparts whose results were more questionable.
2. If an investigator's finding replicated, his or her reputation and ability were enhanced (which is rather obvious).
3. If the finding did not replicate, the original investigator's externally perceived ability and ethical behavior decreased somewhat (but definitively more so if he or she criticized the replication study as opposed to agreeing that the original result *might* be wrong or if he or she performed a self-replication in self-defense).

So Simmons might have been on to something when he suggested that John Bargh's online tantrum constituted a textbook case in how *not* to respond to a disconfirming replication. And Ebersole et al. were even kind enough to provide a case study of an exemplary response to a disconfirming replication by Matthew Vees.

> Thank you for the opportunity to submit a rejoinder to LeBel and Campbell's commentary. I have, however, decided not to submit one. While I am certainly dismayed to see the failed attempts to reproduce a published study of mine, I am in agreement with the journal's decision to publish the replication studies in a commentary and believe that such decisions will facilitate the advancement of psychological science and the collaborative pursuit of accurate knowledge. LeBel and Campbell provide a fair and reasonable interpretation of what their findings mean for using this paradigm to study attachment and temperature associations, and I appreciated their willingness to consult me in the development of their replication efforts. (p. 5)

And, as our survey authors opined, a response such as this will, if anything, gain the respect of one's peers.

The Fetterman and Sassenberg Survey (2015)

This survey presented one of four different scenarios to 279 published scientists. The methodology involved in the two scenarios of most interest to us here were described as follows:

> Participants were told to think about a specific finding, of their own, that they were particularly proud of (self-focused). They then read about how an independent lab had conducted a large-scale replication of that finding, but failed to replicate it. Since the replicators were not successful, they tweaked the methods and ran it again. Again, they were unable to find anything. The participants were then told that the replicators published the failed replication and blogged about it. The replicators' conclusion was that the effect was likely not true and probably the result of a Type 1 error. Participants were then told to imagine that they posted on social media or a blog one of the following comments: "in light of the evidence, it looks like I was wrong about the effect" (admission) or "I am not sure about the replication study. I still think the effect is real" (no admission). (p. 4)

(The other two scenarios were practically identical except the description of the replication involved a well-known study published by a prominent researcher to whom the two alternate comments were ascribed.)

The basic results were similar to those of the Ebersole et al. findings. Namely (a) that scientists tend to overestimate the untoward effects of negative replications and (b) these effects are less severe if the original investigators "admit" that they may have been wrong. The authors accordingly speculate that such an admission might repair some of the reputational damage that is projected to occur in these scenarios.

These two surveys produced a plethora of different results that are not discussed here so, as always, those interested in these issues should access the full reports. The Ebersole et al. survey, for example, was able to contrast differences between the general public and scientists with respect to several issues (e.g., researchers were considerably more tolerant of researchers who did not replicate their own research than the general public and were also more appreciative [at least in theory] of those who routinely performed "boring" rather than "exciting" studies). Similarly, Drs. Fetterman and Sassenberg were able to study the relationship between some of their scenarios and whether or not the respondents were in the reproducibility or business-as-usual research camps.

Both teams transparently mentioned (a) some of the weaknesses of employing scenarios to explain real-life behaviors, (b) the fact that radically different results could be produced by minor tweaks therein, and (c) that the admission of such problems doesn't mitigate their very real potential for themselves creating false-positive results. One potential problem with the interpretation of these surveys, at least in my opinion, is the implication that real-life scientists would be better off admitting that they were (or might have been) wrong when they may have actually believed their original results were correct. (This is ironic in a sense, given Ebersole et al.'s nomination of Matthew Vees's strategy as an exemplary alternative to going into "attack mode" following the failure of one of his study's to replicate since Dr. Vees did not suggest or imply that his initial results might be wrong. Nor did he imply that his replicators were wrong either—which is not necessarily contradictory.)

Of course anyone can quibble about scenario wordings and suggest minor tweaks thereto (which is surely a weakness of scenarios as a scientific tool in general). In truth no one knows whether the results of such studies reflect "real-life" behaviors, reactions, or even if the same results (or interpretation thereof) might change over time. But as imperfect as surveys and scenarios are, the two just discussed at least provide the best assurances available that replication failures, while disappointing, disheartening, and perhaps enraging, are not as bad as they seem to their "victims" at the time. Time may not heal all wounds, but it almost surely will blunt the pain from this one and have very little impact on a career. At least barring fraudulent behavior.

Two Additional Strategies for Speeding the Healing Process Following a Failure to Replicate

The first was proffered by John Ioannidis who, based on meta-analytic evidence, argued that the vast majority of prospective cohort studies indicating a positive relationship between a wide variety of different food intakes and mortality risk probably constituted false positive results.

> The nutritional epidemiology community includes superb scientists. The best of them should take ownership of this reform process. They can further

lead by example (e.g., by correcting their own articles that have misleading claims). Such corrections would herald high scientific standards and public responsibility. (Ioannidis, 2018, p. 970)

At first glance this plea may seem somewhat Pollyannaish and destined to be ignored, but with a bit of thought it may be excellent advice for preempting a negative replication of one's work regardless of discipline. Many investigators probably suspect at least one of their published studies to represent a false-positive result. So, rather than hunkering down in the dark in the hope that some reproducibility miscreant with nothing better to do will decide to replicate said study, why not replicate it oneself (perhaps involving one's original co-authors)?

If the study replicates, great! If it does not, it will represent another publication and label the original author as a principled, 21st-century scientist.

The second strategy was offered by Jarrod Hadfield (2015), an evolutionary biologist, who proposed a strategy for implementing Ioannidis's plea. Professor Hadfield suggested that since investigators often "continue to collect data on their study systems that could be used to validate previous findings. [Thus a]llowing authors to publish short (two paragraph) addenda to their original publications would lower the costs of writing and submitting replication studies, and over time these addenda may reduce the stigma associated with publishing false positives and increase transparency."

He was also reported by Forstmeier, Wagenmakers, and Parker (2017) as somewhat iconoclastically suggesting

> that researchers running long-term studies [to which longitudinal studies stretching over decades, such as the Framingham Heart Study, would qualify] should publish addenda to their previous publications, declaring in a one-page publication that their original finding did or did not hold up in the data of the following years (after the publication), and comparing the effect sizes between the original data and the newer data. This would be a quick way of producing another publication, and it would be enormously helpful for the scientific field. This may also relax the feeling of stigma when something does not hold up to future evaluation. Admitting a failure to replicate could actually be perceived as a signal of a researcher's integrity and be praised as a contribution to the scientific community." (p. 1960)

Which Provides a Conveniently Lead-In to Chapter 9

To this point, some derivation of the word "publish" has been mentioned more than 200 times in several different contexts—many suggesting various biases due to the various actors involved in the process or inherent to the process itself. It seems natural, therefore, that these factors should be examined in a bit more detail from the perspectives of (a) identifying some of their more salient characteristics that impede the replication (and the scientific) process, (b) facilitating the prevalence of false-positive results in the scientific literature, and, of course, (c) examining a number of suggestions tendered to ameliorate these problems. All of which constitute the subject matter of Chapter 9.

References

Bargh, J. A., Chen, M., & Burrows, L. (1996). Automaticity of social behavior: Direct effects of trait construct and stereotype-activation on action. *Journal of Personality and Social Psychology, 71*, 230–244.

Bruns, S. B., & Ioannidis, J. P. A. (2016). P-curve and p-hacking in observational research. *PLoS One 11*, e0149144.

Carney, D. R., Cuddy, A. J. C., & Yap, A. J. (2010). Power posing: Brief nonverbal displays affect neuroendocrine levels and risk tolerance. *Psychological Science, 21*, 1363–1368.

Carney, D. R., Cuddy, A. J. C., & Yap, A. J. (2015). Review and summary of research on the embodied effects of expansive (vs. contractive) nonverbal displays. *Psychological Science, 26*, 657–663.

Chivers, T. (2019). What's next for psychology's embattled field of social priming. *Nature, 576*, 200–202.

Dominus, S. (2017). When the revolution came for Amy Cuddy. *New York Times Magazine*, Oct. 18.

Doyen, S., Klein, O., Pichon, C., & Cleeremans, A. (2012). Behavioral priming: It's all in the mind, but whose mind? *PLoS ONE, 7*, e29081.

Ebersole, C. R., Axt, J. R., & Nosek, B. A. (2016). Scientists' reputations are based on getting it right, not being right. *PLoS Biology, 14*, e1002460.

Fetterman, A. K., & Sassenberg, K. (2015). The reputational consequences of failed replications and wrongness admission among scientists. *PLoS One, 10*, e0143723.

Forstmeier, F., Wagenmakers, E-J., & Parker, T. H. (2017). Detecting and avoiding likely false-positive findings: a practical guide. *Biological Reviews, 92*, 1941–1968.

Gelman, A., & Fung, K. (2016). The power of the "power pose": Amy Cuddy's famous finding is the latest example of scientific overreach. https://slate.com/technology/2016/01/amy-cuddys-power-pose-research-is-the-latest-example-of-scientific-overreach.html

Hadfield, J. (2015). There's madness in our methods: Improving inference in ecology and evolution. https://methodsblog.com/2015/11/26/madness-in-our-methods/

Ioannidis, J. P. A. (2018). The challenge of reforming nutritional epidemiologic research. *Journal of the American Medical Association, 320*, 969–970.

Klein, R., Ratliff, K. A., Vianello, M., et al. (2014). Investigating variation in replicability: A "many labs" replication project. *Social Psychology, 45*, 142–152.

Ranehill, E., Dreber, A., Johannesson, M., et al. (2015). Assessing the robustness of power posing: No effect on hormones and risk tolerance in a large sample of men and women. *Psychological Science, 26*, 653–656.

Simonsohn, U., Nelson, L. D., & Simmons, J. P. (2014). P-curve: A key to the file drawer. *Journal of Experimental Psychology: General, 143*, 534–547.

Vinkers, C. H., Tijdink, J. K., & Otte, W. M. (2015). Use of positive and negative words in scientific PubMed abstracts between 1974 and 2014: Retrospective analysis. *British Medical Journal, 351*, h6467.

Weingarten, E., Chen, Q., McAdams, M., et al. (2016). From primed concepts to action: A meta-analysis of the behavioral effects of incidentally-presented words. *Psychological Bulletin, 142*, 472–497.

Yong, E. (2012). Bad copy. *Nature, 485*, 298–300.

PART III
STRATEGIES FOR INCREASING THE REPRODUCIBILITY OF PUBLISHED SCIENTIFIC RESULTS

9
Publishing Issues and Their Impact on Reproducibility

The vast majority of the studies or opinions cited to this point have been published in peer reviewed journals. Historically, scientific publishing has taken many forms and is an evolving process. In past centuries, books were a primary means of communicating new findings, which is how James Lind announced his iconic discovery regarding the treatment and prevention of scurvy in 1753 (*Treatise of the Scurvy*). Which, of course, was ignored, thereby delaying a cure being adopted for almost half a century.

Gradually scientific journals began to proliferate, and as many as a thousand were created during Lind's century alone. Peer reviewed journals now constitute the primary medium for formally presenting new findings to the scientific community, acting as repositories of past findings, and forming the foundation on which new knowledge is built.

So, obviously, the issue of scientific reproducibility cannot be considered in any depth without examining the publishing process itself. Especially since over half of published studies may be incorrect.

Publishing as Reinforcement

The adage of "publish or perish" is a time-honored academic expression, often rued as a professional curse and more recently as a contributor of false-positive results. Nevertheless, peer reviewed publications are the firmly entrenched coin of the scientific realm as surely as money is for investment banking.

Unfortunately, the number of publications to which a scientist's name is attached has sometimes becomes a compulsive end in and of itself, especially when it governs how some scientists, institutions, and funding agencies judge themselves and others. This is a compulsion reflected by number 10 of our inane scientific policies (Chapter 3) in which a surprising number of individuals publish one paper every 5 days (Ioannidis, Klavans, & Boyack, 2018)—some of which are never cited and probably never even read by their co-authors.

However, with all of its deficiencies as a metric of scientific accomplishment, publishing is an essential component of the scientific enterprise. And, in an era in which traditional scientific book publishing appears to be in decline, academic journal publishing has evolved relatively quickly into a multibillion-dollar industry.

Whether most scientists approve of the direction in which publishing is moving is unknown and probably irrelevant. Certainly nothing is likely to change the practice of publishing one's labors as a reinforcement of behavior or the narcotic-like high that occurs when investigators receive the news that a paper has been approved for publication. (Of course a "paper," while still in use as an antiquated synonym for a research report, is now digitally produced and most commonly read online or downloaded and read as a pdf—all sans the use of "paper.")

So let's take a quick look at the publishing process through the lens of scientific reproducibility. But first, a few facts that are quite relevant for that purpose.

The Scope of the Academic Publishing Enterprise

Every 2 years the National Science Foundation (NSF) produces a report entitled *Science and Technology Indicators*, which is designed to encapsulate the scientific institution as a whole. As of this writing, the current version (National Science Board, 2018) is an impressive and exhaustive volume running to well over a thousand pages, dealing with the national and international scientific enterprise as a whole including, among other things, the number of science, technology, engineering, and mathematics (STEM) personnel, PhDs awarded, patents granted, and, of course, scientific reports published.

Most relevant to our purposes here are its data on the last indicator, which, for 2016 alone (the most recent year covered in the 2018 publication) totaled a mind-boggling 2,295,608 published entries, a figure that has been increasing at a rate of 3.9% *per year* for the past 10 years. And while a 46.5% decadal increase in the number of publications is astonishing and bordering on absurd, some scholars consider the number of publications as a positive, planet-wide indicator of scientific progress. (It should be mentioned that Björk, Roos, and Lauri [2009] take issue with this projection and argue that 1,275,000 is a more accurate figure, but both estimates are mind-boggling.)

As shown in Table 9.1 (abstracted from table 5-6 of the original NSF report) only 18% of these publications emanated from the United States, and the majority (54%) of the 2 million-plus total involved disciplines not or only spottily covered in this book (e.g., engineering, the physical sciences, mathematics, computer sciences, agriculture). However, this still leaves quite a bit of publishing activity in almost all recognized disciplines.

Psychology alone contributed more than 39,000 publications, with the other social sciences accounting for more than three times that number. How many of these included p-values are unknown, but, given the modeling studies discussed earlier, psychology alone undoubtedly produces thousands of false-positive results per year with the other social sciences adding tens of thousands more based on their combined 2016 output of more than 120,000 publications. So let's hope that Ioannidis (2005) was wrong about *most* disciplines being affected by his pessimistic modeling result. Otherwise, we're

Table 9.1 Numbers of scientific publications in 2016

Discipline	World	United States	European Union	China
Total publications, all disciplines	2,295,608	408,985	613,774	428,165
Engineering	422,392	50,305	89,611	123,162
Astronomy	13,774	3,272	5,524	1,278
Chemistry	181,353	20,858	41,123	53,271
Physics	199,718	27,402	50,943	42,190
Geosciences	130,850	20,449	33,758	30,258
Mathematics	52,799	8,180	15,958	8,523
Computer sciences	190,535	26,175	52,785	37,076
Agricultural sciences	50,503	4,908	12,275	9,376
Biological sciences	351,228	73,208	92,066	59,663
Medical sciences	507,329	119,833	149,761	56,680
Other life sciences	27,547	9,816	7,979	852
Psychology	39,025	14,314	12,889	1,278
Social sciences	121,667	29,447	49,102	4,262

talking about hundreds of thousands (perhaps more than a million) of false-positive scientific reports being produced each year worldwide.

The Current Publication Model

While everyone reading this book probably already knows the basics of the subscription-based 20th-century journal publication process, let's review a few of its main components in order to examine how it has changed in this increasingly digital age (and may even effectively disappear altogether in a few decades), what some of its limitations are, and some rather radical suggestions for its improvement (or least alteration). If readers will forgive a brief self-indulgence, I will begin this attempt by detailing some of my experiences as an editor-in-chief as an example of these relatively rapid changes.

In 1978, an unfortunately now-deceased faculty colleague and I decided to establish a peer reviewed program evaluation journal dedicated to the health sciences. In those pre-email days all correspondence was done by mail (usually on an IBM Selectric typewriter somewhere) hence communication with reviewers, the editorial board, and authors involved enumerable trips back and forth to the post office. After a few issues in which we two academics served as publishers, printers, mailers, promoters, editors, solicitors of articles (at least in the journal's early days), selectors (and solicitors) of peer reviewers, and collators of the latter's often disparate and contradictory reviews, all while maintaining additional back-and-forth correspondence with authors regarding revisions or reasons for rejections ad nauseam, we were quite happy to sell the journal for a whopping $6,000 each to Sage Publications (making it that publisher's *second* peer reviewed, professional journal).

Four decades later, Sage now publishes more than a thousand journals and is dwarfed by other journal publishers such as Elsevier and Wolters Kluwer. All communications now employ the internet rather than the postal service, all of which pretty much encompasses the evolution of academic publishing prior to exclusively online publishers—that and the price changes, with the average library subscription for a scientific journal now being around $3,000 per year plus the substantial (sometimes exorbitant) publishing fees levied by many journals on authors for the privilege of typesetting their labors and distributing them to (primarily) institutional subscribers. (The actual number of legitimate [i.e., non-predatory] peer reviewed scientific journals is actually not known but it is undoubtedly well over 25,000.)

The Peer Review Process

The hallmark of scientific publishing involves the use of other scientists (i.e., peers) to review their colleagues' work in order to determine its significance, quality, and whether or not it should be published. Practices vary for the peer review process, but typically once a manuscript is submitted for publication someone gives the manuscript a quick read (or perusal) to ensure its appropriateness for the journal in question.

Who this "someone" is varies with the size of the journal and the resources available to it, but typically three peer reviewers are solicited who hopefully are conversant with the topic area in question. Assuming that these individuals comply with the request, the manuscript is forwarded to them accompanied by instructions, a checklist of some sort (hopefully), a time table for their response (often not met), and a request for comments accompanied by a bottom-line decision regarding publication or rejection. Unfortunately, after all of this time and effort, it is not unusual for three disparate decisions emanating from the process such as one reviewer decreeing "publish with minor revisions," another suggesting that the authors make the requested revisions followed by another round of reviews, and the third rejecting the paper out of hand.

Fortunately for everyone concerned, the entire process is now conducted online, but, as described later, the process can take an inordinate of time and is becoming increasingly unwieldy as the number of journals (and therefore submitted papers) proliferate. The present output of 2,295,608 annual papers, for example, could theoretically require 6,886,824 peer reviews if every submission was published and only three reviewers were used. However, if the average journal accepts only half of its reviewed manuscripts (a large proportion typically accept less than 10% of submissions), the peer review burden jumps to north of 10 million reviews, not counting resubmitted rejections to other journals.

Thus, the bottom line of even online publication is that the peer review process has become impossibly overwhelmed and the quality of those reviews has become extremely suspect since many lower tiered journals are forced to take whoever they can get (which often equates to reviewers who have little or no expertise regarding what they are reviewing). And things are likely to get worse as more and more scientists attempt to publish ever greater numbers of studies.

However, to be fair, the peer review process was always a few orders of magnitude short of perfection. Until a few years ago, no one had come up with a better solution, and it did seem to work reasonably well, especially for the larger and more prestigious journals.

True, recently artificial intelligence aids have been developed that may *slightly* facilitate handling the increasing glut of manuscripts to be reviewed. These platforms can now perform cursory supplemental tasks, as succinctly described by Heaven (2018) including checking for problematic statistical or procedural anomalies (e.g., ScholarOne, StatReviewer), summarizing the actual subject matter of an article rather than relying on a quick perusal of its abstract (which one sage describes as "what authors come up with five minutes before submission" [courtesy of a marketing director in the Heaven paper]), and employing automated plagiarism checks. But regardless of the sophistication achieved by these innovations, the heavy lifting will remain with scientists as long as the current system exists.

Like every other scientific topic, a substantial literature has grown up around the shortcomings of peer review. While this literature can't be done justice here, one of the most thorough and insightful discussions of the entire publishing process (accompanied by potential solutions) must surely be Brian Nosek and Yoav Bar-Anan's (2012) essay entitled "Scientific Utopia: I. Opening Scientific Communication."

However, prior to considering some of these quite prescient suggestions, let's consider the following either amusing or alarming study, depending on one's perspective.

Who's Afraid of Peer Review?

John Bohannon (2013)

Actually, this paper is as much about the dark side of predatory open-access journals as it is about the peer review process but let's continue to focus our attention upon the latter. Financed by *Science* magazine (which perhaps not coincidentally is definitely not an open-access journal and may have had a conflict of interest here), John Bohannon wrote a completely fake article under a fictitious name from a fictitious university ballyhooing the promise of an also fake cancer drug. (Several versions of the paper were prepared, and the language was purposefully distorted by translating it into French via Google Translate and then back again into English. However, all versions involved a fictitious drug that had not gone to trial, and the article was purposefully filled with so many obvious flaws that [according to the author] no competent reviewer would recommend its acceptance.)

The article was then submitted to 304 open-access journals and was accepted by more than half of them, all with no notice of the study's fatal methodological flaws. As one example, the *Journal of Natural Pharmaceuticals* (published by an Indian company which owned 270 online journals at the time but has since been bought by Wolters Kluwer, a multinational Netherland publishing behemoth with annual revenues of nearly $5 billion) accepted the article in 51 days with only minor formatting changes requested. Nothing was mentioned concerning the study flaws.

For the exercise as a whole, the author reported that

> The paper was accepted by journals hosted by industry titans Sage and Elsevier. The paper was accepted by journals published by prestigious academic institutions such as Kobe University in Japan. It was accepted by scholarly society journals. It was even accepted by journals for which the paper's topic was utterly inappropriate, such as the *Journal of Experimental & Clinical Assisted Reproduction*. (p. 61)

Incredibly, only *PLoS One* (the flagship journal of the Public Library of Science and much maligned by John Bargh) rejected the paper on methodological grounds. One of Sage's journals (*Journal of International Medical Research*) accompanied its acceptance letter with a bill for $3,100. (For many such acceptances, Bohannon sent an email withdrawing the paper due to an "embarrassing mistake.")

Perhaps the main conclusion that can be drawn from this iconoclastic "survey" is that everything in science, as in all other human pursuits, can be (and often is) gamed. Other examples designed to expose the problems associated with peer review abound. For example, MIT students used SCIgen, a computer program that automatically generates gobbledygook papers, to submit papers that somehow got through the peer review process; a recent group of bizarre fake articles and authors were published in small peer review journals with titles such as "Human Reactions to Rape Culture and Queer Performativity at Urban Dog Parks in Portland, Oregon" (Wilson [Retracted, 2018] published in *Gender, Place, & Culture: A Feminist Geography Journal*) and Baldwin [Retracted, 2020] "Who Are They to Judge?" and "Overcoming Anthropometry Through Fat Bodybuilding" (published in the journal *Fat Studies* [yes, this and the feminist geography journal are actual journals]); and, of course, the iconic hoax by physicist Alan D. Sokal, who published a completely nonsensical article allegedly linking the then post-modernism fad with quantum physics in the non-peer reviewed *Social Text* (1996) (https://en.wikipedia.org/wiki/Alan_Sokal).

One problem with both the publishing and peer review processes lies in the number of publishing outlets available to even borderline scientists, which means that just about anything can be published with enough perseverance (and money). Another lies in the mandate that journals (especially subscription-based ones) are typically required to publish a given number of articles every quarter, month, or in some cases week. This dilutes the quality of published research as one goes down the scientific food chain and may even encourage fraudulent activity, such as the practice of authors nominating actual scientists, accompanied by fake email addresses (opened solely for that purpose and sometimes supplied by for-profit companies dedicated to supporting the process). Springer, for example (the publisher of *Nature* along with more than 2,900 other journals and 250,000 books), was forced to retract 107 papers from *Tumor Biology* published between 2010 and 2016 due to fake peer reviews. Similarly, Sage Publications was forced to retract 60 papers (almost all with the same author) due to a compromised peer review system. And these are known and relatively easily identified cases. We'll probably never know the true extent of these problems.

On a lighter note, Ferguson, Marcus, and Oransky (2014) provide several interesting and somewhat amusing examples of peer review fraud along with potential solutions. (Cat Ferguson, Adam Marcus and Ivan Oransky are the staff writer and two co-founders, respectively, of *Retraction Watch*, an extremely important organization designed to track retracted papers and whose website should be accessed regularly by all scientists interested in reproducibility.)

Examples supplied by the group include:

1. The author asks to exclude some reviewers, then provides a list of almost every scientist in the field.
2. The author recommends reviewers who are strangely difficult to find online.
3. The author provides gmail, Yahoo, or other free e-mail addresses to contact suggested reviewers, rather than email addresses from an academic institution.
4. Within hours of being requested, the reviews come back. They are glowing.
5. Even reviewer number three likes the paper (p. 481). [In my experience three-for-three uncritically positive reviews are relatively uncommon.]

Predatory (Fake) Journals

As mentioned, another downside of the publishing process involves the proliferation of predatory journals that will publish just about anything for "a few dollars more." We can blame this one on the inexorable movement toward open access, online publishing which ironically was originally motivated by the most idealistic of goals: making scientific knowledge open to everyone.

Regardless of who or what is to blame (if anyone or anything other than the perpetrators themselves), most of these journals appear to be located in India and China but often pretend to be located in the United States—although the latter contributes its share of home-grown outlets as well. Besides scamming gullible, desperate, and/or inexperienced investigators, predatory journals' characteristics include no true peer review system (just about every submission is accepted), no subscription base, a title that neither the submitting investigators nor their colleagues have ever heard of, websites replete with misinformation, and impact factors approaching or including zero.

Declan Butler (2013) provides an excellent overview of this problem in an article entitled "The Dark Side of Publishing," as well as suggesting several strategies for identifying suspect journals.

- Check that the publisher provides full, verifiable contact information, including an address, on the journal site. Be cautious of those that provide only web contact forms.

- Check that a journal's editorial board lists recognized experts with full affiliations. Contact some of them and ask about their experience with the journal or publisher since sometimes these journals simply list prominent scientists without their knowledge.
- Check that the journal prominently displays its policy for author fees.
- Be wary of email invitations to submit to journals or to become an editorial board member. [This one is tricky since legitimate journals sometimes use email correspondence to solicit manuscripts for a special issue or to contact potential board members based on recommendations from other scientists.]
- Read some of the journal's published articles and assess their quality. Contact past authors to ask about their experience. [This, too, has a downside since some of the authors may be quite proud of the fact that their articles were accepted with absolutely no required revisions.]
- Check that a journal's peer review process is clearly described, and try to confirm that a claimed impact factor is correct.
- Find out whether the journal is a member of an industry association that vets its members, such as the Directory of Open Access Journals (www.doaj.org) or the Open Access Scholarly Publishers Association (www.oaspa.org).
- Use common sense, as you would when shopping online: if something looks fishy, proceed with caution. (p. 435)

But Is an Actual Paradigmatic Publishing Change Actually Needed?

Certainly just about every reproducibility methodologist mentioned here would agree that some changes are desperately needed, although the "paradigmatic" adjective might appear unnecessarily radical to some. However, suggested change of any sort often appears radical when first proposed, so let's consider a few of these suggested changes here—some radical, some paradigmatic, and some simply sensible.

And who better to present these suggestions than Brian Nosek and one of his like-minded colleagues via the following forward-thinking essay.

Scientific Utopia I. Opening Scientific Communication

Brian Nosek and Yoav Bar-Anan (2012)

This long, comprehensive article clearly delineates the problems bedeviling publishing in peer reviewed scientific journals followed by proposed solutions for each. Ideally it should be read in its entirety by anyone interested in reforming the current system, but, for present purposes, what follows is an encapsulated version of its authors' vision of both some of the problems with the current system and their potential solutions.

First the problems:

1. The long lag between submission of an article and its appearance in print (perhaps averaging close to 2 years);
2. The astonishing subscription costs to university libraries;
3. The inaccessibility of scientific articles to anyone not affiliated with a subscribing institution;
4. *The odd arrangement of scientists turning over ownership of their articles (for which they are not paid) to publishers to print their work in one of the publisher's journals* (emphasis is added since the arrangement truly is odd and because scientists or their institutions typically have to *pay* these publishers for the honor of accepting said ownership.);
5. The myriad problems with the peer review system;
6. The static nature of a journal article that, once printed, stays printed and remains unchanged while science itself is a fluid and ever-changing process (this is also quite odd since, among other absurdities, author-initiated errata are often published in later journal issues instead of being corrected in the original digital version of the article. And to add insult to injury, authors are often charged for their errata or their requested withdrawal [i.e., retraction] of an article);
7. And, finally, the limited amount of space in printed journals, making it only possible to communicate what the investigators and reviewers consider essential information (but almost never enough information to allow a study to be replicated).

Needless to say Drs. Nosek and Ban-Anan have potential solutions for each of these shortcomings.

Fully embracing digital communication and completely abandoning journal issues and paper copies. Of the 25,000 or so scientific journals, the majority are online only. (Actually no one knows for sure just how many journals exist, although Björk et al., 2009, reported that there were 23,750 in 2006, which was over a decade ago.) Today almost everyone reads journal articles online or downloads pdfs to read at their leisure. Paper copies of all but the most widely read journals are disappearing from academic libraries so part of this suggestion is well on its way to being implemented. As for journal issues, little would be lost and precious time gained if the increasingly popular practice of making the final version of papers available to subscribers online in advance of the completed issue were to simply replace the issue system itself. The PLoS model, for one, already does this by making papers freely available as soon as they are accepted, suitably revised, and copyedited.

Going to a totally open access model in which all scientists (whether they work at universities, for private industry, or in their own basements) can access *everything* without costs. Of course the obvious problem with this is that *someone* has to pay the costs of copyediting and other tasks, but going to an exclusively digital model (and possibly, ultimately, a nonprofit one) should greatly reduce costs. Nosek and Bar-Anan suggest that most of these costs should be borne by scientists, their funding agencies, or their institutions (perhaps augmented by advertisers).

Again the PLoS open-access, purely digital model is presented as one example of how this transition could be made. Another example is the National Institutes of Health (NIH)'s PubMed Central (http://www.ncbi.nlm.nin.gov/pmc/), which attempts to ensure open access to all published reports of NIH-funded research. The greatest barriers to the movement itself are individual scientists, and a number of suggestions are made by the authors to encourage these scientists to publish their work in open-access journals. The major disadvantage resides in the inequitable difficulty that unfunded or underfunded scientists will have in meeting publication costs (which are considerable but are often also charged by subscription outlets) although some open-access journals theoretically reduce or even waive these fee if an investigator has no dedicated funding for this purpose.

Publishing prior to peer review. Citing the often absurd lag between study completion and publication, the authors suggest that "Authors prepare their manuscripts and decide themselves when it is published by submitting it to a repository. The repository manages copyediting and makes the articles available publicly" (p. 231).

Examples of existing mechanisms through which this is already occurring are provided, the most notable being the previously mentioned decades-old and quite successful arXiv preprint repository (https://en.wikipedia.org/wiki/ArXiv), followed by a growing group of siblings (some of which allow reviewer comments that can be almost as useful as peer reviews to investigators). An important subsidiary benefit for scientists is avoiding the necessity of contending with journal- or reviewer-initiated publication bias.

In this model, preprints of manuscripts are posted without peer review, and under this system, the number of submissions for arXiv alone increased by more than 10,000 *per month* by 2016. Most of the submissions are probably published later in conventional outlets, but, published or not, the repository process has a number of advantages including

1. Allowing papers to be revised at the scientists' pleasure, thus becoming a sort of living document (relatedly, it might be wise for even preprint registries to require some disciplinary-appropriate version of Simmons, Nelson, and Simonshohn's "21-word solution");
2. Aiding scientists in keeping up with advances in the field much more quickly than waiting on the snail-paced conventional publishing system;
3. Establishing the *priority* of a discovery since the preprinted paper can be posted almost immediately after the discovery is made instead of waiting months or years in the conventional journal system (this would also reduce investigator paranoia that someone might steal his or her discovery or idea for one);
4. Greatly reducing publication bias since (a) no one can reject the preprint because it wasn't accompanied by a favorable p-value and (b) the process increases investigator incentives to deposit their negative studies in a public repository rather than in their unrewarded file drawers;
5. Preventing costly redundancies by allowing other investigators to build on work that they would otherwise have to perform

themselves (of course, replications would remain an important scientific activity, just a more selective one);
6. And, in some cases (where professional comments are permitted), providing a valuable form of peer review preceding formal journal submissions of manuscripts.

However, there is a moral here, and it is that even extremely promising innovations such as preprint repositories that appear to be available free of cost must be financed somehow. And, as an ironical example of this, the Center for Open Science (COS) repository (ironic since Brian Nosek serves as its executive director) has announced that, as of 2019, it will begin charging fees to the 26 other organizations' repositories that it hosts since its projected costs for the year 2020 of $260,000 can no longer be covered by grant funding. And, as only one example, the INA-Rxiv alone, which now receives more than 6,000 submissions per year, will be faced with $25,000 in annual fees and accordingly has decided to leave the COS repository (Mallapaty, 2020).

Making peer review independent of the journal system. Here the authors really hit their stride by suggesting the creation of general or generic peer review systems independent of the journals themselves. In this system "instead of submitting a manuscript for review by a particular journal with a particular level of prestige, authors submit to a review service for peer review . . . and journals become not the publisher of articles but their 'promotors'" (p. 232).

This process, the authors argue, would free journals from the peer review process and prevent investigators from the necessity of going through the entire exercise each time they submit a paper following a rejection. Both the graded results of the reviews and the manuscript itself would be available online, and journal editors could then sort through reviewed articles according to their own quality metrics and choose which they wished to publish. More controversially, the *authors also suggest that there would be no reason why the same article couldn't be published by multiple journals.* (How this latter rather odd strategy would play out is not at all clear, but it is a creative possibility and the suggestion of employing a single peer review process rather than forcing each journal to constitute its own, while not unique to these authors, is in my opinion as a former editor-in-chief absolutely *brilliant*.)

> *Publishing peer reviews.* Making a case for the important (and largely unrewarded) contribution made by peer reviewers, the authors suggest that peer reviews not only be published but also not be anonymous unless a reviewer so requests. This way reviewers could receive credit for their scientific contributions since, as the authors note, some individuals may not have the inclination, opportunity, or talent for *conducting* science but may excel at evaluating it, identifying experimental confounds, or suggesting alternative explanations for findings. In this scenario, reviewers' vitas could correspondingly document this activity.
>
> It might also be possible for "official" peer reviewers to be evaluated on both the quality of their reviews and their evaluative tendencies. (Many journal editors presently do this informally since some reviewers inevitably reject or accept every manuscript they receive.) Listed advantages of these suggestions include the avoidance of "quid pro quo positive reviewing among friends" as well as the retaliatory anonymous comments by someone whose work has been contradicted, not cited, or found not to be replicable by the study under review.
>
> *Continuous, open peer review.* Peer reviews are not perfect and even salutary ones can change over time, as occurs when a critical confound is identified by someone following initial review and publication or a finding initially reviewed with disinterest is later found to be of much greater import. The authors therefore suggest that the peer review process be allowed to continue over time, much as book reviews or product evaluations do on Amazon.com. To avoid politically motivated reviews by nonscientists, a filter could be employed, such as the requirement of an academic appointment or membership in a professional organization, in order to post reviews. (This latter suggestion could be buttressed by the creation of an interprofessional organization—or special section within current professional organizations—devoted to the peer review process.)

Naturally, there are other insightful suggestions for reforming the publishing and peer review process not mentioned in this seminal article, but this one certainly constitutes the most comprehensive discussion of possible reforms of which I am aware. One such unmentioned innovation that has witnessed a degree of implementation follows. *"As is" peer review.* Eric Tsang and Bruno Frey (2007) have proposed an "as is" review process that they argue should, among other things, shorten the review process for everyone involved and reduce "the extent of intellectual prostitution" occurring when authors accede to sometimes inappropriate revisions by reviewers less knowledgeable than themselves.

The authors also note that while peer reviews in management journals (their discipline) were originally a page or less in length, reviews had mutated to eight or more single-spaced pages by the middle of the 1980s, followed by editorial requirements that authors supply point-by-point responses thereto. All of which were often followed by more reviews and more responses, the totality of which sometimes exceeded the length of the original article.

Tsang and Frey accordingly proposed a process (some variant of which has subsequently been implemented by a number of journals) that would involve the reviewers' providing feedback as currently practiced but then restricting their bottom-line decision to only an accept or reject option (not the dreaded "resubmission for further review" based on minor or major revisions). Then, in the authors' words,

> Based on the referees' recommendations, and his or her own reading of the manuscript, the editor makes the decision to accept or reject the manuscript. If the editor accepts the manuscript (subject to normal copy editing), he or she will inform the authors accordingly, enclosing the editorial comments and comments made by the referees. It is up to the authors to decide whether, and to what extent, they would like to incorporate these comments when they work on their revision for eventual publication. As a condition of acceptance, the authors are required to write a point-by-point response to the comments. If they refuse to accept a comment, they have to clearly state the reasons. The editor will pass on the response to the referees. In sum, the fate of a submitted manuscript is determined by one round of review, and authors of an accepted manuscript are required to make one round of revision. (pp. 11–12)

There are possible variations on this, as well as for all the proposals tendered in this chapter for reforming the publication and peer review process. All have their advantages, disadvantages, and potential pitfalls, but something has to be changed in this arena if we are to ever substantively improve the reproducibility of empirical research.

Other Publishing Issues Impacting Reproducibility

Retractions of Published Results

So far we haven't discussed the retraction of erroneous research findings in the detail that the process deserves since its proper application has the potential of reducing the prevalence of irreproducible results. However, although the handling of retractions is an important component of the publication process, they appear to have often been treated as an unwelcomed stepchild.

In my opinion (and one which I feel confident is shared by Adam Marcus and Ivan Oransky of Retraction Watch [https://retractionwatch.com/]), many journal editors and their publishers are several steps beyond passive aggression when it comes to handling retractions. In addition to the already mentioned tendency for journals to require publication fees from authors, many charge an additional fee for anyone who wishes to retract a study, correct a mistake, or even alert readers to an egregious error not reported by the original authors.

A ridiculous example of this reluctance on the part of journals to acknowledge the existence of such errors is provided by Allison, Brown, George, and Kaiser (2016), who recount their experiences in alerting journals to published *errors* in papers they were reviewing for other purposes. They soon became disenchanted with the process, given that "Some journals that acknowledged mistakes required a substantial fee to publish our letters: we were asked to spend our research dollars on correcting other people's errors" (p. 28).

Of course, some retractions on the part of investigators reflect innocent errors or oversights, but apparently most do not. Fang, Steen, and Casadevall (2012), for example, in examining 2,047 retractions indexed in PubMed as of 2012 found that 67.4% were due to misconduct, fraud, or suspected fraud. And what is even more problematic, according to the Retraction Watch website, some articles are actually cited more frequently *after* they are retracted than before.

Exactly how problems such as this can be rectified is not immediately clear, over and above Drs. Marcus and Oransky's continuing Herculean efforts with Retraction Watch. One possibility that could potentially put a dent in the problem, however, is to send a corrective email to any investigator citing a retracted article's published results, perhaps even suggesting that he or she retract the citation.

Should Scientists Publish Less Rather Than More?

Undoubtedly some should, but who is to decide who and how much? Many of the Utopia I article's recommendations would probably result in a significant increase in per scientist published outputs, and whether or not this is desirable is open for debate.

Brian Martinson (2017) makes a persuasive case for some of overpublication's undesirable consequences, and few scientists would probably disagree with it (at least in private).

> The purpose of authorship has shifted. Once, its primary role was to share knowledge. *Now it is to get a publication* [emphasis added]—"pubcoin: if you will. Authorship has become a valuable commodity. And as with all valuable commodities, it is bought, sold, traded and stolen. Marketplaces allow unscrupulous researchers to purchase authorship on a paper they had nothing to do with, or even to commission a paper on the topic of their choice. "Predatory publishers" strive to collect fees without ensuring quality. (p. 202)

However, many fewer would agree with his solution of giving scientists a "lifetime word limit" which Martinson himself freely admits might have a number of negative consequences. Alternately, Leif Nelson, Joseph Simmons, and Uri Simonsohn (2012), in responding to the Nosek and Bar-Anan "Utopia" paper, floated a one paper per year alternative (a solution not without its own advantages and drawbacks, but which is probably as equally unlikely to be implemented as Martinson's suggestion).

More importantly, however, the Nelson et al. response also makes a strong case against relaxing the publication process to the point where just about *everyone* can publish just about *anything*—a *possible* outcome if some of the more radical Nosek and Bar-Anan proposals were to be implemented.

> Bad papers are easy to write, but in the current system they are at least *somewhat* [emphasis added] difficult to publish. When we make it easier to publish papers, we do not introduce good papers into the market (those are already going to be out there); we introduce disproportionately more bad papers. (p. 292)

And then, of course, there is always John Ioannidis and colleagues' astonishing documentation that "thousands of scientists publish a paper every five days."

But Must Publishing Be the Only Coin of the Realm?

Professors Nosek and Bar-Anan have a tacit answer for this question along with just about everything else associated with publishing. Namely (a variant of which has been suggested by others as well), that some "scientists who do not have the resources or interest in doing original research themselves can make substantial contributions to science by reviewing, rather than waiting to be asked to review" (p. 237).

In a sense all scientists are peer reviewers, if not as publication gatekeepers, at least for their own purposes every time they read an article relevant to their work. So why not officially create a profession given over to this activity, one accompanied by an official record of these activities for promotion and tenure purposes? Or, barring that, an increased and rewarded system of on-line reviews designed to discourage methodologically unsound, unoriginal, or absurd publications.

Alternately, there are presently multiple online sites upon which one's comments regarding the most egregious departures from good scientific practice can be shared with the profession as a whole and/or via one-on-one correspondences with the authors themselves. From a scientific perspective, if institutionalized as a legitimate academic discipline, the hopeful result of such activities would be to improve reproducibility one study and one investigator at a time.

There are, of course, many other options and professional models already proposed, such as Gary King's 1995 recommendation that scientists receive credit for the creation of datasets that facilitate the replication process. In models such as this scientists would be judged academically on their performance of duties designed to facilitate the scientific process itself, which could include a wide range of activities in addition to peer reviewing and the creation of databases. Already existing examples, such as research design experts and statisticians, are well established, but the list could be expanded to include checking preregistered protocols (including addendums thereto) against published or submitted final reports.

And, of course, given the number of publications being generated in every discipline there are abundant opportunities for spotting questionable research practices (QRPs) or errors in newly published studies. Perhaps not a particularly endearing professional role or profession, but letters to the offending journals' editors and/or postings on websites designed for the specific purpose of promulgating potential problems could be counted as worthy, quantifiable professional activities. And, of course, the relatively new field of *meta-science* is presently open for candidates and undoubtedly has room for numerous subspecialties including the development of software to facilitate all of the just-mentioned activities. This is an activity which has already produced some quite impressive results, as described next.

Already Existing Statistical Tools to Facilitate These Roles and Purposes

Taking the search for potential errors and misconduct in published research as an example, there are a growing number of tools available to facilitate this purpose. In addition to ScholarOne, StatReviewer, and p-curve analysis already mentioned, other approaches for identifying *potential* statistical abnormalities exist—three of which are described clearly in a very informative article by Bergh, Sharp, and Li (2017) titled "Tests for Identifying 'Red Flags' in Empirical Findings: Demonstration and Recommendations for Authors, Reviewers, and Editors." A sampling of some other indirect (but creative approaches) for spotting statistical abnormalities included the following:

1. An R-program (statcheck) developed by Epskamp and Nuijten (2015) which allows extracted p-values to be recalculated to spot abnormalities, possible tampering, and errors based on reported descriptive statistics. Using this approach Nuijten, Hartgerink, van Assen, Epskamp, and Wicherts (2016) identified a disheartening number of incorrectly reported p-values in 16,695 published articles employing inferential statistics. As would be expected by now, substantively more false-positive errors than negative ones were found.
2. Simulations such as bootstrapping approaches (Goldfarb & King, 2016) for determining what would happen if a published research result were to be repeated numerous times, with each repetition being done with a new random draw of observations from the same underlying population.
3. A strategy (as described by Bergh, Sharp, Aguinis, & Li, 2017) offered by most statistical packages (e.g., Stata, IBM SPSS, SAS, and R) for checking the accuracy of statistical analyses involving descriptive and correlational results when raw data are not available. Using linear regression and structural equation modeling as examples, the authors

found that of those management studies for which sufficient data were available and hence could be reanalyzed, "nearly one of three reported hypotheses as statistically significant which were no longer so in retesting, and far more significant results were found to be non-significant in the reproductions than in the opposite direction" (p. 430).
4. The GRIM test (Brown & Heathers, 2016) which evaluates whether or not the summary statistics in a publication are mathematically possible based on sample size and number of items for whole (i.e., non-decimal) numbers, such as Likert scales.
5. Ulrich Schimmack's "test of insufficient variance" (2014) and "z-curve analysis" (Schimmack & Brunner, 2017) designed to detect QRPs and estimate replicability, respectively.

All of which could be a most propitious hobby or turn out to be an actual scientific discipline. By way of illustration, consider the following study emanating from an anesthesia researcher's hobby (or passion) for improving reproducibility in his chosen field.

Data Fabrication and Other Reasons for Non-Random Sampling in 5087 RCTs in Anesthetic and General Medical Journals

John B. Carlisle (2017)

Recently highlighted in a *Nature* article (Adam, 2019), Dr. Carlisle arises before dawn to let his cat out and begins entering published experimental anesthesia data into a spreadsheet that will eventually be analyzed for suspicious values—the presence of which he has the temerity to inform the editors of those journals in which the offending articles appear. A process, incidentally, that has resulted in the identification of both fraudulent investigators and numerous retractions.

The study being described here involved (as its title suggests) more than 5,000 anesthesia studies from six anesthesia and two general medical journals (the *Journal of the American Medical Association* and the *New England Journal of Medicine*). These latter two journals were most likely added because anesthesia research appears to be the medical analog to social psychology as far as suspicious activities are concerned. Hence

> Dr. Carlisle may have wanted to ascertain if his profession did indeed constitute a medical outlier in this respect. (According to the *Nature* article, four anesthesia investigators [Yoshitaka Fujii, Yuhji Saitoh, Joachim Boldt, and Yoshihiro Sato] eventually had 392 articles retracted, which, according to Retraction Watch, dwarfs psychologist Diederik Stapel's 58 admitted data fabrications: https://retractionwatch.com/2015/12/08/diederik-stapel-now-has-58-retractions/.)
>
> Carlisle's analysis involved 72,261 published arithmetic means of 29,789 variables in 5,087 trials. No significant difference occurred between anesthesia and general medicine with respect to their baseline value distributions, although the latter had a lower retraction rate than the former. And in agreement with just about all of the authors of this genre of research, Dr. Carlisle was quite explicit in stating that his results could not be interpreted as evidence of misconduct since they could also be functions of "unintentional error, correlation, stratified allocation and poor methodology."
>
> He did implicitly suggest, however, that more investigators should join him in this enterprise since, "It is likely that this work will lead to the identification, correction and retraction of hitherto unretracted randomised, controlled trials" (p. 944).

And the *New England Journal of Medicine* obviously agrees since it has announced that it will be applying this technique to future submissions.

Relatedly, Bolland, Avenell, and Gamble (2016) applied a variant of Dr. Carlisle's approach via a meta-analysis of 33 problematic trials investigating elderly falls (i.e., problematic with respect to baseline data "involving [uncharacteristically] large numbers of older patients with substantial comorbidity, recruited over very short periods"). The results of that analysis being that

> [o]utcomes were remarkably positive, with very low mortality and study withdrawals despite substantial comorbidity. There were very large reductions in hip fracture incidence, regardless of intervention (relative risk 0.22, 95% confidence interval 0.15–0.31, $p < 0.0001$ ... that greatly exceed those reported in meta-analyses of other trials. There were multiple examples of inconsistencies between and within trials, errors in reported data, misleading text, duplicated data and text, and uncertainties about ethical oversight. (p. 1)

So Is It Time for Publishers' to Step Up?

It is past time, regardless of how the journal system evolves over the next few decades. Hopefully the *New England Journal of Medicine*'s tentative step in following John Carlisle's lead is only a precursor to more hands-on actions by journals to ensure the integrity of what they publish. Perhaps the reproducibility crisis's expanding profile will facilitate such actions, supplemented by continuing efforts by scientists such as Dr. Carlisle and the determined efforts of the myriad contributors to this scientific initiative. For it is important to remember that scientific publishing is not solely in the hands of the publishing industry CEOs and CFOs or the editors-in-chief. Rather it is a symbiotic process involving multiple other actors, the most important of which are scientists themselves.

Methodological recommendations in these regards have been tendered by a number of the reproducibility advocates mentioned in recent chapters so there is no need to restate them here. However, given the descriptions of statistical/empirical aids for ensuring the validity of empirical results just mentioned, let's revisit the aforementioned Bergh et al.'s "Red Flags" article that very succinctly reminds us of the multiple roles and responsibilities for ensuring reproducibility from a statistical perspective in the initial phase of the publishing process:

First author's responsibilities:

1. Include such values as coefficient estimates, standard errors, p-values in decimals, and a correlation matrix that includes means, standard deviations, correlations [including those between covariate and outcome], and sample sizes.
2. "Describe all data-related decisions such as transformed variables and how missing values and outliers were handled."
3. "Attest to the accuracy of the data and that the reporting of analytical findings and conclusions."

Next, the editor's responsibilities:

1. Ensure that all of the previous disclosure requirements are satisfied.
2. Require "authors to attest that their findings are based on the reported data and analytical findings; indicate that findings will be confirmed through retesting if article receives a conditional acceptance."

3. "Amend manuscript evaluation form sent to reviewers to include a check of the expanded data disclosure reporting requirements and for consistency between disclosure, analysis, hypotheses, and conclusions."
4. "Retest findings using Tests 1 [ensuring the accuracy of p-values] and 3 [verifying study results based on matrices of descriptive statistic] after a conditional acceptance is awarded and before a final acceptance is reached." (Obviously, the editors themselves won't do this but will task an employee or contractor to do so.)

And, of course, the peer reviewers' responsibilities:

1. "Confirm that data reporting is complete and meets expanded disclosure requirements (permitting the tests described above)."
2. "Assess relationships between the data, findings, and interpretation of hypotheses to ensure consistency." (All dicta are courtesy of table 5, p. 122—exact quotations are so noted, others are paraphrased.)

Plus a few from Philip Bourne and Alon Korngreen (2006):

1. Do not accept a review assignment unless you can accomplish the task in the requested timeframe—learn to say no.
2. Avoid conflict of interest [e.g., cronyism].
3. As a reviewer you are part of the authoring process [which means you should strive to make whatever you review a better paper].
4. Spend your precious time on papers worthy of a good review.
5. Write clearly, succinctly, and in a neutral tone, but be decisive.
6. Make use of the "comments to editors" [i.e., comments designed to help the editor but not to be shared with the author] (pp. 0973–0974).

A Final Publishing Vision that Should Indirectly Improve Reproducibility

Reading the following brief article when it was first published would probably have been viewed as absurd by most social scientists dabbling in computational research (and completely irrelevant for those not involved therein). Today, however, in the context of the reproducibility crisis, it resonates as downright prescient for improving all genres of scientists' day-to-day empirical practice.

> **What Do I Want From the Publisher of the Future?**
>
> Philip Bourne (2010)
>
> Written by the editor of *PLoS Computational Biology*, in some ways this editorial picks up where the just discussed article by Nosek and Bar-Anan leaves off even though it was published 2 years earlier. The paper would be greatly improved by a few detailed examples, but it isn't fair to criticize a brief editorial targeted at computational scientists on this basis since undoubtedly almost everyone in that discipline could supply several such examples from their own experience.
>
> So, in support of many of Drs. Nosek and Bar-Anan's suggested changes, consider the following rhetorical question posed by Dr. Bourne:
>
>> After all our efforts at producing a paper, very few of us have asked the question, is journal X presenting my work in a way that maximizes the understanding of what has been done, *providing the means to ensure maximum reproducibility* [emphasis added] of what has been done, and maximizing the outreach of my work? (p. 1)
>
> Dr. Bourne's answer to this question is a resounding "no," consonant with Nosek and Bar-Anan's (2012) observation:
>
>> Authors are so happy to have their submission accepted that they blissfully sign a copyright transfer form sent by the publishers. Then publishers recoup their investment by *closing* [emphasis added] access to the articles and then selling journal subscriptions to the scientists and their institutions (individual articles can be purchased for $5 to $50 depending on the journal). [Actually, in my experience it is a rare article that can be obtained as cheaply as $5.] In other words, the funding public, universities, and scientists who produced and pay for the research give ownership of the results to publishers. Then, those with money left over buy the results back from the publishers; the rest are in the dark. (p. 228)

> Now back to Professor Bourne, who suggests that, in addition to what is normally included in a scientific article (i.e., the stored data and its documentation—recalling that his specialty is computational research where usable shared data is usually a publication *requirement*), journals (or third parties associated therewith) should also publish what he calls the relevant laboratory's *workflow*—which constitutes a detailed record of everything pertinent to the creation of the published or deposited article.

Naturally, such a workflow might be quite different for experimental work involving humans and animals. However, every laboratory and every study should keep a detailed workflow comprising every aspect of a study from the initial idea to the details of the literature review to descriptions of any meetings or correspondence with investigators or assistants to the writing of the regulatory and/or funding proposals. Pilot studies should also be recorded in detail, including accrued data as well as for those aborted along the way. And, of course, every aspect of the actual study would be available therein, including

1. The final registered protocol;
2. The approved institutional review board (IRB) or institutional animal care and use committee (IACUC) proposal (including correspondences regarding it and any amendments submitted thereto);
3. Any procedural glitches and/or minor "tweaking" of study conditions, outcome variables, and analytic decisions; and
4. All correspondences with journal editors and peer reviewers.

While Dr. Bourne's vision of a computational workflow would probably not be quite this comprehensive, he provides some universal advantages of the process based on correctable inefficiencies experienced in his own lab.

1. The intellectual memory of my laboratory is in my e-mail folders, themselves not perfectly organized. This creates a hub-and-spoke environment where lab members and collaborators have to too often go through me to connect to each other.
2. Much of our outreach is in the form of presentations made to each other and at national and international forums. We do not have a good central repository for this material; such a repository could enable us to have a better understanding of what other researchers are doing.

3. While we endeavor to make all our software open source, there are always useful bits of code that languish and disappear when the author leaves the laboratory.
4. Important data get lost as students and postdoctoral fellows leave the laboratory. (p. 2)

Now certainly most workflows will probably never be archived by a publisher. However, there is no reason why individuals cannot and should not keep better, even compulsively detailed, *personal* records of their research for at least two very rather obvious reasons.

First, comprehensive workflows will greatly facilitate the replication process because even if a request for information from an independent replicator of one's study is stressful, it is advantageous for the original investigator that the replication be performed with maximum fidelity. (And, has been mentioned previously, a direct replication is seldom possible based on nothing more than the information typically available in the methods section of a typical research article.)

Second, when investigators are young and have conducted only a few studies, their memories may be sufficient to reconstruct the exact procedures employed in their studies. However, with time and its many other competitors for cerebral storage, earlier memories begin to deteriorate—a phenomenon for which I could serve as an expert witness.

A parable: As a bizarre hypothetical example, let's return to our hypothetical graduate student of long and many pages ago and his obviously fictitious series of 20+ incremental experiments (all negative) designed to develop a prose study skill capable of producing superior learning to reading and rereading a passage when *time on task was held constant*. As we know, to his everlasting shame, he never attempted to publish this rather Quixotic effort since none of the experiments reached statistical significance (recalling that publication bias of this sort is a very old phenomenon, as the graduate student would be by now had he ever existed).

Suppose further that within a couple of decades he had realized the folly of his ways because, by then, his results appeared to have significant implications for a time-on-task theory he was working on. A theory, in fact, that not coincidentally explained his earlier failures to produce an efficacious method of study and could be validated by a relatively simple series of experiments.

However, there was no possibility that he could recall sufficient details concerning even a few of the studies, much less all of them (or even how many he had conducted). And, of course, practically no one in those days kept detailed paper-based records of procedures, detailed protocols, or data for decades, especially following the institutional moves that often accompany such time intervals—and especially not for unpublished studies. (Hence, even the file drawer constitutes an insufficient metaphor for this problem.)

Today, however, after the Digital Revolution, there is really no excuse for such behavior. There is also no excuse for not sharing all of the information contributing to and surrounding a scientific finding, published or unpublished. And that just happens to constitute the subject of Chapter 10.

References

Adam, D. (2019). The data detective. *Nature, 571*, 462–464.
Allison, D. B., Brown, A. W., George, B. J., & Kaiser, K. A. (2016). A tragedy of errors: Mistakes in peer reviewed papers are easy to find but hard to fix. *Nature, 530*, 27–29.
Baldwin, R. (2018) Retracted Article: Who are they to judge? Overcoming anthropometry through fat bodybuilding, *Fat Studies, 7*, i–xiii.
Bergh, D. D., Sharp, B. M., Aguinis, H., & Li, M. (2017). Is there a credibility crisis in strategic management research? Evidence on the reproducibility of study findings. *Strategic Organization, 15*, 423–436.
Bergh, D. D., Sharp, B. M., & Li, M. (2017). Tests for identifying "Red Flags" in empirical findings: Demonstration and recommendations for authors, reviewers, and editors. *Academy of Management Learning and Education, 16*, 110–124.
Björk, B.-C., Roos, A., & Lauri, M. (2009). Scientific journal publishing: Yearly volume and open access availability. *Information Research, 14*(1), paper 391.
Bohannon, J. (2013). Who's afraid of peer review? *Science, 342*, 60–65.
Bolland, M. J., Avenell, A., & Gamble, G. D. (2016). Systematic review and statistical analysis of the integrity of 33 randomized controlled trials. *Neurology, 87*, 1–12.
Bourne, P. E. (2010). What do I want from the publisher of the future? *PLoS Computational Biology, 6*, e1000787.
Bourne P. E., & Korngreen, A. (2006). Ten simple rules for reviewers. *PLoS Computational Biology, 2*, e110.
Brown, N. J., & Heathers, J. A. (2016). The GRIM test: A simple technique detects numerous anomalies in the reporting of results in psychology. *Social Psychological and Personality Science*, https://peerj.com/preprints/2064.pdf.
Butler, D. (2013). The dark side of publishing. The explosion in open-access publishing has enabled the rise of questionable operators. *Nature, 435*, 433–435.
Carlisle, J. B. (2017). Data fabrication and other reasons for non-random sampling in 5087 randomised, controlled trials in anaesthetic and general medical journals. *Anaesthesia, 72*, 944–952.

Epskamp, S., & Nuijten, M. B. (2015). Statcheck: Extract statistics from articles and recompute p values. R package version 1.0.1. http://CRAN.R-project.org/package=statcheck

Fang, F. C., Steen, R. G., & Casadevall, A. (2012). Misconduct accounts for the majority of retracted scientific publications. *Proceedings of the National Academy of Science, A, 109*, 17028–17033.

Ferguson, C., Marcus, A., & Oransky, I. (2014). Publishing: The peer review scam. *Nature, 515*, 480–482.

Goldfarb, B. D., & King, A. A. (2016). Scientific apophenia in strategic management research: Significance tests and mistaken inference. *Strategic Management Journal, 37*, 167–176.

Heaven, D. (2018). AI peer reviewers unleashed to ease publishing grind. *Nature, 563*, 609–610.

Ioannidis, J. P. A. (2005). Why most published research findings are false. *PLOS Medicine, 2*, e124.

Ioannidis, J. P. A., Klavans, R., & Boyack, K. W. (2018). Thousands of scientists publish a paper every five days, papers and trying to understand what the authors have done. *Nature, 561*, 167–169.

King, G. (1995). Replication, replication. *PS: Political Science and Politics, 28*, 444–452.

Mallapaty, S. (2020). Popular preprint sites face closure because of money troubles. *Nature, 578*, 349.

Martinson, B. C. (2017). Give researchers a lifetime word limit. *Nature, 550*, 202.

National Science Board. (2018). *Science and engineering indicators 2018*. NSB-2018-1. Alexandria, VA: National Science Foundation. (www.nsf.gov/statistics/indicators/)

Nelson, L. D., Simmons, J. P., & Simonsohn, U. (2012). Let's publish fewer papers. *Psychological Inquiry, 23*, 291–293.

Nosek, B. A., & Bar-Anan, Y. (2012). Scientific Utopia I: Opening scientific communication. *Psychological Inquiry, 23*, 217–243.

Nuijten, M., Hartgerink, C. J., van Assen, M. L. M., Epskamp, S., & Wicherts, J. (2016). The prevalence of statistical reporting errors in psychology (1985–2013). *Behavior Research Methods, 48*, 1205–1226.

Schimmack, U. (2014). The test of insufficient variance (TIVA): A new tool for the detection of questionable research practices. https://replicationindex.wordpress.com/2014/12/30/the-test-ofinsufficientvariance-tiva-a-new-tool-for-the-detection-ofquestionableresearch-practices/

Schimmack, U., & Brunner, J. (2017). Z-curve: A method for estimating replicability based on test statistics in original studies. https://replicationindex.files.wordpress.com/2017/11/z-curve-submission-draft.pdf

Sokal, A. D. (1996). Transgressing the boundaries: Toward a transformative hermeneutics of quantum gravity. *Social Text, 46–47*, 217–252,

Tsang, E. W., & Frey, B. S. (2007). The as-is journal review process: Let authors own their ideas. *Academy of Management Learning and Education, 6*, 128–136.

Wilson, H. (2018). Retracted Article: Human reactions to rape culture and queer performativity at urban dog parks in Portland, Oregon. *Gender, Place & Culture, 27*(2), 1–20.

10
Preregistration, Data Sharing, and Other Salutary Behaviors

The purpose of this chapter is to discuss some of the most powerful, effective, and painless strategies for lowering the prevalence of irreproducible scientific findings in a bit more depth. So hopefully I will be excused for a bit of redundancy here given the crucial role preregistration and data sharing play in reducing the opportunities for poorly trained or inadequately acculturated or even unscrupulous scientists to engage in research driven by questionable research practices (QRPs).

For while it is unfortunate that these individuals have been denied (or have chosen to ignore) educational or principled mentoring opportunities, it is absolutely necessary that the continued production of false-positive results in the scientific literature be arrested. And it is highly unlikely that the continuing promulgation of explicit edicts regarding the behaviors in question will be capable of trumping these individuals' perceived (or real) need to continue to publish such findings.

So, in way of review, what are needed are universal requirements for

1. The protocols of all empirical research to be preregistered accompanied by amendments detailing any procedural changes occurring during the conduct thereof, and
2. All data accruing from these protocols to be deposited in an accessible, discipline-approved registry (i.e., not a personal website) and available to all scientists, along with all the code required to permit it to be quickly run as a check against the published descriptive and inferential results.

However, as will be illustrated shortly, this, too, will be insufficient if

1. The preregistered protocols are not compared point-by-point to their published counterparts (preferably involving a standardized methodological and statistical checklist), and
2. The registered data are not reanalyzed (using the available code) as a descriptive check on the demographics and as an inferential check on the primary p-values reported.

But who is going to do all of this? The easy answer is the multi-billion dollar publishing industry, but even though its flagship, the *New England Journal of Medicine*, has joined Dr. Carlisle in checking distributions of the baseline values of submitted randomized controlled trials (RCTs), this does not mean that other journals will embrace this strategy (or the considerably more effective and time-consuming suggestion just tendered of reanalyzing the actual data).

However, since this industry is built on the free labor of scientists, why not create a profession (as alluded to in Chapter 9) devoted to checking preregistered protocols against the submitted research reports and actually rerunning the key analyses based on the registered data as part of the peer review process? These efforts could be acknowledged via (a) a footnote in the published articles; (b) journal backmatter volume lists, in which peer reviewers and authors are often listed; (c) or even rewarded as a new form of "confirming" authorship. (And, naturally, these behaviors would be acknowledged and rewarded by the participants' institutions.)

And why is all of this necessary since replication is the reproducibility gold standard? The easy answer is that the number of studies presently being published precludes replicating even a small fraction thereof given the resources required. (Not to mention the fact that some studies cannot be replicated for various reasons.)

So, if nothing else, with a study that passes these initial screenings involving the original protocol and the registered data (and assuming it also passes the peer review process), considerable more confidence can be had in its ultimate reproducibility. This evidence, coupled with the perceived importance of a study, will also help inform whether a replication is called for or not. And if it is, the registered information will make the replication process considerably easier—for not only is it the gold standard for reproducibility, its increased practice provides a powerful disincentive for conducting substandard research.

But if a study protocol is preregistered why is it necessary to check it against the final published product? Partly because of scientists' unique status in society and the conditions under which they work. Unlike many professions, while scientists typically work in teams, their individual behaviors are not supervised. If fact, most senior scientists do not actually perform any of the procedural behaviors involved in their experiments but instead rely on giving instructions to research assistants or postdocs who themselves are rarely supervised on a day-to-day basis—at least following a training run or two.

Of course scientists, like other professionals such as surgeons, are expected to follow explicit, evidence-based guidelines. But while surgeons are rewarded for positive results, their operating procedures can also be investigated based on an egregious and expected negative result (or a confluence thereof via malpractice suits or facing the families of patients who have died under their watch)—*unless* they have rigorously followed evidence-based, standard operating (not a pun) procedures. And while scientists also have standard operating procedures, as documented in guidelines such as the CONSORT and ARRIVE statements, their carrots and sticks are quite different. The "sticks" have historically come in the form of little more than an occasional carrot reduction, with the exception of the commission of outright fraud. (And some countries even permit investigators who have committed egregious examples of fraud to continue to practice and publish.) The carrots come in the form of tenure, direct deposit increases, and the esteem (or jealousy) of their colleagues, which in turn requires numerous publications, significant external funding, and, of course statistically significant research results.

All of which *may* have resulted in the social and life sciences waking up to the reproducibility crisis after a prolonged nap during which

1. The scientific version of standard operating procedures involved avoiding some of the QRPs listed in Chapter 3 while ignoring many of the others; hence,
2. Scientific training had typically underplayed the problematic nature of these practices; while
3. There has been increased pressure from the publishing industry to keep research reports brief in order to include as many articles as possible in each issue; with
4. Little or no cultural imperative for change for any of the preceding via the "if it isn't broken don't fix it" paradigm.

But, of course, "it" is broken, and the traditional emphasis on intensive practice as the primary mechanism for teaching the skills necessary for the continuity of the scientific professions is not sufficient. At least, not if this intensive practice involves the continuance of traditional research behaviors. Thus the most important message emanating from the reproducibility movement is that many (not all) of the traditional ways in which science has been conducted must change and that the most important mechanisms to drive this change involve *required* transparency in the conduct of the research practices which *precede* the collection of a study's first data point. (Or, for investigations involving existing databases, transparent steps that *precede* inferential analyses.)

But barring constantly monitored surveillance equipment in the laboratory, which isn't a bad idea in some cases (as recommended by Timothy Clark [2017]), what could accomplish such a task? Hopefully the requisite behaviors are obvious by now, but let's examine these behaviors' absolute essentials, their requisite policies, and their current levels of implementation in a bit more detail.

Preregistration of Study Protocols

The primary purpose of the preregistration of research protocols is to prevent the real and simulated methodological abuses detailed in previous chapters. The Simmons, Nelson, and Simonsohn "Hot Potatoes" study (2011), for example, would have most likely been accepted for publication in the past as initially written sans any disclaimers. However, even in the "good old days," had the study's hypotheses, experimental conditions, primary outcomes, sample size justification, inclusion/exclusion criteria, and analytic approaches (such as the handling of missing data and the use of covariates and blocking variables) been preregistered on an accessible website prior to running the first participant and that preregistered document compared to the submitted publication by either a peer reviewer or an employee of the publisher, then the submitted manuscript would have surely been summarily rejected.

However, even if no officially registered protocol–manuscript comparison had taken place, the process itself would serve a number of useful purposes.

1. The very *threat* of third-party preregistration evaluations broadcasting any discrepancies on social media might be a significant deterrent in and of itself;
2. The preregistration document could serve as a reference to conscientious investigators in the preparation of their final manuscripts by reminding them of exactly what they had originally proposed and what actually transpired in the conduct of the study—especially when there is a significant delay between study initiation and manuscript preparation;
3. Since changes and amendments to the document following study commencement are often necessary, the preregistration process and the highly recommended laboratory workflow can operate together synergistically to keep track of progress and facilitate memories.

In the long run, then, the very existence of a thorough preregistration document accompanied by amendments should serve to encourage investigators to transparently list major changes in the published document occurring between the originally proposed and completed study. This, in turn, will facilitate publication by stealing peer reviewers' and critics' thunder and, in some instances, turn a potential QRP into a non sequitur. And finally, the very threat of future methodological criticisms might make some investigators more careful in the conduct of their studies and their subsequent writing style.

The Initiation of the Preregistration Movement

The most substantive move toward requiring preregistration as a condition of publication came from journal editors themselves in 2004, after several years of sporadic lobbying in the form of articles by methodologists and, in some cases, professional associations. In that year the International Committee of Medical Journal Editors (ICMJE), which is comprised of the most prestigious journals in the field (including the *New England Journal of Medicine* and the *Journal of the American Medical Association*) announced that, as of July 1, 2005, preregistration of all clinical *trials* would become a prerequisite for publication in all of the organizations' affiliated journals (De Angelis, Drazen, Frizelle, et al., 2004).

Interestingly, the primary rationale for instituting this policy appeared to be to vitiate publication bias, which has a somewhat different genesis (and an even more important implication) in medicine as opposed to the social sciences. Namely, it was undertaken to prevent research sponsors such as the pharmaceutical industry (second only to the federal government as a source of medical research funding) from concealing the presence of selected (presumably negative) trials that could potentially "influence the thinking of patients, clinicians, other researchers, and experts who write practice guidelines or decide on insurance-coverage policy" (De Angelis et al., 2004, p. 1250).

The early requirements for registries were actually rather modest and could be of the investigators' choosing as long as they were (a) free of charge and accessible to the public, (b) open to all prospective registrants, and (c) managed by a not-for-profit organization. The requirements for the preregistrations of the clinical trials themselves, while far from onerous, suggested an acute awareness of many of the QRPs listed in Chapters 3 and 4.

> There must be a mechanism to ensure the validity of the registration data, and the registry should be electronically searchable. An acceptable registry must include at minimum the following information: a unique identifying number, a statement of the intervention (or interventions) and comparison (or comparisons) studied, a statement of the study hypothesis, definitions of the primary and secondary outcome measures, eligibility criteria, key trial dates (registration date, anticipated or actual start date, anticipated or actual date of last follow-up, planned or actual date of closure to data entry, and date trial data considered complete), target number of subjects, funding source, and contact information for the principal investigator. (p. 1251)

The social sciences eventually followed suit. The Open Science Project undoubtedly provided the greatest impetus and, as of this writing, has more than 8,000 registered studies from a wide variety of disciplines. There are also an unknown but large number of US and international research registries in a bewildering number of disciplines—*clinicaltrials.com* being the largest with more than 300,000 registered trials representing more than 200 countries. (The World Health Organization's International Clinical Trials Registry Platform is also huge, listing approximately 200 participating countries.)

Epistemologically, the most important functions served by the preregistration of trials may be the potential of the process to operationally differentiate between hypotheses (a) generated a posteriori based on a completed a study's data and (b) those generated a priori and constituting the actual rationale for the study in the first place. Obviously a great deal more confidence can be had in the latter as opposed to the former—perhaps most effectively illustrated metaphorically by James Mills (1993) in his classic article "Data Torturing" ("If the fishing expedition catches a boot, the fishermen should throw it back, not claim that they were fishing for boots" [p. 1198]) and Andrew Gelman and Eric Loken's (2014) previously discussed seminal article (in a "garden of forking paths, whatever route you take seems predetermined" [p. 464]).

This genre of ex post facto behavior is surely one of the most frequently engaged in of QRPs (our number 15 in Chapter 3), one of the most insidious, and more often than not accompanied by several related undesirable practices. It is also a leading cause of irreproducibility and publication bias. And, like most QRBs, its end results (artifactually produced and incorrect p-values < 0.05) are reinforced by publication and peer review policies.

However, while preregistration is an important preventive measure for ex post factor hypothesizing, the practice has a long history in the social sciences. So let's take a quick trip back in time and review a small piece of the history of this particular QRP—and in the process perhaps help explain why it is so difficult to eradicate.

HARKing: Hypothesizing After the Results Are Known

Norbert Kerr (1991)

This is a classic and even-handed discussion of what Brian Nosek and colleagues (2018) would later refer to as "postdiction" and to which Professor Kerr bestowed the acronym "HARKing" before concluding that its disadvantages far outweighed its few advantages and hence was completely contraindicated unless an unpredicted-unexpected finding was clearly labeled as such. However, an almost equally interesting aspect of the article is its illustration of how widely accepted the practice apparently was in 1991. Even such a methodological luminary as Daryl Bem is quoted as giving the following advice to students and new researchers in a book chapter entitled "Writing the Empirical Journal Article." (The 1987 book itself was designed to mentor the soon to be practicing generation of researchers and was aptly titled *The Compleat Academic: A Practical Guide for the Beginning Social Scientist.*)

Sample Bem quotes regarding HARKing include the following:

> There are two possible articles you can write: (1) the article you planned to write when you designed your study or (2) *the article that makes the most sense now that you have seen the results* [emphasis added]. They are rarely the same, and the correct answer is (2) . . . the best journal articles are informed by the actual empirical findings from the opening sentence. (pp. 171–172)

Or,

> The data may be strong enough to justify recentering your article around the new findings and subordinating or even ignoring your original hypotheses. . . . If your results suggest a compelling framework for their presentation, adopt it and make the most instructive findings your centerpiece. (p. 173)

Now it's easy to bash Daryl Bem today, but this sort of attitude, approach, or orientation toward ensuring publication through either writing a research report or conducting the research seemed to be generally acceptable a quarter of a century ago. This is better illustrated via an unpublished survey Professor Kerr conducted in 1991 (with S. E. Harris), in which 156 behavioral scientists were asked to estimate the frequency that they suspected HARKing (somewhat broadly defined) was practiced in their discipline. A majority reported the belief that this set of behaviors was actually practiced more frequently than the classic approach to hypothesis testing. Finally, Professor Kerr advanced a number of untoward side effects of HARKing including (in his words)

1. Translating Type I errors into hard-to-eradicate theory;
2. Propounding theories that cannot (pending replication) pass Popper's disconfirmability test;
3. Disguising post hoc explanations as a priori explanations (when the former tend also be more ad hoc, and consequently, less useful);
4. Not communicating valuable information about what did not work;
5. Taking unjustified statistical license;
6. Presenting an inaccurate model of science to students;

> 7. Encouraging "fudging" in other grey areas;
> 8. Making us less receptive to serendipitous findings;
> 9. Encouraging adoption of narrow, context-bound new theory;
> 10. Encouraging retention of too-broad, disconfirmable old theory;
> 11. Inhibiting identification of plausible alternative hypotheses;
>
> And last but definitely not least:
>
> 12. Implicitly violating basic ethical principles. (p. 211).
>
> This absolute gem of an article is reminiscent of Anthony Greenwald's (1975) previously discussed classic. One almost feels as though the social sciences in general (and the reproducibility crisis in specific) are in some sort of bizarre time loop where everything is repeated every few decades with little more than a change of terminology and an escalating number of publishing opportunities.

The Current State of the Preregistration Movement

While Anthony Greenwald and Norbert Kerr are hard acts to follow, someone must carry on the tradition since few practicing researchers read or heed what anyone has written or advocated decades in the past. And certainly Brian Nosek (and his numerous collaborators) is eminently qualified to assume that mantle as witnessed by the establishment of the Open Science Collaboration, which advocates preregistration and provides a multidisciplinary registry of its own, and a number of instructional articles on the topic (e.g., Nosek & Lakens, 2014) as well as the following aptly titled article.

> **The Preregistration Revolution**
>
> Brian Nosek, Charles Ebersole, Alexander DeHaven, and David Mellor (2018)
>
> As just about everyone interested in scientific reproducibility agrees, one of the most effective ways of discouraging scientists from disguising

postdiction as prediction is to require them to publicly register their primary hypotheses and approaches prior to data collection. However, Professor Nosek and his co-authors are quick to emphasize, as did Kerr before them, that both prediction and postdiction are integral parts of science and that "Preregistration does not favor prediction over postdiction; its purpose is to make clear which is which" (p. 2602).

The authors make a clear distinction between generating a hypothesis and then testing it with data versus "hypothesizing after the results are known" (with a well-deserved nod to Professor Kerr). And while the untoward effects of this form of illogical reasoning and faulty empirical practice have already been illustrated in several ways, this particular article provides a more succinct description of the underlying logic behind this most serious and pervasive of artifacts.

> It is an example of circular reasoning––generating a hypothesis based on observing data, and then evaluating the validity of the hypothesis based on the same data. (p. 2600)

The authors then go on to list nine challenges that can be associated with the preregistration process. The first seven will be discussed here because they are seldom (if ever) mentioned in the context of preregistration, although the description here does not do them justice and they are better read in their entirety in the original article.

1. *Changes to procedure during the conduct of the study.* Probably the most common example of this occurs in clinical trials when recruitment turns out to be more difficult than anticipated. To paraphrase a shock trauma investigator I once worked with, "The best prevention for spinal cord injury is to conduct a clinical trial requiring the recruitment of spinal cord injured patients. Diving and motor cycle accidents will inevitably almost completely disappear as soon as study recruitment begins." Which, of course, sometimes unavoidably results in readjusting the study's originally proposed sample size. But this is only one of a multitude of other unanticipated glitches or necessary changes to a protocol that can occur after the study begins. As the authors suggest, transparently reporting what these changes were and the reason that they were necessitated will go a long way toward salvaging a study and making it useful, *unless*

said changes were made after looking at the data (with the exception of the next "challenge").

2. *Discovery of assumption violations during analysis.* This one is best avoided by pilot work, but distribution violations may occur with a larger and/or slightly different sample than was used in the pilot study process. The authors suggest that a "decision tree" approach be specified in the preregistration regarding what analytic steps will be taken if the data do not fit the preregistered analytic plan (e.g., non-normality or missing values on a prespecified covariate). However, some unanticipated problems (e.g., a ceiling effect or basement effect with regard to the outcome variable) can be fatal, so it should be noted that the options presented deal only with the violation of statistical assumptions. By the same token, all experienced investigators have a pretty thorough knowledge of what *could* possibly occur during the course of a study in their fields, hence Lin and Green's (2016) suggestion that common genres of research adopt SOPs which can be copied and pasted into a preregistration document to cover possible discrepancies between published and prespecified analysis plans.

3. *Analyses based upon preexisting data* and (4) *longitudinal studies and large, multivariate databases.* These two research genres involve uses of preregistration that are seldom considered. For example, the utility of blinding is well-established in experimental research but it can also apply to longitudinal databases in the form of generating and registering hypotheses prior to data analysis. When this isn't practical perhaps a reasonable fallback position would be to include those relationships which the investigators have already discovered and reported in a preregistration document while transparently reporting them as such in the published analysis, accompanied by an alpha adjustment of 0.005. (Which isn't as onerous for large databases as it is for experiments.) And, of course, non-hypothesized relationships found to be of interest should also be similarly reported. (These latter suggestions shouldn't be attributed to the Nosek team since they are my opinions.)

5. *Running many experiments at the same time.* Here, the authors described a situation in which a "laboratory acquires data quickly, sometimes running multiple experiments per week. The notion of pre-registering every experiment seems highly burdensome for

their efficient workflow" (p. 2603). Some might find this scenario a bit troublesome since it is highly unlikely that said laboratories publish all of these "multiple experiments per week" hence the trashing of the non-significant ones might constitute a QRP in and of itself. In their defense, the authors suggest that this is normally done "in the context of a methodological paradigm in which each experiment varies some key aspects of a common procedure" (p. 2603). So, in this case, the authors describe how a preregistration can be written for such a program of research as a whole, and any promising findings can then be replicated. However, for experiments conducted in this manner which do not simply vary "some key aspects of a common procedure," one wonders what happens to all of the "negative" findings. Are they never published? At the very least, they should at least be mentioned in the published article and recorded in the laboratory's official workflow.

6. *Conducting a program of research.* This one is a bit like the previous challenge but seems to involve a series of separate full-blown experiments, each of which is preregistered and one eventually turns out to be statistically significant at the 0.05 level. The authors make the important point that such an investigator (reminiscent of our hypothetical graduate student's failed program of research) should report the number of failures preceding his or her success since the latter is more likely to be a chance finding than a stand-alone study. Few investigators would either consider doing this or adjusting the positive p-value based on the number of previous failures. But they probably should and, perhaps in the future, will.

7. *Conducting "discovery" research with no actual hypotheses.* In this scenario, similar in some ways to challenges (3) and (4), researchers freely admit that their research is exploratory and hence may see no need to preregister said studies. The authors conclude that this could be quite reasonable, but it is a process fraught with dangers, one of which is that it is quite possible for scientists to fool themselves and truly believe that an exciting new finding was indeed suspected all along (i.e., "the garden of forking paths"). Preregistration guards against this possibility (or others' suspicions thereof) and possesses a number of other advantages as well—serving as an imperfect genre of workflow for investigators who do not routinely keep one.

> The authors conclude their truly splendid article by suggesting that the preregistration movement appears to be accelerating, as illustrated by the numbers of existing research registries across an impressive number of disciplines and organizations. They also list resources designed to facilitate the process, including online courses and publishing incentives, while warning that the movement still has a long way to go before it is a universal scientific norm.

So why not use this article to draw a red line for reproducibility? I have suggested tolerance for investigators such as Carney and Cuddy because they were simply doing what their colleagues had been doing for decades. Some have even partially excused Daryl Bem by saying that his work met minimum methodological standards for its times, but let's not go that far since there is really no excuse for pathological science.

But surely, at some point, enough becomes enough. Perhaps a year after the publication date of the preceding article (2018) could constitute such a red line. A zero tolerance point, if you will, one beyond which it is no longer necessary to be kindly or politically correct or civil in copious correspondences to offending editors and investigators, nor on posts on social media regarding studies published beyond this point in time that ignore (or its authors are ignorant of) the precepts that have been laid down from Nosek et al.'s 2018 declaration of a "Preregistration Revolution." And this is not to mention the myriad publications that surfaced around 2011–2012 or Anthony Greenwald's 1975 classic.

But isn't this a bit Draconian, given the progress we're making (or is something missing)? There is definitely something missing, even from medicine's inspired edicts listing the registration of clinical trials as a publication *requirement* in its highest impact journals after 2004—and even after this became a legal requirement for some types of clinical trials in 2007, with the law being expanded in 2017.

The first problem, as documented by a number of studies, involves the lack of compliance with preregistration edict. The second reflects the far too common mismatch between the preregistered protocol and what was actually published.

Mathieu, Boutron, Moher, and colleagues (2009) illustrated the necessity for ameliorating both of these problems in a study designed to compare the key elements present in preregistrations with their published counterpart.

Locating 323 cardiology, rheumatology, and gastroenterology trials published in 10 medical journals in 2008, these investigators found that only 147 (46%) had been adequately registered *before the end of the trial* despite the International Committee of Medical Journal Editors 2004 edict. And almost equally disheartening, 46 (31%) of these 147 compliant articles showed a discrepancy in the primary outcome specifications between the registered and published outcomes. And definitely most problematically, 19 (83%) of the 23 studies for which the direction of the discrepancies could be assessed were associated with statistically significant results. These published versus preregistration discrepancies were distributed as follows:

1. The introduction of a new primary outcome in the article ($N = 22$ studies);
2. The registered primary outcome was not even mentioned in the article ($N = 15$, which borders on the unbelievable);
3. The primary outcome morphing into a secondary one (i.e., either a secondary outcome or an unregistered variable becoming the primary outcome in the published article; $N = 8$);
4. A secondary outcome morphing into a primary one ($N = 6$); and/or
5. A discrepancy in the timing of the primary outcome between the two sources ($N = 4$).

(Note that some of the 46 articles had more than one of these QRPs.)

In an interesting coincidence, in the same publication year as this study, Ewart, Lausen, and Millian (2009) performed an analysis of 110 clinical trials and found the same percentage (31%) of primary outcomes changed from preregistration to published article, while fully 70% of the registered secondary outcomes had also been changed.

And, 2 years later, Huić, Marušić, and Marušić (2011) conducted a similar study comparing a set of 152 published RCTs registered in ClinicalTrials.gov with respect to both completeness and substantive discrepancies between the registry entries and the published reports. As would be expected by now, missing fields were found in the preregistrations themselves as well as substantive changes in the primary outcome (17%). Progress from 31% perhaps?

Now granted this is a lot of number parsing, but for those whose eyes have glazed over, suffice it to say that while the ICMJE initiative was a seminally exemplary, long overdue, and bordering on revolutionary policy for a tradition-bound discipline such as medicine and undeniably better than nothing, it was, however, quite disappointing in the compliance it elicited several years after initiation.

Naturally all of the investigators just discussed had suggestions for improving the preregistration process, most of which should sound familiar by now. From Mathieu and colleagues (2009):

> First, the sponsor and principal investigator should ensure that the trial details are registered *before* [emphasis added] enrolling participants.
>
> Second. the comprehensiveness of the registration should be routinely checked by editors and readers, especially regarding the adequate reporting of important items such as the primary outcome.
>
> Third, editors and peer reviewers should systematically check the consistency between the registered protocol and the submitted manuscript to identify any discrepancies and, if necessary, require explanations from the authors, and
>
> Finally, the goal of trial registration could [*should*] be to make available and visible information about the existence and design of any trial and give full access to all trial protocols and the main trial results. (p. 984)

The conclusions for Huić et al., on the other hand, were a combination of good and bad news.

> ICMJE journals published RCTs with proper registration [the good news] but the registration data were often not adequate, underwent substantial changes in the registry over time and differed in registered and published data [the *very* bad news]. Editors need to establish quality control procedures in the journals so that they continue to contribute to the increased transparency of clinical trials. (p. 1)

In summary: While the movement toward preregistration is an inspired and profound boost to scientific reproducibility, it should in no way be voluntary. It should be required in some form or another for every published study—not just in medicine but in all disciplines aspiring to be taken seriously. From a journal's perspective, this process might entail the following steps:

1. Since manuscripts are now submitted online, the first item on the submission form should include a direct link to the *dated* preregistration

document, and the submission process should be summarily terminated if that field is missing. Any major deviations from the preregistration document should be mentioned as part of the submission process and the relevant declaration thereof should be included in the manuscript. (Perhaps as a subtitled section at the end of the methods section.)
2. At least one peer reviewer should be tasked with comparing the preregistration with the final manuscript using *a brief checklist that should also perfectly match the required checklist completed by the author in the preregistration document*. This would include the specification of the primary outcome, sample size justification, identity of the experimental conditions, analytic approach, and inclusion/exclusion criteria.
3. Any unmentioned discrepancies between the authors' rendition and the completed manuscript should either result in a rejection of the manuscript or their inclusion being made a condition of acceptance.
4. Readers should be brought into the process and rewarded with the authorship of a no-charge, published "errata discovery" or "addendum" of some sort since these could constitute fail-safe candidates for both checking and enforcing this policy. (This might eventually become common enough to expel the personal onus editors seem to associate with publishing errata or negative comments.)

Now, of course, these suggestions entail some extra expense from the journals' perspective but academic publishers can definitely afford to hire an extra staff member or, heaven forbid, even pay a peer reviewer an honorarium when tasked with comparing the preregistered protocol with the manuscript he or she is reviewing. (After all, publishers can always fall back on one of their chief strengths, which is to pass on any additional costs to authors and their institutions.)

On a positive note, there is some evidence that compliance with preregistration edicts may have begun to improve in the past decade or so—at least in some disciplines. Kaplan and Irvin (2015), for example, conducted a natural experiment in which large (defined as requiring more than $500,000 in direct costs) National Heart, Lung, and Blood Institute-funded cardiovascular RCTs were compared before and after preregistration was mandated by clinicaltrials.gov.

Unlike preregistration requirements announced by journals or professional organizations, this one has apparently been rigorously enforced since Kaplan and Irvin found that 100% of the located 55 trials published after

2000 were registered as compared to 0% prior thereto. (Note the sharp contrast to other disciplines without this degree of oversight, as witnessed by one disheartening study [Cybulski, Mayo-Wilson, & Grant, 2016] which found that, of 165 health-related *psychological* RCTs published in 2013, only 25 [15%] were preregistered.)

Perhaps equally surprisingly (and equally heartening), the 2015 Kaplan and Irwin cardiovascular study also found a precipitous drop in publication bias, with trials published prior to 2000 reporting "significant benefit for their primary outcome" in 17 of 30 (57%) studies versus 8% (or 2 of 25) after 2000 ($p < 0.0005$). The authors attributed this precipitous drop in positive findings to one key preregistration requirement of the ClinicalTrials.gov initiative.

> Following the implementation of ClinicalTrials.gov, investigators were required to prospectively declare their primary and secondary outcome variables. Prior to 2000, investigators had a greater opportunity to measure a range of variables and to select the most successful outcomes when reporting their results. (p. 8)

Now while such a dramatic drop in positive results may not be particularly good news for sufferers of heart disease, their families, or congressional oversight committees, it constitutes a large step back from the rampant publication bias discussed earlier and hopefully a major step forward for the reproducibility movement for several reasons.

1. In mature disciplines such as cardiovascular research, dramatic new findings *should* decrease over time as the availability of low-hanging fruit decreases.
2. Clinical RCTs are expected to adhere to the CONSORT agreement, and strict adherence thereto precludes the false-positive producing effects of most (if not all) of the QRPs listed in Chapter 3.
3. Well-funded National Institutes of Health (NIH) clinical trials are also associated with adequate sample sizes—hence adequate statistical power—which is not the case in the majority of social science experiments.

And lest it appear that social science experiments are being ignored here with respect to changes from preregistration to publication, the National Science Foundation (NSF)-sponsored Time-sharing Experiments for the

Social Sciences (TESS) provides an extremely rare opportunity for comparing preregistered *results* with published *results* for a specialized genre of social science experiments. The program itself involved embedding "small" unobtrusive interventions (e.g., the addition of a visual stimulus or changes in the wording of questions) into national surveys conducted for other purposes. Franco, Malhotra, and Simonovits (2014) were then able to compare the unpublished results of these experiments with their published counterparts since the NSF required not only the experimental protocols and accruing data to be archive prior to publication, but also the *study* results as well.

The authors were able to locate 32 of these studies that had been subsequently published. They found that (a) 70% of the published studies did not report all the outcome variables included in the protocol and (b) 40% did not report all of the proposed experimental conditions. Now while this could have been rationalized as editorial pressure to shorten the published journal articles, another study by the Franco et al. team discovered a relatively unique wrinkle to add to the huge publication bias and QRP literatures: "Roughly two thirds of the reported tests [were] significant at the 5% level compared to about one quarter of the unreported tests" (p. 10). A similar result was reported for political science studies drawn from the same archive (Franco, Malhotra, & Simonovits, 2017).

And if anyone needs to be reminded that registry requirements alone aren't sufficient, one extant registry has even more teeth than ClinicalTrials.gov. This particular registry is unique in the sense that it potentially controls access to hundreds of billions in profit to powerful corporations. It is also an example of a registry developed by a government agency that closely examines and evaluates all submitted preregistrations before the applicants can proceed with their studies, as well as the results after the trials are completed.

That honor goes to the US Food and Drug Administration (FDA). The FDA requires that positive evidence of efficacy (in the form of randomized placebo or active comparator trials) must be deposited in its registry and that this proposed evidence must be evaluated by its staff before a specific medical condition can be approved for a specific diagnosis.

To fulfill this responsibility, the agency's registry requires a prospective protocol, including the analysis plan, the actual RCT data produced, and the results thereof in support of an application for either marketing approval or a change in a drug's labeling use(s). FDA statisticians and researchers then review this information to decide whether the evidence is strong enough to warrant approval of each marketing application. Such a process, if implemented properly, should preclude the presence of a number of the QRPs described in Chapter 3.

Alas, what the FDA does not review (or regulate) are the *published* results of these trials that wind up in the peer reviewed scientific literature. Nor does it check those results against what is reported in its registry. But what if someone else did?

Selective Publication of Antidepressant Trials and Its Influence on Apparent Efficacy

Erick Turner, Annette Matthews, Eftihia Linardatos, et al. (2008)

The authors of this study took advantage of available information from the FDA approval process to evaluate the possibility of bias in pharmaceutical industry-funded research. The studies under review were all randomized, placebo-controlled evaluations of 12 antidepressant agents approved by the FDA between 1987 and 2004.

Of the 74 FDA-registered studies, 38 were judged by the FDA as being positive and 37 (97%) of these were published. Of the 36 trials that were judged to be *negative* ($n = 24$) or *questionable* ($n = 12$), only 14 (39%) wound up being published.

So far all we have is more unnecessary evidence of publication bias since positive results were almost two and a half times more likely to be published. The next finding, however, is what should be of primary interest to us here.

Of the FDA-judged positive studies published, all (100%) were presented as positive in the final report. (No surprise here since pharmaceutical companies aren't likely to suppress good news.) However of the 12 questionable studies, the 6 that were published all appeared in the published record as *positive* (i.e., a miraculous 50% improvement in effectiveness between the time the FDA reviewed the trial results and the time they appeared in print). Of the 8 published negative trials, five (64%) of the ineffective drugs somehow, magically, became effective as a function of the publication process.

Other, equally disturbing finding, reported were

> The methods reported in 11 journal articles appeared to depart from the pre-specified methods reflected in the FDA reviews.... Although for each of these studies the finding with respect to the protocol-specified primary outcome was non-significant, each publication highlighted a positive results as if it were the primary outcome. [Sound familiar by now?] The non-significant results of the pre-specified primary outcomes were either subordinated to non-primary positive results (in two reports) or omitted (in nine). (p. 255)

And,

> By altering the apparent risk–benefit ratio of drugs, selective publication can lead doctors to make inappropriate prescribing decisions that may not be in the best interest of their patients and, thus, the public health. (p. 259)

Another team conducted a study around the same time period (Rising, Bacchetti, & Bero, 2008) employing different FDA studies and reported similar results—including the propitious changes from registry to journal articles. However, in 2012, Erick Turner (with Knowepflmacher & Shapey) basically repeated his 2008 study employing antipsychotic trials and found similar (but less dramatic) biases—hopefully due to Erick's alerting pharmaceutical companies that someone was watching them, but more likely due to the greater efficacy of antipsychotic drugs. (Or perhaps placebos are simply less effective for patients experiencing psychotic symptoms than in those with depression.)

One final study involving an often overlooked preregistration registry: we don't conceptualize them in this way, but federally mandated institutional review boards (IRBs) and institutional animal care and use committees (IACUCs) are registries that also require protocols containing much of the same basic information required in a preregistration.

The huge advantage of these local regulatory "registries" is that it is illegal for any institution (at least any that receive federal funding) to allow the recruitment of participants (human or animal) for any research purposes without first submitting such a protocol for approval by a committee designated for this purpose. More importantly most of these institutions are quite conscientious in enforcing this requirement since federal research funding may be cut off for violations. So, obviously, such registries would constitute an excellent opportunity for comparing regulatory protocols with their published counterparts, but for some inexplicable reason IRB and IACUC records are considered proprietary.

However, occasionally investigators are provided access to selected IRBs for research purposes. Chan, Hrobjartsson, Haahr, and colleagues (2004), for example, were able to obtain permission from two Danish IRBs to identify 102 experimental protocols submitted between 1994 and 1995 that had subsequently been published. Each proposed protocol was then compared to its published counterpart to identify potential discrepancies in the treatment or the specified primary outcome. The identified changes from application to publication in the pre-specified primary outcomes were that some (a) magically became secondary outcomes, (b) were replaced by a secondary outcomes, (c) disappeared entirely, or (d) regardless of status, the outcomes used in the power calculations required by the IRB protocols differed from those reported in the published articles (which were necessitated by the previous three changes).

Of the 102 trials, 82 specified a primary outcome (it is "puzzling" that 20 did not). Of these, 51 (62%) had made at least one of the four just-mentioned changes. And, not surprisingly, the investigators found that

> The odds of a particular outcome being fully reported were more than twice as high if that outcome was statistically significant. Although the response rate was relatively low, one of the most interesting facets of this study was a survey sent to the studies' authors. Of the 49 responses received, 42 (86%) actually "denied the existence of unreported outcomes despite clear evidence to the contrary." (p. 2457)

So while we don't conceptualize them as such, federally mandated IRBs and IACUCs are actually preregistration repositories for all studies involving human or animal participants that basically contain all of the necessary information required for preregistered protocols. (Sometimes a proposed study is granted exemption from IRB review, but the reasons for such decisions are also on file.)

While these committees differ among institutions from being overly officious to operating with a wink and a nod, there is no reason why all proposals couldn't (a) collect the requisite information on the same standardized checklist form suggested earlier and (b) register that information on either their institutional websites or a national registry. Obviously, such a process would elicit massive hues and cries from everyone—investigators, administrators, and their lobbyists citing both financial and privacy issues—but so what?

After all, shouldn't an IRB that is instituted for research participants' protection also protect those participants from squandering their time and effort on the incessant production of fallacious research results? Also, since no participant identifiers are contained in an IRB application, no privacy concerns can emanate from them except for study investigators. And anonymity is the last thing investigators want since they gladly affix their names to the resulting published articles.

While the following three potential objections may smack of once again raising strawmen, they probably need to be addressed anyway.

1. *Cost*: Admittedly in research-intensive institutions the IRB-IACUC regulatory process would require at least one additional full-time staff member to upload all approved proposals and attached amendments (or at the very least the completed standardized checklist just suggested) to a central registry. (Or perhaps someone could write a program to do so automatically.) Just as obviously, some effort (preferably automated) will be necessary to ensure that none of the required fields is empty. However, research institutions receive very generous indirect costs (often exceeding 50% of the actual research budget itself) so there is adequate funding for an integral research function such as this.
2. *Release date*: The proposal (or possibly simply the minimal information containing hypotheses, primary outcomes, experimental conditions, sample size, study design, and analytic approach), while already uploaded and dated, could be released only upon submission of the final manuscript for publication and only then following the journal's decision to submit it to the peer review process. This release process could be tightened by the principal investigator granting access to the proposal (and its amendments) only to the journal to which it is submitted. However, any such restrictions would be lifted once the manuscript had been published or after a specified period of time. (Both the submitted manuscript and the published article would have to include a direct link to the archived IRB proposal or checklist.)
3. *Amendments to regulatory proposals*: As mentioned, IRBs and IACUCs differ significantly in their degree of oversight and conscientiousness.

My familiarity with IRBs extends only to those representing academic medical centers, which are probably more rigorous than their liberal arts counterparts. However, any serious IRB or IACUC should require dated amendments to a proposal detailing changes in (a) sample size (increased or decreased) along with justifications thereof, (b) experimental conditions (including changes to existing ones, additions, and/or deletions), (c) primary outcomes, and (d) analytic approaches. All such amendments should be attached to the original protocol in the same file, which would certainly "encourage" investigators to include any such changes in their manuscripts submitted for publication since their protocols would be open to professional scrutiny.

The same procedures should be implemented by funding agencies. All federal (and presumably most private) funders require final reports which include the inferential results obtained. While these are seldom compared to the original proposal prior to simply being checked off and filed away by bureaucrats (probably often without even being read), they, too, should be open to public and professional scrutiny.

A final note: preregistration of protocols need not be an onerous or time-consuming process. It could consist of a simple six- or seven-item checklist in which each item requires no more than a one- or two-sentence explication for (a) the primary hypothesis and (b) a justification for the number of participants to be recruited based on the hypothesized effect size, the study design, and the resulting statistical power emanating from them. The briefer and less onerous the information required, the more likely the preregistration process and its checking will be implemented.

Data Sharing

Universal preregistration of study protocols and their comparison to published manuscripts may be the single most effective externally imposed strategy for decreasing the presence of false-positive results. However, the universal requirement for sharing research data isn't far behind.

While data sharing is important in all fields, the main emphasis here will be on data generated in the conduct of discrete experimental or correlational research studies rather than the mammoth databases produced in such fields as genomics, physics, or astronomy which tend to be already archived, reasonably well-documented, and available for analysis. Also excluded are large-scale surveys or datasets prepared by the government or major foundations. (Naturally it is important that all code, selection decisions, variable transformations, and analytic approaches used in any published works be available in archived preregistrations for any typed of empirical study.)

Donoho, Maleki, Shahram, and colleagues (2009) make the argument that "reproducible computational research, in which all details of computations—code and data—are made conveniently available to others, is a necessary response to this credibility crisis" (p. 8). The authors go on to discuss software and other strategies for doing so and debunk excuses for not doing so. Or, as Joelle Pineau observed regarding artificial intelligence (AI) research (Gibney, 2019): "It's easy to imagine why scientific studies of the natural world might be hard to reproduce. But why are some algorithms irreproducible?" (p. 14) Accordingly, at the December 2019 Conference on Neural Information Processing Systems (a major professional AI meeting) the organizing committee asked participants to provide code along with their submitted papers followed by "a competition that challenged researchers to recreate each other's work." As of this writing, the results are not in, but the very attempt augurs well for the continuance and expansion of the reproducibility movement both for relatively new avenues of inquiry and across the myriad classic disciplines that comprise the scientific enterprise. Perhaps efforts such as this provide some hope that the "multiple-study" replication initiatives discussed in Chapter 7 will continue into the future.

At first glance data sharing may seem more like a generous professional gesture than a reproducibility necessity. However, when data are reanalyzed by separate parties as a formal analytic replication (aka *analytic reproduction*), a surprisingly high prevalence of errors favoring positive results occur, as discussed in Chapter 3 under QRP 6 (sloppy statistical analyses and erroneous results in the reporting of p-values).

So, if nothing else, investigators who are required to share their data will be more likely to take steps to ensure their analyses are accurate. Wicherts, Bakker, and Molenaar (2011), for example, found significantly more erroneously reported p-values among investigators who refused to share than those who did.

So, the most likely reasons for those who refuse to share their data are

1. The data were not saved,
2. The data were saved but were insufficiently (or too idiosyncratically or not at all) documented and hence could not be reconstructed by the original investigators or statisticians due to the previously discussed inevitable memory loss occurring with time,
3. The investigators planned to perform and publish other analyses (a process that should have also been preregistered and begun by the time the original article was published, although a reasonable embargo period could be specified therein),
4. Proprietary motives, in which case the study probably shouldn't have been published in the first place, and
5. Several other motives, some of which shouldn't be mentioned in polite scientific discourse.

However, data sharing, unlike preregistration, possesses a few nuances that deserve mention. First of all, the creation of interesting data often requires a great deal of time and effort, so perhaps it is understandable that an investigator resents turning over something so precious to someone else to download and analyze in search of a publication. Perhaps this is one reason that Gary King (1995) suggested that data creation should be recognized by promotion and tenure committees as a significant scientific contribution in and of itself. In addition, as a professional courtesy, the individual who created the data in the first place could be offered an authorship on a publication if he or she provides any additional assistance that warranted this step. But barring that, data creators should definitely be acknowledged and cited in any publication involving their data.

Second, depending on the discipline and the scope of the project, many principal investigators turn their data entry, documentation, and analysis over to someone else who may employ idiosyncratic coding and labeling conventions. However, it is the principal investigator's responsibility to ensure that the data and their labeling are clear and explicit. In addition, code for all analyses, variable transformations, and annotated labels, along with the data themselves, should be included in a downloadable file along with all data cleaning or outlier decisions. All of these tasks are standard operating procedures for any competent empirical study, but, as will be discussed shortly, compliance is far from perfect.

The good news is that the presence of archived data in computational research sharing has increased since the 2011–2012 awakening, undoubtedly facilitated by an increasing tendency for journals to delineate policies to facilitate the practice. However, Houtkoop, Wagenmakers, Chambers, and colleagues (2018) concluded, based on a survey of 600 psychologists, that "despite its potential to accelerate progress in psychological science, public data sharing remains relatively uncommon." They consequently suggest that "strong encouragement from institutions, journals, and funders will be particularly effective in overcoming these barriers, in combination with educational materials that demonstrate where and how data can be shared effectively" (p. 70).

There are, in fact, simply too many important advantages of the data sharing process for it to be ignored. These include

1. It permits exact replication of the statistical analyses of published studies' as well as a resource to pilot new hypotheses; since data on humans and animals are both difficult and expensive to generate, we should glean as much information from them as possible.

And, relatedly,

2. It permits creative secondary analyses (including combining different datasets with common variables) to be performed if identified as such, thereby possibly increasing the utility of archived data.

But for those who are unmoved by altruistic motives, scientific norms are changing and professional requests to share data will continue to increase to the point where, in the very near future, failing to comply with those requests will actually become injurious to a scientist's reputation. Or, failing that, it will at the very least be an increasingly common publication requirement in most respectable empirical journals.

However, requirements and cultural expectations go only so far. As Robert Burns put it, "The best laid schemes o' mice an' men gang aft agley," or, more mundanely translated to scientific practice by an unknown pundit, "even the most excellent of standards and requirements are close to worthless in the absence of strict enforcement".

"Close" but not completely worthless as illustrated by a study conducted by Alsheikh-Ali, Qureshi, Al-Mallah, and Ioannidis (2011), in which the first 10 original research papers of 2009 published in 50 of the highest impact scientific journals (almost all of which were medicine- or life sciences-oriented) were reviewed with respect to their data sharing behaviors. Of the 500 reviewed articles, 351 papers (70%) were subject to a data availability policy of some sort. Of these, 208 (59%) were not completely in compliance with their journal's instructions. However, "*none of the 149 papers not subject to data availability policies made their full primary data publicly available* [emphasis added]" (p. 1).

So to be even moderately effective, the official "requirement for data registration" will not result in compliance unless the archiving of said data (a) precedes publication and (b) is checked by a journal representative (preferably by a statistician) to ensure adequate documentation and transparent code. And this common-sense generalization holds for funding agency requirements as well, as is disturbingly demonstrated by the following teeth-grating study.

A Funder-Imposed Data Publication Requirement Seldom Inspired Data Sharing

Jessica Couture, Rachael Blake, Gavin McDonald, and Colette Ward (2018)

The very specialized funder in this case involved the Exxon Valdez Oil Spill Trustee Council (EVOSTC), which funded 315 projects tasked with collecting ecological and environmental data from the 1989 disaster. The Couture et al. team reported that designated staff attempted to obtain data from the 315 projects (a funding requirement), which in turn resulted in 81 (26%) of the projects complying with some data.... No data at all were received from 74% of the funded entities (23% of whom did not reply to the request and 49% could not be contacted).

Unfortunately, although the EVOSTC reported funding hundreds of projects, "the success of this effort is unknown as the content of this collection has since been lost [although surely not *conveniently*]." But, conspiracy theories aside, the authors do make a case that a recovery rate of 26% is not unheard of, and, while this may sound completely implausible, unfortunately it appears to be supported by at least some empirical evidence, as witnessed by the following studies reporting data availability rates (which in no way constitutes a comprehensive or systematic list).

1. Wollins (1962) reported that a graduate student requested the raw data from 37 studies reported in psychology journals. All but 5 responded, and 11 (30% of the total requests) complied. (Two of the 11 investigators who did comply demanded control of anything published using their data, so 24% might be considered a more practical measure of compliance.)
2. Wicherts, Borsboom, Kats, and Molenaar (2006) received 26% of their requested 249 datasets from 141 articles published in American Psychiatric Association journals.
3. Using a very small sample, Savage and Vickers (2009) requested 10 datasets from articles published in *PLoS Medicine* and *PLoS Clinical Trials* and received only 1 (10%). This even after reminding the original investigators that both journals explicitly required data sharing by all authors.
4. Vines, Albert, Andrew, and colleagues (2014) requested 516 datasets from a very specialized area (morphological plant and animal data analyzed via discriminant analysis) and received a response rate of 19% (101 actual datasets). A unique facet of this study, in addition to its size, was the wide time period (1991 to 2011) in which the studies were published. This allowed the investigators to estimate the odds of a dataset becoming unavailable over time, which turned out to be a disappearance rate of 17% per year.
5. Chang and Li (2015) attempted to replicate 61 papers that did not employ confidential data in "13 well-regarded economics journals using author-provided replication files that include both data and code." They were able to obtain 40 (66%) of the requisite files. However, even with the help of the original authors, the investigators were able to reproduce the results of fewer than half of those obtained. However, data sharing was approximately twice as high for those journals that required it as for those that did not (83% vs. 42%).
6. Stodden, Seiler, and Ma (2018) randomly selected 204 articles in *Science* to evaluate its 2011 data sharing policy, of which 24 provided access information in the published article. Emails were sent to the remaining authors, of which 65 provided some data and/or

> code, resulting in a total of 89 (24 + 65) articles that shared at least some of what was requested. This constituted a 44% retrieval rate, the highest compliance rate of any reviewed here, and (hopefully not coincidentally) it happened to be the most recent article. From these 89 sets of data, the investigators judged 56 papers to be "potentially computationally reproducible," and from this group they randomly selected 22 to actually replicate. All but one appeared to replicate, hence the authors estimated that 26% of the total sample may have been replicable. [Note the somewhat eerie but completely coincidental recurrence of this 26% figure.]

All six sets of authors provided suggestions for improvement, especially around the adequacy of runnable source code. This is underlined in a survey of 100 papers published in *Bioinformatics* (Hothorn & Leisch, 2011), which found that adequate code for simulation studies was "limited," although what is most interesting about this paper is that the first author serves (or served) as the "reproducible research editor" for a biometric journal in which one of his task was to check the code to make sure it ran—a role and process which should be implemented by other journals. (Or perhaps alternately a startup company could offer this role on a per-article fee basis.)

Despite differences in methodologies, however, all of these authors would probably agree with Stodden and her colleagues' conclusion regarding *Science*'s data sharing guidelines (and perhaps other journal-unenforced edicts as well).

> Due to the gaps in compliance and the apparent author confusion regarding the policy, we conclude that, although it is a step in the right direction, this policy is insufficient to fully achieve the goal of computational reproducibility. Instead, we recommend that the journal verify deposit of relevant artifacts as a condition of publication. (p. 2588)

A disappointment, perhaps due to an earlier finding by the Stodden team (2013) which found a 16% increase in data policies occurring between 2011 and 2012 and a striking 30% increase in code policies in a survey of 170 statistically and computationally oriented journals.

But does universal access to data affect reproducibility? No one knows, but, speculatively, it should. If data are properly archived, transparently documented, and complete (including variables not presented in the official publication and all variable transformations), the availability of such data may present a rare roadblock on Robert Park's path (straight, not "forked") from "foolishness to fraud."

How? By making it possible for other scientists or statisticians to ascertain (a) if more appropriate alternate analyses produce the same result and (b) if certain QRPs appeared to be committed. The process would be exponentially more effective if the universal requirement for data archiving could be coupled with the preregistration of study hypotheses and methods. And that may be in the process of occurring.

At the very least, journals should require all empirical research data to be permanently archived. It is not sufficient to "require" investigators to share their data upon request because too many untoward events (e.g., multiple job and computer changes, retirement, and dementia) can occur over time to subvert that process, even for investigators with the most altruistic of motives.

Journals should also not accept a paper until the archived data are checked for completeness and usability. Furthermore, a significant amount of money should be held in escrow by funders until the archived data are also checked. In one or both cases, an independent statistician should be designated to check the data, code, and other relevant aspects of the process as well as personally sign off on the end product of said examination. To facilitate this, the archived code should be written in a commonly employed language (a free system such as R is probably preferable, but software choices could be left up to investigators as long as they are not too esoteric). Everything should also be set up in such a way that all results can be run with a mouse click or two.

While short shrift has admittedly been given to purely computational research and reproducibility, this field is an excellent resource for suggestions concerning computational reproducibility. Sandve, Nekrutenko, Taylor, and Hovig (2013), for example, begin by arguing that ensuring the reproducibility of findings is as much in the self-interest of the original investigators as it is to the interests of others.

> Making reproducibility of your work by peers a realistic possibility sends a strong signal of quality, trustworthiness, and transparency. This could increase the quality and speed of the reviewing process on your work, the chances of your work getting published, and the chances of your work being taken further and cited by other researchers after publication. (p. 3)

252 INCREASING SCIENTIFIC REPRODUCIBILITY

They then offer 10 rules for improving the reproducibility of computational research that are largely applicable to the replication of all empirical research (and especially to the issue of sharing reproducible data). The explanations for these rules are too detailed to present here, and some are applicable only to purely computational efforts, hence, hopefully, I will be forgiven for my rather abrupt abridgments and selections. Hopefully also, the original authors will excuse the verbatim repetition of the rules themselves (all quoted passages were obtained from pp. 2–3), which apply to all types of complex datasets).

Rule 1: For every result, keep track of how it was produced. This basically reduces to maintaining a detailed analytic workflow. Or, in the authors' words: "As a minimum, you should at least record sufficient details on programs, parameters, and manual procedures to allow yourself, in a year or so, to approximately reproduce the results."

Rule 2: Avoid manual data manipulation steps. In other words, use programs and codes to recode and combine variables rather than perform even simple data manipulations *manually*.

Rule 3: Archive the exact versions of all external programs used. Some programs change enough over time to make exact replication almost impossible.

Rule 4: Version control all custom scripts. Quite frankly, this one is beyond my expertise so for those interested, the original authors suggest using a "version control system such as Subversion, Git, or Mercurial." Or, as a minimum, keep a record of the various states the code has taken during its development.

Rule 5: Record all intermediate results, when possible in standardized formats. Among other points, the authors note that "in practice, having easily accessible intermediate results may be of great value. Quickly browsing through intermediate results can reveal discrepancies toward what is assumed, and can in this way uncover bugs or faulty interpretations that are not apparent in the final results."

Rule 7: Always store raw data behind plots. "As a minimum, one should note which data formed the basis of a given plot and how this data could be reconstructed."

Rule 10: Provide public access to scripts, runs, and results.

Of course, while all of these rules may not be completely applicable for a straightforward experiment involving only a few variables, their explication does illustrate that the process of data archiving can be considerably more complicated (and definitely more work intensive) than is generally perceived by anyone who hasn't been involved in the process. Most journals will require only the data mentioned in a published article to be archived, and that may be reasonable. But again, someone at the journal- or investigator-level must at least rerun all of the reported analyses using only the archived information to ensure that the same results are identically reproduced for tables, figures, and text. Otherwise it is quite improbable that said results will be reliably reproducible.

Of course, sharing one's analytic file doesn't guarantee that the final analysis plan wasn't arrived at via multiple analyses designed to locate the one resulting in the most propitious p-value. As a number of statisticians note, and both Steegen, Tuerlinckx, Gelman, and Vanpaemel (2016) and our favorite team of Simonsohn, Simmons, and Nelson (2015) empirically illustrate, different analytic decisions often result in completely different inferential results. And while such decisions are capable of being quite reasonably justified a posteriori, one of our authors has previously reminded us that in a "garden of forking paths, whatever route you take seems predetermined."

Ironically, the Steegen et al. team illustrated this potential for selectively choosing an analytic approach capable of producing a statistically significant p-value by using the study employed by Andrew Gelman to illustrate his garden of forking paths warning. (The study—and a successful self-replication thereof—it will be recalled was conducted by Durante, Rae, and Griskevicius (2013) who "found" that women's fertility status was influenced both their religiosity and political attitudes.) The Steegen team's approach involved

1. Employing the single statistical result reported in the Durante et al. study,
2. Constructing what they termed the "data multiverse," which basically comprised all of the *reasonable* coding and transformation decisions possible (120 possibilities in the first study and 210 in the replication), and then
3. Running all of these analyses and comparing the p-values obtained to those in the published article.

The authors concluded that investigator data processing choices are capable of having a major impact on whether or not significant p-values are obtained in an observational or experimental dataset. For the studies employed in their demonstration the authors suggest that

> One should reserve judgment and acknowledge that the data are not strong enough to draw a conclusion on the effect of fertility. The real conclusion of the multiverse analysis is that there is a gaping hole in theory or in measurement, and that researchers interested in studying the effect of fertility should work hard to *deflate* the multiverse. The multiverse analysis gives useful directions in this regard. (p. 708)

The Simonsohn team (whose work which actually preceded this study) arrived at the same basic conclusions and provided, as is their wont, a statistical approach ("specification-curve analysis") for evaluating the multiple results obtained from these multiple, defensible, analytic approaches.

Both the Steegen et al. (2016) and the Simonsohn et al. (2015) articles demonstrate that different analytic approaches are in some cases capable of producing both statistically significant and non-significant results. And certainly some investigators may well analyze and reanalyze their data in the hope that they will find an approach that gives them a p-value < 0.05—thereby suggesting the need for a new QRP designation to add to our list or simply providing yet another example of p-hacking.

However, the recommendation that investigators analyze and report all of the "reasonable scenarios" (Steegen et al., 2016) or "multiple, defensible, analytic approaches" (Simonsohn et al., 2015) is, in my opinion, probably going a bridge too far. Especially since the first set of authors found an average of 165 possible analyses in a relatively simple study and its replication. So perhaps the group should have stuck with their advice given in their iconic 2011 article which involved (a) the preregistration of study analysis plans and (b) reporting results without the use of covariates.

Materials Sharing

Types of experimental materials vary dramatically in form and portability from discipline to discipline. In the types of psychology experiments replicated via the Open Science Collaboration or the Many Labs initiative these range from questionnaires to graphic-laden scenarios along with explicit (hopefully) standardized instruction to participants and research assistants. These are easily shared via email, Drop Box, or some other internet delivery system, and compliance with requests for such information doesn't appear to be especially problematic, as witnessed by the degree of cooperation afforded to the above-mentioned multistudy replication initiatives.

Other setups in other disciplines can be more idiosyncratic, but it is difficult to imagine that many requests to visit a laboratory to inspect equipment or other apparatuses would be denied to interested investigators. For most bench research, the equipment used is often fairly standard and purchasable from a commercial supplier if the replicating lab doesn't already possess it. For studies employing sophisticated and expensive equipment, such as the functional magnetic resonance imaging (fMRI) studies described earlier, investigators should include all of the relevant information in their published documents or in an accessible archive.

In those instances where cooperation is not provided, the published results should be (and are) viewed with the same suspicion by the scientific community as afforded to unpublished discovery claims. The cold fusion debacle constituted an unusual example of this in the sense that the specifications for the apparatus were apparently shared but not the procedures employed to generate the now infamous irreproducible results.

Material sharing 2.0: Timothy Clark (2017), an experimental biologist, suggests taking the sharing of materials and procedures a step farther in a single-page *Nature* article entitled "Science, Lies and Video-Taped Experiments." Acknowledging the difficulties, he succinctly presents the following analogy: "If extreme athletes can use self-mounted cameras to record their wildest adventures during mountaintop blizzards, scientists have little excuse not to record what goes on in lab and field studies" (p. 139).

Perhaps his suggestion that journals should require such evidence to be registered (and even used in the peer review process) may *currently* be unrealistic. But the process would certainly (a) facilitate replication, (b) serve as an impressive and time-saving teaching strategy, and (c) discourage scientific misconduct. And there is even a journal partly designed to encourage the process (i.e., the *Journal of Visualized Experiments.*)

And Optimistically: A Most Creative Incentive for Preregistration and Data/Material Sharing

At first glance the reinforcement strategy for principled research practice about to be discussed may smack of the simplistic stickers used as reinforcements for preschoolers. However, the Open Science Network should never be underestimated, as witnessed by its advocacy of "badges" awarded to published studies for which the investigator has pledged (a) the availability of data, (b) experimental materials (which, as just mentioned, can obviously vary quite dramatically from discipline to discipline), and/or (c) an archived preregistered protocol.

The potential operative components of this simple reinforcement strategy are that paper badges affixed to a research article may potentially

1. Encourage other scientists to read the accompanying article since its results promise to be more reproducible, citable, and perhaps even more important;
2. Encourage colleagues and scientists interested in conducting secondary analyses of data or performing replications to not only read and cite the article but possibly provide its author(s) with collaborative opportunities. Piwowar, Day, and Fridsma (2007), for example, found that the citation rate for 85 cancer microarray clinical trial publications which shared *usable* research data was significantly higher compared to similar studies which did not do so;
3. Identify the badged authors as principled, careful, modern scientists; and
4. Potentially even increase the likelihood of future acceptances in the journal in which the badged article appears.

Interestingly, an actual evaluation of the badge concept has been conducted (Kidwell, Ljiljana, Lazarević, et al., 2016). Beginning in January of 2014, the high-impact journal *Psychological Science* was somehow cajoled into providing its authors with "the opportunity to signal open data and materials if they qualified for badges that accompanied published articles." (The badges were supplied by the journal's editorial team if the authors so requested and gave some "reasonable" evidence that they were indeed meeting the criteria.)

The evaluation itself consisted of (a) a before-and-after comparison of reported offers for data sharing prior to that date, (b) a control comparison of the percentage of such offers in four other high-impact psychological journals that did not offer badges, and (c) an assessment of the extent to which these offers corresponded to the actual availability of said data.

The results were quite encouraging. In brief,

1. From a baseline of 2.5%, *Psychological Science* reported open data sharing increased to an average of 22.8% of articles by the first half of 2015 (i.e., in slightly over 1 year after the advent of badges);
2. The four comparison journals, on the other hand, while similar at baseline to *Psychological Science*, averaged only 2.1% thereafter (i.e., as compared to 22.8% in *Psychological Science*);
3. With respect to actual availability of usable data, the results were equally (if not more) impressive, with the *Psychological Science* articles that earned badges significantly outperforming the comparison journals. Perhaps a more interesting effect, however, involved a comparison of the availability of *usable* data of *Psychological Science* articles announcing availability with badges versus the *Psychological Science* articles announcing availability but *without* badges. For the 64 *Psychological Science* articles reporting availability of data archived on a website or repository, 46 had requested and been awarded a data sharing badge while 18 had not. Of those with a badge, 100% actually had datasets available, 82.6% of which were complete. For those who announced the availability of their data but did not have a badge, 77.7% ($N = 14$) made their data available but only 38.9% ($N = 7$) of these had complete data. And, finally,
4. The effects of badges on the sharing of materials were in the same direction as data sharing, although not as dramatic.

Now granted, these numbers are relatively small and the evaluation itself was comparative rather than a randomized experiment, but the authors (who transparently noted their study's limitations) were undoubtedly justified in concluding that, "Badges are simple, effective signals to promote open practices and improve preservation of data and materials by using independent repositories" (p. 1).

And that, Dear Readers, concludes the substantive subject matter of this book, although the final chapter will present a few concluding thoughts.

References

Alsheikh-Ali, A. A., Qureshi, W., Al-Mallah, M. H., & Ioannidis, J. P. (2011). Public availability of published research data in high-impact journals. *PLoS ONE, 6*, e24357.

Bem, D. J. (1987). Writing the empirical journal article. In M. Zanna & J. Darley (Eds.), *The compleat academic: A practical guide for the beginning social scientist* (pp. 171–201). Mahwah, NJ: Lawrence Erlbaum Associates.

Chan, A. W., Hrobjartsson, A., Haahr, M. T., et al. (2004). Empirical evidence for selective reporting of outcomes in randomized trials: Comparison of protocols to published articles. *Journal of the American Medical Association, 29*, 2457–2465.

Chang, A. C., & Li, P. (2015). Is economics research replicable? Sixty published papers from thirteen journals say "usually not." Finance and Economics Discussion Series. http://dx.doi.org/10.17016/FEDS.2015.083

Clark, T. D. (2017). Science, lies and video-taped experiments. *Nature, 542*, 139.

Couture, J. L., Blake, R. E., McDonald, G., & Ward, C. L. (2018). A funder-imposed data publication requirement seldom inspired data sharing. *PLoS ONE, 13*, e0199789.

Cybulski, L., Mayo-Wilson, E., & Grant, S. (2016). Improving transparency and reproducibility through registration: The status of intervention trials published in clinical psychology journals. *Journal of Consulting and Clinical Psychology, 84*, 753–767.

De Angelis, C. D., Drazen, J. M., Frizelle, F. A., et al. (2004). Clinical trial registration: A statement from the International Committee of Medical Journal Editors. *New England Journal of Medicine, 351*, 1250–1252.

Donoho, D. L., Maleki, A., Shahram, M., et al. (2009). Reproducibility research in computational harmonic analysis. *Computing in Science & Engineering, 11*, 8–18.

Durante, K., Rae, A., & Griskevicius, V. (2013). The fluctuating female vote: Politics, religion, and the ovulatory cycle. *Psychological Science, 24*, 1007–1016.

Ewart, R., Lausen, H., & Millian, N. (2009). Undisclosed changes in outcomes in randomized controlled trials: An observational study. *Annals of Family Medicine, 7*, 542–546.

Franco, A., Malhotra, N., & Simonovits, G. (2014). Underreporting in psychology experiments: Evidence from a study registry. *Social Psychological and Personality Science, 7*, 8–12.

Franco, A., Malhotra, N., & Simonovits, G. (2017). Underreporting in political science survey experiments: Comparing questionnaires to published results. *Political Analysis, 23*, 306–312.

Gelman, A., & Loken, E. (2014). The statistical crisis in science: Data-dependent analysis—a "garden of forking paths"—explains why many statistically significant comparisons don't hold up. *American Scientist, 102*, 460–465.

Gibney, E. (2019). This AI researcher is trying to ward off a reproducibility crisis. *Nature, 577*, 14.

Greenwald, A. G. (1975). Consequences of prejudice against the null hypothesis. *Psychological Bulletin, 82*, 1–20.

Hothorn, T., & Leisch, F. (2011). Case studies in reproducibility. *Briefings in Bioinformatics, 12*, 288–300.

Houtkoop, B. L., Wagenmakers, E.-J., Chambers, C., et al. (2018). Data sharing in psychology: A survey on barriers and preconditions. *Advances in Methods and Practices in Psychological Science, 1*, 70–85.

Huić, M., Marušić, M., & Marušić, A. (2011). Completeness and changes in registered data and reporting bias of randomized controlled trials in ICMJE journals after trial registration policy. *PLoS ONE, 6,* e25258.

Kaplan, R. M., & Irvin, V. L. (2015). Likelihood of null effects of large NHLBI clinical trials has increased over time. *PLoS ONE, 10,* e0132382.

Kerr, N. L. (1991). HARKing: Hypothesizing after the results are known. *Personality and Social Psychology Review, 2,* 196–217.

Kerr, N. L., & Harris, S. E. (1998). *HARKing-hypothesizing after the results are known: Views from three disciplines.* Unpublished manuscript. Michigan State University, East Lansing (not obtained).

Kidwell, M. C., Ljiljana B. Lazarević, L. B., et al. (2016). Badges to acknowledge open practices: A simple, low-cost, effective method for increasing transparency. *PLoS Biology, 14,* e1002456.

King, G. (1995). Replication, replication. *PS: Political Science and Politics, 28,* 443–499.

Lin, W., & Green, D. P. (2016). Standard operating procedures: A safety net for pre-analysis plans. *Political Science and Politics, 49,* 495–500.

Mathieu, S., Boutron, I., Moher, D., et al. (2009). Comparison of registered and published primary outcomes in randomized controlled trials. *Journal of the American Medical Association, 302,* 977–984.

Mills, J. L. (1993). Data torturing. *New England Journal of Medicine, 329,* 1196–1199.

Nosek, B. A., Ebersole, C. R., DeHaven, A. C., & Mellor, D. T. (2018). The preregistration revolution. *Proceedings of the National Academy of Sciences, 115,* 2600–2606.

Nosek, B. A., & Lakens, D. (2014). Registered reports: A method to increase the credibility of published results. *Social Psychology, 45,* 137–141.

Piwowar, H. A., Day, R. S., & Fridsma, D. B. (2007). Sharing detailed research data is associated with increased citation rate. *PLoS ONE, 2,* e308.

Rising, K., Bacchetti, P., & Bero, L. (2008). Reporting bias in drug trials submitted to the Food and Drug Administration: Review of publication and presentation. *PLoS Medicine, 5,* e217.

Savage, C. J., & Vickers, A. J. (2009). Empirical study of data sharing by authors publishing in PLoS journals. *PLoS ONE, 4,* e7078.

Sandve, G. K., Nekrutenko, A., Taylor, J., & Hovig, E. (2013). Ten simple rules for reproducible computational research. *PLoS Computational Biology, 9,* e1003285.

Simmons, J. P., Nelson, L. D., & Simonsohn, U. (2011). False-positive psychology: undisclosed flexibility in data collection and analysis allows presenting anything as significant. *Psychological Science, 22,* 1359–1366.

Simonsohn, U., Simmons, J. P., & Nelson, L. D. (2015). Specification curve: Descriptive and inferential statistics on all reasonable specifications. Manuscript available at http://ssrn.com/abstract=2694998

Steegen, S., Tuerlinckx, F., Gelman, A., & Vanpaemel, W. (2016). Increasing transparency through a multiverse analysis. *Perspectives on Psychological Science 11,* 702–712.

Stodden, V., Guo, P., & Ma, Z. (2013). Toward reproducible computational research: An empirical analysis of data and code policy adoption by journals. *PLoS ONE, 8,* e67111.

Stodden, V., Seiler, J., & Ma, Z. (2018). An empirical analysis of journal policy effectiveness for computational reproducibility. *Proceedings of the National Academy of Sciences, 115,* 2584–2589.

Turner, E. H., Knowepflmacher, D., & Shapey, L. (2012). Publication bias in antipsychotic trials: An analysis of efficacy comparing the published literature to the US Food and Drug Administration database. *PLoS Medicine, 9,* e1001189.

Turner, E. H., Matthews, A. M., Linardatos, E., et al. (2008). Selective publication of antidepressant trials and its influence on apparent efficacy. *New England Journal of Medicine, 358,* 252–260.

Vines, T. H., Albert, A. Y. K., Andrew, R. L., et al. (2014). The availability of research data declines rapidly with article age. *Current Biology, 24,* 94–97.

Wicherts, J. M., Bakker, M., & Molenaar, D. (2011). Willingness to share research data is related to the strength of the evidence and the quality of reporting of statistical results. *PLoS ONE, 6,* e26828.

Wicherts, J. M., Borsboom, D., Kats, J., & Molenaar, D. (2006). The poor availability of psychological research data for reanalysis. *American Psychologist, 61,* 726–728.

Wollins, L. (1962). Responsibility for raw data. *American Psychologist, 17,* 657–658.

11
A (Very) Few Concluding Thoughts

So where does the reproducibility initiative go from here? Certainly no one knows, although the easy answer is for reproducibility methodologists and advocates to continue their past and current efforts and continue the process of mentoring the next generation of researchers in their image, as well as including reproducibility concepts in all of their teaching activities (formal and informal).

Unfortunately, a precept in one of the several disciplines in which I have worked (preventive and health-seeking behaviors) was that "knowledge alone is inadequate to ensure the adoption of salutary behaviors and the avoidance of harmful ones."

However, this precept was not meant to imply that health education is worthless—only that the resulting knowledge is an insufficient condition for the adoption of such behaviors. So, in somewhat of a generalization leap, why shouldn't this precept also apply to increasing the reproducibility of scientific findings?

Educational Interventions

In the present context, the preventive behaviors in most dire need of adoption involve the responsible conduct of reproducible research. Hence, if nothing else, all students in all disciplines need to be taught the key importance of recognizing and avoiding questionable research practices (QRPs) along with the basics (and importance) of the replication process.

There are now many excellent resources available to anyone charged with this type of instruction. One resides in accessible articles on the topic which haven't been discussed previously such as those by Asendorpf, Conner, De Fruyt, and colleagues (2013) and by Munafò, Nosek, Bishop, and colleagues (2017)—both of which provide excellent overviews of the reproducibility process and suggestions for teaching its precepts to students and implementing them in practice.

With respect to the educational process, Munafò et al. suggest (in addition to a formal course, which is more common in the social sciences than their life and physical counterparts) that

> The most effective solutions [for both students and faculty] may be to develop educational resources that are accessible, easy-to-digest... web-based modules for specific topics, and combinations of modules that are customized for particular research applications). A modular approach simplifies the process of iterative updating of those materials. Demonstration software and hands-on examples may also make the lessons and implications particularly tangible to researchers at any career stage . . . [such as] the Experimental Design Assistant (https://eda.nc3rs.org.uk) supports research design for whole animal experiments, while *P*-hacker (http://shinyapps. org/apps/p-hacker/) shows just how easy it is to generate apparently statistically significant findings by exploiting analytic flexibility. (p. 2)

While both the Asendorpf et al. (2013) and the Munafò et al. (2017) articles are too comprehensive to abstract here and their behavioral dicta have been discussed previously, each deserves to be read in its entirety. However, in buttressing the argument that methodological and statistical resources should be sought after by (and available to) all *investigators*, the latter recommends a model instituted by the CHDI Foundation (which specializes in research on Huntington's disease). Here, a committee of independent statisticians and methodologists are available to offer "a number of services, including (but not limited to) provision of expert assistance in developing protocols and statistical analysis plans, and evaluation of prepared study protocols" (p. 4).

Of course, students can and should be brought into such a process as well. In the conduct of science, few would argue that hands-on experience is one of the, if the not the most, effective ways to learn how to *do* science. Hence the Munafò et al. paper describes a resource designed to facilitate this process in psychology available under the Open Science framework umbrella called the Collaborative Replications and Education Project (https://osf.io/wfc6u/), in which

> A coordinating team identifies recently published research that could be replicated in the context of a semester-long undergraduate course on research methods. A central commons provides the materials and guidance

to incorporate the replications into projects or classes, and the data collected across sites are aggregated into manuscripts for publication. (p. 2)

The Asendorpf paper also strongly advocates student replication projects and suggests what a reproducibility curriculum should emphasize, such as (a) the methodological basics of both single and multiple experimentation, (b) transparency, (c) best research practices (primarily the avoidance of QRPs), and (d) the introduction of students to online resources for preregistration of protocols and registries for archiving data. (A number of other investigators also advocate student replication projects; see Frank & Saxe, 2012; Grahe, Reifman, Hermann, et al., 2012; Jekel, Fiedler, Torras, et al., 2019—the latter summarizing how a student replication initiative [the *Hagen Cumulative Science Project*] can be practically implemented based on its track record, which, as of this writing, has produced 80+ student replications.)

In addition, regardless of discipline, all science students (undergraduates through postdocs) should be exposed to the Consolidated Standards of Reporting Trials (CONSORT; http://www.consort-statement.org/), along with one or more of its relevant 20+ specialized extensions (including one on social science experimentation). As mentioned previously, CONSORT is the gold standard of publishing guidelines, but a plethora of others exist for almost all types of research such as animal studies (ARRIVE), observational studies (STROBE), diagnostic studies (STARD), quality improvement efforts (SQUIRE), systematic reviews and meta-analyses (PRISMA), and so forth. All are accessible through EQUATOR (Enhancing the Quality and Transparency of Health Research: http://www.equator-network.org/), and, while most target health-related inquiry, a bit of translation makes them relevant to all empirical disciplines employing human or animal participants. Students should also be directed to the exemplary, extremely informative, huge (and therefore somewhat intimidating) Open Science Framework website (https://osf.io/), as well as the National Institutes of Health (NIH) Rigor and Reproducibility initiative (https://www.nih.gov/research-training/rigor-reproducibility).

Most importantly of all, however, mentorships should be formalized and supplemented because sometimes some of the most "successful" (i.e., sought after) research mentors in an institution may also be the most adept at obtaining $p < 0.05$ driven irreproducible results. For, as Andrew Gelman (2018) succinctly (as always) reminds us,

The big problem in science is not cheaters or opportunists, but sincere researchers who have unfortunately been trained to think that every statistically "significant" result is notable.

And a big problem with un-training these "sincere" researchers involves how dearly we all hold on to something we have learned and internalized. So the best educational option is probably to keep the sanctity of obtaining p-values at all costs from being learned in the first place. If not by one-on-one mentoring, at least by providing graduate students, postdocs, and even advanced undergraduates with both conventional educational opportunities as well as online tutorials.

So What Is the Future of Reproducibility?

Burdened by the constantly regretted decision to employ Yogi Berra as a virtual mentor, I cannot in good faith answer this question directly. Instead, a pair of alternative scientific scenarios will be posited to allow others far wiser than I to choose the more likely of the two.

In one future, the reproducibility initiative will turn out to be little more than a publishing opportunity for methodologically oriented scientists—soon replaced by something else and forgotten by most—thereby allowing it to be reprised a few decades later under a different name by different academics.

In this future, publication bias will remain rampant and journals will continue to proliferate, as will the false-positive results they publish. After acknowledging the importance of reproducibility, most scientists will soon grow bored with its repetitively strident, time-consuming precepts and simply redouble their efforts in search of new publishing opportunities.

No harm done and few if any benefits achieved—at least as far as the social sciences are concerned, since they are cyclical rather than cumulative in nature anyway. And as for psychology, well, perhaps Paul Meehl's 1990 description of his discipline decades ago will continue to be prescient for future decades as well.

> Theories in the "soft areas" of psychology have a tendency to go through periods of initial enthusiasm leading to large amounts of empirical

investigation with ambiguous over-all results. This period of infatuation is followed by various kinds of amendment and the proliferation of ad hoc hypotheses. Finally, in the long run, experimenters lose interest rather than deliberately discard a theory as clearly falsified. (p. 196)

In the other alternative future, the avoidance of QRPs will be added to the professional curricula to reduce the prevalence of false-positive results in all the major disciplines. Practicing researchers will become enculturated into avoiding problematic practices and rewarded for engaging in salutary ones—partly through peer pressure, partly because they have always wanted to advance their science by producing credible findings (but were stymied by conflicting cultural/economic forces), partly because committing QRPs has become increasing difficult due to publishing requirements and increased professional scrutiny, and partly because teaching is part of their job description and they have come to realize that there is nothing more important for their students and mentees to learn than the basic principles of conducting reproducible science.

But for those investigators who refuse or are unable to change? They'll gradually fade away and be replaced by others for whom acculturation into the reproducibility paradigm won't be an issue because it has already occurred. And the most conspicuous result of this evolution will be scientific literatures no longer dominated by positive results but replaced by less scintillating, negative studies suggesting more fruitful lines of inquiry.

So Which Alternative Future Is More Likely?

If Las Vegas took bets on the two alternative futures, the smart money would undoubtedly favor the first. After all, conducting reproducible research is monetarily more expensive and entails extra effort and time. Thirty percent more time as estimated by one mathematical biologist—who nevertheless suggests that achieving this valued (and valuable) scientific commodity is not insurmountable.

> Reproducibility is like brushing your teeth. . . . It is good for you [and] . . . once you learn it, it becomes a habit. (Irakli Loladze as quoted by Monya Baker, 2016, p. 454)

While 30% more time and effort is probably an overestimate, there will also surely be other reproducibility edicts and strategies suggested in future that will add time and effort to the research process. Some of these have already been proposed, such as "multiverse analyses," and these will most likely fall by the wayside since they require too much time and effort and actually violate some well-established practices such as employing only the most defensible and discipline-accepted procedures.

Undoubtedly, if the reproducibility initiative persists, new facilitative roles and the expansion of existing ones will also be adopted. And almost certainly, if the prevalence of QRPs and misconduct persists (or increases), some of these will gain traction such as (a) required use of institutionally centralized drives for storing all published data *and* supporting documentation (hopefully including workflows) or (b) institutions requiring the perusal of papers by an independent scientist prior to submission, with an eye toward spotting abnormalities. Existing examples of the latter include Catherine Winchester's (2018) "relatively new reproducibility role" at the Cancer Research UK Beatson Institute and some institutions' use of outside firms to conduct reproducibility screening in response to one or more egregiously fraudulent incidents (Abbott, 2019).

Perhaps even the increased awareness that a cadre of scientists is searching, finding, and promulgating published examples of irreproducibility results may encourage their colleagues to avoid the deleterious effects of QRPs. Meta-researchers, even with the considerable limitations of their approach, are already beginning to play a growing role in both increasing the awareness of substandard methodologies and tracking reproducibility progress over time.

And progress is being made. Iqbal, Wallach, Khoury, and colleagues (2016), for example, in analyzing a random sample of 441 biomedical journal articles published from 2000 to 2014 found a small but positive trend in the reporting of a number of reproducibility and transparency behaviors over this time interval. However, as the authors' noted, the continuance of such studies plays an important role in tracking the effects of the reproducibility initiative over time.

> By continuing to monitor these indicators in the future, it is possible to track any evidence of improvement in the design, conduct, analysis, funding, and independence of biomedical research over time. (p. 9)

Similarly, a more recent meta-scientific tracking study (Menke, Roelandse, Ozyurt, et al., 2020) similarly found small but positive gains from 1997 to either 2016 or 2019 involving six methodological indicators (e.g., randomization, blinding, power analysis) and six indicators related to the provision of sufficient information on the biological materials employed and that are essential for replication purposes (e.g., antibodies and cell lines). Unfortunately, the results were not especially impressive for several of these key indicators.

1. Blinding was mentioned in 2.9% of articles in 1997 and in only 8.6% of articles in 2016;
2. Any mention of a power analysis rose from a jaw-breaking 2.2% in 1997 to 9.9% in 2016;
3. Description/provision of the actual organisms employed (e.g., mice, human cancer cells) improved from 21.1% to 22.0% over the 10-year span; and
4. The identity of the cell lines witnessed an increase of less than 3% (36.8% to 39.3%).

Still, improvement is improvement, and all 12 behaviors are important from a reproducibility perspective: methodological rigor indicators because their presence dramatically impacts the prevalence of false-positive results; key biological materials because they are often a necessary condition for replication in this line of research.

This particular study is of special interest for two other reasons. First, it employs a QRP identification instrument (SciScore) that could be employed with certain modifications by other researchers in other disciplines. Second, the data generated could serve as an adjunct to education, best-practice guideline standards (e.g., CONSORT and ARRIVE), and publishing mandates as well as an evaluation tool thereof. The authors also provide an interesting case study of this genre of evaluation (as did the Kidwell et al. (2016) badges study discussed in Chapter 10).

It is worth repeating that the evidence presented so far suggests that education, professional guidelines, and unenforced mandatory best-practice publishing edicts, while better than nothing, are far from the most important drivers of reproducibility. That distinction belongs to journal editors and peer reviewers who are willing to enforce reproducibility behaviors, preferably coupled with a growing number of online resources to facilitate compliance.

The Open Science Framework website is an impressive example of the latter; in addition, as discussed by Menke et al., one of the most useful resources for bench researchers is undoubtedly the Resource Identification Portal (RRID) which facilitates the reporting RRID identifiers in all published research employing in vivo resources. These identifiers are essential for the replication of many if not most such findings, and the RRID portal allows interested scientists to ascertain the specific commercial or other sources for said resources associated with their identifiers—and thus provides the capacity to obtain them.

Using antibodies as an example, Menke and colleagues found that, by 2019, 14 of the 15 journals with the highest identification rates participated in the RRID initiative. The average antibody identification rate for this 14-journal cohort was 91.7%, as compared to 43.3% of the 682 journals that had published at least 11 antibody-containing articles. Not proof-positive of the causal effects of the RRID initiative, but the fact is that the journal impact factor of the 682 journals was unrelated to this indicator (numerically the correlation was negative) and certainly buttressed by the 2016 decree by the editor of *Cell* (Marcus et al., 2016)—one of the most cited journals in all of science and definitely the most cited and prestigious journal in its area.

The decree itself was part of an innovation consisting of the Structured, Transparent, Accessible Reporting (STAR) system, which not only required the implementation of the RRID initiative but also required that the information be provided in a mandated structured Key Resources Table along with standardized section headings that "follow guidelines from the NIH Rigor and Reproducibility Initiative and [are] aligned with the ARRIVE guidelines on animal experimentation and the Center for Open Science's Guidelines for Transparency and Openness Promotion (https://cos.io/top/)" (Marcus et al., 1059).

Not surprisingly the compliance rate for reporting the requisite antibody information in the eight *Cell* journals was even higher than the other seven journals (93.6% vs. 91.7%) included in Menke et al.'s previously mentioned top 15 journals (see table 5 of the original article). If other journal editors in other disciplines were this conscientious, there is little question which of the two posited alternative futures would result from the reproducibility initiative as a whole.

Of Course, There Are Other Alternative Futures

Perhaps it is equally likely that neither of the preceding alternative futures will occur. Instead, it may be that something in between will be realized or some completely unforeseen paradigmatic sea change will manifest itself. But, as always, in deference to my virtual mentor, I will defer here.

In all fairness, however, I cannot place the blame for my reticence solely upon my mentor for the simple reason that the reproducibility field as a whole is in the process of changing so rapidly (and involves so many empirical disciplines with unique challenges and strategies to meet those challenges) that no single book could cover them all in any detail. So the story presented here is by necessity incomplete, and its ending cannot be told for an indeterminate period of time.

However, I have no hesitation in proclaiming that the reproducibility initiative represents an unquestionable *present-day success story* to be celebrated regardless of what the future holds. And we all currently owe a significant debt to the dedicated investigators, methodologists, and statisticians chronicled here. Their contributions to improving the quality and veracity of scientific inquiry over the past decade or so deserve a place of honor in the history of science itself.

References

Abbott, A. (2019). The integrity inspectors. *Nature, 575*, 430–433.

Asendorpf, J. B., Conner, M., De Fruyt, F., et al. (2013). Recommendations for increasing replicability in psychology. *European Journal of Personality, 27*, 108–119.

Baker, M. (2016). Is there a reproducibility crisis? *Nature, 533*, 452–454.

Frank, M. C., & Saxe, R. (2012). Teaching replication. *Perspectives in Psychological Science, 7*, 600–604.

Gelman, A. (2018). The experiments are fascinating. But nobody can repeat them. *Sciences Times.* https://www.nytimes.com/2018/11/19/science/science-research-fraud-reproducibility.html

Grahe, J. E., Reifman, A., Hermann, A. D., et al. (2012). Harnessing the undiscovered resource of student research projects. *Perspectives in Psychological Science, 7*, 605–607.

Iqbal, S. A., Wallach, J. D., Khoury, M. J., et al. (2016). Reproducible research practices and transparency across biomedical literature. *PLoS Biology*, e1002333.

Jekel, M., Fiedler, S., Torras, R. A., et al. (2019). How to teach open science principles in the undergraduate curriculum: The Hagen Cumulative Science Project. http://www.marc-jekel.de/publication/teaching_hagen/

Kidwell, M. C., Ljiljana B., Lazarević, L. B., et al. (2016). Badges to acknowledge open practices: A simple, low-cost, effective method for increasing transparency. *PloS Biology, 14*, e1002456.

Marcus, E., for the Cell team. (2016). A STAR is born. *Cell, 166*, 1059–1060.

Meehl, P. E. (1990). Appraising and amending theories: The strategy of Lakatosian defense and two principles that warrant using it. *Psychological Inquiry, 1*, 108–141.

Menke, J., Roelandse, M., Ozyurt, B., et al. (2020). Rigor and Transparency Index, a new metric of quality for assessing biological and medical science methods. bioRxiv http://doi.org/dkg6;2020

Munafò, M. R., Nosek, B. A., Bishop, D. V. M., et al. (2017). A manifesto for reproducible science. *Nature Human Behavior, 1*, 1–9.

Winchester, C. (2018). Give every paper a read for reproducibility. *Nature, 557*, 281.

Index

For the benefit of digital users, indexed terms that span two pages (e.g., 52–53) may, on occasion, appear on only one of those pages.

Tables and figures are indicated by *t* and *f* following the page number

abstracts, 22, 198
academic disciplines, 73
academic respectability, 114
Academy of Management, 67
access to data, 252
acculturation, 72
acupuncture, vii
addiction research, 24
advertising, scholarship, 245, 246
aesthetics, 100
affective perspective, 133–34
agricultural sciences
    number of scientific publications per year, 195–96, 195*t*
    positive publishing, 20
Alberts, Bruce, 153–54
allergy trials, 26
The All Results Journals, 28
alpha level, 39, 41, 52, 61–63, 102–3
alternative medicine, vii, 24, 25
American Association for the Advancement of Science, 113
American Psychiatric Association, 249
American Psychological Association, 67
Amgen, 153–54, 156*t*, 169
analysis
    data-dependent, 91–92
    meta-, 117, 263
    multiverse, 253, 266
    power, 77, 267
    power/sample size, 87
    secondary, 101–2
    specification-curve, 254
    statistical, 76–77, 212–13
    z-curve, 213
analytic replications, 135, 245

anesthesia research
    non-random sampling in, 213–14
    publication bias in, 24
animal magnetism, 112
Animal Research: Reporting of in Vivo Experiments (ARRIVE), 86–87, 224, 263
animal studies
    blinding in, 87
    post hoc deletion of participants, 78–79
    publication bias in, 23
    publishing guidelines for, 263
    QRPs and, 86–88
antidepressant trials, 24, 240–41
a posteriori hypothesizing, 59
archives
    preprint, 30, 205
    rules for improving, 252
ARRIVE (Animal Research: Reporting of in Vivo Experiments), 86–87, 224, 263
artificial intelligence (AI) aids, 198
artificial intelligence (AI) research, 245
arXiv preprint repository, 30, 205
as is peer review, 207–8
authorship
    confirmation of, 223
    purpose of, 210
author's responsibilities, 215
automated plagiarism checks, 198

badges, 256–57
Baker, Monya, 265
Banks, George, 67–70
Bar-Anan, Yoav, 198, 203–7, 211
Bargh, John, 173–78, 184, 199

Bausell, Barker, 119
Bausell, R. B., 9–10
Bayer Health Care, 154–55, 156*t*, 169
Bazell, Robert, 125
Begley, Glenn, 153–54, 155, 169–70
behavioral planning, 173–78
behavioral priming, 174–76
behavioral sciences, 2
bell-shaped, normal curve, 40*f*, 40
Bem, Daryl, 62, 112–18, 126, 138, 141, 180, 228–29, 234
Benjamin, Daniel, 93–94
bias
   confirmation, 80–81, 100–1
   implicit, 91–92
   publication (*see* publication bias)
   systematic, 41–42, 41*t*
Bioinformatics, 250
biological sciences
   number of scientific publications per year, 195–96, 195*t*
   publication bias in, 20, 24, 25
   replication initiatives, 155, 156*t*
Biological Technologies Office (DARPA), 170
biomedical research, 266–67
bioRxiv, 30
Blake, Rachael, 248–50
blinding, 83–84, 109, 267
   in animal studies, 87
   in fMRI studies, 99
   follow-up experiments, 175–76
   methodological improvements, 175
Bohannon, John, 198–99
Boldt, Joachim, 213–14
bootstrapping, 212
Bourner, Philip, 217–18
Box, George, 43–44
Boyle, Robert, 15
brain volume abnormalities, 24
Brigham Young University, 120
British Journal of Psychology, 114–15
British Medical Journal, 83–84
Bruns, Stephan, 181
Burger, Jerry, 142
Burns, Robert, 7, 247
Burrows, Lara, 174

Burt, Cyril, 81
business, 20
Butler, Declan, 201–2

Cal Tech, 120–21
Camerer, Colin, 165–66
Campbell, Donald T., 3, 4, 5
cancer research
   preclinical, 158
   publication bias in, 23, 24, 25
   replication initiatives, 155, 156*t*
Cancer Research UK Beatson Institute, 266
cardiovascular research, 238
Carlisle, John B., 213–14, 215
Carney, Dana, 180, 181–82
Carroll, Lewis, 113–14
case control studies, 101–2
CDC (Centers for Disease Control and Prevention), 103–4
Cell, 268
Center for Open Science (COS), 155, 206, 268. *See also* Open Science Framework (OSF)
Centers for Disease Control and Prevention (CDC), 103–4
Chabris, Christopher, 93–94
Chambers, Chris, 31, 140
CHDI Foundation, 261–64
checklists, 87–88
chemistry, 127, 195–96, 195*t*
Chen, Mark, 174
child health, 24
China, vii, 24, 195, 195*t*,
chrysalis effect, 67–70
Claerbout, Jon, 245, 246
clairvoyance, 110
Clark, Timothy, 225, 255
class sizes, small, 104–6
Cleeremans, Axel, 174–76
Clever Hans (horse), 110
clinical research
   misconduct in, 57
   preregistration of trials, 226–28
   publication bias in, 20, 25
   published versus preregistration discrepancies, 235

RCTs (*see* randomized controlled trials [RCTs])
  sample sizes, 238
clinicaltrials.com, 227
ClinicalTrials.gov, 238
close (direct) replications, 135–38
  advantages of, 139–40
  combined with extension (aka conceptual) replication, 144–45, 145*f*
  hypothetical examples, 143–47
Cochrane Database of Systematic Reviews, 26
Cockburn, Iain, 157–58
code availability, 250, 251
cognitive science, 24, 160
Cohen, Jacob, 4
cold fusion, 80–81, 109–10, 119–22, 120*f*, 255
  lessons learned, 123–25
Collaborative Replications and Education Project (OSF), 136–37, 156*t*, 159–61, 165–66, 227, 254–55, 262–63
communication
  digital, 204
  online journalism, 177, 182
  scientific (*see* scientific journals)
complementary and alternative medicine, 25
computational research, 245
  number of scientific publications per year, 195–96, 195*t*
  rules for improving reproducibility, 252–53
conceptual (aka differentiated, systematic) replications, 138–40, 143–47, 145*f*
conference abstracts, 22
conference presentations, 30
confirmation bias, 80–81, 100–1
conflicts of interest, 25–26, 48
Consolidated Standards of Reporting Trials (CONSORT), 47, 83, 224, 263
continuous, open peer review, 207
control procedures, 75, 77–78, 82, 109
correlation coefficients
  high, in fMRI studies, 95–99
  maximum possible, 95–96
  voodoo correlations, 100, 101
Cortex, 31
COS (Center for Open Science), 155, 206, 268., *See also* Open Science Framework (OSF)
costs, 157–58, 243
Couture, Jessica, 248–50
COVID-19, vii
craniology, 80–81
credit, due, 59
Crick, Francis, 85
criminology trials, 26
crowdsourcing, 164
Cuddy, Amy, 178–83
culture, 6–7
current publication model, 196–97

damage control, 173
  case study 1, 173–78
  case study 2, 178–83
  exemplary, 184
DARPA (US Defense Advanced Research Projects Agency), 170
data
  alteration of, 68–69
  availability of, 248–50
  fabrication of, 57, 213–14
  missing, 79
  registration requirements for, 248
  rules for improving manipulation of, 252
  selective reporting of, 59
data analysis, 60–66, 254
databases
  large, multivariate, 232
  secondary analyses of, 101–2
data cleaning teams, 103
Data Coda, 181
data collection
  deletion or addition of data, 68
  missing data, 79
  rules for, 64
  undisclosed flexibility in, 60–66
data-dependent analysis, 91–92
data dredging, 81

data mining, 21, 44–45
data multiverse, 253
data sharing, 244–57
  advantages of, 247
  funder-imposed data publication requirements and, 248–50
  guidelines for, 250
  incentive for, 256–57
  most likely reasons for those who refuse, 246
  suggestions for improvement, 250, 251
data storage, 252, 266
data torturing, 81
decision tree approach, 232
DeHaven, Alexander, 230–34
Delorme, Arnaud, 118
design standards, 82
diagnostic studies, 263
Dickersin, Kay, 15
differentiated, systematic (conceptual) replications, 138–40
digital communication, 204, 211–12
Directory of Open Access Journals, 202
direct (aka close) replications, 135–38
  advantages of, 139–40
  combined with extension (aka conceptual) replication, 144–45, 145f
  hypothetical examples, 143–47
Discover Magazine, 177
discovery research, 233
disinterest, 4
documentation, 266
  of laboratory workflows, 218–19
  personal records, 219
Dominus, Susan, 179, 181–82, 183
Doyen, Stéphane, 174–76
Dreber, Anna, 165–66
drug addiction research, 24
due credit, 59

Ebersole, Charles, 183–84, 185–86, 230–34
economics
  data availability, 249
  experimental, 156t, 165–66
  publication bias in, 20, 24
  of reproducibility in preclinical research, 157–58

editors, 29, 215–16
educational campaigns, 28
educational interventions, 261–64
educational research
  20th-century parable, 16–18
  publication bias in, 24
effective irreproducibility, 156–57
effect size, 33, 41, 160
  prediction of, 40
  and true research findings, 46
ego depletion effect, 138
elife, 155
Elsevier, 196–97, 199
English-language publications, 25
engrXiv, 30
epidemiology
  irreproducibility in, 101–7
  limits of, 102
  publication bias in, 24, 25
epistemology, 134
EQUATOR (Enhancing the Quality and Transparency of Health Research), 263
errors, 209
  opportunities for spotting, 211–12
  statistical tools for finding, 212–14
  Type I (*see* false-positive results)
ESP (extrasensory perception), 110, 118
ethical concerns, 28–29
ethos of science, 4–5
European Prospective Investigation into Cancer, 76
European Union, 195, 195t
EVOSTC (Exxon Valdez Oil Spill Trustee Council), 248–49
expectancy effects, 75
experimental conditions, 65
Experimental Design Assistant, 262
experimental design standards, 82
experimental economics, 165–66
experimental procedures, 77–78, 82
external validity, 3–6, 139
extrasensory perception (ESP), 110, 118
Exxon Valdez Oil Spill Trustee Council (EVOSTC), 248–49

Facebook, 182
false-negative results, 39, 41–42, 41t

false-positive psychology, 60–66
false-positive results
  arguments against, 49
  definition of, 39
  detection of, 28
  epidemics of, 53
  genetic associations with general intelligence, 93–94
  modeling, 39, 41t, 50t
  probabilities of, 41–42, 41t
  reason for, 43–48
  simulations, 61
  statistical constructs that contribute to, 39–40
Fanelli, Daniele, 20, 56, 128
Fat Studies, 199–200
feedback, 208
Ferguson, Cat, 200
Fetterman and Sassenberg survey (2015), 185–86
financial conflicts of interest, 25–26, 48
findings
  irreproducible (*see* irreproducible findings)
  negative, 18–19 (*see also* negative results)
  outrageous, 126–27
  true, 46
fiscal priorities, 72–73
Fisher, Ronald, 3
fishing, 81
Fleischmann, Martin, 119, 120–21
flexibility in data collection and analysis, undisclosed, 60–66
Food and Drug Administration (FDA), 152, 239–40
footnotes, 223
Forsell, Eskil, 165–66
Framingham Heart Study, 101–2
fraud, 26, 81, 125–26, 199–201
Freedman, Leonard, 157–58
Frey, Bruno, 207–8
Fujii, Yoshitaka, 81, 213–14
functional magnetic resonance imaging (fMRI) studies, 99–101, 255
  high correlations in, 95–99
  publication bias in, 24
funding, 30, 74, 97, 248–50

Fung, Kaiser, 181–82
future directions, 215–20, 264–69

Galak, Jeff, 115–17, 180
gaming, 33, 77, 199–200
Gardner, Howard, 92
gastroenterology research, 24
Gelman, Andrew, 91–92, 181–82, 183, 228, 263–64
Gender, Place, & Culture: A Feminist Geography Journal, 199–200
gene polymorphisms, 44–45
General Electric, 122
genetic association studies
  false-positive results, 43–48, 93–94
  with general intelligence, 93–94
genetic epidemiology, 24
Georgia Tech, 120–21
geosciences
  number of scientific publications per year, 195–96, 195t
  positive publishing, 20
Giner-Sorolla, Roger, 100
Gonzalez-Mule E., Erik, 67–70
Gould, Jay, 92
Greenwald, Anthony, 10, 52–54, 230
GRIM test, 213
Guidelines for Transparency and Openness Promotion (COS), 268

Hadfield, Jarrod, 186–87
Hagen Cumulative Science Project, 263
Hall, R. N., 109
hard sciences, 128
HARKing (hypothesizing after the results are known), 106, 228–30
Harris, Christine R., 49, 95–99
Harris, Richard, 147
Harris, S. E., 229
Hartshorne, Joshua, 165
Harwell Laboratory, 120–21
health sciences, 2
heavy water, 119
herbal medicine, vii
Herbert, Benjamin, 93–94
Hill, Austin Bradford, 3
Holzmeister, Felix, 166
homeopathy, 113

human studies, 78–79
Hume, David, 124–25
hydroxychloroquine, vii
hyping results, 85
hypotheses
　changing, 80
　post hoc dropping or adding of, 69
　QRPs regarding, 80
　after results are known
　　(HARKing), 106, 228–30
　reversing or reframing, 69
hypothesis tests
　alteration of data after, 68–69
　deletion or addition of data after, 68

IACUCs (institutional animal care and use
　committees), 22, 241–44
IBM SPSS, 212–13
ICMJE (International Committee of
　Medical Journal Editors), 226, 236
IISPs (inane institutional scientific
　policies), 71–74
immunology, 20
implicit bias, 91–92
inane institutional scientific policies
　(IISPs), 71–74
INA-Rxiv repository, 206
Incredibility-Index, 141
independent replications, 143–47
independent scientists, 266
information sharing, 220., *See also* data
　sharing
institutional animal care and use
　committees (IACUCs), 22, 241–44
institutionalization, 73
institutional review boards (IRBs), 22, 241–44
institutions
　fiscal priorities, 72–73
　inane institutional scientific policies
　　(IISPs), 71–74
　requirements for promotion, tenure, or
　　salary increases, 74
insufficient variance: test of, 213
intelligence-genetic associations, 93–94
internal validity, 3–6
International Clinical Trials Registry
　Platform (WHO), 227

International Committee of Medical
　Journal Editors (ICMJE), 226, 236
investigators, 29., *See also* scientists
　acculturation of, 72
　disciplinary and methodological
　　knowledge of, 72
　educational interventions for, 261–64
　how to encourage replications
　　by, 147–48
　mentoring of, 72, 263
　reputation of, 183–84
Ioannidis, John P.A., 10, 43–48, 95, 181,
　186–87, 210
IRBs (institutional review
　boards), 22, 241–44
irreproducible findings *See also*
　reproducibility crisis
　approaches for identifying, 131
　behavioral causes of, 7
　case studies, 91
　costs of, 157–58
　damage control, 173
　effectively irreproducible, 156–57
　publication and, 18–21
　QRP-driven, 91
　scientific, 18–21
　strategies for lowering, 222
　warnings, 10

James, William, 180
Johnson, Gretchen, 104–6
Jones, Stephen, 120
journalism, online, 177, 182
*Journal of Articles in Support of the Null
　Hypothesis*, 28
*Journal of Experimental & Clinical Assisted
　Reproduction*, 199
*Journal of Experimental Psychology:
　Learning, Memory, and
　Cognition*, 159
*Journal of International Medical
　Research*, 199
Journal of Natural Pharmaceuticals, 199
*Journal of Negative Observations in Genetic
　Oncology*, 28
*Journal of Personality and Social
　Psychology*, 52, 112, 114–15, 159

Journal of Pharmaceutical Negative Results, 28
Journal of the American Medical Association (JAMA), 25, 47, 213–14, 226
Journal of Visualized Experiments, 255
journals
    backmatter volume lists, 223
    control procedures, 236
    current publication model, 196–97
    devoted to nonsignificant results, 28
    editors, 29, 215–16
    letters to the editor, 211–12
    medical, 226
    open-access, 198–99
    peer-reviewed, 193, 203–7
    predatory (fake), 198–99, 201–2
    published versus preregistration discrepancies, 234–35
    requirements for publication in, 81, 216–20, 226
    scientific, 193
    Utopia I recommendations for, 203–7, 210

Kamb, Sasha, 153–54
Karr, Alan, 103
Kerr, Norbert, 106, 228–30
King, Gary, 211, 246
King, Stephen, 134
Klein, Olivier, 174–76
Kobe University, 199
Krawczyk, Michal, 76–77

laboratory workflows, 218–19
Lakens, Daniël, 31
Langmuir, Irving, 109, 110–12, 118, 119, 123
Laplace, Pierre-Simon, 124–25
large, multivariate databases, 232
LeBoeuf, Robyn A., 115–17
letters to the editor, 211–12
life sciences research, 128, 195–96, 195*t*
Linardatos, Eftihia, 240–41
Lind, James, 193
Loken, Eric, 91–92, 228

Loladze, Irakli, 265
longitudinal studies, 101–2, 232

magnetic resonance imaging (MRI) studies *See* functional magnetic resonance imaging (fMRI) studies
magnetism, animal, 112
management studies, 212–13
Many Labs replications, 136–37, 156*t*, 165–66, 254–55
    Many Labs 1, 156*t*, 161–62, 177
    Many Labs 2, 156*t*, 162
    Many Labs 3, 156*t*, 162–63
    methodologies, 163–65
Marcus, Adam, 200, 209
Martinson, Brian, 210
materials: sharing, 254–55
    incentive for, 256–57
    suggestions for, 255
materials science, 20
Matthews, Annette, 240–41
McDonald, Gavin, 248–50
medical research
    misconduct, 57
    non-random sampling in, 213–14
    number of scientific publications per year, 195–96, 195*t*
    preclinical studies, 152–58, 156*t*
    prerequisites for publication, 226
    publication bias in, 24
    replication initiatives, 152–58, 156*t*
MedRxiv, 30
Meehl, Paul, 264
Mellor, David, 230–34
mentorships, 72, 263
Merton, Robert, 4–5
Mesmer, Franz, 112
meta-analyses, 117, 263
MetaArXiv, 30
metacognitive myopia, 100
meta-research, 21, 266
    publication bias in, 25, 26
    survival of conclusions derived from, 32
meta-science, 21
microbiology, 20
Milgram, Stanley, 142

Mills, James, 228
misconduct, 26
　prevalence of, 57
　statistical tools for finding, 212–14
missing data, 79
MIT, 120–22
modeling false-positive results
　advantages of, 49
　examples, 42
　Ioannidis exercise, 43–48
　nontechnical overview, 39, 41*t*
modeling questionable research practice effects, 60–66
molecular biology-genetics, 20
morality, 129, 178–83
mortality, 104–6
MRI (magnetic resonance imaging) studies *See* functional magnetic resonance imaging (fMRI) studies
Muennig, Peter, 104–6
multicenter trials, 104–6
multiexperiment studies, 112–13, 232–33
multiple sites, 164
multiple-study replications, 152, 156*t*, 168*t*
multivariate databases, large, 232
multiverse analyses, 253, 266

National Death Index, 104–6
National Heart, Lung, and Blood Institute (NHLBI), 237
National Institute of Mental Health (NIMH), 97
National Institutes of Health (NIH), 74, 83, 238
　PubMed Central, 204
　Resource Identification Portal (RRID), 268
　Rigor and Reproducibility initiative, 263
National Science Foundation (NSF)
　Science and Technology Indicators, 194–95
　Time-sharing Experiments for the Social Sciences (TESS), 238–39
Nature, 76–77, 87, 97, 128, 166, 200, 213–14, 255

Nature Human Behavior, 52
Nature Neuroscience, 97
negative results, 7, 18–19
　conference presentations, 30
　false-negative results, 39, 41–42, 41*t*
　publishing, 18, 30
　reasons for not publishing, 19
　unacceptable (but perhaps understandable) reasons for not attempting to publish, 19
Negative Results in Biomedicine, 28
Nelson, Leif D., 60–66, 115–17, 141–42, 180, 210
NeuroImage, 97
neuroimaging, 24
neuroscience-behavior, 20
New England Journal of Medicine (NEJM), 25, 47, 213–14, 215, 223, 226
New Negatives in Plant Science, 28
New York Times Magazine, 179, 181–82
NIH *See* National Institutes of Health
non-English publications, 25
non-random sampling, 213–14
nonsignificant results, 85
normal, bell-shaped curve, 40*f*, 40
Nosek, Brian, 31, 140, 159, 177, 183–84, 185–86, 198, 202–7, 211, 228–29, 230–34
not reporting details or results, 59
NSF *See* National Science Foundation
nuclear fusion, 119
null hypothesis
　prejudice against, 52–54
　probability level deemed most appropriate for rejecting, 53
　statistical power deemed satisfactory for accepting, 53
Nutrition, Nurses' Health Study, 76

obesity research, 24
O'Boyle, Ernest, Jr., 67–70
observational studies
　large-scale, 103
　publication bias in, 25
　publishing guidelines for, 263
　rules for reporting, 65
online science journalism, 177, 182
open-access publishing, 198–99, 204

Open Access Scholarly Publishers Association, 202
open peer review, continuous, 207
Open Science Framework (OSF), 134, 149, 230, 256, 268
    cancer biology initiative, 169–70
    Collaborative Replications and Education Project, 136–37, 156*t*, 159–61, 165–66, 227, 254–55, 262–63
    methodologies, 163–65
    publishing guidelines, 263
OPERA (Oscillation Project Emulsion-t Racking Apparatus), 127
operationalization, 138
Oransky, Ivan, 200, 209
orthodontics, 24
Oscillation Project Emulsion-t Racking Apparatus (OPERA), 127
OSF *See* Open Science Framework
outcome variables, 75

palladium, 119
paradigmatic shift, 2
Parapsychological Association, 118
parapsychology, 112, 117–18
Park, Robert (Bob), 10, 99–100, 101, 119, 125–26
parsimony principle, 9–10
partial replications, 142–43
Pashler, Harold, 49, 95–99
pathological science, 109
    criteria for, 123–25
    examples, 111–12
    lessons learned, 123–25
Payton, Antony, 93
p-curves, 181, 212
pediatric research, 24
PeerJ, 30
peer review, 28, 29–30, 197–208
    as is, 207–8
    continuous, open, 207
    dark side of, 198–99
    fake, 200
    fake articles that get through, 198–200
    flawed systems, 72
    fraudulent, 200–1
    guidelines for, 65–66
    independent, 206
    publishing, 207
    publishing prior to, 205–6
    shortcomings, 198–99
peer review aids, 87–88, 198
peer-reviewed journals, 193
    problems bedeviling, 203
    Utopia I recommendations for, 203–7, 210
peer reviewers, 211, 216
personality studies, 95–99
personal records, 219
Perspective on Psychological Science, 137, 148–49
P-hacker, 262
p-hacking, 80
pharmaceutical research
    misconduct in, 57
    publication bias in, 20, 24, 26, 240–41
physical sciences, 20, 25, 128
physics
    cold fusion, 80–81, 109–10, 119–22, 120*f*, 255
    number of scientific publications per year, 195–96, 195*t*
    positive publishing, 20
    replications, 127
Pichon, Clora-Lise, 174–76
pilot studies, 79, 218
Pineau, Joelle, 245
plagiarism checks, automated, 198
plant and animal research
    data availability, 249
    positive publishing, 20
PLoS Clinical Trials, 249
PLoS Medicine, 249
PLoS ONE, 114–15, 177, 199
PLoS ONE's Positively Negative Collection, 28, 87
political behavior, 24
Pom, Stanley, 119, 120–22
Popham, James (Jim), 143
positive publishing, 20., *See also* publication bias
positive results, 7, 20–21, 22–23, *See also* publication bias
    false-positive results (*see* false-positive results)
postdiction, 228–29
post hoc analysis, 127
power, statistical, 33, 39–40, 41
power analysis, 77, 87, 267

Preclinical Reproducibility and Robustness Gateway (Amgen), 28, 153–54, 156t, 169
preclinical research
　approaches to replication, 156–57
　economics of reproducibility in, 157–58
　publication bias in, 24
　replication initiatives, 152–58, 156t
predatory (fake) journals, 201–2
　open-access journals, 198–99
　strategies for identifying, 201–2
predatory publishers, 210
prediction markets, 165–66
prediction principle, 10
prejudice against null hypothesis, 52–54
preprint archives, 30, 205
preprint repositories, 30, 205–6
preregistration
　advantages of, 241–42
　of analysis plans, 103
　benefits of, 238
　challenges associated with, 231–33
　checklist process, 244
　of clinical trials, 226–28
　control procedures, 236
　current state, 230–44
　functions of, 228
　incentive for, 256–57
　initiation of, 226–44
　via IRBs and IACUCs, 241–44
　and publication bias, 25, 238
　purposes of, 225–26
　Registered Reports, 31
　of replications, 133
　requirements for, 222, 227, 236–37, 241
　revolution, 230–34
　in social sciences, 228
　of study protocols, 82, 117–18, 165–66, 225–26
　suggestions for improving, 236–37, 244
preregistration repositories, 241–44
Presence (Cuddy), 179
press relations, 30
prevention trials, 26
probability
　mean level most appropriate for rejecting null hypothesis, 53
　of possibly study outcomes, 41–42, 41t

professional associations, 81
Project STAR, 105
promotions, 74
pseudoscientific professions, 73
psi, 112–14, 115–17, 126
PsyArXiv, 30
PsychDisclosure initiative, 87–88
Psychological Science, 87–88, 114–15, 159, 256–57
psychology, 112–18
　cognitive, 160
　data availability, 249
　experimental, 159–65
　false-positive results, 49, 52, 60–66
　number of scientific publications per year, 195–96, 195t
　publication bias, 20, 23, 24, 264–65
　questionable research practices (QRPs), 57
　Registered Reports, 31
　replication failures, 115–17
　replication initiatives, 156t
　reproducibility, 49, 159–61
　significance rates, 20
　social, 160
publication bias, 11, 15, 53, 56
　definition of, 15
　documentation of, 21–26
　effects of, 18
　evidence for, 26–27
　factors associated with, 25–26
　future, 264
　implications, 27
　inane institutional scientific policy (IISP), 71–72
　preregistration and, 25, 238
　steps to reduce untoward effects of, 30
　strategies helpful to reduce, 28–29
　systematic review of, 26
　topic areas affected, 23–24
　vaccine against, 31–33
　what's to be done about, 28–30
Public Library of Science (PLoS), 199, 204
public opinion, 30
public relations
　case study 1, 173–78
　case study 2, 178–83

damage control, 173
exemplary, 184
publishers, 217–18
publishing, 2, 188
  20th-century parable, 16–18
  author's responsibilities, 215
  chrysalis effect in, 67–70
  current model, 196–97
  editors, 29, 215–16
  funding agency requirements
    for, 248–50
  guidelines for, 263
  initiatives for improving, 7
  limits on, 210
  number of scientific publications per
    year, 194–96, 195t
  open-access, 204
  peer reviewers' responsibilities, 216
  positive, 20 (*see also* publication bias)
  prerequisites for journal
    publication, 226
  prior to peer review, 205–6
  as reinforcement, 193–94
  and reproducibility, 18–21, 193
  requirements for journal
    articles, 81, 216–20
  retractions, 199–200, 209
  scope of, 194–96
  selective publication, 240–41
  statistical tools for finding errors and
    misconduct in, 212–14
  suggestions for opening scientific
    communication, 203–7
  Utopia I recommendations
    for, 203–7, 210
  value of, 211–12
  vision for, 216–20
  word limits, 210
publishing negative results, 30
  benefits of, 18
  dedicated space to, 28
  reasons for not publishing, 19
  in traditional journals, 28
  unacceptable (but perhaps
    understandable) reasons for not
    attempting to publish, 19
publishing nonsignificant results
  educational campaigns for, 28
  journals devoted to, 28

publishing peer reviews, 207
publishing positive results *See also*
    publication bias
  prevalence of, 20–21, 22–23
"publish or perish" adage, 193
PubMed Central (NIH), 204, 209
pulmonary and allergy trials, 26
p-values, 33, 42, 160, 181
  under 0.05, 77
  adjusted, 76
  definition of, 39
  inflated, 43
  QRPs regarding, 61, 76–77
  rounding down, 76–77
  R-program (statcheck) for
    recalculating, 212

QRPs *See* questionable research practices
quality improvement efforts, 263
questionable research practices (QRPs)
  case studies, 91
  and effect size, 40
  effects of, 56
  irreproducible results from, 91
  modeling, 60–66
  multiple, 81–82
  observance of, 59
  opportunities for spotting, 211–12
  partial list of, 74–85
  prevalence of, 58t, 56–60
  simulations, 61–63
  z-curve analysis for, 213

randomisation, 87
randomized controlled trials
    (RCTs), 46–47, 238
  control procedures, 236
  guidelines for, 83
  multicenter, 104–6
  non-random sampling in, 213–14
  publication bias in, 24, 25
  small class sizes and mortality
    in, 104–6
  survival of conclusions from, 32–33
recordkeeping
  laboratory workflow, 218–19
  personal records, 219
registered replication reports
    (RRRs), 148–50

Registered Reports, 31–33
registration *See also* preregistration
  of data, 248
  registry requirements, 227, 239–40
regulatory proposals, 243–44
reliability, 161
replicability crisis *See* reproducibility crisis
replication failure, 109–10
  case study 1, 173–78
  case study 2, 178–83
  damage control, 173
  effects of, 185
  exemplary response to, 184
  rates of, 169
  reasons for, 134
  and reputation, 183–84, 185
  strategies for speeding healing after, 186–87
replication studies
  analytic, 135
  conceptual (aka differentiated, systematic), 138–40, 143–47, 145f
  design of, 133
  direct (aka close), 135–38, 139–40, 143–47, 145f
  Ebersole, Axt, and Nosek (2016) survey, 183–84, 185–86
  exact, 135
  exemplary response to, 184
  extensions, 141–42, 143–47, 145f
  Fetterman and Sassenberg survey (2015), 185–86
  how to encourage, 147–48
  hypothetical examples, 143–47
  independent, 143–47
  multiple-site, 164
  multiple-study initiatives, 152, 156t, 168t
  need for, 126–27
  of outrageous findings, 126–27
  partial, 142–43
  pathological science with, 109
  preregistered, 133
  process, 7, 133
  recommendations for, 134
  Registered Reports, 31, 148–50
  requirements for, 133–34
  results, 168, 168t
  self-replications, 143–47
  survey approach to tracking, 165
reporting not quite QRP practices, 85
repository(-ies)
  preprint, 30, 205–6
  preregistration, 241–44
reproducibility, 5–6, 265., *See also* irreproducible findings
  arguments for ensuring, 251
  author's responsibilities for ensuring, 215
  causes of, 157–58
  economics of, 157–58
  editor's responsibilities for ensuring, 215–16
  educational interventions for increasing, 261–64
  estimation of, 159–61
  future directions, 264–65
  peer reviewers' responsibilities for ensuring, 216
  preclinical, 157–58
  of psychological science, 159–61
  publishing issues and, 193
  rules for improving, 252–53
  screening for, 266
  strategies for increasing, 191
  value of, 6
  vision to improve, 216–20
  z-curve analysis for, 213
reproducibility crisis, vii, 1, 261
  arguments against, 49
  background and facilitators, 13
  strategies for decreasing, 191
reproduction *See also* replication studies
  analytic, 245
reproductive medicine, 24
reputation
  Ebersole, Axt, and Nosek (2016) survey, 183–84, 185–86
  Fetterman and Sassenberg survey (2015), 185–86
  "publish or perish" adage and, 193
  replication failure and, 183–84, 185
research *See also specific disciplines*
  20th-century parable, 16–18
  chrysalis effect in, 67–70
  design standards, 82

discovery, 233
educational interventions for, 261–64
funding, 30, 74, 97
glitches and weaknesses in, 75
longitudinal studies, 101–2, 232
misconduct, 57
multiexperiment studies, 112–13
negative studies, 18–19
not quite QRPs but definitely irritating, 85
pathological, 109
post hoc deletion of participants or animals, 78–79
preregistration of (*see* preregistration)
programs, 233
protocols, 117–18, 225–26
questionable practices (*see* questionable research practices [QRPs])
replication (*see* replication studies)
standards, 81
statistical tools for finding errors and misconduct in, 212–14
study limitations, 85
suggestions for improvement, 64–66
research findings
irreproducible (*see* irreproducible findings)
negative, 18–19 (*see also* negative results)
outrageous, 126–27
true, 46
research publishing *See* publishing
research results *See* results
Resource Identification Portal (RRID), 268
respectability, academic, 114
results
false-negative, 39, 41–42, 41*t*
false-positive (*see* false-positive results)
hyping, 85
irreproducible (*see* irreproducible findings)
negative (*see* negative results)
nonsignificant, 85
positive, 7, 20–21, 22–23 (*see also* publication bias)
reproducibility of (*see* reproducibility)
retractions of, 209, 213–14
rules for improving, 252
selective reporting of, 75–76
spinning, 85
retractions, 209, 213–14
Retraction Watch, 200, 209
Rhine, Joseph Banks, 110–11
Rosenthal, Robert, 29
R-program (statcheck), 212–13
RRID (Resource Identification Portal), 268
RRRs (registered replication reports), 148–50

Sagan, Carl, 124–25
Sage Publications, 196–97, 199, 200
Saitoh, Yuhji, 213–14
salary increases, 74
sample size, 46, 87, 238
sampling, 100, 213–14
SAS, 212–13
Sato, Yoshihiro, 213–14
Schachner, Adena, 165
Schimmack, Ulrich, 141–42, 213
Schlitz, Marilyn, 118
ScholarOne, 198, 212
scholarship advertising, 245, 246
Schooler, Jonathan, 137
science, 1, 7.–8, *See also specific fields*
ethos of, 4–5
hierarchy of, 128
pathological, 109
snake oil, 119, 121–22
voodoo, 10, 100, 101, 119
Science, 97, 114–15, 159, 166, 198–99, 249–50
Science and Technology Indicators (NSF), 194–95
science journalism, 177, 182
scientific journals, 114, 193., *See also specific journals*
current publication model, 196–97
library subscription rates, 196–97
number of publications per year, 194–96, 195*t*
Utopia I recommendations for, 203–7, 210
scientific publishing *See* publishing
scientific results *See* results

scientists *See also* investigators
  independent, 266
  peer reviewers, 211
  "publish or perish" adage and, 193
  reputation of, 183–84
  word limits, 210
SCIgen, 199–200
SciScore, 267
scurvy, 193
secondary analyses, 101–2
self-replications, 143–47
sensationalism, 97
sharing data, 244–57
sharing information, 220
sharing materials, 254–55
  incentive for, 256–57
  suggestions for, 255
significance, statistical, 33
Simcoe, Timothy, 157–58
Simmons, Joseph P., 60–66, 115–17, 141–42, 180, 183, 210
Simonsohn, Uri, 60–66, 141–42, 180, 183, 210
simulations, 212, 250
single-nucleotide polymorphisms (SNPs), 44–45, 93
skepticism, 4
Slate Magazine, 116, 181–82
snake oil science, 119, 121–22
SNPs (single-nucleotide polymorphisms), 44–45, 93
SocArXiv, 30
Social Cognitive and Affective Neuroscience, 97
social media, 182
Social Neuroscience, 97
social psychology, 160
social sciences, 2, 127
  fMRI studies, 95–99
  general studies, 166–67
  number of scientific publications per year, 195–96, 195*t*
  positive publishing, 20
  preregistration, 228
  publication bias, 24, 25
  registered studies, 227
  replication initiatives, 156*t*
Social Text, 199–200
soft sciences, 128

Sokal, Alan D., 199–200
specification-curve analysis, 254
spinning results, 85
Springer, 200
SSRN, 30
staff supervision, 81
standards, 81
Stanley, Julian C., 3, 4, 5
Stapel, Diederik, 213–14
STAR (Structured, Transparent, Accessible Reporting) system, 268
Stata, 212–13
statcheck (R-program), 212
statistical analysis, 76–77, 212–13
statistical power, 33, 39–40, 41
  definition of, 39–40
  QRPs regarding, 77
  satisfactory for null hypothesis, 53
statistical significance, 33, 39
  artifactual, 76
  comparisons that don't hold up, 91–92
  definition of, 39
  example, 42
  QRPs regarding, 77–78, 79–80
statistical tools, 212–14
StatReviewer, 198, 212
Steinbach, John, 7
Sterling, T. D., 21–22
stroke research, 24, 26
Structured, Transparent, Accessible Reporting (STAR) system, 268
subjective timing, 176
supervision, staff, 81
surveys
  comparisons between, 59
  for tracking replications, 165
systematic, differentiated (conceptual) replications, 138–40
systematic bias, 41–42, 41*t*
systematic reviews, 263

Taubes, Gary, 121–22, 125–26
Ted Talks, 179
telepathy, 110
Tennessee Class-Size study, 104–6
tenure, 74
TESS (Time-sharing Experiments for the Social Sciences), 238–39
Texas A&M, 120–21, 122

Thun, Michael, 102–3
Time-sharing Experiments for the Social Sciences (TESS), 238–39
timing, subjective, 176
traditional Chinese medicine, vii
traditional journals, 28
trends, 48
tritium, 122
true-positive results, 44
Trump, Donald, vii
truth: half-life of, 32
Tsang, Eric, 207–8
Tumor Biology, 200
Turner, Erick, 240–41
Twitter, 182
Type I error *See* false-positive results

Uhlmann, Eric, 167
United States
  number of scientific publications per year, 195, 195t
  publication bias in, 25
universalism, 4
universal requirements, 222
University of Utah, 119, 120
US Defense Advanced Research Projects Agency (DARPA), 170
Utah, 109–10

validity
  external, 3–6, 139
  internal, 3–6
variables, 69, 75
variance, insufficient, 213
Vaux, David, 76–77

Vees, Matthew, 184, 186
verbal overshadowing, 137
version control, 252
Vickers, Andrew, vii
vision for publishing, 216–20
volumetric pixels (voxels), 98
voodoo science, 10, 100, 101, 119
voxels (volumetric pixels), 98
Vul, Edward, 95–99

Wall Street Journal, 121
Ward, Colette, 248–50
Watson, James, 85
White Queen, 113
Wilde, Elizabeth, 104–6
William of Occam, 9–10, 113, 124, 139–40
Winchester, Catherine, 266
Winkielman, Piotr, 95–99
Wiseman, Ritchie, 114–15
Wolters Kluwer, 196–97, 199
Wonder Woman, 181
word limits, 210
workflows, 218–19, 266
World Health Organization (WHO), 227

Xi Jinping, vii

Yale University, 121–22
Yogi of Bronx, 10
Yong, Ed, 114–15, 140, 177
Young, Stanley, 103
YouTube, 179

z-curve analysis, 213

CELEBRATING
50 YEARS

Texas A&M University Press
publishing since 1974

NORSEMEN DEEP IN THE HEART OF TEXAS

◊ ◊ ◊

Number Thirty-one
Tarleton State University Southwestern Studies in the Humanities
DEBORAH M. LILES, GENERAL EDITOR

Norsemen Deep in the Heart of Texas
NORWEGIAN IMMIGRANTS, 1845–1900

◊ ◊ ◊

Gunnar Nerheim

Texas A&M University Press
COLLEGE STATION

COPYRIGHT © 2024 BY GUNNAR NERHEIM
All rights reserved
First edition

∞ This paper meets the requirements of ANSI/NISO Z39.48-1992 (Permanence of Paper). Binding materials have been chosen for durability.

LIBRARY OF CONGRESS CATALOGING-IN-PUBLICATION DATA

Names: Nerheim, Gunnar, 1949– author.
Title: Norsemen deep in the heart of Texas : Norwegian immigrants, 1845–1900 / Gunnar Nerheim.
Other titles: Norwegian immigrants, 1845–1900 | Tarleton State University southwestern studies in the humanities ; no. 31.
Description: First edition. | College Station : Texas A&M University Press, [2024] | Series: Tarleton State University southwestern studies in the humanities ; number thirty-one | Includes bibliographical references and index.
Identifiers: LCCN 2023040293 | ISBN 9781648430220 (hardcover) | ISBN 9781648430879 (ebook)
Subjects: LCSH: Norwegians—Texas—History—19th century. | Norwegian Americans—Texas—History—19th century. | Norwegian Americans—Texas—Bosque County—History—19th century. | Texas—Emigration and immigration—History—19th century. | BISAC: HISTORY / United States / State & Local / Southwest (AZ, NM, OK, TX) | HISTORY / United States / 19th Century
Classification: LCC F395.N6 N47 2024 | DDC 908.93/982076409034—dc23/eng/20230905
LC record available at https://lccn.loc.gov/2023040293

UNLESS OTHERWISE INDICATED, ALL PHOTOGRAPHS ARE FROM THE AUTHOR'S PERSONAL COLLECTION.

CONTENTS

◊ ◊ ◊

Acknowledgments xi

Introduction 1

CHAPTER 1

Johan Reinert Reiersen and the First Norwegian Migration Chain to Texas

13

CHAPTER 2

Chain Migration from Southern Norway to East Texas and Four Mile Prairie

36

CHAPTER 3

Cleng Peerson, the Father of Norwegian Immigration, Goes to Texas

58

CHAPTER 4

Why Did Immigrants from Hedmark County in Eastern Norway End Up in East Texas?

84

CHAPTER 5

Finding Virgin Land in the Bosque Valley

103

CHAPTER 6

More Norwegian Settlers on the Indian Frontier

125

CHAPTER 7

Living among Transplanted Hillbillies and Slaveowners

145

CHAPTER 8

Right or Wrong—My Texas!
Norwegians in the Confederate Army during the Civil War

170

CHAPTER 9

Norwegians in the Confederate Army after the Conscription Law

188

CHAPTER 10

A Growing Norwegian Colony in Bosque County,
despite Indian Threats

231

CHAPTER 11

A System for Indentured Servants from Norway

250

CHAPTER 12

Harder to Climb the Agricultural Ladder

273

CHAPTER 13

A Transplanted Community

298

CHAPTER 14

Conclusion

329

APPENDIX 1

Norwegians in Texas in 1860

341

APPENDIX 2

The Wealthiest Men in Bosque County, Texas, 1860

343

APPENDIX 3

Slave Owners in Bosque County, Texas, 1860

344

APPENDIX 4

Number of Sheep in Some Texas Counties, 1880–1900

349

Notes 351

Bibliography 395

Index 423

A gallery of photos follows page 216

To Inger Kari

◊ ◊ ◊

ACKNOWLEDGMENTS

◊ ◊ ◊

THE FIRST TIME I heard about early Norwegian immigrants in Texas was at a conference of the Norwegian-American Historical Association at Luther College, Decorah, Iowa, in June 2011. One of the presenters mentioned Clifton College in Bosque County, Texas. I had never heard of the college nor of the place. But if there was a Norwegian college in Texas, might this mean that many Norwegian immigrants lived in Texas?

After my return to Norway I googled both names. My ignorance changed to mild interest. I learned that the city of Clifton had given itself the name "the Norwegian capital of Texas." The Cleng Peerson Memorial Highway, FM 219, goes through Clifton from north to south. Bosque County, I found out, was almost literally "Deep in the Heart of Texas." The "father of Norwegian immigration," Cleng Peerson, lived in Bosque County among Norwegian neighbors the last ten years of his life and died there in 1865. Cleng Peerson is a brand name in Norway; even people who know nothing about Norwegian emigration recognize his name. But why did he go to Texas?

At the time I was Vice-Dean of Research at the Faculty of Arts and Humanities at the University of Stavanger and formally in charge of the organization of the PhD education at the faculty. I was contacted by the Norwegian Consul General in Houston at the time, Jostein Mykletun. Could I set up a meeting at the University of Stavanger in December 2011 about cooperation between the Bosque Memorial Museum in Clifton, Bosque County, Texas, on the one hand, and Tysvær municipality and the University of Stavanger, Norway, on the other? The Consul General wanted to present plans for a future Cleng Peerson Institute in Clifton.

Cleng Peerson grew up in Tysvær, across the fjord north of where I live, and I wholly agreed that Tysvær municipality should be contacted as soon as possible. The Consul General knew I was more knowledgeable about Texas than most Norwegians. My interest in the history of Texas was kindled around 1990 when I was doing research for a book on the history of the Norwegian oil industry. To get a better understanding of the early history of the offshore industry in the Gulf of Mexico, I visited the libraries at the University of Houston and Texas A&M University at College Station on two occasions. During the first visit, I bought for one dollar the classic book by John Stricklin Spratt, The Road to Spindletop: Economic Change in Texas, 1875-1901, at a used bookstore on Westheimer Road in Houston. I learned a lot from reading that book. Randolph B. Campbell's Gone to Texas: A History of the Lone Star State was published in fall 2003. I read it like a novel you cannot put down.

I was already curious about the history of Norwegian immigrants in Texas at the meeting with the Consul General in December 2011. He asked me at the end of the meeting if I could find time to visit Clifton and talk to the people involved in the plan for the Cleng Peerson Institute. I found an opportunity to make a detour to Texas ahead of an American Society of Environmental History conference in Madison, Wisconsin, four months later.

On Sunday, March 14, 2012, at high noon I arrived outside the Bosque Memorial Museum to meet the director, Dr. George Larson. During the next 48 hours, I met friendly and hospitable people connected with the Bosque Museum and the planned Cleng Peerson Institute. It all started with a lunch on the porch of the Omenson House at Norway Mills, followed by a visit to Our Savior's Lutheran Church in Norse, the Cleng Peerson monument, the Questad place, and the Rock Church. Maybe most important to me, and certainly for this book, were the hours I spent in the museum archives. I found material I had never dreamed of finding. I copied as much as I could within my time limits and brought the copies home with me.

I had discovered a treasure trove of very interesting material on the last years of Cleng Peerson and his fellow antebellum Norwegians in Texas. What should I do with it? "Cleng Peerson and early Norwegian immigration to the United States" is a field no sensible Norwegian historian will touch with a stick. So many Norwegians are fully convinced they already know that story. They might have read the novels of Alfred Hauge, or the work of Professor Ingrid Semmingsen, a historian. Norwegian-Americans interested in Norwegian immigration history might also have read the works of Professors Rasmus B. Anderson and Theodore Blegen, or the more recent work of Odd Lovold. Should I wade into a minefield I had sworn I would never go near? In the course of 48 hours, I became obsessed with the unknown history of Cleng Peerson and Norwegian immigrants in Texas. That obsession has stayed with me and motivates me to this day. I began to gather material and wrote the first tentative drafts in summer and fall 2012, in between my normal tasks. From summer 2016, when I retired from my position as professor of modern history at the University of Stavanger, until summer 2018, I worked full time on this book. The manuscript was revised for publication between March and June 2020.

I have written many books in my life. The material I have had to dig up to write this book is maybe the most complicated I have ever encountered. On the other hand, the project has given me a lot of joy, day in and day out.

Many have helped me during the research process. George Larson gave me full access to the library and archives at the Bosque Memorial Museum. The Cleng Peerson Research Library at the museum has a very good collection of archives and books about Norwegian immigration in the Southwest. Shirley Dahl, the librarian at the Cleng Peerson Research Library, played a central role in the professionalization of the library. I have enjoyed the best of help from her.

During my first visits at the Cleng Peerson Research Library, William Calhoon was very helpful. During three visits at the Bosque County Historical Commission in Meridian I received excellent assistance, first from Ruth Crawford and later from William Calhoon.

Two visits to the Texas Collection at Baylor University gave me important insights into the larger context of the relationship the Norwegians had with Waco. On both occasions I was served in an expert and friendly manner by Amie Oliver, Tiffany Sowell, and Brian Simmons. In one chapter I used extensively copies of the original Secretary Books at Our Savior's Lutheran Church, Norse. Patsy Lund, Shirley Bronstad, Gayle Squyres, and Patsy Squyres opened the doors to their archives. Both they and I were delighted that I was able to understand the excellent Norwegian prose in the minutes.

One of the most important sources in my work throughout has been the book *Norge i Texas*, published by Odd Magnar Syversen and Derwood Johnson in 1982. George Larson put me in contact with Judge Derwood Johnson in Waco. In the early stages Derwood read drafts and encouraged me in my endeavor to understand the Norwegians in the social and political context where they settled.

A Norwegian participant at an Annual Meeting of the Texas State Historical Association might seem somewhat out of place. At my first TSHA meeting in Corpus Christi in 2015, my wife and I were introduced to Lindsay T. Baker of Tarleton University and Jim Kearny, University of Texas, Austin. Since that first meeting I have had the pleasure of discussing my topic with both on several occasions. Kearney has done research and published on the history of the first German immigrants, and Baker has done research and published on the history of the first Polish immigrants in Texas. Both presented papers on German and Polish immigrants in Texas at the seminar on the "Legacy of Cleng Peerson" at the Bosque Museum in October 2015. In the final stages of writing, Baker gave me invaluable support to help me find the energy I needed to complete the manuscript.

Lee Parsons of Helotes, Bexar County, who is researching the history of the Christian Grøgaard family and their children in East Texas, has shared his research with me for several years now. Because of his work I have been able to understand better the difference between the life lived by Norwegian immigrants in East Texas and that of the immigrants in Bosque County, Central Texas. Dale Orbeck Van Sickle of Austin, a great-great granddaughter of Ovee Colwick, has shared with me primary research on the history of the Cleng Peerson/Ovee Colwick farm in Bosque County. Thanks to her work, some of my conclusions about the last years of Cleng Peerson are built on a more solid foundation than earlier researchers had available.

I, as well as my wife, have been on several research trips to Bosque County. On many occasions, last in February 2020, we have had the pleasure of being invited to dinners at the house of Virginia and Aubrey Richards. Those visits have meant more to me and to us than either of them can imagine.

Since fall 2015 our home away from home has been the "annex" at the home of Judy and Charlie Blue in Clifton, Bosque County. Every time I plan a new study tour, I send them an email and ask if we can stay for some days in the annex. The answer is always yes. Staying with Judy and Charlie is always a joy, but also a challenge. When the day is done, we are usually invited for a glass of wine or dinner. Sooner or later during our conversations we end up discussing and comparing ways of life, culture, and politics in Norway and the United States. It is not easy to compare Norwegian thinking and politics with thinking and politics in the United States and Texas. I hope these discussions have helped me become more tolerant.

Dean Tor Hauken at the Faculty of Arts and Education at the University of Stavanger shared my enthusiasm for the esoteric Texas project and always supported me. At the end of 2013 I left my position as vice-dean of research at the faculty and became professor of modern history at the Institute of Arts and Languages. The leader of the institute, Professor Odd Magne Bakke, now the dean of the faculty, has supported me wholeheartedly until this day.

When my colleague Nils Olav Østrem at the University of Stavanger heard about the new source material I had found at the Bosque Museum, Texas, in 2012, he urged me to write about it. He has done extensive research and published on the history of emigration from Rogaland County and had also published an article about Cleng Peerson. Every time I returned from a research trip to Bosque County, he came by my office to inquire about what I had unearthed this time. Both Østrem and our colleague Olav Tysdal have read and commented on some of the chapters about Cleng Peerson and his way to Texas.

Gunleif Seldal has been doing important historical research in primary sources on the first organized emigration from Norway in 1825. He has deep knowledge about the history of the *Restauration*, the background of the people who crossed the Atlantic aboard the sloop, and what happened to them later in life. He has shared his research with me unconditionally. His help has been much appreciated.

During my research over many years, I have met numerous people in Texas and elsewhere who have supported me. Not wishing to forget anyone, I thank them all for sharing their thoughts with me.

I am in eternal debt to two persons who have read every chapter of this book repeatedly and until completion. My good friend over many years, Arild Oma Steine, former chief information director at the Norwegian oil company Statoil (today Equinor), has performed the role of "qualified reader" meticulously. His main question during the reading of every chapter has been: Does the author follow the path he has said he will follow, or is he letting himself get sidetracked by interesting side paths? His feedback and suggestions have always been welcome and constructive.

This book is dedicated to my wife, Inger Kari. She has supported me in good and bad days since 1972. Inger Kari has enjoyed both my topic and the research into the lives of ordinary Norwegians who settled in Texas. Countless are the times over the last years when I have told her a story about orphans, mothers dying in childbirth, both parents dying from tuberculosis, relatives taking responsibility, etc., with her ending up with tears in her eyes, and I as well. These occasions were always good reminders of the importance of empathy and passion when writing history. It is a great blessing to be a good friend with the one you love and to share and discuss all intellectual matters together.

It is always hard to translate good advice into good prose. I have tried to do so to the best of my ability and ask the reader to bear with me.

—RANDABERG, NORWAY, 2023

NORSEMEN DEEP IN THE HEART OF TEXAS

◊ ◊ ◊

Introduction

NORWAY IS A FOREIGN LAND to Texans, and Texas is a foreign land to Norwegians. Neither in Norway nor in Texas has there been any awareness of the many Norwegian immigrants in antebellum Texas. This book, about the first Norwegian immigrants in Texas, tells the story of Norwegian "plain folks" who settled "Deep in the Heart of Texas," to cite a well-known Norwegian song (originally American). This is a book about chain migration, adaptation, and assimilation. Specific Norwegian social and cultural traits were adapted to an environment and culture very different from the places where the immigrants grew up. Norwegian traditions brought to Texas were treasured among the immigrants and their children for decades and tolerated by their neighbors in Texas.

A SHORT OVERVIEW OF NORWEGIAN EMIGRATION TO THE UNITED STATES BEFORE THE CIVIL WAR

Organized emigration from Norway to the United States began in 1825. Cleng Peerson, who was born in 1782 in the municipality of Tysvær in Rogaland County, western Norway, led the settlement of Norwegians on American soil.[1] Quakers in the town of Stavanger sent him to the United States in 1821 to explore the conditions for emigration and to locate land. He reported back to his sponsors in Norway in 1824, returned to New York, and bought land from the Pulteney Estate in Western New York before the end of the year.

On July 4, 1825, forty-five men, women, and children and a crew of seven left the city of Stavanger on the small sloop *Restauration*. The journey of the sloop across the Atlantic to New York City marked the beginning of organized emigration from Norway to the United States, where the families and single men on board hoped to find a Canaan's Land.

After ninety-eight days at sea the *Restauration* arrived at the port of New York. Most of the passengers boarded a steamship from New York to Albany. They arrived at Holley, the harbor on the Erie Canal closest to their land in Orleans County, in early November. The Norwegian immigrants in Western New York did not find the milk and honey they had dreamed of. Their land was overgrown with woods and difficult to clear. During the first years they often suffered and wished they were back in Norway. The immigrants experienced extended illness, and many died in Western New York. In letters to relatives in Norway their disappointment was visible between the lines. For many years, few Norwegians followed in their footsteps.

The first group of organized Norwegian emigrants traveled on the sloop *Restauration* in 1825, reaching New York City on October 9. The emigrants continued north on the Hudson River to Albany, traveling by canalboat on the Erie Canal. The fast-growing city Rochester made a strong impression on them. West of Rochester, at Holley, they disembarked.

The Erie Canal opened in fall 1825, providing white immigrants an enormous area for homesteading that had previously belonged to native Indian groups. Before the arrival of the railroad, the Erie Canal was the most important transport route between New York and Chicago and contributed to city growth. During the building of the Erie Canal and the decades after the canal began operating, the area experienced strong economic growth.

In 1833 Peerson traveled further west, exploring for better land. When he returned to New York, he strongly promoted new areas he had seen in La Salle County, Illinois. According to Theodore C. Blegen, his "glowing reports on conditions in the West found ready acceptance among the New York colonists. In 1834 six families moved to Illinois where they established the Fox River settlement, the second Norwegian settlement in the United States.[2] The Norwegian Fox River settlement became the mother colony for thousands of Norwegian emigrants. Their letters home to family and friends helped build up the emigration "pull." Many Norwegians became infected with the urge to emigrate.[3] Farms were sold and tickets bought for the journey from Norway to Fox River.

The breakthrough for organized Norwegian emigration came in 1836 and 1837. On May 25, 1836, the brig *Norden* left Stavanger with 110 passengers, arriving in New York on July 20. A couple of weeks later, the brig *Den Norske Klippe* left Stavanger for New York with fifty-seven passengers. In spring 1837 another two emigrant ships left Stavanger: the bark *Ægir*, followed some weeks later by the bark *Enigheden*. "In size and influence no other group of immigrants in the first generation of Norwegian immigration can compare with the 343 passengers of these four ships that constituted the bulk of the exodus of 1836 and 1837," Henry J. Cadbury maintained.[4]

Between 1836 and 1845 emigration to the United States had spread to nine Norwegian counties. The total number of emigrants remained small, with an average of 620 persons annually. This was less than 0.5 emigrants per 1,000 Norwegian inhabitants. Between 1846 and 1855 the average number of emigrants grew to 3,200 persons per year, or 2.3 persons per 1,000. A record 6,050 Norwegians emigrated in 1853, and 8,900 crossed the Atlantic in 1861.[5] Between 1836 and 1865 the emigration flow was most intense from Rogaland, the county *Restauration* had departed from in 1825.[6]

The immigration situation was discussed in the introduction to the Eighth Census of the United States. "As our own people, following 'the star of empire', have migrated to the west in vast numbers, their places have been supplied by Europeans, which has modified the character of the population."[7] Ireland topped the list of foreigners in the United States in 1860 with 1.6 million, followed by immigrants from German states with 1.3 million. Next came England (431,692), Canada (249,970), Scotland (108,518), and France (100,870). Immigrants from Norway

Emigration from Norway 1825–1865

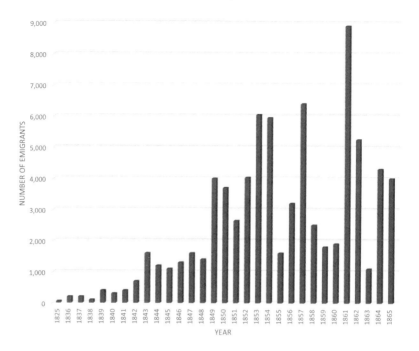

held ninth place. The 1860 census registered 43,996 Norwegians living in the United States, an increase of almost 250 percent compared with the 12,678 Norwegians enumerated in the 1850 census.[8]

Wisconsin was the preferred destination for Norwegian immigrants. In 1860 21,442 Norwegians lived in Wisconsin, followed by 8,428 in Minnesota, 5,668 in Iowa, and 4,891 in Illinois. The future pattern of Norwegian immigration to the Midwest was already in place before the outbreak of the Civil War. Perhaps surprisingly, 715 Norwegians lived in California, but many Norwegians were drawn to the state by the 1849 California gold rush. Comparatively few Norwegians were registered in the state of New York (539) compared with the 326 Norwegians who had settled in Texas.

It is no surprise that German immigrants dominated among foreigners in Texas; 20,553 Germans lived there in 1860. But it might have been unexpected that so many from Poland had gone to Texas (783). Twice as many Norwegians were living in Texas in 1860 as immigrants from Sweden (153) and Denmark (150) combined.

TABLE 1. *The distribution of Norwegian immigrants in the United States in 1860*

State	Number
Wisconsin	21,442
Minnesota	8,428
Iowa	5,668
Illinois	4,891
California	715
New York	539
Michigan	440
Texas	326
Kansas	223
Massachusetts	171
Oregon Territory	159
Missouri	146
Dakota Territory	120

SOURCE: US Census 1860, p. xxviii.

TABLE 2. *Numbers of immigrants living in Texas in 1860*

Country of origin	Number
Germany	20,553
Ireland	3,489
France	1,883
England	1,695
Poland	783
Scotland	524
Switzerland	453
Norway	326
Sweden	153
Denmark	150
Holland	76
Spain	69
Russia	42

SOURCE: Population of the United States in 1860, p. 490.

THE FIRST NORWEGIAN COLONIES IN TEXAS

The first Norwegian colony in Texas was established in Brownsboro in Henderson County, East Texas, by Johan Reinert Reiersen in 1845. By 1850, the largest Norwegian settlement in East Texas was found at Four Mile Prairie on the border between Kaufman and Van Zandt counties. In the mid-1850s a new Norwegian settlement was established on the frontier with the Indians in Bosque County, west of Waco in Central Texas.

The Norwegian settlers in antebellum Texas followed two very distinct migration chains from Norway. The Reiersen chain had two main sources: some small municipalities in Aust-Agder County in southern Norway and a few municipalities in Hedmark County in eastern Norway. Toward the end of his life, Peerson, "the father of Norwegian immigration," became an ardent promotor of Norwegian migration to Texas. Several Norwegians originally from western Norway followed him to Texas from the Norwegian mother colony in La Salle County, Illinois.

MIGRATION CHAINS AND DIASPORAS

In her studies of Scottish migration, Marjory Harper used the expression "the social cement of chain migration."[9] Scots showed a persistent tendency over more than two hundred years to seek out and settle alongside their fellow countrymen, for "most emigrants needed to be reassured, not only of better working conditions, remuneration and prospects, but also that they would not be cutting themselves adrift from all the familiar and comforting associations of home."[10] When Scots were sent out on a reconnaissance mission, they were not only "instructed to choose fertile, well-watered tracts of land, but also to ensure that religious and educational facilities were available within a reasonable distance."[11] Immigrants tended to come in parties. Relatives and friends from the same region followed each other at intervals, and "new arrivals could therefore feel confident that a bridgehead had been established and they were not coming to an entirely alien land."[12]

As information about migration opportunities spread, migration chains not only linked localities across space but also tended to increase the volume of migration from the sending communities. There is a considerable literature on the Scottish diaspora and migration chains.[13] Better wages and working conditions overseas were important, wrote Harper, but for many "the anticipated neighbourliness, cooperation and familiarity of an established Scottish settlement were incentives just as important as material gain and the absence of domineering landlords."[14]

In two pathbreaking articles in the 1880s, E. G. Ravenstein formulated what he called the "laws of migration."[15] Ravenstein noted that Americans were "greater wanderers, less tied to home associations, than are the inhabitants of Europe." He found the phenomenon could be explained by the vast extent of unoccupied land and the great natural resources of the country.[16] Immigrants to the United States were "for the most part content with going no further than the nearest foreign province or the most convenient centre of absorption."[17] When migrants found a place they liked, they tended to remain there and often began recruiting people who shared their values and outlook on life. Most immigrants, whether they had grown up in the United States or Europe, shared the dream of finding cheap and abundant agricultural land.

In many cases, people had very personal reasons for making the big decision to emigrate. Some chose to leave because of personal debt. This was a well-known motive among early Norwegian immigrants.[18] In other cases, family and marital relations had grown complex and were simplified by emigration, which gave everyone a fresh start. People who had experienced great tragedy hoped to ease their pain and find a sanctuary on the frontier, where they could begin anew.

Emigration involved many uncertainties. There was insecurity and risk connected with the Atlantic crossing as well as the inland journey in the destination country.[19] Emigrants often chose as their first destination places where they already had personal contacts. "Families and friends in such locations could provide temporary lodgings, arrange a job, and generally ease the shock of confronting a new society, culture, and economy," wrote Walter Kamphoefner.[20] The people already living there provided support and aid, as well as advice about crops and local markets, whom to trade with, and how to get temporary work as farm laborers.[21] Once a chain of migration had been established, letters and money were sent back along the chain to the home communities the immigrants had left. This in turn encouraged other people to leave home and emigrate to the same place. Chain migration was an important contributor to the establishment of strong ethnic communities.

Immigrants settled in clusters, in a relatively limited number of locations, and among people who spoke their mother tongue.

Systems of chain migration "operated like a transmission belt that brings newcomers from one area to a particular location," wrote Robert Ostergren.[22] The ethnic culture of a group in a country or region in Europe was transferred to specific places in a city or rural area in the United States. "Without having been steered by any organized colonization efforts, one-twelfth of all former Braunschweigers in the whole United States in 1860 lived in one single Missouri county where they made up a quarter of the German population," observed Walter Kamphoefner.

In his book *A Community Transplanted. The Trans-Atlantic Experience of a Swedish Immigrant Settlement in the upper Middle West, 1835–1915*, Ostergren followed more than one hundred migrants who left their homes in the Swedish parish of Rättvik to seek a new life on the raw frontiers of Isanti County, Minnesota, and Clay County, South Dakota. He tried to reconstruct one specific trans-Atlantic chain migration "from its source area in Europe to the place or places where it settled in America as a means of appreciating the totality of the immigrant experience."[23] He wanted to break away from "the long-standing tendency to view emigration and immigration as separate phenomena within particular national contexts. Scholars have often been overly preoccupied with the effects of trans-Atlantic migration on sending or receiving societies, when in fact it may also be viewed as an important link between changing conditions on both sides of the ocean."[24] This author shares Ostergren's intentions. My narrative, however, has been highly influenced by the historical sources available to me and the methods I have chosen to use.

FAMILY HISTORY AS A SOURCE

Local histories and family histories have been important for this study. The book by Odd Magnar Syversen and Derwood Johnson, *Norge i Texas. Et bidrag til norsk emigrasjonshistorie,* published in 1982 by the Stange local historical society in Hedmark County in eastern Norway, has been a treasure trove.[25] Syversen wrote two hundred pages of historical narrative about the Norwegian emigration to Texas, beginning with Reiersen in 1845 and closing with the large influx of emigrants from Hedmark County in the two decades following the Civil War. The last three hundred pages feature the lifelong genealogical work of Johnson as the backbone. Every time I consult *Norge i Texas,* I learn something new.

Syversen did excellent historical research in primary sources in Norway and Texas. Johnson's main interest was genealogy, the study of families through the tracing of their lineages and history. He concentrated on giving the correct name of each person, the date they were born, if they married, the day of their marriage, the name of the spouse, if the couple had children, the name of the children, the day they were born, the day they died, and where they are buried. The same process was repeated for their children in the next generation. The year a person or family first arrived in Texas is the organizing principle in the genealogical part of *Norge i Texas.*

Peder Olsen (Peter Pierson), for example, is found under No. 1853/60, which means he arrived in Texas in 1853. He was born February 27, 1843, and died April 27, 1895. He was the son of the pioneer Ole Pierson, who came to Bosque County in 1854 with the first group of Norwegian immigrants, but there is no information about that. Johnson wrote that Peter was a hornblower in Company E, McCord's Frontier Regiment in the Confederate Army during the Civil War. But there is nothing in the genealogy about the date of his enlistment or that he deserted in May 1864, first to El Paso and later to Santa Fe, where he enlisted in the Union Army. The genealogy informs us that he was a farmer and gives the name of his wife, when she was born and died, and the names of their thirteen children and when they were born, married, and died. It does not tell us that Peter Pierson played a central role in the Norwegian Lutheran Church at Norse and in the larger Norwegian community from the time he returned from the Civil War and until his death. Such information must be found elsewhere.

Elsewhere has often meant *Bosque County: Land and People (A History of Bosque County)*, compiled by the Bosque County History Book Committee, published in 1985.[26] In the introduction to the family histories, the reader is informed that the families in question were invited to write the history of their own family. Some were reluctant to do so; nevertheless, 1,337 family stories were submitted and printed in the book. "Even though valuable genealogical information is present in each presentation, the stories do not comprise a book of genealogy but a book of history."[27] Both claims can certainly be contested.

The family histories in *Bosque County: Land and People* are subjective by design. Nevertheless, the book has been used as an important source. In many cases, the stories did not tell me much at first. But as my understanding of the general context grew, and as I used my understanding of Norwegian history to understand specific Norwegian immigrants in their county context, I began to find gold nuggets. To put it a little differently, despite the subjectivity of the family stories, they have been important pieces in my weaving of the web of collective biography.

The genealogical part of *Norge i Texas* and the family histories in *Bosque County: Land and People* are more trustworthy than most of the histories of specific Norwegian families I have read. Seen from the perspective of a professional historian, the older family histories are generally more trustworthy than the most recent, although there are some fine exceptions.

In the age of the Internet, the number of resources readily accessible to genealogists increase every year, but that does not seem to guarantee that family histories will be more trustworthy. The quality of family histories can vary greatly, depending on the quality of sources, the use of original records, and the place family lore has been given in the narratives.[28] Sometimes secondary sources have been preferred, even if primary sources were available. The writer might have found the information in the primary sources uncomfortable. Those too young might have been provided with a modified age to be allowed to marry or join the armed forces. Birth and marriage dates might have been adjusted to cover pre-wedding pregnancies.[29]

Some compilers use oral sources without the necessary caution. Data recorded soon after the event are usually more reliable than data recorded many years later. Primary sources are records that were made at the time of the event. Secondary sources are records that are made days, weeks, months, or even years after an event.

Family histories that use modern Internet templates have been of very little use as sources of information for this book. The templates seem to give little room for telling stories from the lives lived by the template user's ancestors. I have preferred to use family histories based on primary sources, especially those that document their sources in footnotes or with copies of primary documents in the text.

INSPIRED BY PROSOPOGRAPHY AND THICK DESCRIPTION

This book has been inspired by prosopography. The term can be defined as investigations of common characteristics of a historical group, a kind of collective study of their lives.[30] Prosopography is related to, but distinct from, both biography and genealogy. Family histories based on primary sources are useful, but the aims of prosopography are wider than those of genealogy. Through the study of biographies of a well-defined group, patterns and relationships between actors can be uncovered. In this book, prosopography has been used as a tool to understand the lives of "invisible" Norwegian immigrants in Texas.

The content as well as the method David Hackett Fischer used in *Albion's Seed. Four British Folkways in America* have proved relevant for this study.[31] Fischer studied and discussed the "folkways" of four main groups of immigrants from the British Isles who left distinct regions in Great Britain and Ireland and settled in different regions in the American colonies. He argued that the culture of each of the groups persisted and provided the basis for the modern United States. Fischer tried to explain the origins and stability of a social system that for two centuries had remained "stubbornly democratic in its politics, capitalist in its economy, libertarian in its laws, individualist in its society and pluralistic in its culture."[32]

According to Fischer, folkways are "the normative structure of values, customs and meanings that exist in any culture," which rise from social and intellectual origins. Folkways are "often highly persistent, but they are never static. Even where they have acquired the status of a tradition, they are not necessarily very old."[33]

From 1629 to 1775, North America was settled by four great waves of English-speaking immigrants. Fischer presents and discusses the four regional groups in four long main chapters. The first discusses the the exodus of Puritans from the east of England to Massachusetts (1629–40). The second was the movement of a Royalist elite and their indentured servants from the south of England to Virginia (1649–75). The third was the "Friends" migration, the Quakers, from the North Midlands and Wales to the Delaware Valley (1675–1725). The fourth was a great flight from the borderlands of North Britain and Northern Ireland to the American backcountry (1717–75). Many among them settled in the Appalachian Mountains and along the Indian frontier from Pennsylvania and southward. Members of this group came to have a large impact on the ranching and farming culture in the south and southwest. Both the Virginian slave culture and the folkways of the Appalachian Mountains came to have a strong influence on the people who went to Texas.

This book has also been inspired by the concept of "thick description." The influential social anthropologist Clifford Geertz advocated thick description in ethnographic analysis in his *The Interpretation of Cultures*. Geertz argued that it was one of the main tasks of anthropology to explain societies through specifying social interaction, the actors' interpretations of their actions, and the flux of conceptual structures and meanings, while "thin description" was satisfied with giving a factual account without any interpretation. Thick description not only describes "a particular event, ritual, custom, idea," but also includes commentary and interpretations of those comments and interpretations. The task is "sorting out structures of signification . . . and determining their social ground and import." The social interaction must be described and interpreted in the context where the behavior occurred to make specific behavior meaningful to outsiders. According to Geertz, "what we call our data are really our own constructions of other people's constructions of what they and their compatriots are up to."[34]

The questions historians choose to bring to the empirical material will always be influenced by what is experienced as relevant from today's situation. The past can be explored in the same way we travel to a new and unfamiliar country or city. We may use a map or a guidebook, or we may prefer to wander around and discover things for ourselves. The historian is like a tour guide, trying to explain to a group of travelers what makes the material interesting and relevant based on the knowledge of the actors who created what we study, the people who lived in their own modern time, or like a travel writer who describes their own experiences and discoveries to those unable to visit the places themselves.

CHAPTER I

Johan Reinert Reiersen and the First Norwegian Migration Chain to Texas

◊ ◊ ◊

TIRED OF NORWAY—DREAMING OF A BETTER LIFE IN AMERICA

In January 1843, newspaper editor Johan Reinert Reiersen, who lived in Kristiansand, the largest city in southern Norway, visited his good friends Christian and Thomine Grøgaard in the small town of Lillesand, east of Kristiansand. Their main topic for discussion was emigration to America. Reiersen gave them a vivid account of his almost feverish emotions after having read parts of Captain Marryat's book *A Diary in America with Remarks on its Institutions*, published in London in 1839.

Reiersen wanted to emigrate with his family to the United States.[1] During the spring of 1843, he tried to sell his newspaper and printing office to get the cash he needed to finance the journey. The moment he left, he would "feel like a captive who leaves his narrow prison; I therefore long to start and to say good-bye to Norway, and I think, yes, I am convinced, that caravans will follow in my wake."[2] An important aim for him, he told his friends, was to find the best place for a Norwegian colony. "I would then visit Brazil, Chile, - yes even the entire paradise of California." During his visit, Grøgaard mentioned to him that he knew some influential people in Eide parish, east of Lillesand, who had expressed a strong interest in emigration. Maybe they would be willing to help finance Reiersen's explorations?

On March 12 Reiersen wrote Grøgaard he was convinced that "North America, after all, must become the first place for emigration … Captain Marryat has made me quite enthusiastic about Wisconsin and has infused me with abhorrence for Illinois as a veritable pesthouse."[3] The stories told by Marryat about ague, seasonal fevers, and deaths in Illinois made a very strong impression on Reiersen. Even before he left Norway, he had decided that he would not visit the Norwegian mother colony at Fox River, Illinois.

WHO WAS JOHAN REINERT REIERSEN?

Johan Reinert Reiersen was born on April 17, 1810, in Vestre Moland in Aust-Agder County, southern Norway. He died on September 4, 1864, in Prairieville, Kaufman County, Texas. His father, Ole Reiersen, a teacher and sexton, did not have the economic means to send his son to university. Nevertheless, Johan Reiersen got a far better education than most Norwegians at the time. The family lived in Holt parish, east of the city of Arendal in southern Norway, near the large Næs Jernverk (ironworks). The owner of the ironworks, Jacob Aall, was an influential member of the first Norwegian national assembly at Eidsvold Værk in spring 1814,[4] which wrote the Norwegian Constitution (Grunnloven), signed on May 17, 1814.

Aall had a reputation for paying out of his own pocket for the education of boys showing intellectual promise, and was willing to pay for Johan Reiersen's university education. But the young student disappointed his sponsor. In 1832, after only one year of study at the University of Oslo, Reiersen left for Copenhagen, Denmark. For several years he made a living by translating novels and editing literary journals. He met Henriette Christine Waldt, the daughter of a wealthy merchant, in Copenhagen. They were married in 1836; in October of the same year Henriette gave birth to a son, Oscar. The Reiersen family moved back to Norway in 1838 and stayed with Reiersen's parents at Holt. In 1839 Reiersen procured the necessary funding to establish the liberal newspaper *Christianssandsposten* in Kristiansand.

Christianssandsposten was one of very few newspapers in Norway at the time. Even though the 1814 Norwegian Constitution propagated freedom of speech, newspapers in Norway around 1840 continued to encounter mild censorship from the authorities. Political and economic liberalism had still not had its breakthrough in Norway. Reiersen was in favor of both. In his newspaper he was apt to criticize the abuses and arrogance of civil servants and was proud to have become a constant nuisance in the eyes of conservative state officials. Reiersen felt strongly that the freedom lacking in Norway was already in place in America. He began to publish positive articles about America, its political freedom, and the availability of cheap land for all.

Reiersen established the first colony of Norwegian immigrants, called Normandy, in Henderson County, East Texas, in 1845. The story has been covered in several books and articles about Norwegian emigration to the United States over the years. The best historical work on Reiersen and the Texas colony was written by Frank G. Nelson in his introduction to the English translation of Johan Reinert Reiersen, *Pathfinder for Norwegian Emigrants*, published by the Norwegian-American Historical Association in 1981. Odd Magnar Syversen wrote an interesting chapter on Reiersen in Syversen and Johnson, *Norge i Texas*. Last, but not least, Erik Aalvik Evensen contributed with new primary sources in his Ph.D. dissertation *From Canaan to the Promised Land. Pioneer Migration from Hommedal Parish (Landvik and Eide Sub-Parishes), southern Norway, to St. Joseph, Missouri and East Norway, Kansas*. Evensen discussed the relationship between Reiersen and the group of men from Eide parish who financed most of Reiersen's traveling expenses during his exploratory journey in 1843–44.[5]

The works mentioned above have all been used in this narrative. Several letters Reiersen wrote in 1843 and 1844, published in 1944 by the Norwegian-American historian Theodore C. Blegen under the title: "Johan Reinert Reiersen, Behind the Scenes of Emigration: A Series of Letters from the 1840's,"[6] have never been used before. They shed new light on Reiersen's emotional swings and his impulsiveness when making decisions.

BRAZIL, MISSOURI, OR CALIFORNIA?

The readers of *Christianssandsposten* were well informed about Reiersen's plans to travel to the United States in spring 1843. In an enthusiastic letter to Christian Grøgaard, Reiersen wrote that if interested people found it "desirable for the common good to get a clear and comprehensive idea of circumstances and conditions in the New World, and if they offer me a guarantee of 200 or 300 specie dollar, then my plans are already made. I shall go by steamer from England to New York -- then to lower Canada, through Wisconsin to Missouri, where I believe will be the only region in North America where a Norwegian colony could be founded, next, down the Mississippi to New Orleans, whence I shall go by steamer to Rio de Janeiro. There I shall investigate Brazilian conditions, and this done, I shall return by a merchantman to England or Hamburg."[7]

Either North America or Brazil would be the most interesting destination for emigration. "Between the United States and Brazil a comparison should be made, the advantages and shortcomings of each should be impartially weighed, and the choice ought to depend upon the result of this comparison."[8] At the time, Reiersen was strongly in favor of Brazil. He knew that German-speaking colonies were already flourishing only 80 English miles from Rio de Janeiro in southeastern Brazil at the new town of Novo Freiburgo.[9]

Reiersen was familiar with the German literature on emigration from his time in Copenhagen, and he probably knew that Novo Freiburgo had been settled by 100 Swiss families from the canton of Fribourg in the early 1820s as well as eighty German families. From his reading Reiersen also had formed a distinct conception of the kind of Norwegians he wanted to recruit to follow him abroad. He wanted men to join him, he wrote Grøgaard, who "have some means—several hundred dollars each to spare for a beginning—and who, in the next place, are known as moral, orderly, industrious, and friendly people. The whole society would pay *pro rata* for the most necessary expenses, as for example a saw, with either man or horsepower to cut lumber for houses, a gristmill etc., and would see to it that they have in their midst various professional men."[10] Initially, his emigration society would be an elite society, but after a while others would be allowed to join the colony.

In May 1843, Reiersen sent the Grøgaards some English dictionaries, a phrase book, a grammar, and a reader. His impulsiveness had again taken hold of him. Since his last letter Reiersen has lost his enthusiasm for Brazil and so had changed his plans. He now believed strongly that "it will be most expedient to leave for England—I can go there free with Mr. Howard's lobster smacks—and from there to take the Great Western, by which I can get to New York in 11 to 14 days. The season of the year is so advanced that it is important to make all possible haste."

Grøgaard was skeptical about the cold winters in Wisconsin and Missouri. Reiersen agreed with him. "It seems to me that when one is to change residence and fatherland, one should not choose those which have some of the drawbacks that form contributing causes for the emigration. Every effort should be devoted to finding a place that combines the greatest possible enjoyments with the greatest possible advantages, where the essentials not only for wealth and prosperity but also for happiness and physical well-being are present." The last weeks he had been "amusing myself by reckoning and estimating how one might best and most easily get to California."[11]

Finally, Reiersen succeeded in selling his newspaper and also received a contribution of 300 *riksdaler* in silver to his traveling fund from the Landvik group.[12] Reiersen left Kristiansand in late June, not for New York, but New Orleans.

FROM NORWAY TO NEW ORLEANS, WISCONSIN, TEXAS, AND BACK TO NORWAY

On a hot and humid day in early August 1843, Reiersen arrived in New Orleans. During his stay, he met the consul of the Republic of Texas, who told him that Texas was very eager to attract European settlers as well as a Norwegian colony. Reiersen listened closely to the arguments in favor of Texas but stayed with his original plan to visit some of the Norwegian settlements in the northern states first.

He traveled upriver on a steamboat to St. Louis, Missouri, and continued north on the Mississippi River almost to the border of Wisconsin. He left the steamboat at Galena, Illinois, the main town in the lead mining district in northern Illinois.[13] In a long letter to his friends in Norway, written in January, 1844, he recounted his travels by stagecoach from Galena to Mineral Point in Wisconsin, where he stayed several days to get information from the land office. In the state capital, Madison, he met General Dory, the governor of the territory. "From Madison I traveled about 25 miles west and stopped at Koshkonong Prairie to visit the Norwegian settlements on this and the surrounding prairies within a radius of ten miles."[14]

A week later Reiersen continued by stagecoach through several small villages. He disembarked at Prairie Village to visit the Norwegian immigrant Hans Gasmann, who lived about 18 miles to the northwest. On the way, he learned that the Swedish pastor Unonius and several Swedes, as well as a Dane, had settled there. According to rumors in Norway before Reiersen left, Gasmann had regretted his emigration decision. This was not true, wrote Reiersen. "Gasmann, as well as his wife and the whole family, were in the best of spirits. Far from regretting their decision, they felt satisfied and were happy at having changed countries."

Gasmann had built a temporary log house, a stable, and a barn, as well as a smithy and carpenter shop. He had bought about 1,200 acres of land, cattle, and oxen. "He had also bought a beautiful span of horses and a wagon, in which he and his family had that very day attended the

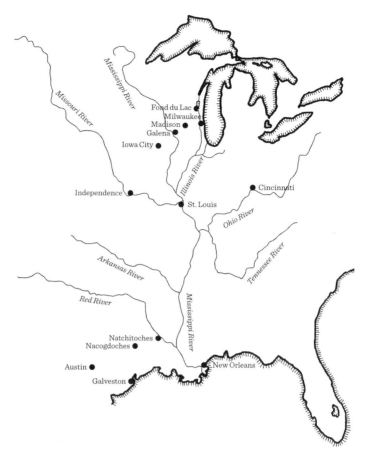

The Norwegian newspaper editor Johan R. Reiersen traveled extensively across the North American continent in 1843 and 1844. In the course of these two years he covered a distance of between 4,000 and 5,000 miles.

dedication of a church in the neighborhood." Several Gasmann family members were employed as craftsmen—a smith, carpenter, builder, wheelmaker, saddle maker, tanner, miller, and sawyer. "Everything accomplished here was the family's own work."[15]

Reiersen was impressed that chance had brought so many educated and wealthy men to this vicinity: "Unonius, Gasmann, Fribert, St. Cyr, and several other Swedes and Norwegians—remarkably enough, the Swedes have settled on the east side of a little lake—Pine Lake—while the Norwegians live on the west shore."

He then continued to Milwaukee, "a town which in seven years has grown to a population of 6,700." He traveled on foot to a place called Muskego, 20 miles south of Milwaukee. According to Reiersen, this settlement of about 2,000 people was the largest Norwegian colony in America at the time. He also visited Port Washington, 30 miles north on Lake Michigan, and ended up at Fond du Lac on Lake Winnebago, where he visited Winnebago.[16]

Reiersen returned to Galena "after a fairly thorough investigation of the natural conditions of Wisconsin" and planned to cross the Mississippi into Iowa and travel as far west as Burlington. Because of drifting ice, however, the Mississippi could not be crossed. He spent Christmas at the home of a blacksmith named Knudtsen from the Norwegian town Drammen.[17] After Christmas, he joined a party of men who crossed the frozen Mississippi River into Iowa Territory, traveling in sleds pulled by horses. His destination was Dubuque, where the main land office in Iowa was located. He met with the governor, Robert Lucas, he wrote, who introduced him to several members of the legislature.

Robert Lucas (1781–1853), born in West Virginia, was a well-connected politician. He belonged to the upper echelons of the Democratic Party in Iowa. After Congress created Iowa Territory in 1838, President Van Buren appointed him Iowa Territorial Governor and Superintendent of Indian Affairs, but Lucas was no longer governor when Reiersen met him. His readers in Norway, however, were hardly concerned with such petty details.[18]

After two days in Dubuque, Reiersen joined a group traveling 25 miles north along the Mississippi River to the Turkey River and even farther up the river to the so-called Neutral Ground.[19] Reiersen then turned south again and traveled through several small towns along the Mississippi as far as Davenport, directly opposite Rock Island, Illinois. From there he traveled west 50 miles to Iowa City, where he wrote his letter.[20]

Reiersen had been traveling for months, and admitted he was exhausted. It was easy to "map out a travel route when sitting at home in one's parlor," where distances had seemed trifling on the map. Despite all his reading and planning, the knowledge Reiersen had acquired before leaving Norway had been incomplete and would remain incomplete. In his January letter, he speculated that his future Norwegian colony should be located somewhere between Wisconsin, Iowa, and Missouri.

Because of the immense transportation challenges, Reiersen gave up his dream of traveling to California. He still planned to visit Texas, though.[21] His last letter from the United States, written in Cincinnati, was dated March 20, 1844. It covered his journey south through Iowa and

Missouri, down the Mississippi, to Red River on the border of Louisiana and Texas, to the Texas capital Austin, where he met with President Sam Houston, to Galveston, back to New Orleans, and north to Cincinnati.

TRAVELING TO MISSOURI WITH HORSE AND BUGGY

Reiersen became acquainted with a young merchant from St. Louis in Burlington, Iowa, who was traveling with his own horses in a buggy. The merchant invited Reiersen to accompany him through the northern part of Missouri to Weston on the Missouri River. Reiersen accepted the invitation and enjoyed the travel through the most densely populated part of Missouri, a distance of 300 miles.

He then went by steamboat down the Missouri to Independence in Osage County. The frontier town Independence made a strong impression on him. The town was one of the most important starting points for ox wagon trains heading to Oregon, California, or Santa Fe in New Mexico. Even though it was very early in the season, February of 1844, two wagon trains were ready to leave. "One party was composed exclusively of merchants, chiefly from St. Louis, on their way to Santa Fe with merchandise. The other was made up of emigrants from all parts of the United States and a few Germans bound for Oregon."[22]

Reiersen traveled by stagecoach from Independence south through Harrisburg, continued to Warsaw on the Osage River, and then east to Jefferson City on the Missouri River (50 miles). He boarded a steamer for St. Louis and further south. He disembarked at Natchez, Louisiana, but forwarded his traveling trunk to New Orleans. "With only a light knapsack, I boarded another boat which went up the Red River to Natchitoches, Louisiana, on the border of Texas. From this town there was a diligence, or stage, to Nacogdoches and San Augustine, Texas." The Texas consul in New Orleans had strongly advised Reiersen to travel through San Augustine.[23] He hired a saddle horse in San Augustine and rode the 80 miles to Austin, the new capital of Texas, located on the Colorado River.

The Texas Congress had just assembled when Reiersen arrived. One of the high points of his journey was meeting President Sam Houston. Houston emphasized that Texas was very interested in recruiting foreign immigrants. "Illiterate peasants mixed with literate political refugees and artisans as Irish, French, English, Scottish, Canadian, Swiss, Scandinavian, Czech and Polish arrivals established colonies in Texas in the 1840s and 1850s," observed Calvert, De León, and Cantrell in their textbook *The History of Texas*.[24] Among the European immigrants to Texas, the Germans were the best known and most numerous. Since their arrival in the 1830s they tended to cluster in ethnic colonies. The first German settlements on the Indian frontier in the Hill Country at New Braunfels and Fredericksburg were established in 1845 and 1846. It has been estimated that by 1860 between 30,000 and 35,000 Germans lived in Texas.[25]

President Houston did not leave Reiersen in any doubt that "Congress would give a colony of Norwegians all the encouragement that could reasonably be expected. Houston doubted that Texas would be admitted to the Union in the near future."[26] With respect to Indian threats on the western frontier, Houston was of the opinion that "the Comanche Indian hostilities" were at an end after their last defeat.

Reiersen stayed two days in Austin, and then traveled by stagecoach through Bastrop and Rutersville to Washington on the Brazos River. He arrived in Houston five days later. On March 7, 1844, Reiersen saw the sea again at Galveston, "the most important trading center in Texas, of almost 4,000 inhabitants."[27] The steamboat *Harry of the West* was just then loading for New Orleans, and Reiersen was among the passengers when the ship left Galveston in the evening of March 9, reaching New Orleans on the morning of March 12.

All ships to Europe had just left New Orleans. Reiersen would have to wait two to three weeks before another ship would sail for Le Havre or Liverpool. Instead of waiting, the impulsive Reiersen decided to travel to New York along the waterways to Cincinnati and Pittsburgh, and then by rail and canal to Philadelphia. In early July 1844 he returned from Le Havre, France, to Arendal, Norway, on the brig *Europa*, owned by the local shipowner O. M. Dannevig of Arendal.

AN IMPRESSIVE AND LONG JOURNEY COMPLETED

In the course of twelve months, Reiersen had completed an impressive, strenuous, and long journey. The observations in his letters to Norway were fresh and to the point. It is worth emphasizing that Reiersen traveled through a lot of frontier country in Wisconsin, Iowa, and Missouri. In the case of Iowa, he visited the eastern parts of the state in the middle of winter three years before Iowa became a state, on December 28, 1846.

He passed through the Norwegian colony in Lee County and made interesting comments about the growing Mormon sect at Nauvoo, on the Illinois side of the Mississippi River. Reiersen was certainly the first Norwegian to travel so extensively in that part of Missouri. His descriptions of the wagon trains in Independence, which were preparing for the long trek across the prairies and through the Rocky Mountains, can compete with the best among such narratives.

Nevertheless, both in his letters and in *Veiviser for Norske Emigranter* (*Pathfinder for Norwegian Emigrants*), it is clear that Reiersen often did not build on his own observations but trusted what he had read, or the sales pitches presented to him by powerful people. His descriptions of Texas clearly belonged to this category. He swallowed uncritically what Houston told him about the land in North Texas and the Indian threat. Reiersen's book was "the most comprehensive Norwegian handbook about America published up to its time," Blegen argued in 1931.[28] It is easy to agree with Blegen's views. The book is still very readable today. Blegen, however, warned the reader that Reiersen sometimes was "not wholly to be depended upon."[29]

According to Blegen, Reiersen had traveled extensively in America "and visited the Norwegian settlements in Illinois and Wisconsin, pushed his inquiries widely in other states, and also made a trip to Texas, where he met Governor Sam Houston." Blegen also gave the reader the impression that Reiersen personally had visited both the Norwegian settlements in Illinois and Wisconsin. This was not the case. He visited several Norwegian settlements in Wisconsin, but never visited the mother colony in La Salle County, Illinois.

Several other authors have also assumed that he visited Fox River: "While visiting the United States in 1843, the Norwegian journalist Johan R. Reiersen, who later founded the first Norwegian settlement in Texas, stopped in the Fox River Settlement," wrote Carlton C. Qualey in 1938 in his book *Norwegian Settlement in the United States*. Qualey included a long citation from Reiersen's book about the Fox River settlement. Frank G. Nelson wrote that Reiersen had visited "the Norwegian settlements in and around La Salle County, Illinois," probably after he wrote his last letter from Cincinnati. "It is impossible to say by what route, but he probably traveled primarily by stagecoach."[30]

Nelson's conjecture is hard to verify. If we can trust Reiersen's letters to Norway, he was never even close to visiting the Norwegian mother colony in Illinois. The distance from Galena to Ottawa, La Salle County, was around 150 miles. If La Salle County had been his destination, he would have left the steamboat at Davenport. Diligent and usually meticulous professional historians have taken for granted that he *must* have visited La Salle County, since this was the most important Norwegian settlement at the time.

Most historians have trusted what Reiersen wrote in chapter 10 of *Veiviser*. Reiersen referred to *Ole Rynning's True Account of America*. The "Norwegian emigrants of 1837 settled in the northern part of Illinois, on the Fox River, and at Beaver Creek where Rynning lived. As a result of unhealthful conditions which caused several deaths at the latter place, the survivors moved to the Fox River settlement near Ottawa, La Salle County, Illinois."[31] In his description of La Salle County, Reiersen followed Rynning's narrative but left the impression that he had also been there.

PUBLICATION OF *VEIVISER* (PATHFINDER) AND REIERSEN'S EMIGRATION

After his return to Norway in July 1844, Reiersen joined his family at Holt, where he wrote his report.[32] In the introduction to the 1981 American version of *Veiviser*, Frank G. Nelson argued that Reiersen returned to Norway with plans going far beyond the writing of his report and the advice he gave his main sponsors, the Landvik group. The completion of the report was "only the first step toward the establishment of a model Norwegian settlement in America. He proposed to avoid the miseries and misfortunes of such places as Muskego through his own leadership."[33]

His book would soon be ready for the printing press, he wrote his friend Grøgaard on October 29, 1844. "Within two weeks I plan to be in Lillesand and shall bring with me a number of copies for distribution among the guarantors." Grøgaard had informed him that the Landvik group had still not made the final decision to emigrate and had not sold their farms. Reiersen could not understand their tardiness. He had recently met the Landvik group face to face and given them "all the explanation and information that they desired." If they had got cold feet and decided against emigration, Reiersen would certainly respect their decision. He could not deny, however, that it would please him if "the uncommonly well-educated people of Eide parish join company with me." Reiersen himself intended to emigrate for good "shortly after New Year's with what little I can scrape together."[34] Would Grøgaard go with him? From the tone in the letter, it is evident that Reiersen very much wanted Grøgaard to be a part of his core group of emigrants.

On December 1, 1844, the printer delivered the first copies of *Veiviser for Norske Emigranter til De Forenede nordamerikanske Stater og Texas* (*The Pathfinder for Norwegian Emigrants to the United States and Texas*). The next day Reiersen sent several copies to his Landvik sponsors with a long letter attached.[35] In his book, he told them, he had to the best of his ability and "with an honest love of truth" presented the results of his investigations.[36] There were good reasons for settling in the northern territories. But one of his main conclusions was that "a region further south might be preferable."

Reiersen's narrative clearly shows that he had built up a special fondness for Texas. Nearly all parts of Texas, he wrote, "have sufficient land for the founding of even the largest colonies." Land along the Trinity River, the Colorado River, and the Brazos River seemed very attractive. Reiersen claimed, mainly trusting the words of others, that East Texas was a region "considered very healthful, and I have found no local causes of disease." He characterized the summer heat as intense, but not oppressive; "because a refreshing, steady breeze from the Gulf of Mexico blows regularly every day."[37]

Neither in his long letter nor in his book did Reiersen come out in favor of one specific locality or place. At the end of the book, he wrote that a "comprehensive plan for a future Norwegian colony in America" had to be discussed with his main sponsors. On Saturday, December 9, Reiersen, Peder Nielsen Kalvehaven, Osuld Enge, and Anders Holte discussed emigration and the location for a new colony. At the end of the day, they formulated and signed a letter to be published in the press, announcing their intention to emigrate the following summer. Reiersen would be their guide to the final destination. All interested persons were invited to join them on the journey to New Orleans. The plan was to sail from Norway before the end of July.

The letter was published in *Christianssandsposten* as a paid advertisement on January 28, 1845. People interested in joining the group were invited to sign up at the house of goldsmith Pettersen in Kristiansand, shopkeeper Bache in Lillesand, sexton Reiersen in Holt parish, and Osuld Enge in Eide parish.

Reiersen had argued strongly in favor of New Orleans as a good point of entry for people wanting to settle in Wisconsin. Reiersen personally leaned more and more in favor of Texas, possibly influenced by what he had read and heard about the most recent German settlements in Texas. They were established by liberal-minded noblemen and professionals who aimed to build a new and better Germany in Texas. Reiersen was thinking along the same lines. He wished to become the founder of a Norwegian colony in Texas with himself as the leader, and with his family and close friends forming the nucleus of that colony.

REIERSEN LEFT NORWAY SECRETLY TO AVOID THE NORWEGIAN AUTHORITIES

In a letter from Reiersen to Grøgaard in early spring 1845, Reiersen expressed disappointment. The income from the sale of his book was far lower than he had expected. The state authorities in Kristiansand had taken steps to force Reiersen to pay unpaid public debt. He had been fined 300 *speciedaler* (the exchange rate between a *speciedaler* and a dollar was close to 1). Reiersen felt as "low-spirited as I can be and am only longing anew to remove myself from the coasts of Norway so as to be able to breathe fresh air."[38] The main reason for his depressive thoughts, though, was his "urgent need of money."

Because of the actions taken by the Norwegian state authorities against him, Reiersen changed the plans he and the Landvik group had agreed on in December 1844. Instead of waiting until the whole emigration group was ready to sail at the end of July, Reiersen decided he and a small group of core people would leave as soon as possible, followed by the large group in July.

Reiersen was convinced that Norwegian officials had singled him out for harassment. His closest friends might have agreed with him. Nelson, however, admitted that Reiersen's reactions to Norwegian state officials did "much to discredit Reiersen in Norway and lend color to the persistent tradition that he was prone to shady if not downright criminal behavior."[39] Contemporary accounts were confused and partisan. Nelson agreed with Reiersen that his old enemies in conservative circles were to blame. They had not forgotten what Reiersen had written about them and regarded him as "a dangerous agitator." Nevertheless, Nelson had to admit that Reiersen probably could have emigrated from Norway without serious trouble if he had paid the fines levied at *Christianssandsposten* before leaving for America in 1843. But Reiersen stubbornly refused to pay. He was innocent, he declared; he had just exercised the freedom of the press.

This situation must be understood in the larger Norwegian context. Norwegian authorities in the 1840s showed a very liberal attitude toward people wanting to emigrate. However, a growing problem was that so many left the country without paying their private and/or public debts. On November 11, 1843, the Norwegian Finance Ministry had appointed a public commission to evaluate the significance and volume of emigration from Norway and to propose new laws to regulate emigration. The Ministry asked the commissioners explicitly to be aware of the habits among emigrants "*til at snige seg bort uden at tilfredsstille sine Creditorer*," to sneak out of the country without satisfying their creditors. The Ministry urged all Norwegian public officials to be meticulous in their collection of outstanding debts before letting anyone emigrate. People were free to emigrate but had to announce their intentions to do so in time for the authorities to process requests from creditors.[40]

Because Reiersen seemed unwilling to pay his debts before leaving, the magistrate in Kristiansand ordered his arrest to make sure he paid them. According to Nelson, Reiersen "felt no moral scruples about refusing to pay fines that he considered a violation of his constitutional rights." Being principled, however, did not help him much.

His friends in Landvik were willing to hide him from the authorities. "Strait-laced citizens though they were in most respects, they must have shared his attitude," Nelson wrote, "as they hid him until he could find passage from Lillesand in April."[41] They kept him hidden on an island until a ship from Grimstad left for Le Havre with Reiersen and a small group of men.

Reiersen was transported to Grimstad in secrecy and carried on board in a large box.[42] Members of his family were already on board—his father Ole was crucial, since he was the man with the most money to invest. His brother Gerhard and his sister Caroline were also members of the first group, as well as Christian Grøgaard, blacksmith Syvert Nielsen Haabesland, and carpenter Ole Stiansen.[43] His wife, children, sister-in-law, mother, and the rest of the family would follow in the summer.

The way Reiersen left Norway left a bitter aftertaste. He had planned to leave as the leader of a new Norwegian colony in America. Instead, he was smuggled on board a ship as a lawbreaker fleeing the authorities. Reiersen was a well-known newspaper editor along the southern coast of Norway. The story of how he had left the country spread like prairie fire from one parish to the next. In addition, old stories from his youth resurfaced. From the publication of *Veiviser* in late 1844 and until Reiersen and his core group emigrated in spring 1845, Reiersen lost legitimacy and credibility among some of the people in southern Norway that he was most eager to recruit.

A NORWEGIAN COLONY IN EAST TEXAS

The small Reiersen group left Le Havre in April on board the *Alisto* and arrived in New Orleans on June 9, 1845. Even though no final agreement had been reached between Reiersen and the Landvik group regarding the location of the new Norwegian colony, the first thing Reiersen did after arrival in New Orleans was to visit the Texas consul: He and his group wanted to settle in Texas. At the Texas consulate, he met an employee named Brown who had participated in the war with Mexico in 1836. After the war, Brown had been awarded a document entitling him to one-third of a league of land in Texas, equal to 1,476 acres. According to Texas law, this document gave him the right to choose land wherever free land could be located. Brown offered to sell the document to Reiersen. It seemed to be a good deal, and he persuaded his father Ole Reiersen to buy it for 500 Texas dollars. The transaction was conducted, and the Texan consul signed as a witness. The consul urged the Norwegian group to continue to Nacogdoches, and he wrote a letter of introduction to Dr. James Starr in that city, asking him to help the Norwegians choose land.[44]

In a postscript to a letter to his wife written a few days later, Ole Reiersen wrote that the big decision had now been taken: They would travel to Texas. All were well, but they found the heat oppressive. The mosquitos were eating them up. "Today we bought a deed giving us the right to choose 1,476 acres in Texas." All necessary legal documents concerning the transaction had been written and signed, and 500 Texas dollars paid.[45]

The merchants of New Orleans swarmed around the Norwegians, eager to supply them with equipment they told them they would need in Texas. Reiersen and his group bought a lot of supplies. They later discovered that most of the equipment could have been acquired at least as cheap elsewhere. On their journey to Nacogdoches, Reiersen bought horses from a previous acquaintance. The seller, however, did not tell him he sold them riding horses. The Reiersen group learned the hard way that riding horses were unfit for pulling a heavily loaded wagon and had to sell the riding horses and buy wagon horses. A lot of money was lost on this hapless horse trade.

Lee Parsons recently did some meticulous historical research in primary sources about what happened to the family of Christian Grøgaard after they arrived in East Texas.[46] Parsons' findings combined with what Odd Magnar Syversen wrote in *Norge i Texas* make it possible to reconstruct most of the chain of events of what happened to the Norwegians in East Texas in fall and winter 1845.

The Reiersen group left New Orleans on the steamboat *The Planter* on June 26 for Natchitoches. On board, Reiersen got to know a doctor from Marshall in Harrison County, who told the Norwegians that employment opportunities would be much better in Marshall than in Nacogdoches. On the doctor's advice the Norwegians changed their plans. At Natchitoches, Johan, his father Ole, and his sister Lina left the steamboat and continued overland to Nacogdoches. The others, including Christian Grøgaard, continued on the smaller steamboat *Man of Kentucky* to Shreveport, and traveled overland from Shreveport to Marshall. The doctor took responsibility for storing most of their equipment for their planned settlement until further notice. He also promised to find good work for the men at Marshall.

Reiersen, his father, and his sister arrived in Nacogdoches on the 4th of July in the middle of lively celebrations. It seemed the whole town as well as people from the surrounding area were present. They were invited to a 4th of July ball in the evening. According to Reiersen, his 16-year-old sister Lina, despite her plain clothing, was the belle of the ball.

They met several important men at the ball. One of them was the German immigrant and merchant George Bondies, who the next day introduced Reiersen to Dr. Starr, who gave the Norwegians excellent help.[47]

James L. Starr was a Yankee immigrant to Texas. He was born in New Hartford, Connecticut, on December 18, 1809, and died on July 25, 1890, in Marshall County, Texas. Starr County was named in his honor. The Starr family had migrated to Franklin County, Ohio, in 1815, where his father died in 1824. James studied at an academy in Worthington, Ohio, got a job as a schoolteacher near Columbus, and taught himself medicine. In 1832 Starr moved to Georgia, where he practiced medicine at McDonough and later at Pleasant Grove. Together with a group of people from Georgia, he and his wife Harriet migrated to Nacogdoches, where they arrived in January 1837.

When the Texas Congress established the General Land Office on December 14, 1837, President Houston selected Starr as president of the Board of Land Commissioners and the receiver of land dues for Nacogdoches County. Two years later, Starr was appointed Secretary of the Treasury of the Republic of Texas by President Mirabeau B. Lamar. Starr played an instrumental role in the design and layout of the streets and the location of the capitol in Austin.[48] He found it too costly to be in politics, resigning in 1840 to return to Nacogdoches and his medical practice.

In addition to his practice as a doctor, Starr was active as a land agent, establishing a land agency in Nacogdoches in 1844 along with Nathaniel C. Amory (1809–64). He was a strong supporter of the annexation of Texas by the United States and was one of the most influential people in Nacogdoches. Starr and Amory remained partners until 1858.[49]

LOCATING LAND IN HENDERSON COUNTY

Starr showed Reiersen a land map and told him that free land could be had 90 miles northwest of Nacogdoches, in an area between the Trinity and Neches rivers. In addition to the right to the 1,476 acres his father had bought in New Orleans, each family head would have the right to choose another 320 acres.

According to Syversen, Reiersen had to wait until a surveyor named Hoffer could go with them.[50] After much archival research, it can be concluded that Syversen got the spelling wrong. The assistant surveyor of Nacogdoches County at the time was Samuel Huffer. He knew the country well, and in the census of 1850, he was listed as living in Henderson County.

Before the Texas Revolution in 1836, very few people of European origin had settled on the land Huffer showed Reiersen. Spanish and Mexican authorities had allowed Cherokee, Shawnee, Delaware, and Kickapoo Indians to live in the Nacogdoches District. In 1827, the Mexican government decided to let the Cherokee Indians, under the leadership of Chief Bowles, settle in the eastern part of what later became Henderson and Van Zandt counties. The Cherokee Indians functioned as a buffer between Mexico and aggressive white American settlers moving to Texas from the north. The government of Texas later validated the claims of the Cherokees but broke the treaty in 1839. War broke out, and during the battle of the Neches on July 15 and 16, 1839, two white men and eighteen Indians were killed, including the 81-year-old Chief Bowles.[51] The battle marked the defeat of the Indians in East Texas. Surviving Indians were forced to abandon their homes and were moved to Oklahoma Territory.

The defeat of the Cherokees opened the area for Anglo-American settlement. White homesteaders who had settled along the Trinity River moved east into areas previously occupied by the Indians. One of the first to settle in this area was Jane Irvine, a widow from Alabama, who migrated with her children to the area before 1835 and built a gristmill at Caney Creek. She owned a Mexican land grant of "a league and a labor." A small community grew up around this mill.[52]

In August 1845 Huffer traveled with Reiersen to land not taken between the Neches and Trinity rivers, land that at the time was still a part of Nacogdoches County. They traveled through East Texas timberland, through wooded, but rolling terrain with sandy loams and sandy soil beneath. The Texas legislature would establish Henderson County on April 27, 1846, in honor of Governor Henderson.

Reiersen chose land for his Norwegian colony along Kickapoo Creek, 92 miles northwest of Nacogdoches. Even though Reiersen had no experience as a farmer, he declared the soil on the chosen land to be very good. He was told good drinking water could be found on the land. There were many creeks and lakes in the area, but also swamps. Cash money was a rarity in Henderson County at the time, and Reiersen had no problems finding a carpenter willing to build two log houses under one roof for 70 dollars. "The place in Henderson County seemed just what they were looking for," wrote Nelson, "a lush prairie with a few low hills and some timber."[53]

RETURNING TO SHREVEPORT AND NEW ORLEANS TO MEET THE FAMILY

After choosing his land, which he named Normandy, Reiersen did not return to Nacogdoches, but rode the 90 miles to Marshall to meet his brother Gerhard, who had established a watchmaker shop there. The blacksmith Haabesland had opened a smithy, and the carpenter Stiansen and Christian Grøgaard were helping him. Grøgaard was not used to manual work. His soft hands were full of blisters, Reiersen commented; Grøgaard soon gave up this work.[54]

It was time for Reiersen to return to New Orleans to meet his family and the first group of immigrants. His brother Gerhard and Stiansen joined him, while Haabesland and Grøgaard remained in Marshall. They arrived in New Orleans at the end of September 1845.[55] "Everything had gone swimmingly so far," Nelson commented. "At this point Reiersen's dream of a model settlement seemed realistically close to fulfillment. To be sure, their property lay some distance from the nearest town and farther still from the nearest steamboat connection with New Orleans." He eagerly looked forward to meeting the group from Eide parish and expected them to contribute considerable energy to his new colony. According to Nelson they shared "Reiersen's own vision, and they had the necessary capital to get started properly."[56]

The American sailing ship *Magnolia* finally arrived on November 17, 1845. Reiersen's wife, children, mother Kirsten, and sister Gina were among the 128 Norwegian passengers on board. Stiansen met his wife Georgine and their children. Five children had died during the Atlantic crossing. Among them was Reiersen's three-year-old daughter and Stiansen's one-year-old daughter, very sad news.

The most bitter pill for Reiersen to swallow was that only his own family and some close friends were willing to follow him to Texas.[57] Not a single person from the Landvik group was on board. He was told that the group had experienced problems with getting ship transport from Norway to Le Havre.

The majority of the Norwegian emigrants on board came from Setesdalen in southern Norway. Reiersen' attempted to persuade them to go with him to his new Norwegian colony in Texas, but they stubbornly refused. Reiersen felt slighted by their attitude and called them stupid inland farmers looking "like a bunch of Indians." Nelson, who often adopted Reiersen's vocabulary, called them "simple mountain farmers, obviously disoriented by their experiences and terrified of the unknown." The crux of the matter, however, was that their leaders showed a hostile attitude toward Reiersen, "at least suspicious of his intentions," wrote Nelson.[58] Reiersen had obviously lost credibility in southern Norway since his emigration in the spring.

Norwegian farmers had a long and ingrained suspicion toward people representing the authorities. Some of the leaders among what Reiersen called "simple farmers" feared that Reiersen in America might want to assume the traditional authoritarian role played by civil servants in Norway. They believed that since Reiersen was better educated than them, he looked upon himself as more civilized, an attitude they resented. They had burned their bridges, sold their farms and brought with them everything they owned from Norway to America in order to be able to make their own decisions. The majority of them boarded a steamboat for St. Louis, and most settled in Wisconsin.[59]

SAVED BY CHARLES VINZENT

Lady luck did not smile much on Reiersen at the end of 1845. The steamboat Reiersen and his family traveled on to return to Normandy went aground and sank. In desperation, Reiersen dived into the water to try to save some of their belongings, and almost drowned. He was rescued by the German immigrant and entrepreneur Charles Vinzent, who saved the Norwegian immigrants not only by bringing the group to Nacogdoches, but also by helping them financially. The meeting with Vinzent marked the beginning of a friendly relationship that was firmly secured with the marriage between Vinzent and Lina Reiersen on September 13, 1847.[60]

In a long letter to his brother Christian in Norway in 1846, Reiersen described the land he had chosen. It was located on a high, rolling country, "overgrown with timber, but so open that in most places one can plow between the trees after first cutting down the smaller ones."[61] The water was generally clean and came from springs. He was firmly convinced that "this is the most healthful region in America." There were no mosquitos, and the ague, "which is more or less present everywhere in western states, is almost unknown." His descriptions of the land, however, were more about what he thought and had heard than what he knew from his own experience. The same was true about his strongly held views on health and sickness. Ague and malaria were rather prevalent in East Texas. The diary of Adolphus Sterne, covering the period September 1840 to early April 1844, and February 1851 to November 1851, is full of descriptions of fever and ague; for example, in 1843 Sterne had a "severe fever all night took medicine today." The next days he hardly had any energy to write, but managed just to "*crawl* from my bedroom to my writing Table."[62]

Reiersen, like many other early settlers in East Texas, was fascinated that cattle could live outdoors the whole year round without being attended to. Corn was the most common crop, but mills for grinding it had not yet been built. Some farmers raised cotton and sugar cane. One of the chief advantages of farming in Texas was that "products will always be well paid for," wrote Reiersen.[63] Wages were high, and land of the best quality was so easy to get that almost every man had his own property. "Consequently, there is a lack of working people here. They can be replaced only by negro slaves."[64] Reiersen showed a very positive attitude toward slave ownership in his letter. He would be interested in renting slaves to work his land, he wrote his brother. But renting a slave was expensive.

Reiersen had seen many "negro slaves" in the short time he had been in East Texas. Some of the nicest people he had met in Texas were slave owners. Within a mere few months after arrival, Reiersen seems to have accepted slavery among his neighbors and in Texas generally. Neither did it bother him that the Indians had been chased away and that he had staked his claim on Indian land. The Reiersen family had settled on land "won only three or four years ago from the Cherokee Indians, who were driven west of the so-called Cross Timbers," he commented. Their land was located about two miles northwest of an Indian trail.

SHATTERED DREAMS? REIERSEN NEVER BECAME THE LEADER OF A NORWEGIAN COLONY IN TEXAS

In summer 1845 Stiansen, a member of the first core group of immigrants, received a letter from his friend, blacksmith O. Røraas in Kristiansand, Norway. Røraas wanted to know the truth about Texas. What were the worst things about Texas compared with Norway? The lack of cash money and having to eat cornbread were the worst things he had experienced in Texas, Stiansen wrote back to Røraas. But he could live with both. At first, he had worked as a carpenter in Harrison County, but his customers were not able to pay until they sold their cotton at the end of the year. On the other hand, it was easy to get enough food to eat and livestock at bargain prices. He and watchmaker Gerhard Reiersen had left Marshall in favor of Nacogdoches, where more money was in circulation. Usually, he earned 2 to 3 dollars a day and sometimes 5 dollars. So far, Stiansen had found class differences in Nacogdoches small compared with what they were used to in Norway. Doctors, businessmen, and farmers all spoke to each other the same way. They did not look down on the lower classes of white people.

Craftsmen such as blacksmiths, shoemakers, and tailors earned good money, Stiansen wrote. He owned 320 acres of land. Government land was to be had for half a dollar per acre.[65] He strongly urged Røraas to emigrate, and Røraas followed his advice. Røraas was on board the *Ancona*, together with his sister and her daughter, when the ship arrived in New Orleans on November 3, 1846. They had traveled with a large group of Norwegians, including Christian and P. G. Reiersen, the brothers of Johan Reiersen. Most of the emigrants on board the *Ancona* came from inland communities in southern Norway. They chose to settle in Normandy (later renamed Brownsboro), Henderson County, where they arrived on December 23, 1846.

THE LANDVIK GROUP NEVER SETTLED IN NORMANDY

Reiersen, however, was still obsessed with the group of people from Landvik, Grimstad. Why were none of them on board the *Ancona*? The members of the group eventually arrived in New Orleans in January 1847 on the *Izette*. None of the 80 Norwegians on board bothered to inform Reiersen of their arrival beforehand, and they took no steps to contact him after their arrival. The group stayed for a while in New Orleans. They were still undecided about where to settle, but agreed that they would not follow Reiersen to East Texas. In the end, the leaders and the majority of the group (around 50 persons) chose to try their luck in Missouri, while a minority (15 persons) chose Wisconsin. They boarded a steamboat for St. Louis in April 1847.[66]

Why did the Landvik group finance Reiersen's pathfinding journey in 1843–44, help him avoid state officials in early 1845, but decide against joining him in Texas? In his doctoral dissertation, Evensen discussed these questions at length. Despite all the new archival material Evensen discovered, he found no documents that gave clear answers to these questions.

Reiersen's report and his knowledge of Texas and the United States had impressed the leaders of the Landvik group. This, however, did not imply that they also accepted Reiersen as their leader. They had their own leadership, men who were influential citizens and active members of the new democratic political movement in Norway. The fact that Reiersen publicly assumed he would be their leader might have become increasingly embarrassing to them. Reiersen had only done an assignment for them, which they had paid for. They might have agreed with many of his political views, but deep down they were still farmers and practical men, and above all they were law-abiding citizens.

They had experienced firsthand how impulsive Reiersen could be when making his decisions—sometimes he risked everything on a whim without thinking through the consequences for himself or others. In spring of 1845, following his principles had been more important to him than respecting Norwegian law. Reiersen was always a person who easily triggered controversy—before he left Norway, and later in Texas.

CHAPTER 2

Chain Migration from Southern Norway to East Texas and Four Mile Prairie

◊ ◊ ◊

FROM AUST-AGDER COUNTY, NORWAY,
TO HENDERSON COUNTY, TEXAS

Most of the Norwegians who emigrated to East Texas between 1846 and 1850 came from farming families in the parishes of Raabyggelaget, a name used since the Viking Age for the communities in the mountain districts of southern Norway. Today these inland municipalities are part of Aust-Agder County. A number of emigrants on board the ship *Ancona* in November 1846 left small and remote valleys in places like Tovdal, Gjøvdal, Åmli, and Froland.

Salve Knudson, from Tovdal, was one of them. According to family history, Knudson contracted "America-fever" in the early 1840s.[1] At that time, Knudson had never heard of Johan R. Reiersen. In the secluded inland municipalities where he and his neighbors lived, they got their knowledge of America and emigration from farmers in Telemark County over the mountains to the northeast. Knudson was born in September 1803 on the Kasim farm at Øvre Ramse in Tovdal parish. The farm had been in the family for generations when he inherited it from his father in 1826. Four years later, in April 1830, he married Asborg Kittelsdatter, born in Mjåland in the neighboring Gjøvdal parish on January 29, 1808. They became the parents of several children. Knud Salve Knudson was the oldest, born on December 1, 1830, followed by Signe Salvesdatter on May 10, 1833, Ketil Oscar Knudson on June 21, 1835, and Targjerd Salvesdatter on September 18, 1837. She died before she was a year old, and another Targjerd was born on June 13, 1839, but died in Le Havre, France, in September 1846. Their youngest child Tollef Knudson was born December 10, 1841 and died May 5, 1843.

Measured by the standards of his neighbors, Knudson was fairly well off as a farmer. Nevertheless, he decided to emigrate in 1843. Organized group migration from Norway had had its breakthrough in 1836 and 1837, mainly affecting communities in northern Rogaland County in western Norway. The number of Norwegian emigrants to America had reached an unprecedented level by 1843. According to Norwegian police accounts 1,451 men, 1,061 women, and 1,428 children emigrated from Norway in 1843. More than 1,800 of them left farms and rural districts in Telemark.[2] From Upper Telemark, and the communities Salve Knudson and others came from, more than 1,000 persons emigrated in 1843.[3]

Salve Knudson had to put emigration plans on hold in 1844 after his wife died at age 36. A year later, on January 1, 1845, he married Signe Tellefsdatter, a cousin of his wife, born in 1805 in Espestøl in Treungen parish. She had inherited a farm, and Knudson and his children moved in with her.

Knudson hardly put down any roots on the new farm. In spring 1846, he decided to join a group of relatives, neighbors, and friends who planned to emigrate along the route Reiersen had suggested in *Pathfinder*.[4] The group of emigrants traveled south across lakes and along rivers to the agreed meeting point at the town of Lillesand on the coast. Reiersen's two younger brothers, Christian (born 1814) and Peder Georg (born 1825), would be their leaders on the journey. Small groups of emigrants from the coastal region east and west of the city of Arendal also met at Lillesand. According to the plan, a German ship would arrive at Lillesand and bring them to Le Havre, France. They waited for weeks, but no German ship showed up. Finally, the emigrant group rented a small sailing boat to bring them west to Kristiansand. This of course led to loss of time and unexpected expenses. The group had to wait another month in Kristiansand until they got passage on the schooner *Den flyvende fisk*, which left Kristiansand on August 29, 1846, and arrived in Le Havre ten days later.[5] They waited another eight days at Le Havre. During that time the Knudson family lost their seven-year-old daughter Targjerd, who was buried there.

Also among the emigrants from Tovdal, Gjøvdal, Åmli, and Treungen was the family of Jørgen Olsen Hastvedt, who had left his farm in Espestøl in Treungen parish. Hastvedt was born on July 29, 1793, in Espestøl and died along the Sabine River in Texas in December 1846. On May 14, 1814, Hastvedt married Torborg Knudsdatter (born May 2, 1792, in Askeland, Gjøvdal). The couple took over the farm of his parents in 1820 but sold it in 1846 when they emigrated with their six children. At the time of emigration, Hastvedt was 53 years old and his wife 54.⁶ Their oldest daughter, Gro Jørgensdatter, born in 1819, had married Jørgen Terjesen Hastvedt in July 1843. They emigrated with the rest of the family and brought with them a small boy, born in 1844, who died soon after arrival in Texas. In early October, during the Atlantic crossing, Gro gave birth to a daughter.

Hastvedt's 14-year-old son, Knud Jørgensen Hastvedt, wrote about the family's emigration much later in his life, and because of his reminiscences much is known about the journey from inland southern Norway to Henderson County in East Texas.⁷ Reading the narrative of Knud Jørgensen Hastvedt in a wider context, it can be concluded that he is usually right about what happened, but sometimes wrong about when it happened. The cotton packet *Ancona* left Havre on September 15 with a large group of Norwegians on board and arrived in New Orleans on November 3, 1846. Gerhard Reiersen (born in 1812), had arrived in East Texas in 1845 as a member of the core group and was waiting for them in New Orleans. He knew his way around New Orleans and had mastered the English language. The immigrants were full of praise for his help and remembered it for the rest of their lives. After two days, the immigrants bound for East Texas boarded a steamboat, which would take them up the Red River, maybe as far as Shreveport. The water level was very low, however, and it was impossible to ascend the rapids.⁸ The group had to disembark at Alexandria and stayed there a couple of weeks to prepare for the overland journey.

The Norwegians soon discovered that there were hardly any roads in East Texas. Depending on the weather and the road, the group traveled between 5 and 12 miles a day. The journey took a month, in part because the Norwegians would not travel on Sunday, the Lord's Day. While still in Louisiana, Jørgen Hastvedt drank river water and got very sick. After crossing the Sabine River, still 40 miles from St. Augustine, he died and was buried along the road. After the burial, the group went on to St. Augustine, and finally arrived at Nacogdoches, where they met some of the Norwegians who had recently settled there.⁹

The Hastvedt family and most of the other immigrants continued to the Reiersen settlement (Normandy) on the Kickapoo Creek, arriving on Christmas Eve, December 24, 1846. The first day after Christmas, Johan Reiersen and his wife came down to pay the new arrivals a visit. Reiersen made them a proposal: "for a fee of two dollars each, he prom-

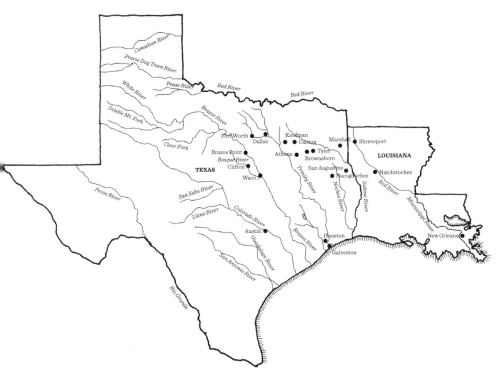

The map shows the most important places where antebellum Norwegian immigrants settled in Texas. The first group of immigrants arrived in New Orleans in 1846. The members of the group traveled by steamboat to Natchitoches, Louisiana, and then headed west into Texas. After crossing the Sabine River, they traveled through San Augustine before they reached Nacogdoches. They settled on land around Brownsboro, Henderson County. After some months, many of the immigrants moved to Four Mile Prairie, between Canton and Kaufman.

In fall 1849, Cleng Peerson visited the Norwegian settlements in East Texas, coming south from Fox River, Illinois. He continued in a northwesterly direction along the Trinity River to visit the Norboe family in Dallas. In spring 1850 he traveled into the region of the Comanches west of Fort Worth, and then changed his path to a southerly direction. He crossed the Brazos River and walked through the Bosque Valley before he arrived at the village of Waco. From Waco he walked back to the Norwegian settlements in East Texas, before returning to Illinois.

ised to provide the newcomers with titles to 320 acres of land at a cost of fifty dollars, which he claimed to be the regular price." Four families accepted the offer, and Knud's mother, recently bereaved Torborg Hastvedt, bought 640 acres. The immigrants were thankful for the help Reiersen offered, but became less happy when they discovered that "the titles cost only thirty dollars; but unfortunately this difference was never credited to us."[10] In this respect Reiersen fit the role of immigration agent: He expected to be paid for services rendered. The problem in this case, however, was that Reiersen did not inform his clients about his fee beforehand. His behavior in money matters earlier in his life are discussed by Syversen in *Norge i Texas*.[11]

Twenty-four-year-old Aslak Nelson Smeland had stayed close to the Hastvedt family during the journey. He was born on August 19, 1822, on the Smeland farm at the upper end of the Gjøvdal Valley west of Åmli and north of Tovdal. Aslak was the second oldest son.[12] He did not get much schooling but showed more aptitude for reading and writing than his fellow students. Before his emigration he had worked as a schoolteacher. It was obvious to fellow immigrants that Smeland and twenty-year-old Margit Jørgensdatter Hastvedt, born on February 27, 1826, were very fond of each other. Smeland was offered a well-paid job as tailor in Nacogdoches, and he decided to stay for a while. On December 25, 1847, Smeland and Margit were married in Smith County, and then moved to Brownsboro.[13] Their daughter Ragnhild was born and died in 1849, and their son Nils Jacob was born on July 1, 1850.

THE NORWEGIAN COLONY AT FOUR MILE PRAIRIE, VAN ZANDT COUNTY

From their arrival at Christmas 1846 until summer 1847, the Norwegian immigrants were content with the land Reiersen had chosen for them in Henderson County. In late summer and early autumn of 1847, however, the Norwegian colony experienced their first attacks of "climate fever" and ague. Four Norwegian families had leased a house from a man who had built a new house on higher ground. They had been warned about living in this house since it was built in a swampy area regarded as unhealthy. The families of Salve Knudson and his brother-in-law, Knud Gundersen, had moved into a detached kitchen on this property. "But with summer came the heat, and with the heat came the mosquitoes that carried the dreaded ague (malaria), a fact unknown to the early settlers."[14] Several of the Norwegians in the house died: Salve's wife, her brother Knud, and the Gundersens' only child. The Norwegian farmers also discovered that the soil around Brownsboro became depleted faster than anyone had expected.

By the end of 1847, many Norwegian immigrants were disappointed with their land and the high risk of contracting serious contagious illness. They left Brownsboro and moved 40 miles northwest to Four Mile

Prairie. The new settlement got its name from a sandy prairie, 10 miles long, which began on the north shore of Cedar Creek in southeastern Kaufman County and continued northeast into Van Zandt County between two streams four miles apart, Cedar Creek on the west side of the prairie and Lazy Fork on the east side.

Very few white people had settled on this prairie. Many years later, H. M. Teel, one of the first settlers in Van Zandt County, gave a vivid description of how the area looked in 1849. Flocks of buffalo had retreated west so were no longer grazing on Four Mile Prairie, but the area was strewn with buffalo horns, heads, and bones from one end to the other. There were no roads, and several herds of wild ponies could be observed. If homesteaders had hides to sell, they had to pack them on horses and travel fifteen miles through the woods to Canton, the county seat.

Four Mile Prairie was part of the Blackland Prairies region and was characterized by tall grasses, mesquite, oak, pecan, and elm trees along the streams. All kinds of wildlife were still abundant. Kaufman County was established on February 26, 1848, and Van Zandt County on March 20, 1848. Two years later Van Zandt County was split again. All land north of the Sabine River became Wood County. Most of the Norwegians lived in the southwestern part of Van Zandt County, but some lived just across the county line in Kaufman County to the west. The first post office in the county was established at Jordan's Saline in 1849. It was named for John Jordan, who had hauled two iron kettles through the dense woods from Nacogdoches and started to manufacture salt in 1845.[15]

The story of the first Norwegian immigrants in Van Zandt County is mentioned briefly in two county histories, W. S. Mills' *History of Van Zandt County*, and Margaret Elisabeth Hall's *A History of Van Zandt County*. Both authors got the story somewhat wrong. According to Mills, "the death of Grogaard, whose expected leadership in the colony had been a factor of importance in winning adherents in Norway, had discouraged prospective immigrants" to settle at Brownsboro.[16] Actually, Grøgaard had not been in Brownsboro or in Van Zandt County; he died of yellow fever early in 1846. The same author maintained that it was Elise Wærenskjold, the "pioneer school teacher" and "very talented writer" and her husband, "J. M. C. W. Waerenskjold, whom she had met and married in Texas the same year, who was also a gifted writer," who gave "the lagging interests in the colonization movement a revitalizing shot in the arm." Neither Elise nor Wilhelm Wærenskjold ever taught school in

Texas. Mills was further of the opinion that it was Johan Reiersen who in early 1848 "founded a second settlement in Van Zandt and Kaufman counties with Prairieville as its nucleus. The Waerenskjolds joined this colony the same year, and in 1850 it received a reinforcement of fourteen families from Norway. This colony was called New Norway, and at times Waerenskjold town." The *History of Van Zandt County, Texas*, published in 1984, repeated the above text verbatim without any citation marks.[17]

In her county history from 1975, Hall argued that it was the "outstanding leader," Johan Reiersen, who "created, located, even nourished a settlement."[18] According to Hall, Reiersen established Four Mile Prairie in southwestern Van Zandt County in 1848. "That same year he also purchased a farm near the town and persuaded three families to do the same. In accordance with state law they obtained homesteads of 640 acres for families and 320 acres for bachelors. Consequently, within two years Johan influenced fourteen more families to join the growing community."[19]

The different versions of Van Zandt County histories seem to follow the mainstream narrative among Norwegian-American historians. They reiterate the myth that Johan Reiersen founded both Brownsboro and Four Mile Prairie. There is no doubt that Reiersen established Normandy (Brownsboro), but he was not the founder of the Norwegian colony at Four Mile Prairie. Several of the Norwegian immigrants became so disgusted and frustrated with the conditions at Brownsboro during their first year in the colony that they decided to move to Four Mile Prairie, while others left East Texas for good and moved to the Midwest.

Reiersen stubbornly continued to argue that the Norwegians themselves were at fault for their sickness and death. In retrospect, he wrote that he had warned the immigrants about the risk of settling on the bottomlands, but they had not heeded his advice. If Reiersen had still wanted to be the colony's leader, he continued to lose credibility in 1847. To uphold his position as the leader of the Norwegian colony, he would have had to follow the other Norwegians from Brownsboro to Four Mile Prairie. He and his family, however, did not move to Four Mile Prairie until the second half of 1850.

CAN WE TRUST THE 1850 CENSUS?

In 1850, Van Zandt County had a population of 1,348. Subsistence farming was the major economic activity, and most of the inhabitants had migrated from states in the Old South: Tennessee (19 percent), Alabama (16 percent), Arkansas (15 percent), Missouri (13 percent), Mississippi (8 percent), Georgia (8 percent), Kentucky (5 percent), North Carolina (4 percent), Virginia (4 percent), and others (8 percent). The Norwegian immigrants constituted a significant minority with 6.25 percent.

Can primary sources shed light on the question of who founded the Four Mile Prairie colony—Reiersen or unnamed "simple Norwegian farmers"? If Reiersen founded Four Mile Prairie in 1848–49, should we not expect to find his name in the census for either Van Zandt or Kaufman County? Most historians would agree that census records belong to the category of primary sources, whereas books written a hundred years later belong to the category of secondary sources. Among historians, primary sources are generally regarded as more dependable than secondary sources. Census data rank high on the scale of documents to trust, although names are often misspelled, and dates can be wrong.

More than fifty Norwegians were listed in the 1850 census for Van Zandt County, but the names of Johan R. Reiersen, his wife, and his children are not listed in either the Van Zandt or Kaufman County census. Is it possible that Reiersen and his family had bought land in Van Zandt or Kaufman County, but had not yet moved there? If fifty Norwegians had already moved from Brownsboro to these two counties, why were not their leader and his family present among them? In fact, Reiersen and his whole family are found in the 1850 census for Henderson County. The conclusion must then be that the Reiersen family still lived in Henderson County when the census was enumerated in 1850.

After having gone through the census for Henderson County for 1850 household by household, it can be concluded that very few Norwegians lived in the county. Only eight were listed in the census, and slightly misspelled: merchant Johan R. Reierson (40 years), his wife Henriette (30), born in Denmark, and their oldest son Oscar Reierson (13), born in Denmark. Nine-year-old John H. was born in Norway as was his brother Christian A., while 1-year-old Charles Reierson was born in Texas. Johan R. Reierson had no assets listed behind his name, while the farmer James L. Reierson (30) and his wife Mary (19) had assets of $320. The farmer William Wærenskjold (27) and his wife Elise Wærenskjold (35) were also listed in the Henderson County census, and their total assets were $480. Thirty-nine-year-old Martha Olson was living in the Wærenskjold household. Twenty-year-old Norwegian Maria was married to the medical doctor Robert L. Mathews, who had total assets of $800.[20]

On the last page the census enumerator concluded: "Free Inhabitants in the County of Henderson amounts to 1157": 193 families lived in 192 households; 628 were males and 529 females. Henderson was a county full of immigrants. The majority of the 192 born in Texas were small children: 174 had been born in Tennessee, 112 in Missouri, 110 in Alabama, 98 in Arkansas, 79 in Kentucky, 69 in Mississippi, 57 in Illinois, 44 in North Carolina, 41 in South Carolina, 38 in Georgia, 28 in Virginia, 18 in Ohio, 18 in Indiana, 14 in England, 8 in Norway, 8 in Denmark, 7 in Louisiana, 7 in Massachusetts, 5 in Delaware, 4 in Germany, 3 in Pennsylvania, 3 in Florida, 3 in Maryland, 1 in Connecticut, 1 in Michigan, 1 in Ireland, 1 in Scotland, and 34 with an unknown birthplace. Among the inhabitants over 20 years of age, 98 were illiterate; they could neither read nor write.

The two richest men in Henderson County in 1850 were the attorney John R. Reagan and the farmer William C. Bobo, with $5,000 in assets each. Their landholdings averaged more than 2,000 acres, "which in East Texas during this period was equivalent to the landholdings of a middle-size planter."[21] Next on the list of wealthy people in the county were the farmers Jarred Erwin with $4,550, followed by G. B. Fancher and John Spikes with $4,000 each, Moses Cavett with $3,300, W. T. Brewer with $3,142, R. B. Thomas $3,000, William L. Gosset $2,000, and E. J. Thompson $2.302. The medical doctor R. W. Gray had assets of $3,000, and the blacksmith Victor Pannel had assets of $2,520.

Both the Reiersen and Wærenskjold families moved to Van Zandt County before the end of 1850. In a letter to Taale Andreas Gjestvang, Løten, Hedmark County, in December 1850 Elise Wærenskjold informed him that she and her husband now had moved from Brownsboro to Four Mile Prairie. She strongly believed that the Norwegians would have "a pleasant settlement here at Four Mile."[22] Johan R. Reiersen had also moved. He lived 2 miles from them, while three other Norwegian families lived in between, and four other families were about to settle. "I can truly say that this area is the most beautiful I have seen in Texas, and I have traveled a good deal," wrote Elise Wærenskjold. According to her, around 200 Norwegians now lived at Four Mile Prairie and Brownsboro.

The 1850 census for Henderson, Kaufman, and Van Zandt counties clearly documents that most Norwegian immigrants before 1850 first settled at Brownsboro, but soon continued to Four Mile Prairie. They felt no moral obligation to stay in Brownsboro to help Reiersen realize his big plans for a Norwegian colony under his leadership. This was also the conclusion of Nelson in his introduction to Reiersen's *Pathfinder*. According to Nelson, Reiersen "continued to move restlessly from one project to another—ran a boardinghouse and store, invested in real estate and founded towns, operated a mill." He seemed to have lost much of his energy and belief in his big project in Texas. The main reason for that, Nelson argued, was his large disappointment in fall 1845, when even "simple Norwegian farmers" stubbornly refused to go with him and build up his Normandy colony. Nelson concluded that "his active crusade for organized immigration was over. His zeal for other public causes, too, ebbed away."[23] Reiersen had fought civil servants in Norway relentlessly; in East Texas he admired "all things American." He was "blind to the injustices of the caste system of the antebellum South" and never criticized the slave system in writing, not even within the "safe confines of the Norwegian community." On the contrary, he became a strong supporter of the "Confederate cause" during the Civil War.

BROWNSBORO, NOT NORMANDY

Reiersen concentrated on his personal business affairs in the 1850s. He had named his colony Normandy, but the place was soon renamed Brownsboro in honor of John "Red" Brown, a wealthy man with considerably more political influence in the community than Reiersen would ever have. Brown was born in Ireland on October 30, 1786, although sources differ about the year and place of birth. He moved to Texas in 1836 and settled near Nacogdoches, where he practiced law and farming, and he probably died in Henderson County in 1852. Brown had represented Nacogdoches County in the House of the Sixth Congress of the Texas Republic (1841–42). After the annexation of Texas into the United States in 1845, Brown was elected to the First Texas Legislature.[24]

Brown was one of the founders of Henderson County. He was notary public in 1848 and operated a ferry across Kickapoo Creek from 1849. He received a license from the state of Texas to build a toll bridge near Normandy. Brown was one of the county commissioners who chose Athens as the county seat in 1850.

THE GRØGAARD FAMILY TRAGEDY

The letters from Johan R. Reiersen in spring 1843 leave no doubt that his friends Christian and Thomine Grøgaard were fascinated by the challenge of emigrating to a wholly new continent. On several occasions Thomine expressed a strong interest in the books about America that Reiersen had told them about. Christian Grøgaard was a member of the core group who arrived in New Orleans in summer 1845.

Both husband and wife had an upper-class background, certainly with respect to education. Christian Grøgaard was the son of Hans Jacob Grøgaard, a well-known Lutheran rector in Vestre Moland parish east of Kristiansand in Vest-Agder County in southern Norway. Hans Jacob Grøgaard was a member of the first Norwegian national assembly Stortinget in spring 1814. He was a typical "Enlightenment minister." He published schoolbooks; an ABC in 1815 and a reader in 1816.[25] Christian was known in the family as "the gifted son." At a young age he learned the scholarly languages Greek and Latin and the living languages French, German, and English, as well as history, geography, mathematics, and physics.

Christian knew that the family expected him to study theology and follow in the footsteps of his father. Every time an important exam came up, he got so nervous that he was not able to perform. In the end, the family had to accept that Christian Grøgaard would never pass his final university exams. He went to sea for a short period in 1815 and was steered in the direction of becoming a merchant.

At the age of 25, in September 1825, Christian married Thomine Ellefsen, the daughter of a ship's captain. Before their emigration in 1845 she gave birth to twelve children; two of them died young.[26] Grøgaard built a large house in the growing port of Lillesand in 1834 on the coast of Aust-Agder County. Despite the general boom in shipping and trade, he did not do so well. His brother Johannes Grøgaard, on the other hand, became one of the leading shipowners and merchants in Lillesand.

Family and social connections helped Christian Grøgaard get appointed sheriff (*lensmann*). After the introduction of the *Municipality Act* in 1837 (*formannskapsloven*), Grøgaard was elected to the city council in Lillesand. The town of Lillesand was small, even for Norway. The census taken in January 1846 counted 571 inhabitants. Kristiansand to the west had 8,349 inhabitants and was the sixth largest city in Norway.

Grimstad, east of Lillesand, had 806 inhabitants and Arendal 3,562.[27] The fortunes of all the coastal towns mentioned depended on the business cycles in the international shipping trade.

Christian Grøgaard belonged to an influential family; his cultural capital was high, and there was nothing wrong with Christian and Thomine's social position in their small urban community. They had problems fulfilling the expectations that went with the role, however, certainly in relation to family income, so husband and wife decided to do something really daring. Some of their closest relatives felt their decision to emigrate was a tragedy.

On September 30, 1845, about three weeks after Thomine Grøgaard left Arendal with her children, her sister-in-law Antonette Grøgaard recounted their departure and her feelings about it in a letter to her sister Anne Marie Müller: "It is such a pain to know that people you love are in distress and need, and not being able to help. It was a terrible sight to see Thomine and her ten children embark on the trip to America without knowing what to do or how to nourish themselves upon arrival. Johannes was very good to them. He has sacrificed significantly for Christian."[28]

Christian and Thomine made their choice to emigrate with open eyes. Thomine and the children crossed the Atlantic on the sailing ship *North Carolina*. According to Syversen and Johnson in *Norge i Texas*, Christian Grøgaard traveled to New Orleans to meet his family either from Marshall, where Reiersen left him in September 1845, or from Nacogdoches. Thomine and the children arrived in New Orleans on December 17, 1845. Grøgaard had decided to settle in Nacogdoches. The two smallest children had been very ill during the Atlantic crossing and never fully recovered. Wilhelm Frederik Krogh Grøgaard, age 4, died before Christmas in 1846, and his 2-year-old brother Johannes Jacob Peter Grøgaard died before the end of January 1847.[29] By then, however, their father was already dead.

THE DEATH OF CHRISTIAN GRØGAARD

Despite the many merchants operating in Nacogdoches, Christian and Thomine had agreed that Christian should try to establish himself as a merchant in town. In summer 1846 he traveled to New Orleans to buy the necessary merchandise. While in New Orleans, he contracted yellow fever. On the steamboat on his way back he became very ill. He left the boat and died at Grand Ecore, Louisiana, 8 miles north of Natchitoches, at a distinctive bluff on the Red River. It was nothing new for a traveler to contract yellow fever in New Orleans and die in Grand Ecore on the way home. Grand Ecore was almost wiped out by a yellow fever epidemic in 1853.

The death of her husband and their two smallest children was a hard blow for Thomine Grøgaard and her remaining children. Thomine's friend Elise Foyn (Tvede) from Lillesand, separated from her husband Svend Foyn, decided to visit Texas in spring 1847. After a long and arduous journey, she arrived in New Orleans on July 10. During her stay, Elise Tvede was informed that Christian Grøgaard had died the year before.

After her arrival in Nacogdoches on July 27, 1847, she was surprised to find Thomine Grøgaard in much better spirits than she had expected. In a letter to Norway she declared: "Friends of Grøgaard will be pleased to hear that his widow is doing quite well by sewing and baking for people." The widow had many children, but it was actually a blessing that the family lived in Texas, Elise Tvede argued, "as it will be much easier for them to make a living here than it would have been in Norway."[30] The two eldest daughters were working for prosperous families "who treated the girls like their own children." Three sons were in school, and two others would begin working with German businessmen in the fall. The Grøgaard family lived outside town in a one-room log cabin, and their living conditions were very spartan compared with the big house the Grøgaard family had left in Norway.

Soon the Grøgaard family was again visited by death. Thomine Grøgaard contracted yellow fever and died in 1848. At the time of her death, her eight children were Anna Maria (21 years), Helene (20), Elize (18), Christopher (13), twins Thomas and Hans Jacob, both 11 years, Emma (9), and Nicholas (6). The three oldest daughters were old enough to make their own living, but the five youngest children would need help. The immigrant community and townspeople of Nacogdoches tried to help them according to the norms and rules of the time.

THE GRØGAARD ORPHANS
IN THE IMMIGRANT COMMUNITY OF NACOGDOCHES

East Texas in the 1840s was a small world, despite large distances between towns and villages. Nacogdoches had 299 white inhabitants in 1847 and

103 slaves. On his journey through Texas in 1852, Frederick Law Olmsted described the town as "compact, the houses framed and boarded." The houses along the road into town stood in gardens and were neatly painted. Some of the old Mexican buildings were still standing but had been put to "the uses of the invading race."[31]

Nacogdoches was smaller than Lillesand, where the Grøgaards came from. Nevertheless, it was one of the oldest settlements in Texas, and most visitors found the town attractive, Olmsted included. Not many European immigrants lived in East Texas before 1850. Most of them knew each other, and the citizens in general knew about each other. It was no coincidence that Johan Reiersen and his fellow travelers had met some of the most prominent foreign citizens in Nacogdoches on July 4, 1845— men like Adolphus Sterne, George Bondies, and Joseph von der Hoya.

Best known among early Germans in Nacogdoches was Adolphus Sterne, born on April 5, 1801, in Cologne, Germany. His father was an Orthodox Jew, his mother a Lutheran. At the age of 16 he was working in a passport office when he learned that he was going to be conscripted for military service. He forged a passport for himself and emigrated to New Orleans in 1817. Sterne found employment with merchants in the city, studied law, and worked as a translator in English, French, Spanish, German, Yiddish, Portuguese, and Latin.

While living in New Orleans, Sterne joined the Masonic Lodge. He worked as an itinerant peddler and merchant with New Orleans as his base until 1824. He traveled as far north as Nashville, Tennessee, and he met and became friends with Sam Houston. Sterne established a mercantile house in Nacogdoches in 1826 but continued to go to New Orleans on business. The normal route was to travel overland to Natchitoches and then by steamboat to New Orleans. In Natchitoches he met Eva Catherine Rosine Ruff, born on June 23, 1809, in Württemberg, Germany. She had immigrated to Louisiana with her family in 1815 and had been orphaned soon afterward when her parents died of yellow fever. The Ruff children were adopted by Placier Bossier and his wife. On June 2, 1828, Sterne and Eva were married. She was a Catholic and persuaded him to change to Catholicism, but his diary clearly documents that he probably looked upon himself more as a deist along Freemason lines or as a freethinker. On the other hand, he enjoyed attending good revival meetings.[32]

Sterne was a strong supporter of the movement for Texas independence, and in 1835 he traveled to New Orleans to recruit an army. He personally financed two companies, later known as the New Orleans Grays. Sterne supported most of Houston's programs during the Texas Republic, except his Indian policy. In the battle of Neches, on July 16, 1839, Sterne commanded a company of militia, and he was strongly in favor of expelling the Cherokees from East Texas. On February 19, 1840, Sterne became postmaster of Nacogdoches. He served as deputy clerk and associate justice of the county court and became a Justice of the Peace in 1841.[33] In 1847, the year Sterne was elected to represent Nacogdoches in the House of Representatives of the Second Legislature, three females and two slaves lived in his house. He owned substantial amounts of land, estimated to be about 16,000 acres at the time of the 1840 census. Most of the daily comments in his diary describe matters related to land, his own and that of others. Sterne died in New Orleans on March 27, 1852.

Joseph T. von der Hoya and his family had emigrated together with three of his brothers from Damme in Oldenburg in northwestern Germany in 1836.[34] He was born on July 18, 1811, in Damme and died in Nacogdoches on May 21, 1896. The Hoya family settled on a farm 5 miles south of Nacogdoches. His brother Fritz von der Hoya worked as a merchant and is frequently mentioned in the diary of Adolphus Sterne.

Several stores were in operation in Nacogdoches in the 1840s, owned by P. F. Renfro, George Bondies, Sampson Shepherd, J. H. Muckleroy, and Charles Chevaillier. In the 1850 census, twelve heads of households were listed as merchants.[35] One of them was Charles Chevaillier, who sold "summer clothing and general goods." Chevaillier looms large in Sterne's diary. He was a wealthy merchant and landowner in Nacogdoches, and according to the tax roll for Nacogdoches in 1837 he owned 2,800 acres of land evaluated at $4,514. He was more affluent than Sterne, and for years Sterne owed him money. On January 5, 1841, Sterne "settled and paid his account to Chevallier up to date—(thank god)," and ten years later, on April 8, 1851, he again noted in his diary that he had settled his debt with Chevaillier. For the first time in ten years, he was finally "out of that man's debt, and if god lets me live I shall remain so forever."[36] On the other hand, Sterne frequently did business with Chevaillier, not least to pay off his debts.

Both Chevaillier and Sterne enjoyed parties and dancing. On August 12, 1841, a ball was held in the evening at "Mr. Chevallier's—was very much amused," and three weeks later there was another at Chevaillier's house, "given by Captain English—all good humored and well pleased." On May 12, 1842, Chevaillier again invited him to a "Dancing Party" and Sterne danced till near morning, a "fine party." Then as now people played several roles in society. On Monday September 26, 1842, the Probate Court was in session and Sterne noted that he had "made an arrangement of an amicable nature (final) between the Brothers von der Hoya think it a meritorious act to Keep *Brothers* from fighting when their difficulties can be settled Brotherly, Mr. John Durst & Mr. Chevallier had a falling out and nearly came to blows—bad that—don't like it." But two months later, on Thursday November 17, 1842, Sterne again attended a fine party at Chevaillier's.[37]

The ebb and flow of the fortunes of the businessmen of Nacogdoches often followed the same rhythm. On February 17, 1844, Sterne wrote in his diary that the brick maker and brick layer Wortham had arrived in Nacogdoches. He was to build a house for Fritz Hoya and one for Chevaillier, and if "all Ends meet I'll build one myself."[38]

Five men and four slaves lived in the Chevaillier household in 1847. At the time of the 1850 census Chevaillier was 37 years old, and he had assets of $30,000. His 19-year-old wife Sarah also lived in the household, as well as two Norwegians—Mariah Gregory (25) and her brother Thomas (14), in other words two of the Grøgaard children.

The merchant George Bondies was born on February 2, 1813, in Schleswig, Denmark, and died on July 7, 1894, in Galveston. He was listed as 36 years old in 1850 and had assets of $10,000. His wife Catharina (20) was born in Tennessee. Two small children lived in the household—4-year-old F. Gerrison from Denmark and 2-year-old T. Rohte from Germany. Two clerks were working for Bondies: J. Federson (32) from Denmark and Christopher Graygard from Norway. Bondies obviously had a soft heart for orphans, not only for Christopher Grøgaard.

The 1847 census for Nacogdoches showed that the Norwegian Gerhard Reiersen lived in the town, as did the blacksmith O. Røraas from Kristiansand. Three females lived in the Røraas household. In the household of James H. Starr, whom Johan Reiersen had met on July 4, 1845, lived two men, five women, and three slaves.[39]

INDENTURE CONTRACTS FOR THREE GRØGAARD BOYS

Lee Parsons has done painstaking historical searches for primary sources on the history of the Grøgaard family in Texas.[40] He discovered that the three oldest Grøgaard brothers were deeded to employers on the same day, March 30, 1849, in the Nacogdoches County Court. Most of the following narrative on the Grøgaard children builds on Parsons' work.

An indenture contract was signed by Sidney Maury Orton, chief justice of the County Court of Nacogdoches, between Christopher Grøgaard and the merchant George Bondies. It stated that the orphan Conrad Claus Christopher Grøgaard had been "placed and bound" according to the law with George Bondies "to live after the manner of an apprentice until he shall obtain the age of twenty-one years." Bondies promised that he would "teach and instruct or cause to be taught and instructed the said apprentice the Art, Mystery and Occupation in which said Bondies is engaged," as well as the "science of Bookkeeping and give him a good Common English Education and when he attains the age of twenty-one the sum of two hundred dollars in cash."[41] Parsons went through Bondies' business day book from October 1851 to April 1853 and found that he took good care of Christopher, providing him with shoes, five pairs of socks, pants, a blanket coat, $75 cash allowance, two pairs of boots, suspenders, a tie, a whisk brush, a fine comb, a silk handkerchief, and a bottle of cologne.[42] Two years after Christopher Grøgaard completed his indenture with Bondies in 1858, he died of pneumonia in Shreveport.

Thomas Fasting Grøgaard signed an indenture contract with Charles Chevaillier. The contract specified that he should learn marketing and bookkeeping. Charles Chevaillier died on December 30, 1852, at the age of 40. Little is known about what happened to the indenture contract with Thomas Fasting Grøgaard after his death. But at the end of the 1850s, Grøgaard had moved to Shelbyville to help his twin brother, Jake, run the Vinzent-Reiersen store.[43] In 1861, he was listed on the tax rolls of Cass County.

His twin brother, Hans Jacob Grøgaard (Jake), born December 24, 1836, had signed an apprentice indenture with Albert Aldrick Nelson, "to learn surveying." Nelson was born in Milford, Massachusetts, on May 15, 1814. After a dramatic career as a sailor on New England whaling ships between 1833 and 1838, he migrated to Texas.[44] Nelson was one of the first professional surveyors to settle in the Republic of Texas. He surveyed for the Nacogdoches Land District, which covered parts of East and Northeast Texas and held the post of elected county surveyor for almost fifty years. Nelson was an active Democrat, was also mayor of Nacogdoches, and was a friend of such public figures as Sam Houston, Thomas J. Rusk, and Adolphus Sterne. For a while he lived with the Sterne family.[45] On October 9, 1845, he married Jane Caroline Simpson, the daughter of a prominent East Texas family, and they had seven children. He died at his home in Nacogdoches on September 25, 1892.[46]

Jake Grøgaard did not stay long with Nelson. He was not very fond of mathematics, and in summer 1850 he persuaded Nelson to transfer his indenture to the farmer Raiford Fulghum. He did not like the farming apprenticeship with Fulghum any better, and abandoned this indenture contract as well. He moved into the household of his sister Helen, who had married Peder Georg Reiersen and lived in Shelbyville.[47] At the age of 15, Jake was appointed assistant postmaster in Shelbyville. He trained as a merchant with his brother-in-law and acquired managerial and marketing skills. He soon showed an aptitude for the cotton brokerage business. When his sister and brother-in-law moved to Mt. Enterprise around 1856, Jake was left in charge of the Shelbyville store; a year later he also moved to Mt. Enterprise.

The youngest brother, Nicolai Christian Keyser Grøgaard, who changed his name to Nicolas Christian, was born on September 22, 1840 and was taken into a foster home with a well-to-do family in Rusk County. According to Parsons, he had at least two foster care providers.[48] None of them were poor. They gave Nicolas a very good education, including studies at Larissa College. The students met in a log cabin but received "a classical education befit for community leaders," James Smallwood wrote. "Overseen by the Presbyterians, the school experienced steady growth through the 1850s,"[49] with 190 students by 1856–57. In a letter from Elise Wærenskjold to Norway on March 25, 1860, she commented that Nicholas Grøgaard "understands English, Greek and Latin, but not a word of Norwegian."[50]

The Grøgaard brothers lived in well-to-do households in East Texas and assimilated into East Texan society much faster than most Norwegians. Cultural assimilation can be defined as a process where immigrants adapt their language and culture to those of a larger and more dominant group. Full assimilation occurs when new members of a society become indistinguishable from the members of the majority group. Different stages of acculturation can be singled out—adaptation of language and customs, norms and values, social integration, and identification.[51]

The first generation of foreign-born immigrants is usually less assimilated to American culture than the second generation, while the third generation might be fully integrated into the dominant American culture. Fully assimilated immigrants not only integrated the culture of their new country, but also put their original cultural heritage, consciously or unconsciously, in the background. The Grøgaard brothers received a better education than most other settlers in East Texas at the time. They acquired important cultural codes from their host families, which certainly differed from the cultural codes in the Norwegian colony at Four Mile Prairie. Slaves and slavery were part of their daily lives. The Grøgaard brothers were well assimilated in East Texas society by the time the Civil War broke out.

CHARLES VINZENT, CAROLINE AMALIE REIERSEN, AND THE NORWEGIANS

Several of the Grøgaard children and other Norwegian immigrants established close relations with the German entrepreneur Charles Vinzent. According to Ralph Wooster, "one of the most versatile East Texas entrepreneurs was Charles Vinzent, 34-year old Rusk County merchant, who maintained a tin shop, tannery, blacksmith shop, saddler, and wool cordery in 1850." Vinzent employed twenty-one workers in his different businesses with an annual turnover estimated at $44,000.[52] He had $7,800 invested in his enterprises and was listed with assets of $22,000. Vinzent owned ten slaves. In the 1850s, before Vinzent moved to California with his family, he also manufactured wagons, buggies, and furniture. For a while he operated four stores outside Mt. Vernon, which sold his manufactured products.

The brothers Edward and Charles Vinzent were German immigrants born in 1813 and 1816 into a well-to-do family in the village of Ovelgönne in the Grand Duchy of Oldenburg in northwestern Germany. While studying at the gymnasium they read a letter in the newspaper about Texas, written by Friedrich Ernst, who had founded the town Industry in Austin County, Texas. This letter made a big impression on the two teenagers and many others and led to considerable emigration from the Oldenburg region to Texas. The Vinzent brothers emigrated in 1839 and settled in Rusk County in 1840, where Charles came up with a plan for recruiting Oldenburgers to the area. He parceled large land grants into smaller lots of 80 acres, which Oldenburgers with small means might find attractive. In 1843, he returned to Oldenburg to recruit emigrants.

Charles was the driving force in their business. This was reflected in an agreement the two brothers signed on October 6, 1851. Their joint estate was valued at $50,000, and it was divided in three parts with two belonging to Charles and one to his brother Edward. Before Edward died in 1854, the estate was valued at $120,000, of which three-fourths was ascribed to Charles and one-fourth to Edward.[53] Charles Vinzent established the Mount Enterprise Male and Female Academy in 1851, and for five or six years he paid the deficit every year. Vinzent also established a close business relationship with the large plantation owner Julian Devereaux and served as a cotton broker for Devereaux in the early 1850s. Devereaux was an important customer, and Vinzent sold him everything from an oven to ink. He was delighted when Vinzent was able to bring him newspapers.[54] Caroline, the younger sister of Johan Reiersen, was born on May 7, 1829, in Holt, Norway, and died 1882 in Oakland, California. Vinzent had known the Reiersen family since January 1846, after the steamboat they were traveling on sank in the Red River above Natchitoches. The Reiersen family survived but lost all their possessions, and was stranded in the cane breaks until another steamboat picked them up and brought them to Shreveport. It was there that Charles Vinzent got to know the Norwegians. He gave them shelter and clothing and transported all of them to Normandy.

Charles Vinzent married Caroline Amalie Reiersen on September 13, 1847. Elise Tvede found that Caroline Reiersen had made an excellent match. Vinzent was a "very kind man as well as industrious and wealthy," she wrote on January 5, 1848, in a letter from Nacogdoches to Taale Andreas Gjestvang in Løten, Norway. Vinzent had not had his own household before marriage but wanted to plant a garden and has "all kinds of fruit trees from Germany, among them some vines that apparently thrive here."[55]

Vinzent employed several Norwegian immigrants at his workshop in Mt. Enterprise. One of them was the rather eccentric wheelwright Reier Olsen Roa, from Næs Ironworks, Holt. Vinzent's brother-in-law Carl Reiersen had arrived from Wisconsin and was working for Vinzent as a carpenter. Another brother-in-law, Christian Reiersen (born 1814), operated a general merchant store in Cherokee County. He was hired by Vinzent to operate his first store in Larissa. Christian Reiersen died as early as 1853. The youngest Reiersen brother, Peder Georg Reiersen (born 1825), was operating the Vinzent store in Pulaski on the Sabine River in Panola County in 1848. He married Helen Grøgaard on October 4, 1849, and lived in household 287 in Panola County in 1850. No other Norwegians lived in that county in 1850. By March 1851, the couple had moved to the Vinzent store in Shelbyville, Shelby County.[56]

In 1855, Charles Vinzent began to sell off his real estate. Lee Parsons located fifteen sales transactions between 1855 and 1861 signed by Vinzent. After 1857, when Vinzent moved with his family to Oakland, California, he gave power of attorney to someone else.[57] Peder Georg Reiersen and his wife moved to Mt. Enterprise in Rusk County in 1857, where he probably took over an interest in the remaining Vinzent business. In 1859, Peder Georg Reiersen owned three lots in Mt. Enterprise, one horse, and three slaves.[58] As of 1860, Jake Grøgaard and 6-year old Mathilda Reiersen, the daughter of his brother who had lost her mother, also lived in the household.

SEVERAL EARLY NORWEGIAN IMMIGRANTS IN EAST TEXAS SETTLED IN TOWNS

The overwhelming number of Norwegian immigrants in Texas before the Civil War were farmers. They lived in Norwegian colonies and spoke Norwegian among themselves. The discussion of town life in Nacogdoches and the indenture contracts of the Grøgaard orphans have given us a broader understanding of the life of Norwegians in East Texas. Many of them were, in fact, town dwellers. The Grøgaard children, none of whom were taken care of by Norwegian families, quickly became integrated in the day-to-day interactions and exchanges of their small Texan town. But Norwegian families and single men were also integrated into the East Texas town life before 1860. Immigrants with education or trained as crafts- and tradesmen in Norway had skills to sell. They earned much more in towns than in the countryside.

More Norwegian immigrants settled in East Texas towns than anyone has previously been aware of. In his history of Smith County, Jim Smallwood wrote about the Norwegian Tom Albertson, who settled in the county seat Tyler. "The town baker Tom Albertson added an international flavor to early Tyler. Originally from Norway, Terje (Tom) and his brother Eilif came to East Texas in 1856, first settling in Normandy

(Brownsboro) in Henderson County. They soon moved to Tyler. On a corner of the square, Tom's bakery was the first in town. He sold quality bread, cake, beer and ginger ale, but his specialty was gingerbread, a delicacy from Scandinavia, which he exported to various towns. Even in faraway Dallas, people clambered for this specialty. The Albertsons would remain in Tyler and in the coming decades continued to be businessmen and boosters."[59]

CHAPTER 3

Cleng Peerson, the Father of Norwegian Immigration, Goes to Texas

◊ ◊ ◊

JUST BEFORE CHRISTMAS IN 1886, the dedication and unveiling of the Cleng Peerson monument took place in the graveyard of Our Savior's Lutheran Church at Norse in Bosque County, Texas. The monument was dedicated to "Cleng Peerson, the Father of Norwegian Emigration to America." It contained the following inscription in both Norwegian and English: "Cleng Pierson, the Pioneer of Norse Emigration to America. Born in Norway, Europe, May 17, 1782. Landed in America in 1821. Died in Texas, December 16, 1865. Grateful Countrymen in Texas Erected this to his Memory."[1]

In his book *The First Chapter of Norwegian Immigration*, first published in 1895, Rasmus B. Anderson characterized Peerson as "thoroughly unselfish in his character, and he devoted his life largely to the service of his countrymen. While he never had what might properly be called a home after he left Norway, he spent his time and his scanty means in getting homes for others." Anderson was not in doubt: "His great services to Norwegian immigration deserve to be remembered and appreciated, and with all his eccentricities and shortcomings his countrymen will look upon him as a benefactor to his race and as an honest and benevolent man."[2]

Anderson's positive view was soon contested by the Norwegian-American George T. Flom, who in 1907 characterized Peerson as a "shiftless" person, a "vagabond," and a "scoundrel." Not until his old age in Texas did he "yield to the monotony of a settled life." Peerson did not take an active part in the development of the Norwegian community in Texas, Flom wrote. He showed "no active interest in its progress. In a settled community he alone was unsettled; he was never able to gather himself together into concentrated action and prolonged effort in a definite cause or undertaking. A vagabond citizen he died in poverty. The only activity we associate with his name is the adventurous wanderings of his youth."[3]

In an article about Cleng Peerson and Norwegian immigration in the *Mississippi Valley Historical Review* in 1921, Theodore C. Blegen argued that it was Peerson who "led the way in the settlement of the Norwegians on American soil." In his view "thousands of natives of Norway and their descendants now occupying happy and luxurious homes in the Fox River Valley owe their prosperity and happiness in part at least to the leadership and efforts of that remarkable man."[4]

Toward the end of his life Peerson became an ardent promotor of Norwegian immigration to Texas and established a strong migration chain from western Norway to the mother colony in La Salle County, Illinois. From 1850 to his death in December 1865 he lived among Norwegian immigrants on the frontier in Central Texas. New material about Peerson has come to the surface in the last decades, and much of this material throws light on Peerson's last years in Texas and his influence on Norwegian immigration to Texas. This chapter focuses on his early work as a pathfinder for Norwegian immigrants in the 1820s and 1830s, and especially his last years in Texas from 1850 on.

ORGANIZED IMMIGRATION FROM NORWAY BEGAN IN 1825

Cleng Peerson was born in Tysvær, Rogaland County, western Norway, in summer 1782.[5] On April 2, 1807, he married the widow Ane Katrine Sælinger, 34 years older than he. She lived at a cotter's place under the farm Kindingstad on the island of Finnøy, Rogaland County. Peerson was her fifth husband.[6] In 1821, Quakers in the town of Stavanger gave Peerson the assignment to travel to New York to explore the conditions for foreign immigrants. The Quakers in Norway and Stavanger were few, but they felt persecuted by the Lutheran state church and wanted the religious freedom known in the United States.[7]

Peerson was not a Quaker but had Quaker sympathies. Between 1821 and 1824 he built up a good relationship with Quaker groups in New York State, especially the Quakers in Farmington in Western New York, located about 25 miles southeast of Rochester.[8] The Farmington Quakers were the "pioneers of all Quakerism in western New York, and also they were the first white men to bring their families into the vast forest of New York State west of Seneca Lake for the purpose of transforming Seneca's wilderness hunting ground into a white man's homeland."[9] Groups of Quakers also settled in nearby Macedon and Palmyra.[10] The

Quakers were active participants in the economic boom connected with the building of the Erie Canal.

Peerson returned to Norway in 1824 and gave a positive report to his sponsors on the prospects for emigration to Western New York. His sponsors decided in favor of emigration and gave Peerson a new assignment: to locate and buy land for them and do other necessary preparations before their arrival in New York the following summer. In a letter to Norway, dated December 20, 1824, Peerson wrote that he had executed important parts of his assignment. "It is well known that Cleng Peerson traveled from the Quaker settlement of Farmington, Ontario County, to Geneva," wrote Richard Canuteson in 1954, "where he purchased land for the prospective settlers from Joseph Fellows, subagent under Robert Troup for the Pulteney Estate."[11] The Pulteney Estate was selling vast amounts of land in the region. The land commissioner in Geneva received him cordially, wrote Peerson. Fellows had promised to aid the Norwegian group as much as he could when they arrived. "We arrived at an agreement in regard to six pieces of land which I have selected, and this agreement will remain effective for us until next fall."[12]

The letter from Peerson in December 1824 is crucial for our understanding of what happened before the first group of Norwegian emigrants left Norway in the summer of 1825, as the decisions made on their behalf before their arrival defined a subsequent path of dependency. Peerson bought land only a short distance away from the Erie Canal, the most dynamic economic region in the United States at the time.[13] The region was also a hothouse for religious revivals and new religious sects.

THE SLOOPER *RESTAURATION* LEFT NORWAY ON THE 4TH OF JULY 1825

Forty-five men, women, and children and a crew of seven emigrated from the city of Stavanger in western Norway on the small sloop *Restauration*, hoping to find a Canaan's Land in America. The journey of the *Restauration* across the Atlantic to New York City marked the beginning of organized emigration from Norway to the United States.

The Atlantic crossing and the dramatic events following the arrival of the emigrants in New York City have been covered by a number of historians and novelists over the years. The *Restauration* arrived at the port of New York on October 9, 1825, after 98 days at sea. "Cleng

Peerson met the immigrants at New York, as he had promised," wrote Blegen, "and the connections that he had already made in that city stood the party in good stead."[14]

Many stories about the *Restauration* and its passengers are inconsistent. Gunleif Seldal, who has done extensive research in primary sources on the history of the preparations for the voyage of *Restauration*, the voyage itself, the background of the passengers on board, and their new life in New York, observed that a lot of what had been written was "erroneous, copying has been extensive, myths abound."[15] Rasmus B. Anderson presented a list of the passengers on board the sloop in his book *The First Chapter of Norwegian Immigration* in 1895. Since then many researchers have put in tremendous amounts of time trying to verify or refute the names and numbers of the emigrants on board. Seldal concluded that the existing primary sources from 1825 can be trusted to a higher degree than documents written decades later. According to the health certificate from the magistrate in Stavanger, there were 52 passengers on board. In his report to the Norwegian government, the bishop reported that there were 51 passengers, and US customs officials registered 52 passengers. This information is still highly reliable, according to Seldal.[16]

For a vessel of her size, the sloop had far more passengers on board than American law allowed, exceeding the allowed number of passengers by 21 persons.[17] After arriving in New York, the vessel was confiscated, the captain was arrested, and the owners were given a severe fine of $3,150. The land agent Joseph Fellows was present in New York with Peerson to receive the Norwegians. Fellows was an influential man with a large network of people in politics and business he could draw on to help the Norwegians solve their legal problems. A petition was prepared and forwarded to the Secretary of the Treasury. It went all the way up to President John Quincy Adams, who instructed his administration to prepare a pardon with these words: "Let the penalty and forfeitures be remitted on payment of costs so far as The United States are concerned. J. Q. Adams. November 5, 1825."[18]

The *Restauration* was sold at a great loss; the sale brought only $400, less than a fourth of what had been paid for the ship in Norway. This meant that the Norwegians had considerably less capital to finance their new Norwegian colony than planned. Their lack of capital on American soil would come to haunt them later.

The majority of the "Sloopers" boarded a steamship from New York to

Albany on October 21. The last stage of their travel up the Hudson River and west on the Erie Canal began only four days before the official opening of the whole length of the Erie Canal.[19] The group arrived at Rochester and continued west to Holley, the harbor on the Erie Canal closest to their land in Orleans County, where they arrived in early November.

They settled on the land Joseph Fellows and Cleng Peerson had agreed on a year earlier. According to long-held and stubborn traditions in the Norwegian-American immigration literature, each adult received a tract of 40 acres of land and paid 5 dollars an acre. The Pulteney Estate gave the Norwegians the same sales conditions as other buyers:[20] They bought land on credit and would have to pay off all their debts to the Pulteney Estate before they would get title to their land.

According to Rasmus B. Anderson the land was "sold to the Norwegians by Joseph Fellows at five dollars an acre; but as they had no money to pay for it, Mr. Fellows agreed to let them redeem it in ten annual instalments."[21] The original source of this information was Ole Rynning in 1837.[22] Almost all later writers have accepted Rynning's account uncritically, observed Richard Canuteson in 1954.[23]

During his research on the Pulteney papers, Canuteson discovered a map showing that Cleng Peerson was the owner, "presumptively at least, of four tracts, alone or in partnership with someone else: Lot 15, at the mouth of the creek, 78.52 acres; jointly with Nelson (probably either Nels or Cornelius Nelson Hersdal), the east part of Lot 25, approximately two thirds of the 154.89 acres; the south part of Lot 26, possibly three fifths of the 170.44 acres; and the western part, about two thirds, of Lot 27, 153.70 acres."[24] Between 1825 and 1833, very few of the Norwegian immigrants were able to pay even small amounts on their land.[25]

NO MILK AND HONEY FOR NORWEGIANS IN WESTERN NEW YORK

The land Peerson had bought was located in a swampy, forested area where in many places water was four feet deep. "The land was thickly overgrown with woods and difficult to clear," according to Rynning. "Consequently, during the first four or five years, conditions were very hard for these people. They often suffered great need and wished themselves back in Norway."[26] Their American neighbors knew from experience how critical the first year was for a new settler. The most common items pioneers brought with them were tools and seed. Building a shelter for the family had top priority, followed by a log cabin. "Both the lean-to and cabin would be built from materials at hand, and often simultaneously with the clearing of land." The planting of the first crop of corn was crucial. When the settlement was on wooded land, a suitable site would have to be cleared. "Whatever the location and circumstances, corn was the universal crop on the trans-Appalachian frontier. It needed little cultivation."[27]

Most of the Norwegian immigrants did not know what corn was, and they arrived at their colony too late in the fall to get a crop in the ground. For more than a hundred years, generations of settlers had known that: "Without this gift of the Indian, so easy to plant and so adaptable to frontier conditions, so nourishing and with so many uses, the cycle of life and labor on the early frontier would have been different. From the time of his arrival, the main effort of the early settler was directed to the cultivation of the corn crop."

The Norwegians had to buy their food the first winter, and most of them did not have any cash. Many experienced extended periods of illness, and many died during their first years in Western New York.[28]

The Norwegian immigrants had settled in a dynamic economic region, but few among them tried to get employment outside their own circle. Those who had skills needed outside agriculture for employment had large language problems. Most of them did not speak English, and they tended to seek security by clustering together.[29]

For several seasons the first Norwegian immigrants in Western New York worked hard to grow enough food to last through a long and cold winter. In their letters home to Norway, they were unable to hide their disappointment with their new life in the woods. Very few Norwegians followed in their footsteps. The first Norwegian colony in western New York was definitely not a Canaan's Land.

THE ESTABLISHMENT OF THE NORWEGIAN MOTHER COLONY IN FOX RIVER, ILLINOIS

The year 1834 marked the beginning of the breakup of the New York Colony. Again Cleng Peerson "appears in the role of trail blazer, set out on an exploratory journey to the West in search of suitable lands for settlement by the Norwegians," wrote Theodore C. Blegen. "His travels took him into Ohio, across Michigan, and through northern Indiana into Illinois. It has been asserted that he walked along the shore of Lake

Michigan as far north as the present site of Milwaukee. Here he met the founder of that city, the famous fur-trader Solomon Juneau, who in response to an inquiry informed him that Wisconsin land was heavily forested and entirely unsuitable for settlement."[30] It was in spring 1833 that Peerson walked the Indian trail along Lake Michigan between Chicago and Green Bay. The trail passed Solomon Juneau's trading post on the Milwaukee River.[31]

Juneau was still the leading fur trader on the Milwaukee River when Peerson met him. Two years later, in 1835, land sales in Wisconsin skyrocketed. Every steamboat arriving at Green Bay that spring brought speculators from the East. They were hunting for fertile land, mill sites, water-power rights, and land for urban centers. Homesteaders, interested in farm land, followed on the heels of the speculators.[32] Had Peerson arrived three years later, he would have found close to 3,000 persons living in Milwaukee. Peerson, however, trusted the word of Juneau and returned to the small town of Chicago. He then walked in a southwesterly direction until he arrived in the recently organized La Salle County. The Black Hawk War was over. Land speculators were convinced that the Illinois and Michigan Canal would be built through La Salle County.

Blegen conceded that he had no primary sources to document that "Peerson served as the chosen representative of the Kendall settlers." But "his glowing reports on conditions in the West found ready acceptance among the New York colonists, for in 1834 six families moved to Illinois and there established the second Norwegian settlement in the United States—the 'Fox River settlement.'"[33]

There are several versions of how the Norwegian Fox River settlement was established; all are based on secondary sources and indirect evidence. Carlton Qualey wrote about Cleng Peerson that "the characteristic feeling of restlessness seems to have possessed him, reinforced by a conviction that Western New York was perhaps not the most desirable place in America for Norwegian settlement."[34] The best known and most romantic version of why Peerson chose Fox River and La Salle County was presented by the Norwegian-American newspaper editor Knud Langeland. Peerson returned to Norway in 1842–43 in connection with the death of his father. While in Norway, he met groups of people interested in emigration. Langeland was present at one of these meetings. Peerson had told the audience that after having walked for days west-southwest from Chicago he finally came to a hill overlooking the Fox River Valley. "Almost dead of hunger and exhaustion as a result of his long wandering through the wilderness, he threw himself upon the grass and thanked God who had permitted him to see this wonderland of nature. Strengthened in soul, he forgot his hunger and sufferings. He thought of Moses when he looked out over the Promised Land from the heights of Nebo, the land that had been promised to his people."[35] Since Langeland was a prospective settler, Blegen comments in a footnote, Peerson might have designed the tale to give Langeland and other listeners a positive picture of America. Nevertheless, Langeland's story became the main source for numerous accounts of what has been termed "Cleng Peerson's dream."[36]

WHO WAS JOSEPH FELLOWS?

The main argument in this book is that it was Joseph Fellows who made Cleng Peerson aware of the possibilities on the western frontier in Michigan, Wisconsin, and Illinois. When Peerson met Fellows the first time in fall 1824, Fellows was a lawyer and bachelor his own age. He had worked for and with rich and influential people since he was 20 years old.

Joseph Fellows Jr. was born in Redditch, Worcestershire, England, on July 2, 1782. The Fellows family emigrated to New York in 1795. They brought with them five children; Joseph was the eldest, while Lydia, the youngest, was born during the Atlantic crossing. After staying some weeks in New York City, the Fellows family continued to Luzerne County, Pennsylvania. They settled there, and during the next years the family became involved in the coal mining business.[37]

Fourteen-year-old Joseph Fellows Jr. remained in New York. He signed an indenture contract to become a lawyer with the well-known lawyer Isaac L. Kip on June 24, 1796. He was to live in the Kip household and be educated in law, and at the end of his indenture he would become a lawyer. Kip was a descendant of the well-known Dutch Kip family who had lived on Manhattan Island since Holland ruled the colony.[38] During his years in the Kip household Fellows met influential men in American politics and business life, men such as Alexander Hamilton and Aaron Burr. He also got to know Colonel Robert Troup. When Fellows received his certificate to practice as a lawyer in 1803, Troup, who was the main agent of the Pulteney Estate, offered him work as a lawyer for the Estate with a yearly salary of $600.[39]

Troup was very satisfied with Fellows' work during the following years. In 1810 Joseph became the sub-agent at the Geneva office of the Pulteney Estate, and his salary was increased to $2,000 per year. In 1824, after Troup moved from Albany, the state capital, and back to New York City, Fellows took over the management of the Pulteney Estate for the entire area of Western New York. In 1832 he was appointed the main agent of the Pulteney Estate with a yearly salary of $5,500.[40]

The Pulteney Estate allowed Fellows to speculate in land on his own account. Fellows knew that the best land deals in the future would be made in the Midwest. Among Norwegian-American historians, only J. Hart Rosdail has argued that Peerson might have been following the advice of Joseph Fellows. In his book *The Sloopers. Their Ancestry and Posterity*, published in 1961, he discussed several reasons why Peerson went on his exploratory trip in 1833. Peerson might have been sent by the Sloopers to "verify rumours of abundant and cheap land."[41] Or Peerson might have gone "in response to his own well-recognized urges to explore the country; or he may have been sent by the Quaker land agent, Joseph Fellows."[42]

A closer study of the activity of Fellows and his extended family in this period confirms the importance of the connection between Fellows and Peerson. The relationship between Fellows and the Norwegians lasted into the 1830s and beyond. He had excellent contacts with leading politicians and investors and was well acquainted with the political debates on land sales in the Congress and Senate. The election of Andrew Jackson as President of the United States in 1828 represented a political earthquake. Jackson and his supporters worked toward greater democracy for the common man, later known as "Jacksonian Democracy." They believed voting rights should be extended to all white men. Jackson worked to end what he termed the "monopoly" of government by elites. He and his followers also strongly believed in the concept of manifest destiny, that white Americans should settle the American West from the Atlantic Ocean to the Pacific. Westward lay freedom, eastward lay old Europe.[43] The West should be settled by yeoman farmers, men who owned and farmed their own land.

Joseph Fellows knew that the policies introduced by Andrew Jackson would have a negative effect on the profits of the Pulteney Estate and other land projects Fellows was involved in. The best land investments in the future would be in western territories. Plans for the Michigan and

Illinois Canal from Chicago to Ottawa in La Salle County were progressing. Land prices on both shores along this waterway would increase when construction work got under way.

Fellows had known Cleng Peerson since 1824. He knew that Peerson was an intelligent and sharp observer who was skilled in retelling what he observed. There are strong reasons to support Rosdail's suggestion that this time Peerson was not an "advance agent" for fellow Norwegians, but for Joseph Fellows.

THE SETTLING OF LA SALLE COUNTY

The first settlers in La Salle County, Illinois, had grown up in the uplands of the Carolinas, Virginia, Tennessee, or Kentucky. They moved north along the rivers into southern Illinois and had ties with New Orleans and the growing steamboat trade on the Mississippi River.[44] Their settlements had a strong southern character. La Salle was organized as a county in winter 1830–31, and the first election was held in Ottawa on March 7, 1831.[45] During the Black Hawk War in 1831 and 1832, immigration came more or less to a standstill, and homesteaders retreated to stronger settlements to the south,[46] but Joseph Fellows and other land speculators expected western migration to rebound strongly the moment a treaty was in place. The Black Hawk War treaty with the Indians was signed in Chicago on September 26, 1833. The Indians were forced to cede their rights to roughly 1,300,000 acres of land in northern Illinois. A stream of new settlers from the northeastern states arrived in La Salle County in 1833, the year Cleng Peerson visited the county for the first time. Most of them settled close to the proposed route for the Michigan and Illinois Canal along the Illinois River.

The opening of the Erie Canal in 1825 had led to massive migration from the northeast states to the western frontier, as it reduced the average traveling time between Buffalo and Detroit to two and a half days at a cost of about 15 dollars.[47] Most of the good soil in New England, New York, and New Jersey had been taken by 1830. Yankee settlers or settlers with Yankee roots constituted the majority of the first settlers in Michigan,[48] and southern Michigan became known as "Greater New England" or Yankeeland. In her book *The Yankee West. Community Life on the Michigan Frontier*, Susan Gray defined the region as "extending west from New England along roughly longitudinal lines through upstate New York, Ohio's Western Reserve, the southern half of Michigan's lower peninsula, northern Indiana, and Illinois."[49]

The first trickle of settlers into the southern Michigan peninsula and to the land around Detroit began around 1820. The Detroit land office, which opened in 1818, sold 92,332 acres in 1825 and 217,943 acres in 1831. "Then probably under the influence of the Black Hawk War and the epidemics of cholera in 1832 and 1834, there was a gradual decline until 1835 when sales suddenly leaped to 405,331 acres, and in 1836 to nearly one and a half million acres."[50] In the wake of the financial panic in 1837, the land boom halted. The Michigan population increased from 28,004 in 1830 to 212,267 in 1840, a growth of 658 percent.[51]

When the ice broke up on Lake Erie and Lake Michigan in spring 1833, months before the Indian treaty was signed, Cleng Peerson began traveling west. He first went to Michigan, where Joseph Fellows and other members of his family had speculated in land in the early 1830s. Fellows and his brother-in-law Joseph Edward Hill bought large tracts of land in Berrien County, Michigan.[52] A letter from Hill to Fellows, dated Berrien, October 5, 1837, clearly documents their land speculations. Joseph Fellows had bought land near Goshen for 8 dollars an acre. A canal passed through this land, and it was now worth $50 an acre, Hill reported.

But land speculation in Illinois was at least as attractive. Sales in Illinois increased by 600 percent between 1834 and 1835: 354,010 acres were sold in 1834 and 2,096,623 acres in 1835.[53] The Illinois population increased from 157,445 in 1830 to 476,183 in 1840. "Chain migration and group migration ushered many Yankees, other Northerners, and foreigners to Illinois, especially to the northern half—the booming half—stamping that region with unique imprints."[54]

NORWEGIANS SETTLED IN LA SALLE COUNTY IN 1834

Cleng Peerson and the Norwegian immigrants were an insignificant creek in a large westward-flowing migration river. "The arrival of these foreigners gave a forecast of the influence which the northern line of transportation was to have upon the character of the settlements in these counties, for soon there was to be a great throng of foreigners poured through the Chicago gateway upon the prairies."[55] Neither the Norwegians nor many of the other early homesteaders in La Salle County got

Cleng Peerson, along with a stream of immigrants from northeastern states (Yankeeland), moved west after the Black Hawk War. They traveled west on the Erie Canal to Buffalo, New York, crossed Lake Erie by boat, and disembarked in Detroit. From Detroit they walked or rode to a harbor in the southwestern part of Michigan and by boat crossed Lake Michigan to Chicago. When Peerson arrived in Chicago, he first walked north along the shore of Lake Michigan to Milwaukee. He turned there, walked back to Chicago, and continued southwest to Ottawa and the Fox River Valley in La Salle County. He praised the quality of the land in the Fox River Valley, which became the mother colony for Norwegian immigrants.

paper title to the land they settled on in 1833 and 1834. The US Congress passed pre-emption laws at different times. Claimants who had made certain specified improvements received the exclusive right to purchase the land they had claimed for a minimum price of $1.25 per acre.[56]

The following Norwegians moved from Western New York to Fox River in 1834: the families of Gudmund Hougaas, George Johnson, Jacob Anderson, and Andrew (Endre) Dahl as well as Torsten Olson Bjorland, Niels Thoresen Brastad (Nels Thompson), and Cleng Peerson. In spring 1835 came the families of Ole Olsen Hetletvedt (1797–1854) and Daniel (Stensen) Rosdail (1779–1854). Kari Pedersdatter (1787–1846), the widow of Cornelius Nelson, came with her family in 1836. Her brother Cleng had purchased land for her and her children. The Norwegians settled mainly in the northeastern part of La Salle County, in the townships of Miller and Mission, but some lived in Adams, Northville, and Serena.[57]

A new land office in Chicago opened in 1835.[58] As the administrator of the Pulteney Estate in New York, Joseph Fellows knew exactly what each Norwegian settler owed to the Estate. Fellows helped the Norwegian immigrants get out of their old debts in Western New York and ensured they were paid for improvements they had made so they could start over again in Illinois. When the Chicago land office opened, Fellows was present. He bought a lot of land for himself, but he also helped the Norwegians register their new homesteads. On June 17, Cleng Peerson registered 80 acres for himself in Mission Township and 80 acres for his sister Carrie Nelson, and registered another 80 acres of land on June 25. It was, however, land speculators like Joseph Fellows who acquired the lion's share of the land for sale in 1835. The same month Fellows helped the Norwegians, he registered at least 5,000 acres of land around the Fox River settlement in Illinois in his own name.[59]

The connection between Fellows and his family and the Norwegian immigrants was strengthened through the marriage in 1839 between Benjamin Beach Fellows, Joseph's nephew (born in Scranton, Pennsylvania, in 1811), and 17-year-old Marthe Karine Hersdal Nelson (1822–1913), the daughter of "Slooper" Hersdal; Peerson was her uncle. Benjamin Beach Fellows had settled in the town of Mission, La Salle County, on May 1, 1835. When Beach Fellows was elected County Treasurer of La Salle County in 1855, the family moved to Ottawa and lived there the rest of their lives.

GOOD NEWS FROM FOX RIVER:
THE BREAKTHROUGH FOR NORWEGIAN EMIGRATION

In the course of a few years the Norwegian Fox River settlement in La Salle County, Illinois, became the mother colony for thousands of Norwegian emigrants. The letters home to family and friends in Norway from Fox River had a much more optimistic tone than the letters written from Kendall, New York and helped build up emigration "pull." Families and friends in the municipalities and regions in western Norway where the "Sloopers" came from were infected with the urge to emigrate.[60] Farmers sold their farms and bought tickets for their families for the journey from Norway to Fox River.

The breakthrough for organized Norwegian emigration came in 1836 and 1837.[61] On May 25, 1836, the brig *Norden* left Stavanger with 110 passengers on board, arriving in New York on July 20. A couple of weeks later, on June 8, the brig *Den Norske Klippe* left Stavanger with 57 passengers on board, arriving in New York on August 15, 1836. Two more emigration ships left Stavanger in spring 1837. The bark *Ægir* arrived in New York on June 11, followed some weeks later by the bark *Enigheden*, which arrived in New York on September 14, 1837, with 91 emigrants on board, including 23 families.[62] "In size and influence no other group of immigrants in the first generation of Norwegian immigration can compare with the 343 passengers of these four ships that constituted the bulk of the exodus of 1836 and 1837," Henry J. Cadbury argued.[63] Between 1836 and 1845, emigration to the United States had spread to nine Norwegian counties. The total number of emigrants remained small; the average was 620 persons annually. It was during those years, however, that many western Norwegians got "emigration fever."[64]

NORWEGIAN SETTLEMENTS IN LA SALLE COUNTY AND BEYOND

A number of Norwegians arrived in La Salle County in 1836 and 1837, just before and after the US financial crisis broke out in 1837. The financial panic brought immigration almost to a standstill. Work on the canal in La Salle County was suspended in 1839.[65] Norwegians already living in the county tried to help the immigrants who had just arrived as best they could by housing and feeding them. They also gave advice with respect to better opportunities in Illinois, Wisconsin, or Iowa.

Cleng Peerson played a crucial role in the establishment of Fox River as the Norwegian mother colony.[66] Because of the great influx of Norwegian immigrants to Fox River in 1836 and 1837, the immigrants asked Peerson to scout for new land again. This time he traveled south and chose land in Shelby County in northeastern Missouri.[67] In a history of Shelby County, published in 1911, it was noted that "a small colony of Norwegians wandering about the country decided to settle at the headwaters of North River" in 1839.[68] The first group arrived in March 1837. A party of twelve to fourteen persons moved from La Salle County together with Cleng Peerson, Jacob and Knud Anderson Slogvig, and Andrew Simonson.[69] In a letter dated Ottawa, April 30, 1837, John Nordboe wrote that he and his family were ready to move to the new Norwegian colony in Shelby County, Missouri.[70] He sold his farm in La Salle County in 1836 for $400.

A fairly large group who came directly from Norway in 1839 settled in Shelby County. The land in Shelby County, however, soon proved disappointing, and in 1839 several Norwegians moved north to the recently organized Norwegian colony in Sugar Creek in Lee County in southeastern Iowa.[71] Peerson followed the others to Sugar Creek. They found a place that reminded them of home 6 miles into Lee County. However, neither the Shelby County settlement nor the Sugar Creek settlement was a success. A stream of immigrants in the 1840s preferred new Norwegian settlements in Wisconsin.

Primary sources document that Cleng Peerson was back in Norway in the second half of 1842. He wanted to visit his father, who had been sick, but found his father had died in 1841. Peerson participated in the legal process of distributing his inheritance[72] then returned to the United States in 1843. He stayed in Missouri and Iowa until 1847, when he joined the Swedish religious sect of Eric Jansson at the Bishop Hill Colony in Henry County, Illinois.

RELIGIOUS WILDFIRES AT FOX RIVER— NORWEGIANS AND MORMONISM

The Norwegian immigrants who settled in the Fox River Valley were unorthodox with respect to religion, wrote Theodore Blegen. They showed little interest in hiring an ordained minister from the Norwegian state church, and many had a "deep-rooted antagonism to the established church of their native land." Lay preachers filled their spiritual needs, especially those with a background from the Hauge movement in Norway.[73]

Nauvoo, on the eastern side of the Mississippi River in Illinois, became the new Mormon headquarters in the summer of 1839. Subsequently, a remarkable number of Norwegian immigrants in the Fox River Valley, Sugar Creek, Iowa, and Koshkonong in the Wisconsin Territory converted to Mormonism in the early 1840s. In March 1842, the Fox River Norwegians were visited by George P. Dykes, one of Joseph Smith's traveling elders. Within a month he had converted the Norwegian Hauge follower and religious leader Ole Heier from Telemark, the schoolteacher Jørgen Pedersen, and the "Sloopers" Andrew Dahl and Gudmund Hougaas.[74]

Knud (Canute) Peterson, age 18, who had emigrated with his parents in 1837, was baptized a Mormon together with his widowed mother and two of his best friends, Swen and John Jacobs, on August 12, 1842.[75]

When Peterson visited Nauvoo in 1844, he was ordained a "Seventy" in the Mormon hierarchy and called to take a mission in Wisconsin. There he succeeded in baptizing several Norwegians and organized a new branch of the Mormon Church. In January 1843, Hougaas and J. R. Anderson spent three weeks among Norwegians in Lee County, Iowa; ten persons were baptized. Hougaas and Ole Heier also made a missionary tour among Norwegians in Wisconsin.[76]

Norwegian Lutheran Church leaders were not happy about the inroads Mormonism made among the Norwegian settlers. Pastor J. W. C. Dietrichson thundered that the sect was destroying the Lutheran unity.[77] By 1845, nearly a hundred and fifty Norwegians in the settlements had accepted Mormonism, some eighty members in the Fox River colony alone.[78]

Sarah Ann Nelson, the daughter of the Slooper Cornelius Nelson Hersdal, and the second Norwegian child born in America, became a Mormon. She was a member of the six wagons that left Fox River for Utah on April 18, 1849. In Mormon history, the group became known as the Norwegian Company. After battling waist-deep snow in the mountains, the group reached their destination on October 25, in time to be numbered in Utah's first census, along with one Swede and two Danes. Sarah Ann married Canute Peterson on the journey to Utah.

CLENG PEERSON JOINED THE UTOPIAN SWEDISH BISHOP HILL SECT

Cleng Peerson was not attracted by Mormon teaching. Peerson, however, was at this time far from immune to the religious wildfires scorching the area. He was strongly attracted to the new Swedish Bishop Hill sect, established by the Swede Eric Jansson in Henry County, Illinois, 80 miles west of La Salle County. The story of the Bishop Hill commune is another interesting chapter in the rich history of religious utopias in antebellum America. Life during the brief existence of the Swedish colony in northern Illinois was intense and often violent.[79] In his book *Wheat Flour Messiah. Eric Jansson of Bishop Hill*, Paul Elmen traced the spiritual odyssey of Eric Jansson from his youth on a Swedish farm until he was murdered at Bishop Hill in 1850.[80]

Jansson emigrated from Sweden with 400 of his followers in 1846. The economic base of the colony, established in Henry County, Illinois, the same year, was agriculture and handicraft. The colony was organized along communist principles, which Janson argued were laid down by Christ himself. Cleng Peerson sympathized with the communitarian ideals promoted by Jansson to such a degree that he sold all he owned in 1847. He "joined the famous religious-communistic settlement in Henry County, Illinois—the Swedish Bishop Hill Colony," wrote Blegen.[81]

Jansson was in favor of celibacy, and during the first two years at Bishop Hill imposed celibacy on the colony. Suddenly it dawned on him that this practice would mean the destruction of his own sect in the long term; no young people would be recruited. The policy changed in summer 1848, when Janson ordered a wedding boom. On June 25, 1848, Jansson married four couples, followed by three couples on July 2, five couples on July 9, five couples on July 16, twenty-four couples on July 23, and nineteen couples on July 30.[82] Peerson was also on the marriage list at Bishop Hill. On September 3, "Kleng Peerson was married to Miss Maria Charlotte Dahlgren."[83] She was born May 26, 1809, in Sødra Bongsbo in Österunda sokn, Uppsala län, Sweden. Dahlgren was 39 and Peerson 65 when the marriage took place. There is no reason to think that they married for love; their marriage was first and foremost a practical matter ordered by the sect leader Eric Jansson.

The international cholera epidemic reached Bishop Hill in July 1849, brought by a group of recently arrived Norwegian converts. None of the Norwegians died, but after they had recovered, Jansson ordered them out of his colony. According to the Jansson's teachings, true believers would not succumb to physical disease.[84] Nevertheless, between July 22 and the middle of September 1849, an alarming number of colony members contracted cholera and died. "The cholera epidemic was catastrophic for the Bishop Hill Colony, not only because two hundred of the members died, but also because it exposed once and for all the theological weakness of the perfectionist principle," Elmen observed.[85] Both Cleng Peerson and his wife got very sick with cholera that summer. Cleng survived, but his wife died. Together with many other survivors he left the colony. When members had settled at Bishop Hill, they gave all their money to Eric Jansson. None of them got any of their money back when they left.

THE LAST MIGRATION CHAIN ESTABLISHED BY PEERSON—
FROM THE MIDWEST TO TEXAS

Cleng Peerson went to Fox River to recuperate among Norwegian friends. Later that fall he decided to go to Texas. "He had long been interested in Texas as a possible site for a Norwegian settlement," Blegen wrote. Peerson knew about Reiersen and the Norwegian colony in East Texas. He also knew that his old acquaintance John Norboe and his family had moved to North Texas.

Norboe had emigrated from Norway in 1832 and settled in Kendall, Western New York. He and his family were among the first to follow Peerson to Fox River in 1834 and then to the new Norwegian colony in Shelby County, Missouri, in 1837. When many of the Norwegians left Missouri and moved north, Norboe and his family went south and ended up as one of the few foreign settlers in Peter's Colony in Dallas County.

Cleng Peerson was 66 years old when he made the journey to Texas in fall 1849. He hoped that he could once more help locate new and good land for fellow Norwegians. He stayed with the Norwegian immigrants in East Texas and with Norboe in Dallas, and returned to Fox River in spring 1850. Peerson wrote a letter about his experiences in Texas to the Norwegian newspaper editor Knud Langeland, dated August 20, 1850. The letter was published in the newspaper *Democraten* in Racine, Wisconsin, on September 7.

When he arrived in Texas in late fall 1849, Peerson wrote, he first visited the Norwegians who had settled in the eastern part of Texas under the leadership of Johan Reiersen. They were about a hundred in number and lived 30 miles east of the Trinity River. "I do not particularly like their choice of land, but I was astonished to see how much they harvested from it, especially maize, rice, and sweet potatoes. They also have large numbers of cattle and horses."[86]

To get from East Texas to Norboe's place in Dallas, Peerson traveled in a northwesterly direction. At the major river junction on the Trinity River in southwestern Kaufman County, he continued along the West Fork into Dallas County. He found the land good and beautiful, but it lacked the trees and forests Norwegians were so fond of. Peerson traveled on both sides of the Trinity River for "a couple of hundred miles," but chiefly on the west bank. "The larger part of the land is high rolling prairie, rich in rivers and streams, but with little forest land in comparison with the prairie. All kinds of oaks are found and the osage grows eternally. Cypress and red cedar are also found in abundance. The prairie is just as fertile as the bottom land."[87]

Peerson experienced snow in East Texas in December 1849. "We had about an inch of snow on the eighth and ninth of December, and ice to about the thickness of a finger." He found Texas "a glorious place for raising cattle. They feed themselves the whole year through without special supervision." The products he saw cultivated most frequently were maize, wheat, barley, rye, oats, rice, tobacco, hemp and rope, and sweet potatoes. "Sheep thrive exceedingly well."[88] Instead of bunching together in Wisconsin, Minnesota, and Iowa, Peerson argued, Norwegians could find better land and a better climate on the Texas frontier. "God willing," he planned "to leave again for Texas in the month of September."[89]

THE NORBOE FAMILY—NORWEGIAN PIONEERS IN DALLAS COUNTY

During his visit in Texas in 1849–50, Cleng Peerson was skeptical about the quality of the soil in East Texas and the great risk of serious disease. Time would prove him right. He was much more enthusiastic about the land and environment in Dallas County, where John Norboe and his family had settled.

Johannes Pedersen Nordbu, Nordboe, or Norboe emigrated with his family from eastern Norway in 1831/1832. He was born in 1768 at Venabygd, Ringebu parish, Gudbrandsdalen, Norway. His father died in 1779, and Norboe took over the family farm in 1784, including the traditional contract to take care of his mother in her old age. He sold the farm in 1797 for 2,000 *riksdaler*. He left hardly any traces in Norwegian public records until 1815, when he bought the farm Nar-Holo in Ringsaker, Hedmark County, for 3,500 *riksdaler*. Norboe acquired many skills and practiced many trades in addition to farming.[90] He married Kari Knudsdatter in 1814 (born maybe in 1791), and they had four children—Oline Abelone, born February 1, 1816, Peder, born February 8, 1820, Johannes, born February 6, 1822, and Poul Martinus, born May 14, 1825. Norboe was not able to service his debt on the Ringsaker farm and had to leave it in 1825. The family settled in the Larvik district on the western side of the Oslofjord. In an article about Norboe, Arne Odd Johnsen found it unlikely that Norboe would leave Norway with his family and cross the ocean to a new and strange world unless he had very good reasons to do so.[91]

Many among the early Norwegian emigrants traveled to the United States as passengers on iron-exporting ships leaving from Gothenburg, Sweden; Norboe and his family were among them. The schooner they boarded in Gothenburg ran into bad weather in the Bay of Biscay and suffered damages off the Portuguese coast. From Lisbon, the family sailed on the ship *Delta* and arrived in New York on March 12, 1832.[92] The Norboe family had no money and stayed and worked in New York for three months before they continued to the Norwegian colony in Western New York. The immigrants at Kendall helped the family as best they could.

Norboe and his family were the first Norwegian immigrants to settle permanently in North Texas in the 1830s. In 1852, Norboe received a letter from Taale Andreas Gjestvang, from Løten, Hedmark County. It was the first letter he had received from Hedmark in twelve or fourteen years, Norboe commented in his reply. He had never regretted his decision to emigrate to America. It would give him great joy if Gjestvang came to Texas next year to visit and to explore the prospects for emigration. The Norboe family had lived in Dallas County since 1841. Their farms were located along Five Mile Creek, 5 miles south of the city of Dallas and 2 miles east of the Trinity River. The soil on his land was good, the winters mild, and the air healthy.[93] Neither he nor any of his children or grandchildren had been seriously ill. He was now 84 years old and had problems with his hearing and eyesight. Norboe finally prospered, wrote Arne Odd Johnsen: "he secured 1,920 acres for himself and his family—a colossal estate compared to those in Norway." According to Johnsen, Norboe died in Dallas in the early 1860s: "He was then over ninety years of age."[94] Based on available documents, however, it seems more plausible that he died in 1855.

DALLAS COUNTY AND PETERS COLONY

The honor of having founded the city of Dallas is usually ascribed to John Neely Bryan (1810–1877), who settled on the east bank of the Trinity River in 1842. Bryan was born in Fayetteville, Tennessee, attended Fayetteville Military Academy, read law, and was admitted to the Tennessee bar. Around 1833 he moved to Arkansas, where he began as an Indian trader. On a journey south from Fort Smith in summer 1839,

Bryan got his first glimpses of the land along the Trinity River. In November 1841, he put up a crude hut by the Trinity bluffs.⁹⁵

The first trickle of homesteaders arrived in Dallas in 1842. On February 26, 1843, Bryan married 18-year-old Margaret Beeman. They lived in a one-room house of hewn cedar logs. The cabin also served as post office and courthouse and was the first meeting place for the Cumberland Presbyterians in Dallas. Bryan claimed land along the best ford on the Trinity River, and he operated a ferry across the river.⁹⁶

Dallas County was officially formed from portions of Nacogdoches and Robertson counties on March 30, 1846. It had a total population of 2,743 white inhabitants and 207 slaves in 1850, and 8,665 whites and 1,074 slaves in 1860. The percentage of slaves was considerably lower than that for the state of Texas as a whole. By 1860, wheat growing had become a major crop in the county, and gristmills were in operation.⁹⁷

The 1850 census registered two Norboe families in Dallas County. John P. Norboe (82 years) and his wife Katharine (59) owned a farm operated by their sons John Jr. (28), and Powel (Paul Martinus) (25). Two laborers from Illinois also lived on the farm, M. D. L. Gracey (18) and C. G. Gracey (16). His oldest son Peter (30) had his own farm and had married Jane Robinson (20; born in Bond County, Illinois) on July 5, 1849. Jane's 46-year-old widowed mother, Rebecca Robinson, born in North Carolina, also lived in their household.⁹⁸

The Norboe family were the only Norwegian settlers in Peter's Colony, which was founded by W. S. Peters and a group of businessmen from Kentucky. The aim of Peters' Colony was to accelerate immigration to Texas and extend the northwestern frontier into Indian country. Only immigrants from outside Texas could acquire land. In his groundbreaking work on the history of Peter's Colony, Seymour V. Connor classified 1,787 settlers who arrived between 1841 and 1848 as bona fide colonists.⁹⁹ Most of the colonists were citizens of the United States; only 2.3 percent were born in Europe. Among the European colonists, eleven were born in England or Scotland, six in Germany, five in Ireland, and three in Norway. Connor lists the Norboe family under the name Narboe in "Biographical sketches of Colonists."¹⁰⁰

Most of the settlers in Peters' Colony came from southwestern states: 23.1 percent from Tennessee, 17.6 percent from Kentucky, 8.9 percent from North Carolina, 9.9 percent from Virginia, 8.3 percent from Missouri, 5.5 percent from Illinois, and 4.8 percent from Indiana. Very few came from states along the East Coast, either north or south.[101] A majority of the colonists were married, and most of the married men had children. There were very few slaves in the colony; 31 families owned a total of 106 slaves. Not a single colonist was listed in the 1850 census as a planter, with many slaves.[102] According to Connor, the archetypical colonist in Peters' Colony was a small prairie farmer working his own land with the help of his family.

PEERSON TRAVELED WEST ALONG THE INDIAN FRONTIER IN SPRING 1850

The Brazos River begins in the state of New Mexico in the west and runs approximately 1,050 miles from New Mexico through Texas until it reaches the Gulf of Mexico at Freeport in Brazoria County. It is the longest river in Texas.[103]

Before his return to Illinois, Cleng Peerson traveled along the Brazos River for a total distance of about 160 miles. He took considerable risk by traveling on the edge of the Indian frontier in North Texas through virgin land for settlers, where few dared to settle because of the extreme threat of Indian raids.

After Texas became part of the United States in 1845, and during the Mexican War (1846–48), Indian tribes still controlled the land west of Dallas, Waco, Austin, and San Antonio. In October 1844, Governor Sam Houston "successfully negotiated a treaty of peace and commerce with the Comanches and other western tribes at Tehuacana (in modern day McLennan County) and even though the pact produced a time of relative tranquility for the republic, the Indians' marauding never stopped completely."[104] The treaty had declared an end to horse stealing and the practice of taking children captive. In exchange the Texas government promised to furnish the Indians with gunpowder, lead, spears, and other arms needed for hunting. "But the difficulties that lay between the white people of Texas and the prairie Indians were much too large to be overcome by this treaty effort."[105] White explorers continued to push west into unchartered Indian territories. In 1848, under the command of General William J. Worth, the US army ordered fifteen hundred troops to Texas to try to gain control of the borderlands between whites and Indians. The same year the army began to build a line of seven forts stretching north from Fort Duncan at Eagle Pass on the Rio Grande to

Fort Worth on the upper reaches of Trinity River. The stream of settlers continued to press westward, and in 1851 it was decided to build a new string of forts 150 miles west of the 1848 demarcation line.[106] The Texas Congress established Tarrant County on December 20, 1849, with Fort Worth as the county seat.[107] At the time of the census in 1850, 599 whites and 65 slaves lived in Tarrant County.

When Cleng Peerson walked west of Fort Worth in spring 1850, the area was still frequently raided by Indians. But after twenty-five years as a pathfinder on the frontier, Peerson knew well that the search for good and cheap land was accompanied by the risk of meeting Indians. From Fort Worth he walked in a southerly direction along the Clear Fork River until he came to Benbroke Lake. From there he traveled south until he saw the glitter of the water of the Brazos River in the distance. When he crossed Brazos Point, he knew that it was not too far before he would arrive at Fort Graham, established in March 1849 on the east side of the Brazos River at Little Bear Creek.[108] During his travel between Fort Worth and Fort Graham, Peerson noted large stretches of good prairieland.

After leaving Fort Graham, Peerson continued south in the direction of Fort Gates. He explored the land along several tributaries to the Brazos, including the Bosque River, which had several branches to the west. He was now so far south that Indians no longer represented a large threat. From the new village of Waco in the new McLennan County he began walking in a northeasterly direction back to the Norwegian colonies in East Texas, and from there he continued to Shreveport on the Red River. The distance between Waco and Four Mile Prairie was approximately 125 miles.

ORPHANED NORBOE GRANDCHILDREN AND NORWEGIAN IMMIGRANTS FOLLOWED PEERSON TO TEXAS

Cleng Peerson was known as a defender of the rights of children. During his visit in Dallas County, he brought to the Norboe family news about the Norboe grandchildren living in Fox River. They had recently lost both of their parents. In 1833, the Norboes' oldest child and only daughter Oline Abelone had married the 22-year-old "Slooper" Jørgen Johnsen Hesja (American name George Johnson) in Kendall, New York. The couple had followed the Norboe family to Fox River. Oline Abelone gave birth to five children, all born in La Salle County.

Around 1850, US federal authorities built a string of forts from Fort Duncan on the Rio Grande River in the south to Fort Worth in the north. Comanches and Apaches controlled all the land west of this line of forts. Most white inhabitants lived in East Texas and the region of the Mexican Gulf. The Norwegian immigrants who settled in Bosque County lived on the border with the Comanches.

Caroline Johnson was born in Fox River on December 29, 1837, and their second child, Anne, died before 1850. Their eldest son Joseph was born before Elias, who was born in December 1841. Their youngest son, John W. Johnsen, was born on December 25, 1844. Oline Abelone died on February 3, 1846, at the age of thirty. George Johnson remarried, but died during the cholera epidemic in 1849.[109]

During Peerson's stay at the Norboe ranch in Dallas County it was decided that he would return to Illinois and bring the orphaned grandchildren back to their grandparents in Dallas County. It was further decided that their 28-year-old uncle John Norboe Jr. should travel north to Fox River and help Peerson and the grandchildren on their journey back to Texas. In his letter to Gjestvang in February 1852, John Norboe wrote that his son had walked most of the way north to La Salle County, but used steamboat on the return journey.[110]

SOME RECENT NORWEGIAN IMMIGRANTS FOLLOWED PEERSON TO TEXAS IN 1850

When Cleng Peerson returned to Texas from the Midwest with the Norboe grandchildren late in the fall of 1850, a small group of recent immigrants from Rogaland County in Norway joined him. On May 16, 1849, the brig *Favoriten*, owned by the merchant company J. A. Køhler & Co and captained by J. C. Westergård, had left Stavanger with 191 emigrants on board. The merchant company was one of the largest shipowners in Stavanger at the time and *Favoriten* had been delivered from the shipyard in 1848.[111] The brig arrived in New York on June 28. Most of the passengers on board could only afford a ticket for steerage. Among the better-off cabin passengers was the 36-year old carpenter Abraham Ommundsen from Stavanger, his wife Maria Ommundsen (32), and their eight children, the oldest fourteen years and the youngest two months old. Abraham was born August 23, 1812, in Varhaug, Jæren, Rogaland County, and died December 10, 1874, in Henderson County, Texas. The family's first destination was the Norwegian mother colony in La Salle County, but they soon moved to the Norwegian settlement in Lee County, Iowa. However, they found that the best and cheapest land in Lee County had also been taken. The Ommundsen family, who had changed their last name to Brown, decided to travel to Texas with Peerson after hearing his stories about the area.[112]

Another group of emigrants from Stavanger crossed the Atlantic on *Favoriten* the following year. The brig left Stavanger on the same day as the year before, May 16, but arrived in New York on June 26, 1850, two days earlier. On this crossing the brig had 163 steerage passengers and 43 cabin passengers.[113] Most of the passengers came from the Ryfylke region in Rogaland County, especially from the municipality of Hjelmeland, but several came from the island of Karmøy, the Hardanger region, and the town of Egersund in Rogaland County.[114]

Among the passengers from Karmøy were Knud Knudsen and his family. Knudsen was born in 1802, the son of a cotter, in Stangaland, Karmøy, Rogaland County, and died July 11, 1886, in Bosque County, Texas. Knudsen learned the blacksmith trade from his father at a young age and earned most of his income from blacksmithing.[115] On June 16, 1830, Knud married Anna Karine Olsdatter, born in 1810 in Sørstokke, Avaldsnes. They had three children. Ola was born on September 4, 1832 in Nordstokke, Avaldsnes. When he died, on November 6, 1920 in Waco, Texas, he was registered as Ole Canuteson. Knut Andreas was born in 1845 but died before his first birthday. A third son with the same name was born August 27, 1847 in Karmøy and died in 1914 in Hagerman, New Mexico, registered under his American name Andrew C. Canuteson.

Knud Knudsen had two older brothers who lived in Fox River, La Salle County. Halvor Knudson (born 1799) had emigrated to Kendall, Western New York, as early as 1831 and was a member of the first group Cleng Peerson brought with him to Fox River in 1834. His name had been Americanized to Oliver Canuteson. He and his wife Julia had three children, two boys and one girl. Oliver Canuteson wrote a letter to his relatives in Norway in 1837, where he mentioned that he owned 160 acres of land. When Peerson was back in Norway in 1842, he had visited the home of Knud Knudsen near Kopervik, Karmøy. In a letter to the historian Rasmus Anderson, dated Waco, Texas, December 16, 1894, Ole Canuteson wrote that Peerson on this visit had encouraged his father to emigrate. "My father and I transported him a short distance in a boat, to a man that had a son in America."[116] This probably happened in the autumn of 1842, commented Anderson.

On the Knudsen family's journey from New York to Illinois "cholera broke out among us, many died on the way and others died after we arrived," Ole Canuteson wrote in 1894. "My mother was among those who died and was buried on arrival at the Fox River settlement in Illinois." Not long after their arrival, the Knudsen family met Cleng Peerson, "just back from Texas, and on his advice, and on his promise to be our guide, we concluded to go to Texas."[117] Land prices at Fox River were too high; they were ten times higher in 1850 than they had been in 1835. Land sold for $15 to 20 an acre, which was beyond their economic means.[118] Knud Knudsen Americanized his name to Canute Canuteson, the way his older brothers had done. His 18-year-old son Ole married 18-year-old Ellen Malene Endresdatter (Gunderson) in September 1850. She was born February 2, 1832, in Nordstokke, Avaldsnes, Karmøy, and died on December 9, 1908, in Waco, Texas.

CHAPTER 4

Why Did Immigrants from Hedmark County in Eastern Norway End Up in East Texas?

◊ ◊ ◊

IN EARLY JULY 1848, the Norwegian shipowner Lauritz Christian Stephansen announced in the leading Norwegian newspaper *Morgenbladet* in Oslo that his brig *Amerika* would sail directly from Arendal to New Orleans with emigrants that fall. Interested persons were asked to contact him. Stephansen was born in Tjølling, Vestfold County, in 1810 and died in Arendal in 1863. In his youth he was a seaman and was a captain by the time he left the sea in 1832. He established himself as a shipowner with his own shipyard in the dynamic shipping town of Arendal in southern Norway. There he participated actively in town politics and was elected to the Norwegian Parliament in 1857–58. For a while he was the largest shipowner in Arendal. His ships were engaged in the Norwegian trade with Havre in France as well as the cotton trade with New Orleans. Because of the political revolutions in Europe in 1848, shipowners were experiencing problems securing freight from Europe to the United States.

Amerika left Arendal on September 27, 1848, but with only 36 emigrants on board, and arrived in New Orleans on December 1, 1848.[1] Stephansen lost money on this journey. He did not try the emigrant trade again in 1849. In 1850, Stephansen was better prepared and started his marketing earlier. The brig *Amerika* left Arendal on September 9, 1850, with 101 emigrants and arrived in New Orleans on November 7. Most of the passengers came from southern Norway, from small communities like Tovdal, Gjøvdal, Åmli, and Froland in the inland and Tromøy, Holt, and the town Lillesand on the coast. The emigrants followed the Reiersen migration chain from southern Norway to East Texas. Many of them had relatives and friends who had already settled there.

TAALE ANDREAS GJESTVANG HELPED RECRUIT EMIGRANTS FROM HEDMARK COUNTY

Shipowner Stephansen repeated the roundtrip from Arendal to New Orleans in 1851. He announced in the spring that the bark *Arendal* would sail with emigrants to New Orleans in the fall. The ship left Arendal with 89 passengers on board, and 78 of them chose to settle in East Texas. Surprisingly, more than half of them came from Hedmark County in eastern Norway. They were recruited through the efforts of Taale Andreas Gjestvang, Løten, who was born on January 31, 1811, at Kise, Nes, and died on May 7, 1891, in Løten. He was one of the most interesting personalities in the Løten municipality in the nineteenth century, wrote Syversen, and he was an influential figure in Løten.[2] He had bought the large farm Haukstad in 1838. Compared with most farmers at the time, he was exceptionally well read, and he tried to introduce modern agricultural principles in his farming operations. For decades, he operated the post office at Løten. Gjestvang subscribed to several newspapers and encouraged people to borrow them. Moreover, he played a leading role in local politics and was elected *ordfører* (mayor) of the municipality for many years. This was the highest elected position in local politics in Norway then, and still is today.

Gjestvang subscribed to Reiersen's newspaper *Christianssandsposten*. He was well aware of Reiersen's views; he had corresponded with him and read *Pathfinder* when it was published in 1845. Reiersen had almost persuaded him that Texas would be the best location for Norwegian immigrants.[3] Gjestvang had sent a letter to Stephansen just before the brig *Amerika* left in 1850. Could the shipowner ask his captain to buy a map of Texas in New Orleans and bring it back to Norway? Gjestvang had acquired a lot of literature about America, but he had no map of Texas.[4]

Stephansen sensed that Gjestvang might be a prospective customer or agent for his emigration business. The two men began to exchange letters in the second part of 1850. Could Gjestvang help him recruit passengers for the planned Atlantic crossing in 1851, asked Stephansen? Gjestvang answered in the affirmative, and he soon received samples of passenger contracts and printed announcements for the journey. Stephansen asked Gjestvang to make sure that the information was spread in his community. Announcements were read outside the churches in Løten, Romedal, and Vang and printed in the newspaper in the new town Hamar. Gjestvang did help some people to get in contact with Stephansen, but most of the emigrants who signed contracts for transportation between Arendal and New Orleans did so without Gjestvang's assistance. In July 1851, Gjestvang was informed that a group of emigrants from Tynset, east of Løten, were on their way to Christiania. They had signed no travel contracts yet, and Gjestvang persuaded 17 of the 30 emigrants that it would be a better deal for them to travel to Arendal and make the journey directly to New Orleans on the Stephansen bark *Arendal* than to emigrate from Oslo. They had to wait several months in Arendal, however, until the ship arrived. During the waiting time, Stephansen hired some of the men to work at his shipyard.

WHO WERE THE EMIGRANTS FROM HEDMARK COUNTY, AND WHY DID THEY LEAVE?

Because of Syversen's excellent historical research in primary sources, published in *Norge i Texas*, more is known about the background and motivations of the emigrants from Hedmark in 1851 than is usually the case. Several of the emigrants left the municipalities of Løten and Romedal. Most of them were yeoman farmers or had grown up on independent farms. A number of them emigrated to avoid economic ruin or loss of face.[5] There was a downturn in the local economy in the late 1840s. In Løten the harvest of small grains and potatoes during 1843 to 1853 was below average. The farmers depended on cash from the sale of potatoes as raw material for the aquavit distilleries in the region to pay taxes and loans. Some of the farmers had overextended themselves by mortgaging their farms to invest in businesses outside their core activities in agriculture. When their investments failed, they found themselves in dire economic circumstances.

Johan Olsen Brunstad belonged to this last category. He had invested in the brick works Jordal Teglverk in Oslo together with his brother-in-law Peder Eriksen and P. Bakken.[6] They had been led to expect a considerable profit within a short time, but the brick works soon proved unprofitable. Johan Brunstad was born on the large farm Harstad in Romedal, Hedmark County, on February 27, 1816. His parents, Ole Johannesen Harstad (1787–1866) and Margrete Embretsdatter Westgard (born 1793, died October 9, 1821), had grown up in well-to-do families. His father had been in the military and reached the position of quartermaster.

Brunstad's mother died in 1821. During the inheritance proceedings the assets in the house were valued at 2,000 *riksdaler* and the land was valued at 7,000 *riksdaler*. Ole Johannesen had bought the farm from his father in 1817, at a time of inflated farm prices, for 13,333 *riksdaler*. The household had lived in the style they were accustomed to, and their expenses had been higher than their income. The value of the farm did not cover the debts. Most of the creditors, however, were family and kin. When the inheritance proceedings were completed in 1825, Ole Johannesen found himself in a very difficult economic situation.[7] By 1828, he was no longer able to hold on to the farm and sold it for 3,000 Norwegian Spd (*speciedaler*; 1 Spd equals approximately $1) on July 31, 1828. In March 1835, Ole Johannesen bought the Brunstad farm in Åsbygda, Romedal, for 700 Spd plus expenses for a *føderåd* contract for the old parents of the seller.

Ole Johannesen sold the Brunstad farm to his son Johan for 800 Spd in 1844, including a clause stating that Johan would take care of his father for the rest of his days. Johan had married Anna Eriksdatter Furuset on April 3, 1841. Their daughter Margrethe was born on October 25, 1842, and died in Four Mile Prairie, Van Zandt County, Texas, in October–November 1852. Their second daughter, Karen, was born on April 17, 1844 and died in Washington in 1921. Their third daughter, Ellen, was born on July 6, 1846, and died in 1918 in Cranfills Gap, Bosque County, Texas. Their only son, Ole, was born on February 6, 1848, and died in 1931 in Norse, Bosque County. The two youngest girls were Inga (born February 2, 1850, died in Bosque County in 1919) and Gina (born May 14, 1851, died in 1886 in Bosque County).

Because of his losses on the brick works in Oslo and unsuccessful attempts to refinance his debt, Johan Brunstad had to sell his farm in 1848 for 800 Spd.[8] He had grown up in a family with high ambitions, and his prospects for the future did not look very rosy. The Brunstad family joined the large group of people who emigrated from Hedmark to Texas in 1851. In addition to paying their travel expenses, they brought with them some money to buy land in Texas.

Several of the larger farmers at Løten had invested in *Aadals Brug Mekaniske Verksted og Jernstøperi*, one of the first proto-industrial mechanical workshops in Hedmark County, established around 1830.

In the beginning the firm imported Swedish iron from across the border and used a water-powered trip hammer to produce nails and scythes. The works was located at an isolated place along the river Svartelva, Aadal, in Løten parish, a site with easy access to waterpower. By 1850 the operation included an iron foundry and a mechanical workshop producing agricultural implements; it employed 37 workers in 1850, wrote the governor of Hedmark County.[9]

Carl Questad, from the farm Øde-Kvæstad in Løten, was involved with Aadals Brug. He was born as the youngest child on October 27, 1815 on the farm Sæter, Vardal, close to the city of Gjøvik on the western side of Lake Mjøsa, in the affluent family of Søren Engebretsen (1778–1822) and his second wife Gunhild Marie Ketilsdatter Bjørnstad (1792–1875).[10] Engebretsen worked for one of the largest merchant and banking houses in Norway at the beginning of the nineteenth century, Collett & Søn in Oslo. The merchant house had a strong interest in the lumber export business and owned farms with large wood holdings throughout eastern Norway. Collett & Søn took over the farm Sæter in 1807 but sold it in 1811 to their trusted company agent Søren Engebretsen for 3,800 Spd. After the death of her husband in 1822, Gunhild Marie married Even Ellingsen Øverby in 1823. He took over the farm Sæter and sold it in 1836.

The leading man behind Aadals Brug was Ole Olsen Tranberg. He grew up in Vardal on the western side of Lake Mjøsa but moved to Løten on the eastern side around 1820. He married the daughter of the farm Finstad, took over the farm, and soon bought other farms in Løten. He was involved in the development of Aadals Brug, and he introduced the water-powered trip hammer for making scythes and nails. During the next years, several of his relatives from the western side of Mjøsa came to work at Aadals Brug, including his cousin Even Ellingsen and his family. After the sale of the farm Sæter, Even Øverby had 4,000 Spd to invest, bought a farm, and put the rest of the money into the modernization of Aadals Brug. In 1841, however, he had to declare bankruptcy, and Ole Tranberg likewise.[11] Aadals Brug was sold to one of its main creditors, Lieutenant Colonel Lowsow.

Carl Andreas Engebretsen, or Carl Questad, as he was known in Texas, followed his mother Gunhild Marie and stepfather Even Ellingsen to Løten. On November 26, 1841, Carl Questad married Sedsel Olsdatter Ringness, who was born February 22, 1810 in Ringnes, Løten, and died on July 25, 1890, in Norse, Bosque County. In 1846 Carl Questad invested an inheritance from his father in the farm Øde-Kvæstad; he also worked as a blacksmith at Aadals Brug. Because of poor harvests several years in a row, Questad had economic problems.[12] Instead of throwing good money after bad, Questad saved as much of his inherited capital as possible. Like many others in the community, several of them former workers at Aadals Brug, he decided to follow the advice of Gjestvang and emigrate to Texas with his family. He sold his farm to Jacob Qvæstad for 1,500 Spd in 1851, the year the family emigrated. They brought their two children, Even (born October 17, 1842), and Marthe (born February 13, 1848).

The Questad family's emigration group had booked passage on the bark *Arendal*. On their way to Arendal, Questad got a letter from T. A. Gjestvang, who informed him that Jacob Qvæstad had paid the last 295 Spd he owed him for the farm.[13] Gjestvang also requested that Sedsel Questad's brother, Jens Ringness, should send a signed letter to his brother Ole Haneknæ, giving him the lawful right to act on his behalf. Ole Haneknæ had informed Gjestvang that the assets from the sale of the farm Halstenshov would soon be ready for distribution.

Jens Ringness and his family had chosen the same destination as the Questad family. They traveled together from Hedmark to Arendal, crossed the Atlantic on the sailing ship *Arendal* to New Orleans, and continued from New Orleans to Henderson County in East Texas. Jens Ringness was born on November 6, 1802, in Ringnes, Løten, and died in August–September 1871, in Norse, Bosque County. His wife, Kari Jensdatter, was born on August 12, 1822, in Halstenhov, Løten, and died on June 20, 1891, in Norse. They brought with them four children; the oldest child was eight years old when they left, and the youngest, born in August 1851, died before they boarded the ship in Norway.[14]

The 79-year-old mother of Jens Ringness and Sedsel Questad, Marthe Jensdatter Ringnes, traveled with them but died in Four Mile Prairie in 1852, the year they arrived. The Questad family also brought 19-year-old Anne Pedersdatter Questad (born in 1832) to help the family care for the children. Some months after their arrival at Four Mile Prairie, however, she got the "ague." She intended to take the medicine quinine, but mistook a bottle with strychnine for quinine, and died.

The older brother of Sedsel Questad and Jens Ringness, Knud Olson Storløken, was already living in Henderson County. He was born in

Løten on November 25, 1797, and emigrated from Arendal to Henderson County in 1850. He worked as a farmer, blacksmith, and wheelwright. Before his emigration, he had worked at Aadals Brug for several years.[15]

THE JOURNEY FROM ARENDAL TO FOUR MILE PRAIRIE THOUGH THE EYES OF JOHAN BRUNSTAD

A long letter from Johan Brunstad to Norway some months after the Brunstad family's arrival in Texas in 1852 gives detailed information about the Atlantic crossing, the journey from New Orleans to East Texas, their actions during the first months in Texas, and the dreams the writer had for himself and his family. The bark *Arendal* left Arendal in the morning of November 5, 1851. The ship had been built in 1847 in the Stephansen shipyard in Arendal. After only two and a half days, the ship sailed through the English Channel. The coast of the West Indian Island of St. Domingo was sighted on December 11. At the mouth of the Mississippi River, *Arendal* and three other sailing ships were met by a large steamship and guided upstream to New Orleans. On Christmas morning right after sunrise, New Orleans came into sight. The passengers left the ship on the third day after Christmas and boarded a steamboat for Alexandria, where they changed to a smaller steamboat that took them further up the river to Shreveport. From the time they left Norway until they disembarked in Shreveport on January 3, their worst experience on the journey, according to Brunstad, was the bad accommodation on board the steamboat to Shreveport.[16]

In Shreveport, the Brunstad family rented a simple house for a week for $1.50. Brunstad bought oxen, a wagon, and horses, and paid a total of $70. The Norwegian immigrants had language problems but got some help. Later, Brunstad bought another two oxen and a wagon for $84. The group then began the 175-mile journey to Four Mile Prairie. An endless number of tree stumps along the way caused them daily problems. When it rained, it took time to cross creeks and swamps. The group experienced dry and clear weather on most days, but some days it was very cold, with thick ice on shallow water.

The group arrived at Four Mile Prairie on January 23, 1852, and was received with open arms by the Wærenskjolds.[17] The Brunstad family found lodging with Johan Reiersen, "who lived two English miles from the Wærenskjolds." They stayed there for three weeks while mulling over the question of buying or renting land. Prices of land ranged from six bits to one dollar. There was no vacant government land to be had in the vicinity; speculators had bought it all. Brunstad did not find land with a quality he was willing to buy and therefore chose to rent land for the short term. Together with Hendrick O. Dahl, Brunstad rented a field of 18 to 20 acres, which was already plowed and fenced, for $2 per acre. The agreement included the use of the house on the homestead. They planted a third of their rented land with spring wheat and the rest with corn. Brunstad bought five cows, four of them with calves, together with a three-year-old heifer and one ox. He paid $68 for the animals, and another $26 for a grown sow with three pigs, one young sow, and ten one-year-old pigs.[18]

The farming equipment used in Texas did not impress Brunstad. The grass on the prairie reached a man's shoulders, but was only knee- to hip-high in the forest. The soil was light and sandy in many places, but mixed with clay, and the subsoil consisted mostly of clay, and Brunstad thought it would soon be necessary to fertilize the soil. Most of the Norwegian settlers in the vicinity had arrived in the past year or so. Based on what he had been told, Brunstad was optimistic. Most Norwegian immigrants rented from 100 to 640 acres of land. Although they had little money, and many had still not repaid what they owed for their emigration expenses, most felt they had done well, and "their health is on the whole quite good."

Brunstad had not met one Norwegian immigrant who wanted to go back to Norway, with its rigid social system and "aristocratic officials," and he also referred to his own "unfortunate situation" before he left Norway. He often experienced joy and gratitude that "I made the decision to emigrate here."[19] He was convinced that it would be much easier to get ahead in America than in Norway. For a man who had a large family and no cash to draw on, though, the first years in America would be hard, as all necessities were costly. Wærenskjold had told him, however, that cattle and swine were very profitable; every man could have as many as he wanted, whether he owned land or not, and it cost nothing to feed them. Eleven Norwegians, including Brunstad, had recently traveled to Canton, the county seat of Van Zandt County, to become American citizens, renouncing their allegiance to the Norwegian kingdom and King Oscar I. It was a good feeling, wrote Brunstad, to have become an American citizen. Brunstad emphasized that both he and his wife were content with their decision to emigrate. If he and his family could "enjoy good health and buy a piece of good land and build a good house, I hope that after a couple of years to be in such a situation that I could never hope to have in Norway."

LOOKING FOR LAND IN DALLAS AND TARRANT COUNTIES

Only months after their arrival, the more resourceful and daring among the Norwegian immigrants from Hedmark began to look for better land. In summer 1852 several of them embarked on a land hunt to northwest Texas. Their impressions are well documented, since some of the participants wrote letters home to T. A. Gjestvang of Løten. Gjestvang, on his part, made sure the letters were published in the Norwegian newspaper *Arbeider-Foreningernes Blad*. This newspaper represented the early labor movement in Norway. The editor, Marcus Thrane, used the letters to inform unemployed and poorly paid workers in Norway about the new possibilities in America, "the republic of free and independent people."[20] It was probably not a great surprise to the letter writers that Gjestvang published the letters, and this expectation might have influenced the letters' content.

The men from Hedmark explored some of the same land along the Trinity River that Cleng Peerson had explored in 1849–50. Their conclusion was the same as his—the land in North Texas was much better than the land they tilled in East Texas. In a letter from Four Mile Prairie, dated July 20, 1852, published in *Arbeider-Foreningernes Blad* on December 11, 1852, Johan Halvorsen Grimseth wrote that it would be possible to find land in Texas that was "so rich that I do not think any better can be found on this earth, but it is useless to look for that kind of land in the neighborhood of Four Mile." Grimseth was born on December 1, 1825 on the Grimseth farm in Løten.[21] Since it was easy for newcomers to get temporary work at Four Mile Prairie, Grimseth worked as a hired hand while looking for land. It was his intention to leave the Four Mile community "as soon as I can obtain land somewhere else." Reiersen owned some of the best land at Four Mile Prairie, "but even that is far from as good as I have seen in other districts. Mr. Reiersen himself realizes this full well; but when he settled, he had not been here very long."[22]

The exploratory group traveled along the Trinity River in a northwesterly direction, and followed the West Fork through Dallas County and west to Fort Worth in Tarrant County until they came to the river junction with Clear Fork. Where West Fork joined Walnut Creek, they found "the most beautiful and most fertile land that any of us had ever seen; the water was quite crystal clear and there was also a fair amount of useful forest. But this land is completely deserted and altogether unsettled. We looked over many thousands of acres of wild rye, and timothy and grass grew more luxuriantly than it would be possible to get it to grow in Norway."[23]

They visited Fort Worth, where two hundred cavalry men were posted, and continued west on the western side of the river to Fort Belknap, where five hundred men were stationed. "These forts are there for the purpose of keeping the Indians under control. West Fork of Trinity River passed right through this tract of land." At the time, not many people dared to settle on this land because of the Indian threat.

The group rode through parts of Tarrant County, formally established by the Texas legislature on December 20, 1849, with Fort Worth as the county seat.[24] The US Army had moved into Camp Worth on the south side of the Trinity River; the post was officially named Fort Worth on November 14, 1849.[25] The army troops at Fort Worth moved to Fort Belknap on September 17, 1853. The abandoned barracks were soon used as store buildings by pioneer merchants in Fort Worth.[26] In 1850, 599 white persons and sixty-five slaves lived in Tarrant County, but the population grew tenfold during the 1850s. By 1860, the number of whites in Tarrant County had increased to 5,170 and the number of slaves to 850.

While staying in Fort Worth, Grimseth accompanied Reiersen on a visit to the Indians staying outside the fort. "These people were rather unusual creatures for us to see, for I had never before seen Indians. We sat in their tents a long time conversing with them; this was possible because one of them knew the English language quite well. They were greatly distressed at the miserable treatment they had received and because they were driven hither and thither; and they believed that in time they would die from sheer starvation."[27] Grimseth never put down roots in Texas, but moved north to Otter Tail County in Minnesota.

Since Reiersen knew that some of the men were writing letters to Norway, it was important for him to present his views on the quality of the land they had seen. In a letter dated July 27, 1852, published in *Arbeider-Foreningernes Blad* on January 13, 1853, Reiersen opened with a strong political statement: He loved Texas more than he did his old fatherland! In Texas, he was "free and independent among a free people, who are not chained down by any old class or caste system."[28] He admitted, though, that the soil in the area where he had first settled was poor, and the drinking water was of inferior quality. The crops had been disappointing the last two years. The land around Four Mile Prairie had "large and wide prairies - with a scanty supply of timber along the watercourses. My principal objections to this entire rich expanse of land - where every inch is fertile - are the lack of timber and the conglutinate character of the soil after a heavy rain. The water is clear and clean, but very few of the watercourses have enough water for milling purposes during the entire year. For these reasons I did not believe that this district could be recommended for a Norwegian settlement."

It was he, Reiersen claimed, who had suggested that they should explore the land along the West Fork. In the Fort Worth region the group had discovered "a beautiful, elevated plain, which, in turn, is surrounded by rich prairie land." Between thirty and forty families had settled 5 to 6 miles from the fort during the last years, but no settlements were found further west.[29] Johan Grimseth, Johan Brunstad, Karl Kvæstad, and several other younger men were members of the exploratory group, and "all of them became so enthusiastic about the land that they determined to start a settlement there next fall, and the plan ripened into a definite decision later on."

Nevertheless, Reiersen defended his own original choice of land and argued that "all of the Norwegians who came to this country in 1847 from the mountain communities in western Norway have, without exception, become well-to-do people. Most of them possessed very little surplus capital when they arrived; some were without a penny and a few were even in debt for their passage across the ocean. Now, all of them have bought land and have paid for it; they have built good houses; they have sufficient cattle, oxen, and horses for the management of their farms; and they get annual crops so large that there is a surplus to be sold to the newcomers. A few of these men, who had large families and debts of more than one hundred dollars each can now be estimated as being worth more than one thousand dollars per family. They have succeeded even under unfavorable circumstances; and all of them are actually independent farmers, free from any fears of either taxes, mortgages, or foreclosures."[30] This was boasting, pure and simple. Reiersen would never have been able to prove such wealth among the Norwegian immigrants in East Texas at the time. It is certainly hard to document any wealth to speak of among the Norwegian immigrants in the 1850 census.

Reiersen argued that the recent immigrants from Hedmark in eastern Norway had also done well. "Both Karl Kvæstad, Jens Ringnæs, and Johan Brunstad will get fairly good crops this year. The corn crop is excellent all over the country, and corn is sold at from twenty-five to thirty cents per bushel, but the price is expected to rise to fifty cents in the winter. The wheat crop has also been good, and the sweet potato crop is promising."[31]

Brunstad disagreed with Reiersen. He had been optimistic for the first few months but became very disappointed toward summer. In a letter, dated Four Mile Prairie, July 21, 1852, and printed in *Arbeider-Foreningernes Blad* on December 18, 1852, Brunstad maintained that the soil at Four Mile was "poorer in quality than I had believed land of this kind could be in America." He was disgusted that "we have no other water than what we can find in stagnant pools in the brook, where even the pigs are wallowing and bathing."[32] Brunstad saw a lot of good land on the journey along the West Fork to Fort Worth in summer 1852, and he believed his future prospects in Texas would be far better than in Norway, "if only I could succeed in procuring a good piece of land in a vicinity where I could settle down among pleasant and homey neighbors."[33] Brunstad never got a chance to realize his dream. He became seriously ill with malaria a few months later and died, as did his oldest daughter.

AGUE, MALARIA, AND DEATHS AMONG NORWEGIANS AT FOUR MILE PRAIRIE IN 1852

There was an awareness among Norwegians in the Midwest in the early 1850s that land was cheap in Texas. Among the Johan Reinert Reiersen family papers at the Briscoe Center at the University of Texas, Austin, is a letter addressed to "Kling Person" [Cleng Peerson], dated Lee County, Iowa, February 2, 1852. The letter writer, Martha Olson, thanked Cleng for the letter he had sent them, dated May 1, 1851.

Cleng had only been in Texas since December 1850. Martha Olson had lost her husband and wanted to sell her land and move to Texas as soon as possible. Several other Norwegian families were also interested in migrating to Texas, if they could find good farms. Among them were the families of Sjur Persen and Knud Andersen.[34] Could Cleng give them advice about the best place to locate in Texas, the price of land, and the best way to travel to Texas from Lee County, Iowa?[35] The letter was signed "your friends K. Andersen, G. Helliksen and S. Persen, and Martha Olsen." None of the Norwegians mentioned in the above letter ever moved to Texas.

In the late 1850s and later several families and single persons found their way to Texas along the Peerson chain from La Salle County, Illinois, and other Norwegian colonies in the Midwest. From the end of 1852 and the next years the flow was stronger from Texas northward than from the north to Texas. The main reason was illness—severe epidemics and the ague had befallen the Norwegian settlements in East Texas.

In summer and fall 1852, the Norwegian colony at Four Mile Prairie was hit very hard by the ague. Elise Wærenskjold wrote in detail about disease and death among the Norwegian immigrants at Four Mile Prairie in letters to Norway in 1852 and 1853. "Almost all the newcomers have had the fever and several still have it," she wrote to Gjestvang on September 26, 1852. "Brunstad's family, Dahl and his fiancee Christine, and Anderson, but I think they will soon be well. At Qvestad's and Ringnes's they have been quite well and some haven't been touched by the fever. Grimseth, Halvorsen, Gran and his sister, Ballishoel, Anne Qvestad, and Olsen's son, who are all at Reiersen's, have also been ill now and again, but as far as I know they are now well."[36] She then listed the names of a number of Norwegians who had become ill and died. Some people had also died in 1851, and "old" Reiersen was one of them, "but he was more than seventy, and death comes to you wherever you may be. Brun, a man from Stavanger who came here from Iowa and who lived on one of the unhealthiest farms in the old settlement, has also been very unlucky as he lost three children in the northern states and two this summer." Elise Wærenskjold knew that Gjestvang was seriously considering emigrating to Four Mile Prairie. She therefore felt a need to be brutally honest with him about the serious health situation in the community: "I don't want it said that I've been silent about negative aspects."[37]

Malaria, or the ague, was an illness unknown in Norway but was prevalent in the United States at least until 1880. It affected most populated regions, undermined people's health, had a negative impact on the economy, and was one of the major causes of death. Infants, children under the age of five, and people in southern states had the highest risk of contracting malaria. Even though malaria was less deadly than other diseases in the nineteenth century, malaria infections had a debilitating effect on the long-term health of the persons affected.[38] The disease was also called the shakes, bilious fever, or autumnal fever. It was characterized by fluctuations in body temperature, oscillating between bone-cracking chills and frighteningly high fevers. Those who suffered through such bouts gained partial immunity, a process known by settlers as "seasoning."

"Everywhere the new emigrants went," wrote Fiametta Rocco in her book *Quinine. Malaria and the Quest for a Cure That Changed the World*, "over the Allegheny Mountains, across the Great Lakes, along the Grand River, in the vicinity of the Rideau Canal and into upper Canada—malaria would follow." From the southern Atlantic states, malaria was soon reported in Kentucky, Tennessee, Alabama, Mississippi, Arkansas, and northern Louisiana. "By 1850, practically the entire United States constituted one vast expanse of malarious country, except for Maine, the northern portions of Wisconsin and Minnesota."[39]

The ague usually arrived in late summer and early fall. There were few new cases after severe frosts, and in spring and early summer most people were healthy. During high water in the spring, bottoms were flooded and lagoons and low places along the streams were filled with water. When these places were drying out in the hot sun of July and August, they became a breeding ground for mosquitos and disease.

In another letter to Gjestvang, dated November 21, 1852, Elise Wærenskjold wrote that "all the Norwegians have had the fever and every tenth person has died." In the course of two and a half weeks in October and November, seven Norwegian immigrants had died: (1) Anne Questad, who had served as the Reiersen's housekeeper and who in error killed herself with a strong poison. The doctor had been there when she died but had not seen the bottle of poison she had used until she was dead. (2) Johannes Olsen Foss. Elise found his death very sad since he had been a kind and unusually hardworking provider for his family. During his illness, his wife and their four children had also been very ill. He had lacked care as well as medicine, and his neighbors were also ill and could not help. (3) A young unmarried man from West Norway, John Paulsen. (4) Margith Nielsen from West Norway, "an excellent and truly Christian housewife." (5) Jørgen Vehus, an old man from West Norway. (6) Margrethe Brunstad. (7) Johan Brunstad.

It had been a terrible time of death and mourning, Wærenskjold wrote. The burial ground was on Wærenskjold land. Every morning when she saw neighbors approaching, she feared that "they had come to dig a grave, and those who did the digging were often soon in need of a grave themselves." Wærenskjold felt strongly that most of the people who died, "perhaps all, could have been saved if we had only had quinine for the whole summer, but it was usually not available, as quinine rarely came to the store."[40] When the shopkeeper got a new batch of quinine, the Americans were quick to buy. Even though the shopkeeper was Norwegian, he did not make sure that the Norwegians would also get some and was seemingly unconcerned "whether his countrymen lived or died, and for this reason quinine was only available in the store on the day it arrived." The fever had started in June and was still raging in November. "I believe those who have neglected to supply us with this necessary article have been irresponsible. We have physicians here, but they are not very good, nor have they had quinine. Reiersen acquired three bottles and Qvestad and Ringnes one each, but this was too little for so many." She had talked with Anne Brunstad the other day, who still lived at the Reiersen house. She had told Wærenskjold that the family had paid "fifty dollars for a doctor and medicines. I'm sure they would only have needed ten dollars if they had had medicine in time and then they would also have avoided much pain and suffering in body and soul."[41]

Many people had been careless, Wærenskjold wrote. Gjestvang certainly knew how "careless lower class Norwegians are when faced with illness, in particular as concerns how much they eat and drink, which is so important." All who died, she reflected, with the exception of Margrethe Nielsen, "are among those who arrived this year and last year, so it seems that the longer we are here, the better suited we are for the climate."[42] The many deaths had left painful wounds, which would stay with them for a long time. "All the sickness had caused a general lack of confidence so most appear eager to go to the older states if they only could sell their property." The deaths were a large setback for the Norwegian settlement.

The eyewitness information in the Wærenskjold letters is of crucial importance. Family histories of the Brunstad/Rogstad family often state that Johan Brunstad died in early 1853. Based on Wærenskjold's letter from November 21, 1852, it can be concluded that Johan Brunstad was the last to die before she wrote the letter. Elise Wærenskjold is a trustworthy witness and a primary source compared with all other versions written later, as she reported the deaths when they happened.

Because of recurrent epidemics in Henderson County, the Hastvedt and Smeland families had moved to Van Zandt County. In summer and fall 1852, however, Margit Smeland got "the consumptive chills" and died in late September–early October. In his reminiscences many years later, her brother Jørgen Hastvedt wrote that so many died that "we could scarcely do anything but care for the sick and bury the dead. My sister Margit, Aslak Nielson's wife, died of congestive chills, and when I had dug her grave, I was so exhausted that the same sled which brought her to the grave had to take me home. All the illness and the many deaths made us uneasy about the future."[43]

Aslak Nelson, Margit's husband, chose to stay at Four Mile Prairie. In 1853 he married Gunhild Mjaaland, born on July 27, 1829, on the farm Mjaaland in Gjøvdal, Aust-Agder. She came from the same community where Aslak grew up and emigrated in 1850 with her brother, Ole Evensen Mjaaland. Between 1854 and 1866 Gunhild and Aslak had seven children.[44] Aslak continued his work as a shoemaker and tailor, both at Brownsboro and Four Mile Prairie.

The family of his first wife, however, became disillusioned and pessimistic. Relatives in Wisconsin had emphasized in their letters that their health in Wisconsin was good, and that their prospects for the future also looked good. On May 11, 1853, the Hastvedt family and several other Norwegians left Texas for Iowa and Wisconsin. Some of them had recently arrived in East Texas.

THE LAST BATCH OF EMIGRANTS FROM ARENDAL SAILED ON THE *VICTORIA*

News about all the sickness and death in East Texas reached the public in Norway in fall 1852 through a letter Ole Terjesen Nystøl at Four Mile Prairie wrote to his relatives in Åmli, Aust-Agder, dated August 28, 1852. The Nystøl letter was published in Norwegian newspapers as a warning to people who still contemplated emigration to Texas. It was printed on the front page of the *Vestlandske Tidende* on November 10, 1852, and in the leading newspaper *Morgenbladet* in Oslo on November 17. Since their arrival, Nystøl wrote, his family as well as other immigrants had become very disappointed with Texas. The price of grain was high, and they had to travel far to buy it. The ague (*koldfeber*) had been hard on his family since summer 1851, and it had continued on and off until January 1852. The fever returned in summer of that year. Workers were well paid in Texas compared with Norway, but this did not help much since they got sick all the time and had to pay for medicines and help from a doctor.[45]

The Nystøl family had acquired 160 acres of land; they had improved 16 to 17 acres and built fences around it. But a lot of work was needed before the grain could be harvested. Hogs and cattle broke down the fences, and birds picked on the ears of grain until almost nothing was left. Many of the Norwegian settlers in the colony had become disillusioned, Nystøl wrote, and wanted to move to the northern states if only they could sell their land. They had heard that there was no ague in the northern states, and the land was as cheap as in Texas.

The views expressed in this letter came to have a negative impact on Norwegian emigration to Texas in the late 1850s. But at the time the letter was published, in November 1852, a new and large group of emigrants were already gathering in Arendal, waiting for Stephansen's brig *Victoria* to arrive and bring them to New Orleans. Most of the passengers came from Hedmark County, and again Gjestvang was the driving force behind the emigration.

Victoria should have left Arendal in the fall, but the brig did not return in time, and the departure had to be postponed for weeks and months. This created problems for shipowner Stephansen, but also for Gjestvang. Several of the passengers had traveled to Arendal early in the fall, and they became impatient and angry when *Victoria* did not show up as planned. It cost the passengers a lot of money to rent housing and pay all other living expenses day in and day out. Some of the passengers, with M. Aasetqvænen, G. Nygaarden and Erick Mathiesen as their spokesmen, demanded that Stephansen find another ship for them. With the help of the authority of Gjestvang the little rebellion was smothered, and the passengers accepted to wait longer for the *Victoria*.[46] When the ship finally left, on February 10, 1853, with 114 emigrants on board, T. A. Gjestvang and his nephew Ole Gjestvang were among the passengers. Gjestvang had finally decided to go to Texas and see for himself if the land was as good as he had been told. They arrived in New Orleans on March 29, 1853.

Some of the emigrants who arrived in East Texas in early summer 1853 immediately chose to join the Norwegians who prepared to move to the Midwest. Gjestvang spent weeks in the Norwegian settlements in East Texas; he also traveled to Norboe in Dallas County and lived on his ranch for several weeks. He too became very disappointed by the low quality of the soil, the general environment, and the large health risks. He knew Carl Questad and Jens Ringness well and trusted their judgement. They told him that they intended to move away from East Texas as soon as they could find better land elsewhere. After Gjestvang returned home to his farm at Løten, he found it difficult to hide his disappointment. People in the community were used to listening to his advice. and sensed that he was no longer so enthusiastic about promoting Texas.

WHERE IN NORWAY DID THE IMMIGRANTS IN EAST TEXAS COME FROM?

Most of the Norwegian immigrants who settled in Texas came from a few municipalities in Aust-Agder and Hedmark counties. In his analysis of emigration to Texas from Aust-Agder between 1844 and 1859, Odd Magnar Syversen came to the following conclusion: fifty-seven emigrated from Holt parish, forty-two from Åmli, twenty-seven from Tromøy, twenty-five from Gjøvdal, twenty from Tovdal, nineteen from Lillesand, and eighteen from Froland.[47]

According to the report from the highest state official in Hedmark County, *fylkesmannen*, or the council governor, only six persons emigrated from Hedmark in 1846, eleven in 1847, and six in 1848. But in 1849 the numbers increased to forty-three emigrants and in 1850 to 218. This was the year when many bought tickets from shipowner Stephansen and left Arendal for New Orleans and Texas. The highest number of emigrants left in the years 1851, 1852, and 1853, a total of 924 persons.[48] In 1852, when Gjestvang himself joined the emigrant group on the *Victoria*, 329 emigrants left Hedmark County, but settled in other states than Texas. There can be no doubt, however, that the active recruitment campaign conducted by Gjestvang had an impact on the total number of emigrants from Hedmark County during those years. After Gjestvang returned and stopped promoting Texas, this had an immediate effect on emigration, and the council governor commented on this fact in his report.

Although the council governor argued that there was no clear connection between economic downturn and emigration in Hedmark, some parishes experienced economic decline between 1846 and 1850.

The political revolutions in Europe in 1848 had led to a decrease in timber exports. There were also a growing number of cases involving debt and the handling of debt.

The peak years of Norwegian immigration to Texas between 1845 and 1860 were 1846, when sixty-three immigrants arrived, 1850 (129), 1851 (seventy-six), and 1853 (eighty-seven). No immigrants arrived in 1849, 1852, 1855, and 1856, while nine immigrants arrived in 1854, seven in 1857, twenty-three in 1858, nineteen in 1859, and three in 1860. The group of immigrants who crossed the Atlantic on the *Victoria* in 1853 came to mark the end of the beginning of Norwegian immigration to Texas (see Appendix 1).

Where in Texas did the Norwegian immigrants and their families live in 1860? Syversen and Johnson analyzed the 1860 census lists for Texas counties. In some cases, they found the census lists included incorrect information. Persons or families were listed more than once; boys and girls were listed as born in Texas instead of Norway, or Norway instead of Texas. Their conclusion was that 516 Norwegians and Texans with Norwegian-born parents lived in Texas in 1860, and 306 of them had been born in Norway. It does not come as a surprise that the majority lived in Henderson, Van Zandt, and Kaufman counties, or that Norwegian immigrants lived in other neighboring East Texas counties. The greatest surprise is the high number of Norwegian immigrants who lived in Bosque County, west of Waco. When did the Norwegians begin to settle in Bosque County?

CHAPTER 5

Finding Virgin Land in the Bosque Valley

◊ ◊ ◊

IN LATE 1853 AND EARLY 1854, the first Norwegian immigrants settled on the western frontier in the Bosque River valley. The North Bosque River, the longest river branch, begins in north central Erath County, flows through Hamilton County and Bosque County to the east, and is joined by the East Bosque River, which also begins in Erath County. From their confluence, the North Bosque continues into central McLennan County, is joined by the South and Middle Bosque rivers, and flows into Lake Waco.[1] The Bosque River runs through rolling hills, and the dominant vegetation includes live oak and cedar. The word *bosque* is Spanish for "woods" or "woody lands." According to some accounts, the river was named by the Marqués de San Miguel de Aguayo in 1719, while others claim the river got its name from a French trader, Juan Bosquet, who lived with the Tawakoni Indians in the 1770s.

After the Canuteson family settled in Dallas County in winter 1851, Cleng Peerson repeatedly told stories about the land he had seen to the south along the Brazos River in 1849–50. Twenty-year-old Ole Canuteson listened to the Peerson stories. In summer 1852 he decided to explore the country south of Dallas along the Brazos. "In the month of August 1852," Canuteson wrote three decades later, "I started out with a man by the name of Bryant, to search for vacant land, the legislature having just passed an act donating 320 acres to actual settlers."[2] The Indians no longer seemed such a big threat, and Canuteson and Bryant reduced the risk by following the military road between the forts.

The Canuteson families were not disappointed with their land in Dallas County. But the Texas legislature was discussing a new homestead law that was more favorable to settlers than the first Pre-emption Act in Texas from January 22, 1845, had been. The new law would give a homesteader 320 acres of land for free, on the condition that he lived on it for three years and made improvements. According to Thomas Lloyd Miller, this "pre-emption act became in fact a homestead grant when an act approved on February 7, 1853, declared that those who had settled under the provisions of the pre-emption act need not cover the grant with a valid certificate. The grant became an outright gift with only the requirements of occupancy."[3] Seymour V. Connor called the 1853 law the "western world's first homestead law."[4] In 1854, however, another act reduced the amount of land granted under these conditions to 160 acres per person.

Canuteson and thousands of others used the short window of opportunity to acquire cheap land. On their way south, Canuteson and Bryant traveled through Fort Graham, established in March 1849 near the eastern bank of the Brazos River at Little Bear Creek, 14 miles west of present-day Hillsboro. For a short while Fort Graham was the most important post on the upper frontier. It was close to the camps of many Indian groups. When the line of settlement moved west, Fort Graham ceased to be of strategic importance. The fort was still in operation when Canuteson traveled through the area in 1852 but was closed in August 1853.[5]

The two travelers continued south along the military road to Fort Gates. They left the road when they came to the north side of the Bosque River, and found shelter and hospitality "at the house of the then well-known pioneer settler Jewell Everett."[6] The next morning, Jewell's son Francis went with them up the Bosque Valley. When they reached the place where the town of Clifton was later established, they "overtook three men eating their lunch—L.H. Scrutchfield, Jaspar Mabray, and a man by the name of Bell, a brother-in-law of Mabray, that had just come from East Texas to prospect for land on the Bosque."[7] The whole group continued upriver, crossed Meridian Creek, and camped close to where the town of Meridian was later located. "We found a bee tree and killed a turkey and had quite a feast that night." They returned downriver the next morning, and Bryant and Canuteson left the others to visit with William McCurry, a recent settler on Neil's Creek. "Mr. McCurry had just been out with a surveying party conducted by that well known pioneer, George B. Erath." McCurry went with them along Neil's Creek and "pointed out to us vacancies that had been shown him by the surveying party. I concluded to enter land on Neil's Creek, went to Waco and engaged Major Erath to come and survey these lands and found enough to accommodate many more than at first contemplated."[8]

BOSQUE VALLEY AND GEORGE B. ERATH

George Bernard Erath was born in Vienna, Austria, on January 1, 1813, and died in McLennan County, Texas, in 1891. His father was an immigrant from Würtenberg in Germany who did well in the tannery business in Vienna, while his mother was of Greek heritage. Their son was sent to school at the age of 6, and Erath received an education far beyond the normal for a boy of his class.[9] He became a student at the Vienna Polytechnic Institute at the age of 12 and studied there for two years. After his father died, his mother lived in fear that the Austrian authorities would recruit him into the Austrian army and therefore sent him to visit relatives in Rothenburg on the river Neckar in Würtenberg, Germany. His relatives took very good care of him, and even offered to let him take over the family tannery business. But because Erath had already decided that he wanted to emigrate to the United States, his relatives paid for his journey from Germany to New Orleans instead. He left Germany in early April 1832 and sailed from Le Havre in France on April 18 on the American cotton brig *Motion*, which arrived in New Orleans on June 22, 1832. Many of the German emigrants on board continued to Cincinnati, Ohio, and Erath joined them. Some months later, however, he decided to go to Texas, where he arrived in March 1833. Because of his technical schooling in Vienna, he soon found surveying work in Robertson's colony. In March 1836, at 23 years old, he enlisted as a private in Captain Jesse Billingsley's Company C of the Texan Volunteers for service in the Texas Revolution and participated in the battle of San Jacinto.

Around 1840, wrote William C. Pool in his *History of Bosque County*, "rugged individualists who formed the advance guard of settlement along the Milam frontier began to push into the lands adjacent to the South, the Middle, and Main Bosque rivers in the wake of the increasingly successful war parties of rangers raiding and harassing the Indians to herd them farther west."[10] Erath played a crucial role in the settlement of Central Texas. In early winter 1839 he led a company of Texas Rangers northwest to the headwaters of the Bosque River. The main aim was to scout for signs of Indians. There were no signs to see, and the scouting party chose a different route back, which brought them to the banks of the Bosque River. The rangers were struck "with the beauty of the country" and became interested in taking land. Since

Erath was an official Texas surveyor, the group began surveying and continued surveying until they had no food left. Several men registered land claims.[11]

Between 1843 and 1845, Erath represented Milam County in the House of Representatives during the Eighth and Ninth Congresses of the Texas Republic. He was strongly in favor of the annexation of Texas to the United States.[12] Erath continued working as a Texas Ranger and surveyor. He laid out the towns of Waco, Stephenville, and Meridian in Central Texas, and Erath County was named in his honor.

THE PATHFINDER NEIL MCLENNAN AND MCLENNAN COUNTY

The Bosque valley in early 1854 was part of McLennan County, organized in 1850 and named in honor of Neil McLennan, the first permanent settler in the county. He was a Highland Scot born on the Isle of Skye in northwest Scotland on September 2, 1787. His father, John McLennan, and his mother, Catherine MacKinnon, emigrated with their children and other kin to Richmond County in North Carolina in 1800 or 1801. Highland Scots had been settling in the Cape Fear River district in North Carolina since the 1730s. Emigration ships from Scotland landed in Wilmington at the mouth of the Cape Fear River and the migrants followed the river inland, settling along its banks and creeks.[13] By 1800, the land close to the sea had been taken and new immigrants moved farther west. The family had relatives in Richland County, west of Scotland County, and settled there. They spoke Gaelic and were strict Presbyterians.

Neil McLendon (known as McLennan in Texas) married Christain Campbell in 1812. According to *History of Walton County*, Florida, McLendon decided in spring 1820 to seek new land in the Florida Panhandle east of Pensacola. He "longed for a newer and better country—a cattle, hog and sheep range."[14] Florida had recently been taken from Spain by the United States. In December 1817, President Monroe sent General Andrew Jackson to police the Seminole Indians, who had raided settlers in Southern Georgia. Jackson took the Spanish capital Pensacola in May 1818 and established an interim government. The actions of General Jackson put the United States in a difficult position in relation to international law and other nations. But American public opinion was strongly in favor of what Jackson had done. In the end

Spain sold Florida to the United States for $5 million. President Monroe appointed Jackson as the first civil governor of the territory, and Jackson took formal possession of Florida on July 10, 1821.[15]

McLendon, with his wife and family, his brother Lochlin and family, his mother, his unmarried brother-in-law Daniel D. Campbell, and John Folk and his family, traveled from North Carolina by wagon in the direction of Pensacola, Florida, through South Carolina, Georgia, and Alabama. The group of Highland Scots chose land in what in 1828 became Walton County, Florida, located between the Black Water River in the west, Choctawhatchee River in the east, Alabama to the north, and the Gulf of Mexico to the south.[16] McLendon started a strong migration chain of Highland Scots from North Carolina. Almost every year in the 1820s five or six families moved by wagon from their homes in North Carolina to Walton County, Florida. Some immigrants arrived at Pensacola directly from Scotland.[17]

Around 1830 McLendon began to feel crowded again. According to Ravenstein's laws of migration, the first migration step is usually the hardest, and the second and third migration steps are easier. The family decided to move to Texas and was accompanied by the family of his brother Lochlin and three other families.[18] A rather primitive two-mast schooner was built in the course of eight to ten months and was baptized *Euchee* in honor of an Indian chief. The ship left Florida in November 1833. Several weeks later McLendon's relative John McKinnon received a letter from Galveston, Texas, dated February 6, 1834:[19] After a long and perilous trip they had reached Galveston. McLendon reported that he had left his family there while he scouted for land. He had recently sold the *Euchee* for more than it had cost to build.

THE MCLENNANS IN ROBERTSON'S COLONY

The McLennan's (as they were known in Texas) settled in Robertson's Colony. The following narrative of their settling and life in Walton County, Florida, has been based on the *History of Walton County*, published in 1911. The main source for what happened to the group of Scots in Texas during the next years is Volume X of *Papers Concerning Robertson's Colony in Texas*. The stories in the two sources are inconsistent at times. According to the volume in the Robertson papers, the schooner entered the mouth of the Brazos River in early spring 1835, but struck a snag and sank at Fort Bend, but that differs from the story told in the Walton County history, based on McLennan's letter.

The families settled at Sugar Loaf, later called McLennan's Bluff, in 1835.[20] Approximately 75 families lived at Nashville-on-the-Brazos in the late 1830s, but many of them soon moved farther west into frontier areas that later became Bell, McLennan, Coryell, and Bosque counties.[21] Many settlements on the northwestern frontier in Texas were raided by Indians after the war with Mexico in 1835 and 1836, and the McLennan family paid dearly for their willingness to settle on the frontier. In early October 1835 they were attacked by Indians. The Indians came upon Laughlin McLennan "where he was at work making rails in the bottom, and shot him to death with arrows, twenty-five of which were sticking into and through his body when he was found. They then advanced on the house where Aunt Peggy, Laughlin's wife, her three boys and aged mother were awaiting the return of the son, husband and father to dinner, and killed the old lady by a blow on the head with an axe and cast her body into the house which was set on fire. After wantonly destroving (sic) everything they could, killing cows, calves and chickens, they left, taking the distracted mother and her boys with them."[22] The boys captured were John (eight years), Neil (six), and Daniel (four). "The mother was never recovered, but after several years died in captivity. The boys were traded from tribe to tribe and the eldest, John, was returned to his relatives after being with the Indians ten years." McLennan's brother John was killed in an Indian raid in 1838, but the rest of his family escaped. Neil McLennan's family was also attacked but escaped.[23]

McLennan was a member of the Erath expedition of Texas Rangers in 1839, and he was one of the men who registered a land claim on their way back. When Major George B. Erath began surveying land along the Bosque River in May 1845, McLennan was again a member of the group. This time he brought with him a wagon and people to build a house on the South Bosque River. During the summer, he also brought cattle.[24]

McLennan was the first permanent settler west of the Brazos River. The Texas legislature established McLennan County on January 22, 1850, with Waco as the county seat and a modest 72 inhabitants. On March 1, 1849, Erath laid out the first block of the new town on the site of the former Waco Indian village on the Brazos River.[25] Soon a small business district developed. "The majority of McLennan County's settlers before the Civil War were from the southern United States or other regions of Texas. They brought their culture, community, and 'peculiar institution' of slavery with them."[26] Well-educated planters established their cotton-based plantations along the rich bottomlands of the Brazos River. The surrounding prairie land was primarily used for livestock. According to the US census of 1860, McLennan County had a free white population of 3,811, and 270 of them owned 2,395 slaves. The county seat Waco had around 800 inhabitants.

NORWEGIANS CHOSE LAND IN THE FUTURE BOSQUE COUNTY

The land Ole Canuteson located in the Bosque territory in August 1852 was still a part of McLennan County.[27] News of the good land in the Bosque Hills traveled from Dallas County to the Norwegian settlements at Four Mile Prairie. Several of the Norwegians living at Four Mile Prairie were interested, especially the immigrants who had looked for land in Dallas and Tarrant counties in North Texas in summer 1852.

After the harvest in 1853 and the worst of the heat had worn off in the autumn, a group of men including Cleng Peerson, Ole Canuteson, Canute Canuteson, Carl Questad, Jens Ringness, Ole Ween, Anders Bretten, and Ole Pierson went to investigate and claim or buy land in the Bosque valley. They found successive ranges of hills and intervening valleys, each seemingly more charming than the last. They found clean and running water on the land, many springs, an ample choice of trees, and beautiful and pleasing views of valleys and creeks. They liked the land, and each member of the scouting party chose the place he liked best.

Until now, historians have had very scant primary sources available to reconstruct what the Norwegian immigrants did during the weeks in 1853 when they first visited what later became known as the Norse colony in Bosque County. Excellent historical research in primary sources done by Dale Van Sickle in Austin, Texas, has shed new light on the process. Van Sickle's main interest was to find primary sources making it possible to reconstruct the history of the Cleng Peerson/Colwick farm. She discovered primary sources that throw new light on that specific farm, but also on the actions of the whole group.

Ole Canuteson and his father Canute Canuteson each formally requested that surveys of their land be done in the presence of the

McLennan Justice of the Peace, Jasper Mabray, on November 6, 1853. The other members of the group made the same request on the same day, with the exception of Ole Pierson, who signed his request on November 8. County officials helped the Norwegian immigrants transform their handwritten applications into the required legal language. They handed in their application to the county clerk with a request that George B. Erath, the Deputy Surveyor of Milam Land District for McLennan County, should survey the land they had chosen. It would not have been possible for Erath and his surveyor team to do all the work involved in a couple of days. The Norwegians had probably taken days if not weeks to select the land they wanted, and a survey crew had performed initial surveys and written field notes before it all came together in a formalized procedure in early November.

The case of Anders Bretten, or Andris Bretta (as he signed himself on the request for survey), or Andrew Britten (written on the field notes of the survey), or Anders Broten (name of the original grantee on General Land Office records), will be used as a specific example of the general process. Cleng Peerson later took over Bretten's land, and then sold it to Ovee Colwick. Copies of the original documents in different public archives put together by Dale Van Sickle prove beyond any doubt that even though the names are written slightly differently on different occasions, it is indeed the same person. Anders Andersen Bretten was born on July 8, 1813, in Harildstadeie, Romedal, in Hedmark County in eastern Norway. He had arrived in New Orleans on March 29, 1853 on the ship *Victoria* from Arendal in southern Norway. Bretten was a carpenter by trade.

In his field notes, the surveyor wrote that Andrew Britten claimed 313 acres of land on the north side of Neil's Creek, beginning on the northeast corner of a survey made for John Ringness for the northwest corner of this survey. It was surveyed and signed by Erath on November 6, 1853, certified on November 24, 1853, by G. B. Erath, and signed by J. W. Armstrong, District Surveyor on May 24, 1854.[28]

The Ole Canuteson claim was registered "in Milam District, Bosque County, on the north side of Neil's Creek, about 8 miles from its mouth, and about 13 miles due south of Meridian."[29] The claim had to be registered within three years of settlement. The Bosque County Clerk filed the claim for 303 acres on April 10, 1857, and it was approved on November 9, 1857.[30] The claim included one bank of Neil's Creek, which ensured Canuteson enough water for his livestock. Potable water for family use came from a spring nearby. More than a sufficient supply of wood for almost any purpose—buildings, fences, and fuel—was found close to the homestead. The buildings were put up among some large oak trees on a rise in the ground, and most of the land could easily be cultivated.[31] Canute Canuteson took land adjacent to the property of his son. His land was registered and approved on the same dates as those of his son and was measured to be 313 acres.

The decision to move from East Texas to Bosque had been made, but the farmers had to return to their families to make the necessary preparations for the trek from East Texas to Bosque. Last but not least, landowners in East Texas wanted to sell their land and their farms at the best price possible. The first group of Norwegians left East Texas in late 1853 and arrived in Bosque County in early 1854. Canute Canuteson, Ole Canuteson, and Cleng Peerson went back to Dallas County to prepare for the move south to the Bosque valley. This time, Cleng Peerson was the follower, not the leader. From 1854 until his death in 1865 he spent some good years among fellow Norwegians in the Norse district in Bosque County.

The two bachelors, Anders Bretten and Ole Ween, stayed behind at a camp they established at Upper Turkey Creek. Ole Ween, from Løten parish, Hedmark County, also came on the *Victoria* in 1853. The two men did not remain idle during the weeks they waited for the others to return. They found suitable timber and used the tools they had available to build primitive log cabins on their land.

It did not help Bretten much that all formalities concerning his land were in order. In early 1854, while out hunting near Neil's Creek at the head of Turkey Creek, Bretten accidentally shot himself. He did not die immediately. The wound became infected, he contracted typhoid, and died after a long illness.[32] Bretten was the first Norwegian to die in Bosque County.

HOW MANY NORWEGIANS HAD MOVED TO BOSQUE COUNTY?

At the end of April 1854, T. A. Gjestvang, of Løten, Hedmark County, received the news that several Norwegian families and unmarried men had moved to Bosque County. It was indeed very good news, wrote Gjestvang in a letter to Carl Questad in May 1854, that the Questad and Ringness families had moved from East Texas to land along the

Map of Bosque County. The Brazos River is the natural boundary in the east. In the middle flows the Bosque River. The railroad runs through Bosque Valley and the towns Valley Mills, Clifton, and Meridian. The historic Norse district is shown between Clifton and Cranfills Gap. A significant number of Norwegian-Texans still live in this area.

Bosque River. Gjestvang had tried and failed to find Bosque County on his Texas map. It would please him very much if their new colony succeeded.[33] He asked Questad to include a detailed description of the place in his next letter. How many Norwegians had moved to Bosque County, and who were they?

Gjestvang knew most of the new settlers. Several of them had emigrated from Hedmark County. He knew that Carl and Sedsel Questad and their two children, Even and Marthe, had settled in Bosque County. Questad had selected 256 acres in a valley between low cedar mountains above Gary Creek.[34] He placed his house on a high knoll with a panoramic view of beautiful twin mountains to the east. Fields were cleared on the lower land, between small creeks and streams of water. The first house Questad built was a log cabin for himself, his wife, and two children. Cultivation of the land and the sowing of small grains and corn came first. Questad was primarily a farmer, but he had above average skills in several other fields. He had worked as a blacksmith before his emigration, and was also a skilled carpenter, stone mason, and cabinet maker. While working his land he located suitable building materials, trees as well as rocks, so he could build large, solid stone houses. The defense of family and animals against Indian attacks played a central role in the layout of the buildings. To attack the Questad house, Indians would have to ride uphill on two sides before reaching the buildings, and the field of fire was unhindered by grass and small brush.

Carl Questad still owned land in Four Mile Prairie. He had left the sale of the land to Judge Harrison when he moved to Bosque County, and in July 1856 the Norwegian immigrant Sigurd Ørbæk informed him that the land had been sold for $1.00 per acre.[35] Knud Olson, the older brother of Sedsel Questad, who remained at Four Mile Prairie, had asked Ørbæk to inform Questad about the sale

Other settlers Gjestvang knew were Jens Ringness and his family. Ringness had chosen land neighboring that of Anders Bretten.[36] When Gjestvang wrote his letter, Bretten had already died. Another settler, bachelor Ole Larsen Ween, was working hard on the 312 acres of land he had claimed. He was born October 23, 1822, in Grønsveen, Løten, Hedmark, and died in Bosque County in October 1871. The patent date for the land Ween settled on was October 13, 1858 (Abstract 82, Patent 847); it lay on the north side of Gary Creek, not far from the Questad place. Ween had registered land close to Carl Questad on August 23, 1858 (Abstract 872, Patent 211), measured to be 294 acres. Since Ween was a bachelor, he had many daily chores both around the house and in the field. In the US census of 1860, the Ween property was valued at $936 and his personal estate at $528. The census also listed Cleng Peerson as living on the Ween property as well as the Olson family from Karmøy.[37]

Two of the pioneering families in Bosque County, the Jensens and the Piersons, came from southern Norway; Gjestvang knew both families. They had crossed the Atlantic together on board the *Victoria* from Arendal to New Orleans in 1853. Jens Jensen and his family came from Holt parish east of Arendal, and the Ole Pierson family came from Tromøy, an island just outside Arendal. Jens Jensen Oddersland was born in May 1808 in Brovold in Austre Moland and died on August 25, 1889, in Bosque County. In 1834, he married Tonje Knudsdatter (born December 9, 1812, in Austre Moland, died April 15, 1883, in Bosque County). They brought with them seven children, the oldest 18 years of age and the youngest 3. According to family lore, Jens Jensen brought with him 1,000 dollars, a great sum of money at the time.[38] During the short time the family lived in Four Mile Prairie, their son Knud (born August 4, 1837) was fatally shot on the street without any provocation. The law did nothing to catch the murderer, and the Norwegians learned an important lesson. They had emigrated to a country with wholly different attitudes toward violence and law compared with the country they had left. In Bosque County, Jensen settled at Gary Creek near Norse and built a log house with a rock fireplace on his first homestead.

Ole Pederson Songe (Ole Pierson) was born on December 4, 1804, and died on August 26, 1892, in Bosque County. He was married on June 29, 1833, to Anne Helene Olsdatter, who was born March 2, 1812, in Tromøy, and died on April 29, 1899, in Bosque County. Ole Pierson was almost 50 years old when he and his family decided to emigrate to Texas in 1853. The family had seven children; the oldest was 18 and the youngest not yet 2. After their arrival at Four Mile Prairie, their youngest daughter died in mid-July. The family brought some money with them and arrived just in time for Pierson to join the group of men from East Texas seeking land in the Bosque valley. Pierson also settled on Gary Creek, 10 to 11 miles west of Clifton, and the same distance from Meridian. Wood and fresh water were very important to the Norwegian settlers, and Pierson was no exception. A large spring with a waterspout at least two inches wide came out of the creek bank.[39] Live oak, elm, mesquite, hackberry, and pecan grew along the creek, and cedar was plentiful in the hills. Pierson built a house 500 feet from the north bank of Gary Creek, and between the house and the creek there was a clearing where the stock could graze.[40]

THE ARRIVAL OF HENDRICK O. DAHL AND THE BRUNSTADS

A second group of Norwegian settlers from Hedmark arrived in Bosque County in fall 1854. Hendrick O. Dahl, who was born on November 28, 1827, on the farm Kjemstad, Stange, and died January on 13, 1873, in Bosque County, was 24 when he emigrated in 1851 with the group from Stange and Romedal on the ship *Arendal*. Johan and Anne Eriksdatter Brunstad and their six children traveled on the same ship. Anne's niece, 17-year old Christine Furuseth emigrated with the Brunstad's and assisted with taking care of her children.

Johan Brunstad and Hendrick O. Dahl rented land together at Four Mile Prairie. Some months after Brunstad's death in fall 1852, Dahl married Christine. A year later, on February 22, 1854, Christine gave birth to a son, Ole. The couple had already decided to follow the others to Bosque County, but the journey had been postponed because of the newborn. Christine's aunt, Anne Brunstad, and her children joined them on the journey, as did the Norwegian bachelor Berger Rogstad. He had arrived at Four Mile Prairie after the death of Johan Brunstad. Berger Rogstad was born on December 11, 1820, in Elverum, Hedmark County, and died on December 9, 1880, in Bosque County. He had grown up in a wealthy and influential farming family in Elverum. Before the end of 1854, Anne Brunstad married Rogstad.

Anne Brunstad still had some of the money the family had brought with them from Norway, and with it she bought land from William Gary. He had staked a claim in the hills the Norwegians became so fond of, where Gary Creek still bears his name. Gary and his wife had migrated from South Carolina to Bosque County in 1852. At the first election in Bosque County in 1854, Gary was chosen as one of the first five County Commissioners. His land on Gary Creek was registered as Abstract 306, and Anne Brunstad was registered as the patentee of 312 acres. The patent date was August 24, 1858.

Hendrick Dahl settled on 320 acres of land 1 mile north of Gary Creek and about 6 miles southwest of Clifton. He bought the land from Jasper Newton Mabray, one of the first settlers. Mabray was born on July 17, 1822 in Florence, Alabama, and died in Sonora, Texas, on October 27, 1904. He was a soldier in the Mexican War in 1848 and was wounded at Monterrey. He married Mary Ann Hudson in Rusk County, Texas, on November 27, 1851. Before Bosque County was organized, Mabray was notary public in McLennan County, and when Bosque County was organized in 1854, he was elected county clerk.[41] Dahl owned a fine mare that Mabray desired very much. After some serious haggling, Mabray agreed to trade his 320 acres of land in exchange for the mare. On September 19, 1854, Mabray transferred his land to Dahl. The Dahl family moved into the log cabin already present on the Mabray homestead.

HOW CLENG PEERSON GOT HIS LAND FROM THE TEXAS LEGISLATURE

Cleng Peerson, now 70, was not among the land takers in early November 1853. He had owned land several times before but did not show any interest in becoming a homesteader again. He was present, however, and eager to contribute with his knowledge and experience. Peerson had lived in the United States for more than thirty years, after all, and had been a pathfinder for Norwegian settlers more than once. He also knew a lot more about American land laws than most of the men in the scouting party, some of whom had literally just come off the emigrant ship.

Peerson had lived with the Canuteson family in Dallas County for four years. Was it tacitly agreed that this arrangement would continue in Bosque County? A new situation emerged after Anders Bretten accidentally shot himself and died. What would happen to the land Bretten had pre-empted and registered in November 1853? Bretten's land was located between the land of Canute Canuteson and that of Jens Ringness. Bretten was unmarried, and "his heirs and legal representatives being aliens and bound by the statutes of this state from holding land," his land would be open to be claimed by others, since no owner could be found with a right to sell it. Cleng was the only one among the Norwegians who had chosen no land, and so according to Texan law he was eligible to take over Bretten's land.

Cleng Peerson was far from rich, rather the opposite, after losing money during his time at Bishop's Hill. What if Cleng moved on to Bretten's claim, fulfilled the requirements of the pre-emption law, and became the owner? Even if he did not intend to farm, he would get the land for free, and he could sell it and use the money to make life a little easier for himself in his old age. There was already a log cabin on Bretten's land, and the neighbors could help Peerson fulfill the conditions to get ownership of the land. The idea was introduced, mulled over, and discussed. Toward the end of 1854, Cleng Peerson took possession of Bretten's land. There were no heirs to Bretten's land, and according to Texas law, Peerson would have to apply to the Texas Legislature for ownership. The Norwegian immigrants had trouble just communicating everyday meanings in English; none of them had the skills to write the necessary document in the specialized legal language required. Their American-born neighbors, however, who had been elected to political positions on the county level, were more than willing to help and probably were consulted several times about the right phrasing before Peerson delivered the correct application.

In a document dated December 17, 1855, Peerson formally filed a petition to the Texas Legislature for a Pre-emption Patent for the 313 acres of Bretten's land, stating that he had lived on the property from about 1854 and cultivated it after the death of the original pre-emptor. Hendrick O. Dahl and Ole Canuteson met as his witnesses before the Bosque County Justice of the Peace, J. K. Helton, on that date. Jasper N. Mabray, Clerk of the County Court of Bosque County, certified the petition on December 31, 1855.

On August 13, 1856, the Texas Legislature passed "an Act for the relief of Cling Pearson." He was awarded a deed for the land he had "squatted" on. The survey for 320 acres of land for Andries Braten by the Deputy Surveyor of Milam Land District on November 6, 1853, was transferred for free to Cleng Peerson. The Commissioner of the General Land Office was instructed to issue a patent for this land to Cleng Peerson, as "ascribed in the field notes of the said Andries Braten as pre-emptor."[42] On January 26, 1857 the governor of Texas, Elisha Marshall Pease, signed a deed on fine sheepskin for 313 acres of land to Cleng Peerson. The land was located "on the North side of Neill's Creek, about 8 miles from its mouth into the Bosque, and about 12 miles South 5 (degrees) West of Meridian. The land began at the North East corner of a survey made for John Ringnes for the West corner of the Survey."

There has been much speculation in the Norwegian-American historical literature about what kind of processes took place before Governor Pease signed this deed. Was the deed of land a surprise to Cleng Peerson and his Norwegian neighbors? Did he get it because he brought Norwegian immigrants to Texas? In her search for primary sources about the history of the Peerson/Colwick farm, Dale Van Sickle unearthed official documents that showed that the process followed Texan law and was both straightforward and mundane.

RAIDED BY INDIANS A MONTH AFTER ARRIVAL

The Norwegians had grown up among mountains and high hills in Norway. The hills in Bosque County reminded them of Norway. The land had never been touched by any plow. Changes over time were a result of floods and fires. The mountain country was full of brush and underbrush, infested with rattlesnakes, copperheads, mountain lions, and coyotes. Sage grass and other vegetation, undesirable in the eyes of the farmer, grew in the valleys. There was an abundance of grazing opportunities and fresh and clean water. But the nearest marketplace was 50 miles away and Indians were still roaming the country. The Norwegian immigrants had never given much thought to the risk of Indian raids. Their main concern was locating good land in a healthier environment than they had found in the Norwegian colonies in East Texas.

Within the first month after arrival in spring 1854, Indians arrived and raided the homestead of Canute Canuteson. Canuteson had bought five milk cows from Colonel Frazier, who lived in the northwestern part of the county,[43] and had traveled with his neighbor Ole Ween to the Frazier place on the Brazos River to fetch the cows. While they were away, his son Andrew C. Canuteson and two other small boys from a neighboring homestead were outside playing some distance away from the cabin when they spotted Indians approaching. Terribly scared, the boys ran to the cabin as fast as their small legs could carry them. Andrew's stepmother, Berthe Canuteson, was as frightened as the boys. None of them had seen Indians before; they had only heard scary stories about them.

Berthe had enough presence of mind to grab the $500 in silver Canute had put in an old chest and run up into the hills with the boys, where she hid the money. After a while Berthe and the boys snuck back to see if the Indians were still there. They were, so they walked the three miles to the Ole Pierson place, where they spent the night.

Canute Canuteson returned home with the milking cows toward evening. He was shocked to find Indians occupying his house. He saw no sign of his wife and son and thought the worst. He fled to the home of his son Ole Canuteson and spent the night there. When Canute returned to his home the next morning, the Indians were gone. He felt great relief when he found no dead bodies, but feared that the Indians had taken his wife and son with them. They had stolen many of their possessions and destroyed things they were not interested in. Later that day Canuteson was reunited with his wife and son at the Pierson's, and their joy was indescribable.

Several versions of the above story exist among Norwegian-Texans in Bosque County, and some of them are inconsistent. The version above follows the version Theo Colwick told in a letter to Rasmus B. Anderson in 1894. In his master's thesis in 1954, Oris Pierson maintained that it was 900 and not 500 silver dollars.[44] In his autobiography, the Indian fighter and Texas Ranger James Buckner Barry claimed that "None of the Norwegians had a gun, so they knew that if the Indians had decided to raid the entire settlement, they were at the Indian's mercy."[45]

At the time of the Indian raid, Buck Barry lived in Limestone County, but was called in to lead the Indian chase. Bosque County was only a few miles from the frontier line, he commented, but the "Indians had not bothered it for quite an interval." In the 1854 raid, several Norwegian settlers along Gary's Creek "suffered such misfortune when their savage enemy decided to molest their household property and run off their horses and other livestock."[46] According to Barry, Canuteson found that the Indians had carried off his wife and son, "the bed ticks ripped open and feathers scattered all over the place, and $900 in silver, in a sack, was also gone."[47] It is not unusual that such dramatic and vivid stories change somewhat in being told over and over again.

"The Norwegians had not been molested in East Texas, and when they came to Bosque County, they took very little precaution the first year to ward off any Indian attack," wrote Oris Pierson.[48] They had seen no reason to equip themselves with guns, since they had no perception of the risk of being attacked by Indians. They were totally unprepared.

The Norwegians learned from this incident what it meant to have settled among Americans who had grown up on the frontier between whites and Indians. Their neighbors knew firsthand the potentially disastrous consequences of underestimating the unpredictable behavior of raiding Comanche or Kiowa Indians. They had a tradition of accepting the risk of living on the Indian frontier, and likewise a tradition of strong retaliation when Indian attacks occurred.

The Norwegians learned to adapt to this risk. They soon began to build their houses in a way that helped them more easily defend them in case of Indian attack. It took them longer to adapt to the frontier horse and gun culture and to accept they might have to kill another human being to avoid being killed themselves.

PETITION TO THE GOVERNOR ABOUT THE "INDIAN MENACE"

After the 1854 Indian attack the Norwegians' Anglo-American neighbors called in the Texas Rangers. All of the Norwegian settlers attended a meeting at Neil's Creek, where they were introduced to an English term most of them had never heard before—Indian depredations. The word "depredate" means plunder or ravage, and in the nineteenth century, the term "depredations" was frequently used to describe massacres, conflicts, and cruelty inflicted by Indians upon whites.

At the end of the meeting, someone skilled in the language of politicians and lawyers wrote a petition to Texas Governor E. M. Pease about the "Indian menace." "We the undersigned," they wrote, "inhabitants of the Frontier Counties of Bosque and McLennan, would respectfully represent, that our settlements have within the last two or three weeks been repeatedly attacked by several parties or tribes of Wild hostile Indians, who have robbed our houses, killed and stolen our horses, pursued our people and menaced our lives; compelling us to seek temporary security by abandoning our homes and collecting our families together as our only recourse until aid and assistance can be sent for our protection."[49]

The governor and fellow Texans were urged to raise immediately, "with the greatest dispatch, a company of Texas Rangers formed of experienced volunteers, to be maintained for a period of twelve months or as long as the service might require, for the purpose of scouring through the adjoining border country, and to hunt down and punish our Savage Enemies, and to drive them off effectually from our exposed settlements." If such measures should not be taken "before another moon shall have passed, the fate of our devastated homes and murdered Families may be held up to the world as an awful example of the wilful neglect of those in authority whose duty it is to afford protection to the lives and property of ourselves and families."[50] The letter was signed by an impressive list of early settlers—the McCurrys, Saachers, Gandy, Morgan, several Garys, Goodman, Robertson, Thomas, several Everetts, Mabrays, as well as the lawyer J. H. Scrutchfield. All the Norwegians present signed their name on the petition—Ole Canuteson, Canute Canuteson, Ole Pederson, Peter Spangburg, Ole Ween, Carl Questad, Jens Ringness, and Cleng Peerson.

SUCCEEDING AS SUBSISTENCE FARMERS
REPRESENTED THE GOOD LIFE

The first priority of Norwegian settlers was to supply shelter for their families as quickly as possible. The immigrants followed the local tradition and built houses using easily accessible rocks and logs. Neighbors helped each other erect the houses. Jens Jensen built a log house with a rock fireplace, which took less time to build than a stone house. Since the first log houses were built with the help of neighbors, there were variations in notching and chinking techniques. When the need arose, these early log structures were expanded with additions made of stone.⁵¹ The houses were often built snugly along a mountainside or on high ground. "A variety of immigrant groups directly from Europe copied Anglo-American log construction styles after arriving in Texas," wrote Terry G. Jordan in his book *Texas Log Buildings*. He mentioned the buildings of "the Germans of south-central Texas, the Norwegians of the Bosque County hills, the Wends or the Sorbians of Lee County, the early Czech settlers in Austin and Fayette counties, and some of the Irish Catholics in the border area of Goliad, Refugio, and Victoria counties."⁵²

Compared with the general level of mechanical skills among farmers in Norway around 1850, the antebellum Norwegians in Bosque County had skills far above average. A remarkable number among them had practiced the blacksmith's trade. They were able builders of their own houses; they also made their own equipment to build houses and till the land. Their combined skills gave them comparative advantages to help them succeed on the Texas frontier both individually and as a group.

By 1860 most of the first settlers had moved out of their log buildings and into larger and more solid stone houses. They put a lot of thought into the best way to place their houses and outbuildings in the terrain in case of rainstorms and flooding. The Norwegians demonstrated an ingrained sense of the dangers of sudden floods along the creeks. As a rule, they chose to build on higher ground.

Permanent buildings were often erected on a spot with sand, gravel, or clay, where little would grow. The Norwegians learned to respect rattlesnakes and copperheads and developed strategies to avoid them. They took great care to avoid grass around the house, since it was easier to spot rattlesnakes on white gravel. The homes were built to maximize comfort during hot summer days and nights. In Bosque County, a gentle cool breeze came from the Gulf of Mexico; without this breeze the nights might have been terribly hot. To take full advantage, the corners of their houses faced south so that the southern breeze could waft by two sides of the house and expose more rooms to the breeze.⁵³ The newer houses often featured a hall running through the house, and sometimes a traverse hall, which helped the air circulate in summer. In accordance with Norwegian tradition, only one room was heated in winter.

Ole Pierson built a two-story house measuring 40 × 25 feet with walls 24 inches thick. It had two rooms upstairs and two rooms downstairs with porches the length of the house on both floors and a stairway outside. Each room had a doorway opening onto the porch, but there were no doors connecting the rooms. The west downstairs room had a large fireplace. Hand-hewn beams supported the second story and the roof.[54]

Jens Ringness began building a large stone house on his property near Neil's Creek in 1859. In 1860 the census enumerator visited household no. 27, "John Ringless," 57 years old. Four Americans lived in the household in addition to his wife and children. Two of them were carpenters, John E. Brown (38), born in Kentucky, and T. H. Brown (30), born in Missouri. Two women named Brown also lived in the Ringness household—M. V. Brown, 60 years and born in Georgia, and M. I. Brown (29), born in Indiana.[55]

The Dahl family built a two-room house between 1854 and 1856, and added two more rooms during the 1860s.[56] At the time of the US Census in 1860, two Norwegian blacksmiths lived with the Dahl family—50-year-old Ole Arneberg and 22-year-old Ole Wold. Both men had emigrated from Romedal, Hedmark, in 1858 and arrived in Bosque in 1859.[57] The Dahl family house had a scenic view of the winding Gary Creek valley to the southwest and the Neil's Creek valley to the southeast. The buildings were placed on high ground, and the land they intended to clear and cultivate lay between their houses and the stream below.

Carl Questad constructed a vertical type of lime kiln to make mortar from limestone rock that he used in building two solid rock houses and a large stone barn with specially designed holes in the walls for shooting.[58] Limestone rock chiseled to the right dimension was used to build the walls. The first building to go up was a bedroom and living area with a huge fireplace where all the cooking was done. Next, a kitchen building was erected on a slanting hill. Below the hill was a clay tank for storing water. The kitchen room had a "dog-run" at the back, connected to another storage room, and under this room was a kind of cooling room. A little spring of dripping water made its way outside to the big clay tank of water. A winding curve of steps led below to the cellar.[59]

Because of fire hazard, the rooms were built separately from each other. A second bedroom was built close to the main bedroom, with an attic used as sleeping quarters for the work hands. Large and impressive log rafters were placed on the top of the stone walls. Questad let the Swedish entomologist Gustave Belfrage use the upstairs room to live in and as a storage room for his collections of bugs and insects between 1870 and 1879.

A fourth building was erected on the south side of the cluster of houses. Questad used part of this building as a smithy and workshop, while the second part was used in later years to house buggies and carriages. A big bell on the roof of the smithy was used to call workers for meals during the day and in emergencies. The barn was the largest building on the Questad place. It had rooms for grain and corn, and hay was stored on the second floor. The west side of the building had a huge room for farm implements, buggies, and wagons.

Questad hired skilled workers to help him with his building projects. When assistant marshal Allan S. Anderson registered the inhabitants in Bosque County for the 1860 census, Christian Olsen Strand was listed as living in the Questad household, his occupation given as plasterer. Strand was born in Elverum, Hedmark, in 1831. He had been one of the emigrants on board the *Victoria* in spring 1853. After arriving in Galveston, Strand and several others had traveled on the steamboat *Jack Hays* up the Trinity River as far as Magnolia in Anderson County. From Magnolia, they traveled north for 190 miles to the Norwegian settlements in Henderson and Van Zandt counties. Strand remained there for some years, but then moved to Bosque County.[60]

The distance between the Norwegian homesteaders was considerable. From the Ole Canuteson place to the Brunstad/Rogstad family the distance was close to 14 miles, but Jens Ringness lived between the two. From Ole Canuteson to Carl Questad the distance was approximately 7 miles, and no other Norwegians lived between them during the first years. Before the Civil War broke out, other Norwegian families took up land in the open spaces between the first settlers.[61]

BOSQUE COUNTY WAS A GOOD PLACE TO LIVE

The Norwegian immigrants had been subsistence farmers in Norway, and their idea of a good life in Texas was to continue being subsistence farmers. They might have lacked many implements and tools, but this did not bother them to any great extent. Many of them constructed their own implements, like primitive plows and harrows. Sowing was done by hand. The farmer carried a bag of grain fastened with straps around his shoulder and walked back and forth across the field while scattering the grain with one hand. The cradle was used at harvest time. The Norwegians were avid growers of wheat, but also planted oats and barley. They soon learned to raise corn, like all other migrants to Texas, but did not start growing cotton until after the Civil War.

Bosque County was a good place to live, wrote Cleng Peerson in a letter to the sons of his brother in Norway dated September 12, 1860.[62] This is probably the last existing letter ever written by Peerson. He had left Illinois and gone to Texas eleven years ago, he wrote, and he had never been back to Illinois. Despite his advanced age his health was still good. He now lived among Norwegians in Texas, and he was the only one from the municipality of Tysvær, where he grew up, who lived in this community.

The soil in Bosque County was good. Summers could be very hot, and there was little rain. Droughts could have a very negative effect on crops. On the other hand, winters in Bosque County were short. The land was well adapted to cattle, horses, and sheep because of the nourishing grass. The animals hardly needed any tending and they still got fat.[63] It was not as flat in Bosque County as in Illinois, but he liked the valleys and hills with cedar trees and oak. Good grass and oak trees were found on the hilltops. Down in the valleys the land was tilled for wheat, small grains, and maize.

CHAPTER 6

More Norwegian Settlers on the Indian Frontier

◊ ◊ ◊

THE ESTABLISHMENT OF BOSQUE COUNTY was celebrated with a barbecue at the new county seat, Meridian, on July 4, 1854. The event had been widely announced. "Several hundred people came in ox wagons and on horseback from the surrounding country and from Waco, Gatesville, and other points, and pitched camp for the occasion."[1] Land had been donated to establish the county seat. Surveyor George B. Erath was brought in to survey the lots, and when he marked out the last town lot, the surveyor's chain was held by the small hands of Karen Olene and Marie, the daughters of Ole Pierson.

THE EARLIEST SETTLERS IN BOSQUE COUNTY

The Texas legislature had decided that Bosque County should be organized out of McLennan County on February 4, 1854, and the first election in the county was held in October 1854.[2] The citizens could vote for county officials at three places—at the junction of Steele Creek and the Brazos River, at Meridian, and under an old and beautiful live oak tree between the towns of Clifton and Valley Mills. L. H. Scrutchfield was elected judge, P. Bryant sheriff, J. N. Mabray clerk, Isaac Gary assessor and collector, and Archibald Kell treasurer. The new county had 173 taxpayers in 1856, and they owned 1,426 horses, 11,417 cattle, and 213 negro slaves.[3] Most of the pioneers lived south of Meridian in the Bosque Valley and near the confluence of Steele Creek and the Brazos River.

Among the early settlers who played a leading role in the process leading to the establishment of the new county, several had considerable frontier experience as Texas Rangers and members of parties fighting Indians. One of them was the newly elected county judge Lowery H. Scrutchfield. He was born in Nacogdoches in 1824 and died in Bosque County in 1900. His mother, Nancy Pool Scrutchfield, a widow for the third time, had moved with her family in 1835 to Nashville-on-the-Brazos in Milam County, living with her oldest son, John C. Pool. In the 1840s, young Scrutchfield accompanied George Erath on several expeditions as surveyor and Indian scout. From Erath he learned a lot about different Indian tribes—their habits, skills, languages, and cultural patterns. When Erath laid out the townsite of Waco village in 1849, Scrutchfield assisted him. He was among the twenty-one first settlers to build a log cabin in Waco. After Scrutchfield married Nancy Profitt, in 1851, the couple moved to the Bosque territory and settled on the east side of Bosque River several miles north of Valley Mills.[4]

Samuel R. S. Barnes was elected county commissioner. He was born in Tennessee in 1815 and came to Texas in 1835 with two older cousins. They fought in the Battle of Bexar in December that year. In 1837, he joined a Milam ranger company and settled at Little River in Milam County. In May 1845, he accompanied Erath to Waco as assisting surveyor. He also served on several Indian campaigns between 1845 and 1850. Barnes settled on a claim on Steele Creek in 1852, and married Elisabeth Oakes Barton, the widow of Albert Barton, the same year.[5] The Barton family had moved to a farm on the west side of the Brazos River on Steele Creek across from Fort Graham in 1850, and Albert Barton soon began to operate a ferry across the river on the military road between Fort Graham and Fort Gates. Albert Barton drowned in the river in June the same year when his ferry capsized.[6]

It has already been mentioned that Canute Canuteson bought milking cows from James Frazier in spring 1854. Frazier first passed through Bosque County in March 1851 on an assignment to drive 100 head of cavalry horses from Austin to Fort Graham. At the time, nobody lived at the river crossing on the Bosque River where Valley Mills is today. The river was rising fast, and Frazier had to wait three days before it was safe to cross with the horse herd.[7] Two or three miles out on the prairie on the other side he "saw a house of a new settler who I afterwards learned was named Everett." Thomas Ewell Everett had moved to the Bosque territory in 1849 or 1850 to escape from the deadly Tutt-Everett War in Marion County, Arkansas, which raged between 1844 and 1850.[8] The Everett family had settled about 3 miles north of Valley Mills. Everett was born on November 12, 1800, in Barren County, Kentucky, and married Lucinda Hudson in 1824.[9] They had ten children, all of them born before their arrival in Bosque County. Some of them married members of

Goodall and Wood families, who followed in their footsteps from Arkansas to Bosque County. Everett was one of the appointed commissioners who chose Meridian as the county seat, and it was his son who guided Ole Canuteson upriver on his first day in the Bosque territory in 1852.

J. K. Helton, born in Tennessee in 1817, was also among the first settlers. He arrived in Rusk County, Texas, in 1842 and lived in Harrison and McLennan counties before he settled in Bosque County in 1853. Jaspar N. Mabray grew up in Alabama, fought in the Mexican War, and was also an Indian fighter. Mabray settled on the east bank of the Bosque River below the confluence of Neil's Creek in 1852. William McCurry, born in South Carolina, settled on Neil's Creek several miles west of the other Bosque River settlers. William Gary was the upper settler along the Bosque River. He settled on the north side of the river a little above the mouth of Neil's Creek. Several members of the Gary family settled in the area, including Matt Gary, Gaffey Gary, and Isaac Gary.

A FORMER NORWEGIAN ESTATE OWNER ARRIVES IN BOSQUE COUNTY

The Norwegian immigrants settled far to the west, close to the Indian frontier. After the Norwegians came in 1854, only a few of their countrymen found their way to Bosque County before the Civil War. One of the most well-known in his home community in Norway was Poul Poulson. In the genealogical part of *Norge i Texas*, Derwood Johnson noted that Poul Poulsen Vik (Poul Poulson) came to Bosque County from California in 1857. He was born on April 17, 1818, in Vik, Stange, Hedmark, and died in Bosque County on September 17, 1868. Poulsen had gone bankrupt in Norway in 1848, emigrated to California, and was involved in gold digging until "he was reunited with his family in Texas in 1857."[10]

Poulsen grew up as the oldest son in a very rich farming family in Hedmark County. The Poulsen family had owned the substantial Vik estate since Paul Hansen Vik bought the property in December 1782. On October 20, 1841, 23-year-old Poulsen took over the property for 7,000 Spd. and a contract for lifelong sustenance for his father.[11] The farm had more than 120 acres of tilled land, around 470 acres of wooded land, a saw mill, and a grain mill. It contained many houses—a smithy, a stable for nine horses, a small barn, and a large barn with room for 92 cows. The farm also had eight cotter's places. Poulsen had married Oline

Maria Halvorsdatter on June 16, 1840. Their oldest son, Casper Andreas (Casper Andrew Poulson), was born at the Vik farm on March 20, 1841, followed by Halvor Georg (George H. Poulson), born on March 31, 1843, and daughter Betsy Randine (Betsy Elisabeth Poulsen), born on October 14, 1846.

Odd Magnar Syversen has narrated in detail what happened on the Vik estate from the time Poul Poulsen took over until he bought land in Texas in 1857. Soon after his takeover, he invested in a distillery on the farm, valued at 2,500 Spd. Potatoes for the distillery were transported by sailing vessels on Lake Mjøsa from places like Feiring, Toten, Helgøya, and Nes. The distillery was not a success. On February 3, 1848, Poulsen saddled his horse and left for Christiania on the pretext that he would try to refinance the high debt on the farm. He brought with him, however, all the cash he could lay his hands on. In Oslo, he persuaded butcher Gulliksen to buy 320 cows from him for 320 Spd. The same day he sold the same cows to another butcher, Jacob Andersen, for the same amount.[12] On February 9, 1848, he left Oslo without a trace. His horse was later found in Dillingen, south of Oslo, close to the city of Moss.

On February 10, 1848, the authorities temporarily seized his properties. Merchants in Oslo presented claims of 6,522 Spd and 96 skilling and took control of the estate. The Vik farm and all chattels were auctioned off. The Vik land was sold by the bankruptcy estate on December 1, 1848, for 15,000 Spd.to Conrad Vilhelm Hauge.[13] Poulsen's bankruptcy estate was not concluded until April 24, 1856. Total assets for distribution to creditors were 24,312 Spd, while total outstanding debt was 45,105 Spd. The two butchers in Oslo were among the 175 creditors, but did not get a higher percentage than the rest.[14]

Where had Poulsen gone? This question was of great interest to people in the local communities and his many creditors. Poulsen had left his wife and children and escaped to the United States. He traveled through Sweden and Denmark to Hamburg, where he boarded a ship for America. After Poulsen fled to America, the local pastor H. O. F. Heyerdahl wrote in a report that there had been problems in the marriage during the last years. Poulsen had often been away in Oslo for months at a time.

His wife and the children left the farm and moved in with her parents.[15] More than a year later Poulsen sent his wife a letter, dated New York, July 10, 1849. He knew that she and the children were well. If he had success in his projects in the new world, he wrote, it was his intention to return to Norway in 1850 and bring all of them to the United States. Poulsen, however, did not return in 1850 or the next years after that. His wife grew impatient and applied for divorce, which was granted to her.[16] Poulsen then went to California to try his luck in the goldfields and had some luck. When he arrived in the Norwegian colony in Bosque County, he had enough money to buy land and establish a farm. On October 7, 1857, he bought 318 acres of land from J. K. Helton for $450.[17] Poulsen asked his wife and children to join him in Bosque County. In June 1857, Oline Maria and her children received an official letter from the Norwegian authorities permitting them to emigrate from Stange parish.

Poulsen was not the only Norwegian pioneer in Bosque County from Hedmark who had grown up in an affluent environment, but his without a doubt had been the wealthiest home. When Poulson turned up in Bosque County in 1857 and bought land, he implicitly told his neighbors that he, like them, wanted to build up his life again from scratch. He also signaled that he wanted to build his new life among Norwegians from Hedmark County. His neighbors had no quarrels with that. It was up to him to show that he could succeed in his new surroundings.

ALONG THE PEERSON CHAIN FROM THE MIDWEST

In the late 1850s several immigrants from western Norway followed the Peerson migration chain from the Midwest, most of them from the Norwegian mother colony in La Salle County, Illinois. They had their cultural roots in Vestlandet, the coastal counties of western Norway. They spoke a western Norwegian dialect, which was different from the southern and eastern Norwegian dialects of their neighbors in Bosque County.

In contrast to Poul Poulsen, the Joseph Olson family from Karmøy in western Norway brought no money with them and experienced little economic progress during their first years in Bosque County. Joseph Olson Aadland was born in Skudesneshavn, Karmøy, Rogaland County, on April 11, 1811, and died in Bosque County on March 20, 1894. In 1849, at the age of 38, he married Anna Karina M. Ivarson, who was born on November 19, 1821, in Fosenøy, Avaldsnes, and died in Bosque County on October 23, 1917. The family left the island Karmøy with their four children, traveled south to Stavanger, and boarded a ship for

Quebec on May 28, 1858. After seven weeks and four days the family reached Quebec and continued to La Salle County.

After two months in La Salle County, the Olson family left for Texas, traveling through St. Louis and down the Mississippi to New Orleans. There they boarded a ship to Galveston, and from Galveston traveled on the railroad to Natchitoches, Louisiana. That was the end of the easy part. From Natchitoches they traveled by ox wagon, two families to one wagon, including a bunch of children. Their average speed was 6 miles a day, and on November 10, 1858, they arrived on the farm of Canute Canuteson. They felt at home since Canuteson also had emigrated from their home island Karmøy. Canuteson could not house the family in his cabin for long. After they spent nine days at the Canuteson's, Carl and Sedsel Questad let the Olson family live in Questad's old smithy. They moved into the rather primitive log house on November 19, 1858. The bedstead was made of poles, the chairs from oak logs, and the table was a large box. The kitchen utensils were one small copper kettle, a skillet, and a frying pan. Forks and knives were homemade, and spoons were made of horn. Water was fetched in a homemade wooden bucket.[18]

Joseph Olson joined a threshing team in summer 1859. One day while cleaning up around the threshing machine, he got too close to the horses powering the machine. When one horse started to rear up, Olson stepped back into the open gearing of the tumbling rod. His shoe was cut, and he lost his big toe and part of the next one. He could not work for a while, and he was somewhat handicapped ever after. The Olson family needed money badly to buy wheat and rye. Olson had some skills as a shoe cobbler, and he took on such work when it was available. His wife kept busy carding and spinning wool for others, and she kept half of the production as payment for her work. She continued carding, spinning, and weaving for others at least until after the Civil War.

Questad's old smithy was located about a mile from the Questad house, and Sedsel Questad invited Anna Olson to help her with milking morning and evening. Sedsel Questad was "a wonderful good person, her husband likewise," wrote their son Jacob Olson in 1935. In return for helping to milk the Questad cows, the Olson family got all the milk and butter they might want and "then some." In spring 1860, the bachelor Ole Ween allowed the Olson family to live in one room of his house. The deal was that Anna Olson would cook all his meals and he would provide everything he wanted cooked.

Cleng Peerson was a frequent guest at the Ween table, and he often stayed in the Ween house for extended periods. When the US Census was enumerated on July 25, 1860, Cleng Peerson was listed as living in the Ween household. The household consisted of Jo Wilson (46), Anna V. (39), Anna M. (10), Michael (8), Jacob (6), Torborg Serine (4), and Andrew (3 months). The enumerator spelled their names as Wilson instead of Olson.

The Olson family lived with Ole Ween for seven years, until 1867, when Ween returned to Norway and got married. Joseph Olson had pre-empted 160 acres of land adjoining the Ween place in 1862, and he asked two young neighbors, the recently married Nils Swenson and Canute Halvorsen Canuteson, to break 7 acres of land for him. The same year the two young men joined the Confederate Army, and both died during the war. Joseph Olson only had enough money to buy one yoke of oxen, a plow, and a shackly wagon. Not until 1866 did Joseph Olson build a log house, 14 × 15 feet. The family moved into this house in late December and lived there until 1872, when they moved into a two-room stone house.

Their son Jacob lived with his parents until they passed away. His mother, Anna, afraid he would never get married, took an active part in finding him a spouse. She was the aunt of Lovisa Knutiane Knudsen, born on September 6, 1873, in Fosenøya, Avaldsnes, western Norway. When Lovisa was 19, Anna persuaded her to emigrate to Bosque County and marry her 38-year-old bachelor son. Lovisa arrived on April 15, 1892, and the couple married on November 15, 1892. They never had any children and moved to Clifton in 1922, where Lovisa died on March 23, 1930.

The case of the Olson family is a reminder that many of the early Norwegian settlers in Bosque County needed years to build up their farms. The Olson family was poor on arrival and remained poor, at least until the children could help their parents with the farm's daily chores. Their son Jacob had very strong intellectual leanings. He collected books and artifacts throughout his life and corresponded with scientists at leading institutions in the United States. His collection of artifacts became the backbone of the collection at the Bosque Memorial Museum. The Olson family's log cabin from 1866 is still on display outside the museum.

WHY DID CLENG PEERSON TRANSFER HALF OF HIS LAND TO OVEE COLWICK?

In 1857, when Cleng Peerson received his deed of land from the Texas Governor, he was 74 years old, but still in good shape, mentally and physically. What stuck in the minds of historians and others interested in the life of Cleng Peerson during his twilight years in Texas, wrote Th. Blegen in 1921, were some of the views T. Theo Colwick had put on paper in a letter to Rasmus B. Anderson in 1894, which Anderson repeated almost verbatim in his book. Peerson was somewhat eccentric toward the end of his life, wrote Colwick. This was easily forgiven since "he was at the same time very kindly and accustomed to serving others unselfishly without thought of compensation."[19] He characterized Peerson as "a good person with a strong character. He found it easy to tell and recount stories. One never became tired of listening to Cleng." Peerson was in his element when talking about his travels. "Even among the redskins he was liked, they would never wish to do him any harm, quite the contrary, he could come and go when he wished and always could carry out what he undertook. His main task in life was to be a pathfinder, which as is known, he was well suited for." He still had a keen interest in what happened in the community, and many a time "he trudged the long way to the Land Office in Austin to put his countrymen's land affairs in order but he never took a cent in compensation."

Peerson was the owner of 320 acres of land and could do what he wanted with it. Toiling from dawn to dusk as a farmer had never been his dream of a good life. He liked to visit with people, sharing meals with them, sitting on the porch into the night to reflect and tell stories. To feed their families, however, his neighbors had to work from dawn to dusk. Peerson knew that he was most welcome if he arrived at sundown when preparations for the evening meal were underway. An extra chair was hauled to the table along with eating utensils. After the meal, Cleng would share all the news he had picked up lately and tap his fountain of old stories until it was time for bed. The distance between neighbors was considerable, and when someone came to visit, it was expected that they would sleep over.

Land was abundant in Texas in the 1850s and therefore cheap. Perhaps Peerson could find a Norwegian immigrant family in the Midwest willing to take over his land in exchange for *føderåd*, a tradition practiced in Norway since the Middle Ages. When a farm owner transferred his farm to a relative or sold it, the old owner had a right to *"føderåd, folge, kår,"* which usually included a place to live as well as victuals; sometimes a "good funeral" was also included. The price of a farm was lower if a *føderåd* contract was part of the deal. Such rights were written down and registered by the court as a liability on farms in Norway starting in the 17th century and were practiced into the second half of the twentieth century.

Cleng Peerson kept in contact with old friends in La Salle County. One of his friends was the much younger Ovee Rosdail, born on December 17, 1809, in Tysvær. Rosdail came from the same community in Rogaland where Peerson grew up. He was the son of Daniel Steinson, who was born at the farm Kjølvig in Jelsa, Ryfylke in 1779 and died in 1854 in La Salle County, Illinois. Daniel had married Britha Ovesdatter Rossedal, born on the farm Rossedal in 1786. She was the sole inheritor of the Rossedal farm, and after their marriage Daniel Kjølvig changed his last name to Rossedal. The Rossedal family with their five children belonged to the group of pioneer Norwegian emigrants on board the sloop *Restauration* in 1825.[20] In the United States the family surname was changed to Rosdail, and Ove, the second oldest child, began to spell his given name Ovee. He was close to 16 years old when the family left Norway. The Rosdails followed Cleng Peerson from Kendall, New York, to Fox River, La Salle County, Illinois. Ovee Rosdail did well in La Salle County. He held the position of postmaster for some years, was elected Justice of the Peace, and worked as a cabinet maker, blacksmith, farmer, innkeeper, auctioneer, real estate agent, and business advisor.

A cousin of Ovee Rosdail, Ove Torgersen Kjølvig, arrived in La Salle County with his family in 1854, after having recovered from cholera in Quebec. Ove Kjølvig was born in Jelsa, Ryfylke, on March 19, 1825, the son of Kari Ellingsdatter Mehus (born 1780) and Torger Steinson Kjølvig (born 1785, Jelsa, Ryfylke, and died 1854, Jelsa). Torger Kjølvig took over the Kjølvig place after his mother in 1825.[21] The family had several children, but the focus here is on Ove Kjølvig, who married Johanne Margrete Johannesdatter Nådland on July 7, 1853. She was born on March 18, 1831, on the neighboring island Ombo, and died on March 25, 1881, in Norse, Bosque County. Less than a year after their marriage Ove and Johanne decided to emigrate. They left Stavanger on May 6, 1854, on the sailing ship *Urania*. Johanne's younger sister, Bergitte Cecilie Naadland, born in Jelsa on October 7, 1837, traveled with them.[22] Their destination was the Norwegian mother colony in La Salle County, Illinois, where they had family on both sides.

OVE KJØLVIG CAUGHT CHOLERA, BUT SURVIVED

New York City was still the main destination of Norwegian emigrant ships in the early 1850s, but by 1854 most chose to sail to Quebec in Canada.[23] The Kjølvig family was unlucky with their Atlantic crossing; the *Urania* did not arrive in Quebec until July 24, 79 days after the ship left Stavanger. Johanne was very pregnant when they left Norway; since the voyage took so long, she gave birth to their first child out on the Atlantic on July 2, 1854. The boy was baptized Thomas Theodore, and later in life became known in the Norse community as T. Theo Colwick. While on board, Ove Kjølvig contracted cholera, but survived.

During the deadly cholera epidemic in Quebec in 1832, a quarantine station was opened on the island of Grosse Île, about 30 miles down the St. Lawrence River from Quebec City. Arriving ships had to stop there before entering Quebec City, and all passengers who showed signs of illness had to wait out the prescribed quarantine period. Around 38,000 immigrants arrived in Quebec during the cholera epidemic in 1849, and about 60 percent of them died.[24] Another cholera epidemic occurred in 1851 but was mild compared with the one in 1849. The third serious cholera epidemic broke out in Quebec in 1854; according to conservative estimates the total number of deaths was 3,846. Ove Kjølvig stayed at the cholera hospital at Grosse Île for several weeks. His wife stood by him, but their savings dwindled to nothing. Finally, Ove was healthy enough to continue to La Salle County. They expected help from their kin, and help they got.

Ove Kjølvig changed his name to Ovee Colwick. Land prices in La Salle County were too high for Colwick to contemplate buying land there. He worked several years for others. Cleng Peerson contacted Ovee Rosdail about his land, but Rosdail already owned half of it. He now suggested to his cousin Ovee Colwick that he should take over the other half of Peerson's land in exchange for a lifelong contract for sustenance, *føderåd*, for Cleng. Ovee Colwick and his wife found the offer very interesting. To cite from a biographical paper written by Theodore

T. Colwick after the death of his father in July 1895: "Several years of bad crops in conjunction with the high price of land, in Illinois—the outlook or prospect for a man of his limited means of even being able to secure a place of his own being rather gloomy, he concluded to seek a home in another State where land could be obtained on easier conditions. Learning of the abundance of cheap, productive unoccupied and vacant lands in Texas, through his cousin Ovee Rosdail's correspondence with Cleng Peerson, he concluded to brave the dangers and vicissitudes of a frontier life in that State."

In addition to Theo T. Colwick, Ovee and Johanne had two daughters while in La Salle County. Margaret Cecilie Hjelm Colwick was born on December 8, 1855, and Marthe Christina Colwick was born on May 20, 1857. The Colwicks did not begin their journey to Texas until after the birth of their son John Naadland Colwick, on March 21, 1859. In late summer 1859, the Colwick family together with several other Norwegian immigrants left for Texas and Bosque County. The journey to New Orleans took four weeks, followed by a journey by ship to Galveston. They traveled by ox wagon from Houston and arrived at Neil's Creek in Bosque County in mid-September 1859.

THE EMIGRATION PATH OF OVEE COLWICK AND HIS WIFE FROM A FJORD IN JELSA, RYFYLKE, VIA LA SALLE COUNTY, ILLINOIS, TO BOSQUE COUNTY, TEXAS

In a family history written in 1966 by Mary Latimer Colwick, *The Colwicks—Ovee, John and Family*, she wrote that "Ovee became sufficiently successful in the fishing trade to save a modest sum of money, and he resolved to go to the New Land, to America, the land of freedom, liberty and opportunity, of which he had heard many glowing accounts."[25] In another family history, the authors stated that the Colwick couple "heard numerous glowing stories about America and the opportunities and the freedom that the people found there. Due to the hardships while living under Swedish rule and the dangers of Ovee's profession, fishing, the couple made the very difficult choice to leave their home and family for America, the land of opportunity."[26] Freedom, liberty, and opportunity are recurring themes in countless family histories of Norwegian-Texans. In most cases the motives of the emigrants were more personal and complex, as with Ovee Colwick and his family.

The 1966 family history stated that Ovee Colwick "became intimate friends with the trail-blazer and colonizer, Kleng Person, who pioneered Norwegian immigration in America." When Cleng Peerson told the people in La Salle County about the "wonderful state of Texas," the "Colwicks, with others of the settlement, were so impressed with Peerson's account of Bosque county, Texas, they decided to go there as soon as possible to make their permanent home."[27] A company was formed in 1859, "under the leadership of Kleng Peerson," and including the Colwick family and others. They made the journey from La Salle County to Texas.[28] Did this mean that Cleng Peerson had been back in La Salle County after 1850? Countless hours have been spent to verify this conjecture. In the end, it was a letter from Peerson to the sons of his brother in Norway, dated September 12, 1860, that settled the question. Peerson wrote that he never went back to Illinois after 1850. When the Colwick family arrived in Bosque County in 1859, Peerson met them face to face for the first time. Half of the 320 acres of land he had got from the Texas government for free in 1857 he "sold to Ove Rosedal and the other half I gave for free to Ove Kjølvig from Jelsa in return for him taking care of me in my old age."[29]

According to Mary Latimer Colwick, Ovee learned the fishing trade from his father early in life. He acquired his own fishing boat and was able to save money.[30] In his biographical paper, T. Theo Colwick wrote that his father had owned "a small sailing vessel or yacht (of his own) in which he had done considerable business, trafficking in, buying and shipping fish, from place to place, or rather carrying this business between the towns and cities on the coast of Norway, for several years, doing a good and profitable business. While on one of those fishing excursions, off the coast of Norway, his vessel was capsized and sank in a storm, and he very narrowly escaped. He was rescued in an unconscious condition, but revived again and was thus saved from an early watery grave."[31]

The story told by T. Theo Colwick is consistent with our general knowledge of the history of the spring herring fisheries as a driving force in the economic development of Rogaland County in the middle of the nineteenth century. Despite considerable emigration starting in the 1840s, the population of Rogaland County more than doubled between 1820 and 1870, and the population in the towns increased by a factor of eight. The herring fisheries were plentiful in the 1840s, and Ovee Colwick probably did not have strong economic reasons to emigrate.[32] His narrow escape from death at sea might well have been the turning point in the family's decision to emigrate.

On their arrival in Bosque County, the Colwick family did not immediately move to the Peerson land. Their neighbor Canute Canuteson offered to let them live in the log house where his family had lived during the first years after arrival. Colwick began building a house with three rooms on the first floor plus upstairs rooms. In the rear of the house on the first floor was built an 11' × 17' room for Cleng Peerson, which was still known as "Cleng's room" decades after his death.[33] Olivia Colwick, born on December 18, 1861, was the first Colwick child born in their new house in Texas. According to the Eggen family history, Peerson called Olivia "his little girl" and would rock her in his old handmade rocker, which is now housed in the museum in Clifton.[34]

On February 1, 1860, Peerson sold 160 acres, half of his land, to Ovee Rosdail of La Salle County, Illinois, for $200, or $1.25 per acre. The transaction was registered on February 11, 1860.[35] Five months later, on July 17, 1860, Ovee Colwick bought the same 160 acres from his cousin "Ovee Rosdail of Mission, La Salle County, Illinois for 400 dollars." The document was signed by Lockwood, Justice of the Peace in Bosque County.[36] In September 1860, Peerson and the Colwick family moved into the new house. Before the end of the year, on November 26, 1860, Peerson signed over 153 acres of his land to Ovee Colwick for one dollar.[37] The document was signed in the presence of O. Canuteson and T. Danielson. If such a document had been made in Norway, it would have specified that Colwick as the buyer would have taken all responsibilities and outlays in relation to Cleng Peerson for the rest of his life. The document had no such clauses; Texan law did not require it.

NORWEGIAN IMMIGRANTS MORE THAN WELCOME

Several other Norwegians traveled with the Colwick family from Illinois to Texas in 1859. Three children of Oliver Canuteson (Halvor Knudsen, born 1799), the older brother of Canute Canuteson, were members of the group. Oliver Canuteson had emigrated in 1831 from the island of Karmøy, north of Stavanger, to Kendall, New York, where he had married Julia (born ca. 1807). They later followed Cleng Peerson to La Salle County, Illinois. Their first child, Ole Andreas Canuteson, was born on April 29, 1834, in Kendall, New York. His brother Canute E. Canuteson was born in 1836, in Mission Township, La Salle County, and their sister Margaret Julia Canuteson was born on April 9, 1844.

The Canuteson family in La Salle County knew from letters from their uncle Canute Canuteson in Bosque County that land was much cheaper there than in La Salle County. Ole Andreas Canuteson brought his wife and two small children on the trip to Texas. His wife, Ingeborg Thompson, was born in 1834 in the Stavanger region and died in 1870 in Bosque County. They had married in 1855; their two children were Halver Canuteson (born on July 23, 1856 in Illinois) and Mary Ann Canuteson (born in Illinois in 1858). At the time of the 1860 census, Ole Andreas lived with Canute Canuteson. He was listed as a plasterer, but soon settled on 160 acres of land adjacent to his uncle's land at Norman Hill.[38] During their first ten years in Bosque County, Ingeborg gave birth to five children: Julie Helene in 1860, Elisabeth in 1862, Canute in 1864, Andrew in 1867, and Albert (Ingebret) in 1868. After Ingeborg died in 1870, Ole Andreas Canuteson returned to Illinois to work. While living there he met Christina Erickson (born in Norway in 1835, died in Norse in 1891); they married in 1875 and then returned to Bosque County. Christina gave birth to Alfred Severin in 1877 and Cora Olivia in 1879.[39] Both Ole Andreas and his brother Canute E. Canuteson enlisted in the Confederate Army, and Canute died during the war. Their sister stayed in Bosque County through the Civil War but returned to Illinois after the war.[40]

Two Norwegian brothers, Halvor Smeland, born ca. 1838 (Oliver Thompson), and Andreas Smeland, born ca. 1832 (Andrew Thompson), traveled to Texas with the Colwick group. Both brothers joined the Confederate Army. They survived but went back to Illinois after the war. David Lund was also a member of the Colwick group. Little is known about him before the Civil War, but he left interesting letters from his life in the Confederate Army. He died in the war. Ommund (Omen) Omenson, who was born on March 28, 1833, in Skjold, Rogaland County, and died on May 29, 1901, in Bosque County, arrived in La Salle County with his parents and four siblings in 1857. Omenson was the only one in his family who migrated to Texas. He did not have a penny when he arrived in Texas with the Colwick group in 1859, and he did not master English. Omenson had a hard time finding work but ended up tending sheep near Waco. According to a family history he "lived as a shepherd for ten years, saved his wages, and bought his first piece of land."

John Johnson, born Johan Johansen Vatne on May 1, 1817 at the farm Vatne close to the city of Sandnes in Rogaland County, came to Bosque County with his family in 1854. He married Mary Johnson (Ingeborg Maria Gjerulvsdatter), born in 1823 in Øyestad in southern Norway, in 1843. The couple brought with them four girls to Bosque Count: Jørgine, Talette, Jensine, and Perrine.[41] At the time of the 1860 census, Johnson was listed as a stone mason. The 30-year old laborer Neil Canuteson lived with the Johnson household in 1860. He was born Nils Knudsen Underbakke on October 17, 1829, in Suldal, Rogaland County, and died on December 27, 1914, in Norse. Neil Canuteson had come to Texas with the Cleng Peerson group in 1850.[42]

In April 1857 three brothers named Skimland emigrated to the Midwest from the western Norwegian island of Bømlo, north of Haugesund. They were the sons of Reinert Jacobsen and Kari Knutsdatter and grew up on the farm Skimmeland in Finås, Bømlo. The oldest brother, Knud R. Skimland, was born on this farm on November 2, 1830; Elias R. Skimland was born on May 16, 1833; and R. R. (Andrew) Skimland was born on September 10, 1838. Their parents, together with two younger children, emigrated to the Midwest some years later. The Skimland brothers traveled from the Midwest to Bosque County along with the Jacob Olson family in 1858. The brothers served in the Frontier Regiment during the Civil War. Their role in that context will be covered more extensively in the chapter on the Civil War.

In 1860 Ole Wold and Ole Jensen Arneberg arrived in Bosque County from Hedmark County. Ole Wold, or Ole Andersen Slagsvold, was born on May 3, 1839 on the farm Slagsvold, Romedal, Hedmark County, and died in Seattle, Washington, in 1915. He was the son of Anders Hansen Slagsvold and Mari Olsdatter Harstad. He was related to both the Brunstad and Dahl families.

Ole Wold emigrated from Romedal in 1859 with Ole Jensen Arneberg from Arneberg Østre, Romedal, who was born on July 9, 1811, and died in Bosque County in June 1884. The two men crossed the Atlantic to Quebec, traveled through the Midwest, and arrived in Bosque County just in time to be enumerated in the 1860 census. They lived in household 18, with Hendrick Dahl's family, and they were both listed as blacksmiths. Arneberg had married in Norway; on July 24, 1861; Ole Wold married his cousin Karen Brunstad, born in Romedal on April 17, 1844. She died in Seattle in 1921.

THE LAST DAYS OF CLENG PEERSON

During the writing of *The First Chapter of Norwegian Immigration (1821–1840). Its Causes and Results*, Rasmus B. Anderson corresponded with hundreds of persons with intimate knowledge of early Norwegian immigrants in the United States. In Texas, he corresponded about Cleng Peerson and the early settlers in Bosque County, mainly with Elise Wærenskjold and Ole Canuteson, but also with T. Theo Colwick. Anderson presented Reiersen, Elise Wærenskjold, and Ole Canuteson as his main "heroes" in Texas. He leaned heavily on a long letter from Ole Canuteson, dated December 16, 1894. In this letter, Canuteson played down the role of Cleng Peerson. He had stayed with the Canuteson family in Dallas County for three years, and "when we moved to Bosque County in 1854, he came with us, not as a leader then, but as a follower, being too old to undertake leadership anymore."[43]

Ole Canuteson was "the founder of the largest and most prosperous Norwegian settlement in Texas," Anderson argued, and "Ole Canuteson deserves more than passing notice."[44] He then continued to tell the story of how the Canutesons came from Karmøy, western Norway, to Texas, and about the later career of Ole Canuteson as a factory owner in Waco. After reading those passages in the book, T. Theo Colwick wrote a letter to Anderson arguing that Bosque pioneers did not mind that Ole Canuteson had been given such a favorable presentation. But the Norwegians in Bosque knew very well that Canuteson had not been alone in building up the Norwegian settlement.

According to Anderson, Ovee Colwick had got his land from Cleng Peerson for free without doing much in return. T. Theo Colwick disagreed strongly. In a letter to Rasmus B. Anderson, dated November 4, 1895, Colwick stated that the "general tone or tenor and drift of the relation of Kleng Peerson's home with my father, O. Colwick & co.—as it appears in your History—is repulsive to any one as well acquainted with the actual facts as I am."[45] The way Anderson had described Cleng Peerson's late life and his relationship to his father was entirely uncalled for, he wrote. "O. Colwick's transactions with Kleng were not only legitimate business, but as land was cheap (about 50 cents an acre) at that time, and Kleng lived and had his home with us for several years, it was not onesided, but fair & just."[46] Colwick admitted that Peerson had continued to visit neighbors, where he was always welcome and felt at home, but his home was with the Colwicks and that was where Peerson spent most of his time. "Kleng was not only considered as one of our family but was in every way treated as such during the declining and last years of his life when he lived and had his home with us." Ovee Colwick had not only agreed to take care of Peerson for the rest of his life, but actually had done so, caring for him "the balance of his life."

Even though Cleng Peerson was getting old, as late as January 1865, eleven months before his death, he had all his intellectual faculties intact. His word and his signature were still used as proof of honesty in the community. On January 16, 1865, Ole Ween sold sixteen horses to Carl Questad for $560. The signatures of Cleng Peerson and Jens Ringness are found on the sales contract as witnesses.[47]

Cleng Peerson died in the Colwick home on December 16, 1865. Because of the letters from T. Theo Colwick to Rasmus B. Anderson, we know more about Peerson's last night on earth than many other things in his life. Ole Canuteson had written to Anderson that he had been "with him the last hours of his life. I closed his eyes in the long sleep of death."[48] This statement was contested by T. Theo Colwick. Several people had been present the night he died, and each of them took turns sitting at his deathbed, certainly including members of the Colwick family. Colwick distinctly remembered that he and John Ringness had "waked with Kleng the first part of the very night he died."[49]

"The people of this community that are acquainted with the facts," wrote Colwick, earnestly desired that Anderson in later editions would correct the story about Cleng Peerson and the Norwegians in Texas in accordance with the content of his letter.[50] The Anderson book was published in four editions, with the fourth edition published in 1906, but Anderson never changed one word in the text about the Norwegians in Texas.

ADAPTING TO THE RISK OF INDIAN ATTACKS

The many Norwegians who came to Bosque County in the late 1850s were immediately made aware that the county had been experiencing many Indian raids. After their first encounter with Indians shortly after their arrival in spring 1854, there were few Indian raids in 1855 and 1856. But the number of raids increased considerably in early 1857. Each month during the full moon Indian raids were expected nightly.

Some of the homesteaders in Bosque County organized scouting parties along the northern and western borders. A high point called Lookout Mountain near Martin's Gap on the line between Bosque and Hamilton counties, not far from the village of Fairy, was used as a lookout. The scouting party, usually consisting of eight or ten men, had a camp close to Lookout Mountain. Two men at a time acted as sentinels and kept a constant vigil from the top of the mountain, where they could see for many miles. On the day of the inauguration of President Buchanan, March 4, 1857, Francis Marion Kell was one of the sentinels on Lookout Mountain. Looking to the south and west, he detected a band of five Indians leaving the settlement with a bunch of stolen horses. He crawled back over the point of the mountain and gave the alarm to the other scouts below, who included Kell's brother Abe Kell, Bob Renfro, Jim Babb, Allen Anderson, and Ross Cranfills. The father and brother of young Renfro had recently been killed by Indians.[51] All of the scouts mounted their horses and charged after the Indians. One by one the Indians were killed in hand-to-hand combat; only the leader escaped.

The Indian raids continued throughout 1857 and were so numerous, and the number of killings so many, that the citizens in the new counties west of Waco did not dare to "leave their families unprotected by night." On January 2, 1858, Thos Harrison wrote a letter from Meridian, Bosque County, to George B. Erath, his representative in Austin. He insisted that Erath should present a petition from the citizens of Bosque County to the Texas governor as soon as possible. That same day he had seen with his own eyes the body of Peter C. Johnson, "mangled with five wounds inflicted by arrows. The track of Indians that killed Johnson, carried away a large drove of the most valuable horses in the country." He hoped Erath would "serve your constituents & the cause of humanity by increasing the frontier protection?"[52]

It took some time to piece together the different sources of information about the Indian attacks to form a coherent picture, but on January 13, 1858, Texas Governor H. R. Runnels got a letter from John Forbes in which he reconstructed the sequence of the attacks. A marauding band of Indians had entered Palo Pinto, Erath, Comanche, and Bosque counties in December 1857 and separated into three parties. One of the parties operated in the vicinity of Stephenville in Erath County on December 30, 1857, and in a "very bold and audacious manner stole and drove off from thence a number of valuable horses." On the same day, another party attacked the settlements on Besley's Creek, some 20 miles southeast of Stephenville, and stole the horses of Turnbull Barbee and others. While a black slave on the Barbee ranch was out unhobbling some horses, the Indians were suddenly upon him. The Indians shot "some 7 or 8 arrows into the body of Barbee's negro, and left him for dead, but it is supposed the negro will recover."[53] He managed to reach the house and warn his owner.

The Indians continued down the valley, gathering horses as they went. "A citizen of the name Isaac Bean living on the waters of Besley's creek and his negro man was barbarously murdered by them on the same day and the next day a Mr. Johnson a well known citizen who was driving his wagon on the public roads from Meridian, accompanied by his son a lad of about ten years old, was also murdered by them, his little boy missing and supposed to be taken prisoner and it is said that an American female has been taken prisoner and carried off by them."[54] Peter C. Johnson and his 10-year-old son Peter were returning from Waco with a wagon full of supplies. When they were near a flat-topped mountain, later known as Johnson's Peak, the Indians surrounded the wagon, killed the father, and took the boy and a large band of horses in the direction of Clear Fork on the Brazos River.[55] "

The Indians dropped little Peter when they arrived at Clear Fork after taking his coat, hat, and socks, leaving him with nothing but his shirt and pants. No white people lived closer than 50 to 70 miles to the east. Peter lived for four or five days and nights with nothing to eat but grass roots. On the evening of the sixth day he was found by some men looking for cattle. He was thin as a skeleton when they found him.

The third party of Indians had been stealing horses in the vicinity of Meridian and then joined the other Indian bands 6 miles west of Stephenville. They drove the horses between Barton's Creek and the North Leon River, crossing at the headwaters of the Palo Pinto, and continued in the direction of the Indian Reserve on the Clear Fork of the Brazos River. The citizens of Stephenville followed the broad trail for more than 100 miles, but then gave up. The citizens had experienced great losses in the attacks. The Texas governor was urged "to adopt the most prompt and vigorous measures to prevent the recurrence of Indian incursions by an efficient defense of the frontier and for the protection of the lives and property of your fellow citizens."[56]

The 1858 Indian attacks continued for months. "The western tribes were so troublesome during the year 1858 that they interfered with the attendance at the first school established at Meridian," wrote William C. Pool.[57] On Monday, November 22, 1858, the County Court of Bosque County decided to grant a request from the citizens that the court should raise money for food, horse shoeing, and ammunition for a company of twenty-four rangers to protect the citizens from the "depredations of the Indians."

CHAPTER 7

Living among Transplanted Hillbillies and Slave Owners

◊ ◊ ◊

ECONOMISTS USE THE TERM "barrier to entry" to describe factors that prevent a new enterprise from succeeding. First mover advantages are one of the most important forms of barriers to entry. The first Norwegians in Texas were latecomers compared with most of their American neighbors. They were at a disadvantage with respect to language, and they lacked the skills to use the legal system in their favor. It also took them some years to adapt to a different agricultural environment. They had left wet and snowy Norway and ended up in hot and dry Texas. The seasons were also somewhat different. Plowing and sowing began earlier, and consequently harvesting and threshing were also completed weeks earlier than in Norway. The newcomers had to learn to adapt to new kinds of crops such as maize and sweet potatoes. Last, but not least, very few Norwegians brought with them capital to invest in farms on new soil. They were cash poor. In every new county in Texas there were large farmers, stockraisers, and plantation owners who had brought with them considerable capital to invest. This made it possible for them to develop and build up their holdings within a few years after arrival.

HOW DID THE ANTEBELLUM NORWEGIAN IMMIGRANTS DO ECONOMICALLY?

The Norwegian immigrants experienced more barriers to entry than most of their American neighbors. On the other hand, subsistence agriculture was dominant in frontier counties in Texas in the 1850s. Farmers who managed their farms well were treated with respect. The dynamic farmer was the one who improved his land and planted more acres each year. In an article on the economic development in Henderson County in the 1850s, Kenneth Howell cited a resident of a small farming community in the southern part of the county who recalled that "if a man let his farm grow smaller he was no good, but if a farmer wanted to keep in good repute he should make his farm a little larger every year."[1]

The Norwegian immigrants had grown up with this kind of value system, and they also practiced it subconsciously on their homesteads in Texas. Let us then use this value system to measure improvements among Norwegian settlers in Bosque County between 1854 and 1860. There was without a doubt some internal and tacit competition between Norwegians concerning improvement of land and livestock. All of them started out with approximately 320 acres, and none of them had slaves to work their land. At best, some of them had children who were taught to plow at a very young age, and some might have had cash available to pay recently arrived Norwegian immigrants to do labor-intensive work for them, such as breaking new land. Livestock was of high importance in the frontier counties, and the Norwegians soon began to acquire livestock.

Information registered in the Texas Agricultural Census of 1860 makes it possible to evaluate the work done by Norwegian immigrants between 1854 and 1860. The census distinguished between acres of "improved land" and "unimproved land."

The Norwegian farmer Hendrick Dahl ranked highest with respect to improved acres; he was registered with 100 acres of improved land and 290 acres of unimproved land. Next on the list was Poul Poulsen and Carl Questad, each with 80 acres of improved land; Poulsen had 238 unimproved acres and Questad had 266 unimproved acres. Canute Canuteson was registered with 50 acres of improved land and 213 acres of unimproved land, followed by Ole Pierson with 40 acres of improved and 280 acres of unimproved land. Berger Rogstad had only improved 30 acres and still had 290 unimproved acres, followed by John Ringness (20, 300), Bersvend Swenson (18, 143), Ole Canuteson (16, 213), Neill Swenson (15, 145), Jens Jenson (13, 307), and Ovee Colwick (9, 148).[2]

Most of the homesteaders mentioned had lived in the community six years, while a few had arrived recently. Bersvend Swenson (born 1808) had improved 18 acres and his son Neill Swenson (born 1836) had improved 15 acres since their arrival in Bosque County in 1857. The Swenson family had emigrated from Tynset in Hedmark County in 1851 on the ship *Arendal* and settled at Four Mile Prairie.[3] Questad, Ringness, and several others crossed the Atlantic on the same ship. No Norwegians had dared to settle as far west as the Swenson family did in

1857. Ovee Colwick and his family had taken over half of Cleng Peerson's land in fall 1859. It is no surprise that Peerson had cleared very little of his land and that Colwick began almost from scratch. It is more surprising that young Ole Canuteson only had improved 16 acres of land in six years. Becoming a successful farmer was obviously not his main aim in life. He preferred to work in his smithy and to solve mechanical problems. After the Civil War he sold his land and moved to Waco, where he established the first mechanical workshop.

Who among the Norwegians ranked highest with respect to the value of livestock? Again, Hendrick Dahl topped the list, with livestock valued at $1,454, followed by Carl Questad with $1,330, John Ringness with $888, Berger Rogstad with $838, Ole Pierson $820, Ole Canuteson $740, Canute Canuteson $725, Poulsen $500, J. Jenson $340, B. Swensson $270, Neill Swenson $180, and Ovee Colwick with $41.[4]

Neither acres of improved land nor livestock automatically translated into general wealth. The 1860 census enumerator registered both real and personal estate of the individual households. In 1860, Hendrick O. Dahl was the wealthiest Norwegian with a total of $2,894, followed by Canute Canuteson with $2,505, and his son Ole Canuteson with $2,430. Poul Poulsen was registered with $2,400, Jens Ringness with $2,000, and Carl Questad with $1,957. In the agricultural census, neither Canute nor Ole Canuteson loomed large as farmers with much improved land. Nevertheless, they were numbers two and three on the list of Norwegians with the highest total wealth.

Compared with the wealthiest men in Bosque County, the Norwegians could not compete at all. The main source of wealth in Texas in 1860 was ownership of chattel in the form of slaves, not land, so the richest men in Bosque County in 1860 were slave owners. As elsewhere in Texas, wealthy men were elected to political office, and important political decisions were made by the affluent. The Norwegians did not run for political office. They concentrated on developing their farms and hoped to be left in peace, following their own paths (see Appendix 2).

UPPER AND LOWER SOUTH IN TEXAS

The terms Upper South and Lower South are often used to characterize two different social and economic regions in antebellum Texas. The Lower South included almost all of the Gulf Coastal Plain. This

TABLE 3. *Wealth of Norwegians in Bosque County, 1860 (in US dollars)*

Name	Year of arrival	Real estate	Personal estate	Total
Henrick O. Dahl	1854	1,280	1,614	2,894
Canute Canuteson	1854	1,665	840	2,505
Ole Canuteson	1854	1,515	915	2,430
Poul Poulsen	1857	1,600	800	2,400
Jens Ringness	1854	1,000	1,000	2,000
Carl Questad	1854	572	1,385	1,957
B. & Ann Rochester (Rogstad)	1854	640	900	1,540
Ole Pierson	1854	640	885	1,525
Ole Ween	1854	936	528	1,464
Jens Jenson	1854	640	400	1,040
B. Swenson		330	335	665
Neill Swenson		400	230	630
Ovee Colwick	1859	314	41	355

SOURCE: "Heart of Texas Records," Central Texas Genealogical Society, Inc., Waco, vol. XIX, no. 1, 2, 3, and 4.

was the land of cotton and slavery, a land dominated by the plantation type of agriculture.[5] Many of the settlers in this region had grown up in the Chesapeake Bay colonies of Maryland and Virginia and in the tidal swamps of South Carolina and Georgia. The Upper South, according to one definition, included the states of North Carolina, Tennessee, Virginia, Kentucky, and West Virginia. It is sometimes described as the "Yeoman South" in contrast to the "Plantation South." It also included the southern Appalachian Mountains, Ozarks, and Quachita Mountains as well as the plateaus, hills, and basins between the Appalachians and Ozarks, such as the Cumberland Plateau, part of the Allegheny Plateau, the Nashville Basin, and the Bluegrass Basin. The main ethnicities represented among the early settlers in the Upper South were English, Scots-Irish, Scots, and Germans. The agricultural economy of the Upper South was dominated by yeoman farmers, independent freeholders, with few slaves. Cotton was not the major cash crop. "Grains, and especially corn and wheat, formed the backbone of the rural economy, supplemented in certain areas by tobacco and hemp," according to Terry G. Jordan.[6]

Texas was still a part of Mexico when Anglo-American colonization began in the 1820s. The majority of the first settlers in the Austin and de Witt colonies came from the Upper South states along the eastern seaboard of the United States. After the Texas Revolution in 1836, Texans "were quick to give slavery all the guarantees that it had never been afforded by Mexican governments," Randolph B. Campbell observed.[7] During the next decades people from states of the Lower South began a large-scale migration to Texas, settling in the eastern timberlands and south central plains. The new settlers brought their plantation culture and slaves. Their economy, dependent on agriculture, was concentrated first on subsistence farming and cattle and then on production of cotton as a cash crop. They established political institutions along patterns they were familiar with from the Southern states they had left. People with their roots and culture in the Lower South dominated the economy, culture, and politics in the counties along the Gulf of Mexico.

The majority of the population in the prairie counties in north and central Texas west of Dallas came from the Upper South. Before the Civil War, the agriculture of the farmers in north and central Texas was characterized by "1) an emphasis on wheat as a cash crop 2) a food surplus in corn and small grains 3) an unimportance of cotton 4) a scarcity of negro slaves, and 5) a dominance of the horse as a draft animal."[8]

In 1850, 95 percent of 212,592 Texans lived in the eastern two-fifths of the state. Ten years later, although the population had grown to 604,215, the overwhelming majority still lived in the same region. The 1850 census registered more free persons from Tennessee living in Texas than from any other state, and more people from Tennessee lived in Texas than all foreign immigrants combined. People from northern states also migrated to Texas in increasing numbers; 2,855 persons from Illinois were registered as living in Texas in 1850 and 7,050 in 1860. Still, a mere 22,000 persons from northern states had settled in Texas in 1860, while the total number from the Upper South was 101,633 and from the Lower South 100,690.[9]

NORWEGIANS AMONG HILLBILLIES FROM THE UPPER SOUTH

Most of the neighbors of the Norwegian immigrants in Bosque County were born in and migrated from the Upper South. Bosque County had 2,005 inhabitants in 1860, including 293 slaves, who constituted only 15 percent of the population. The county had 88 persons born outside the United States (4.4 percent); most of them were born in Norway. The county had 85 farms with 4,953 acres of improved farm land and 42,546 acres of unimproved farm land. The farms had an estimated value of $156,417, and the value of the livestock was estimated to be $226,260.[10]

The 1860 census registered 312 heads of families; 291 of them were born in Texas, 70 were born in Tennessee, 29 in Kentucky, 26 in Alabama, 25 in Georgia, 21 in South Carolina, 19 in Arkansas, 16 in Missouri, 15 in North Carolina, 14 in Texas, 13 in Mississippi, 10 in Illinois, and 10 in Virginia. Migrants from states such as New York, Iowa, Louisiana, Indiana, Ohio, Vermont, and Maine were also represented. Fifteen heads of families were born in Norway, two in England, three in Ireland, and one in Scotland.

The Norwegians in Bosque County lived among transplanted hillbillies who had grown up with the values and practices of the Upper South. In the United States today, hillbilly might be understood as a derogatory expression. In this context, the term is used to describe the majority of the neighbors of the Norwegians in Bosque County by their geographical roots. People from the Upper South developed a distinct culture, and at times the term is used to specify where they came from as well as their culture. Homesteaders from the Upper South were known to practice small-scale farming, stock raising, and hunting. Their settlement patterns were different from those in the Deep South and the Midwest.[11]

The major stream of immigration southward from Pennsylvania during the eighteenth century consisted mainly of small farmers of Scots-Irish and German descent. By the time of the Revolutionary War they dominated the back country of Virginia and Maryland and had also settled most of North Carolina.[12] The children of population groups that moved into Tennessee and Kentucky in the eighteenth century became settlers in Missouri and Arkansas in the nineteenth century. Peak migration years like 1813–19, 1833–37, and 1853–57 did not coincide with economic depressions, but "with boom times when prosperity was highest and speculation most rife."

"Arkansas was a child of Tennessee, furnishing a convenient outlet for a considerable portion of her surplus population," maintained William O. Lynch.[13] Malcolm J. Rohrbough wrote that "Arkansas was a reincarnation of the American frontier traditionally associated with places like Kentucky after 1775 and Michigan after 1815." The frontier experience and frontier ideals lived on until well into the nineteenth century. "Arkansans hunted, trapped, grazed livestock, and generally pursued a lonely, solitary existence, consistent with their location in the most remote frontier of the west. Amidst their isolation, the people of Arkansas shared a number of common characteristics: a single-minded materialism; a tendency to unite in groups, linked by kinship, for self-protection and their own self-interest; and an inclination to violence widespread among all elements of society that endured."[14]

It was farmers and livestock herders who led the westward movement across the southern frontier, argued Frank L. Ownsley, the historian of the "plain folk." They were usually "landowning farmers and herdsmen, though a small minority were engaged exclusively in other occupations. Their thoughts, traditions and legends were rural." The children of "plain folk" agriculturalists continued to share a distinctive culture. They preferred lonely and rural neighborhoods, and they often lived among kinsmen. They grew food and cash crops in small, fenced fields cleared from the woods, and they let their livestock take care of themselves in unfenced woodlands. "To them the land was, with God's blessings, the direct source of all the necessities of life and of all material riches."[15] They belonged to evangelical protestant churches and attended revival meetings. "As a rule, plain-folk families migrated laterally across the southern frontier, locating familiar terrain, soils, climate, and vegetation in which to reproduce their agricultural practices."[16]

MOUNTAIN FOLK MIGRATED TO THE TEXAS HILL COUNTRY

"By the middle third of the nineteenth century," wrote Terry G. Jordan, "the hills of Appalachia became overcrowded and emigration occurred."[17] Many among the "mountain folk" found a new homeland in the hills of central Texas, known as the Texas Hill Country. They brought to the area their mountaineer culture and "many of the ingredients of the culture and economy of the southern Appalachians and the Ozark-Quachitas."[18] They liked to name places in Texas the way they did in the Appalachians, such as Palo Pinto Mountains, Brady Mountains, Callahan Divide, Cowhouse Mountains, Limestone Hill Country, Bandera Mountains, or Cranfills and Indian Gap. The natural vegetation was less impressive in Texas than in the Appalachians. Most trees were smaller, and only the live oak of the Texas hills could rival the trees of the mountain south. The Western Cross Timbers region was both drier and more exposed to prolonged drought than the Appalachian region farther north.

A significant number of settlers who had grown up in Tennessee, Arkansas, and Missouri had by 1850 settled in Texas Hill counties like Williamson, Travis, Hays, Comal, and Gillespie. Persons born in the Upper South were almost twice as numerous as natives of the Lower South. The hills could certainly not compare with the mountains they grew up with in the Ozarks and the Appalachians, but they were similar enough "to lure the homesick hillsmen who were migrating from Arkansas, Missouri, and Tennessee."[19] Because of the Norwegian homesteaders in Bosque County, Jordan excluded Bosque County from his study, but he might as well have included it. The Norwegians were highly attracted by the hills and the creeks. Bosque County was not Norway, but it reminded them of Norway.

SLAVERY IN TEXAS BEFORE THE CIVIL WAR

Slavery expanded rapidly in Texas in the 1850s. The 1850 census registered 58,161 slaves, or 27.4 percent of the population. During the 1850s the slave population increased to 182,566 slaves, an increase of more than 200 percent; in fact, the number of slaves grew faster than the Texan population. Nineteen Texas counties had 1,000 or more slaves. The large slave counties were found along the Brazos and Colorado rivers and in East Texas around Nacogdoches and St. Augustine north to the Red River. There were 250 plantation owners with more than twenty slaves in East Texas; close to 30 percent of them lived in Harrison County.[20] By 1860, 64 of 105 Texan counties had 1,000 slaves or more. The number of slaves increased with "special rapidity along the middle Trinity from Polk and Walker to Limestone counties."[21] Slaves constituted more than 50 percent of the population in six counties in 1850. Thirteen Texas counties now had populations where more than 50 percent were slaves.

Did slavery pay? Slave prices almost doubled between 1850 and 1860;

owning slaves was a measure of financial success. "Those who owned a few slaves found them valuable in many ways and hoped to acquire more."[22] According to Campbell, the mean value of slaves increased from $440 in the years 1848–52 to $765 in the years 1858–62.[23] During the late 1850s "prime male field hands" aged between 18 and 30 cost $1,200 on average, and skilled slaves such as blacksmiths were often valued at more than $2,000.[24] In comparison, good Texas cotton land could be bought for as little as 6 dollars an acre. The appreciation in slave values was a potential source of profit in a rapidly growing area with much land and little labor. Large cotton plantations were located along the river valleys of eastern Texas and in the coastal counties around Houston and Galveston. Transportation was the main bottleneck. Only cotton grown close to navigable rivers and the coast could easily be brought to market, and slavery was modest in regions with long and complicated transport options.

When the Civil War broke out, slavery was the leading economic and social institution in Texas. Slaveholders, especially planters with more than twenty slaves, constituted the wealthiest class in the state.[25] Campbell concluded his discussion of slavery and profitable agriculture in Texas the following way: "The institution may have helped retard commercialization, urbanization, and industrialization, but it was satisfactory to a great majority of the state's economic leaders. Moreover, it performed numerous functions that, while not appearing in the profit-or-loss column on a balance sheet, were of great benefit to many Texans."[26]

THE IMPACT OF THE LOWER SOUTH ON EAST TEXAS

Half of the population of Texas in the 1850s lived in thirty-two counties east of the Trinity River. The region was bordered by the Sabine River in the east, the Red River in the north, and the Gulf of Mexico in the south. Most of the new settlers came from the states of the Old South: Alabama, Tennessee, Mississippi, Arkansas, Louisiana, and Missouri. Wheat, rice, and tobacco were grown in East Texas mainly for local consumption; the big cash crop was cotton. The counties with the highest slave population were also the counties with the largest exports of cotton bales each year.

The ethnic diversity of the Texas population was far greater than in other states in the South. Large numbers of Germans lived in some of the south central counties west and north of San Antonio. Many Mexican Americans lived in San Antonio and counties further south. Groups of foreign immigrants such as Poles and Czechs settled in towns and farms west of the Trinity River. Very few foreign immigrants settled east of the Trinity before 1860. The small Norwegian colonies in Henderson County and at Four Mile Prairie along the Kaufman-Van Zandt county line in East Texas were an exception to this rule, Ralph Wooster emphasized. The Norwegians in East Texas lived in a region totally dominated by the social practices and institutions of the Lower South. They lived among slave owners and were expected to adapt to the social practices of their neighbors.

MCLENNAN COUNTY WAS DOMINATED
BY THE SLAVE CULTURE OF THE LOWER SOUTH

Norwegian immigrants in East Texas, but also in Bosque County, were very much aware that they had settled in a slave state. The issue of slavery was less acute on the frontier in Bosque County than in East Texas. Few among their American neighbors owned slaves. The slave owners they knew personally had maybe one or two, and none of them were plantation owners with more than twenty slaves.

The Norwegians felt uncomfortable about slavery. The topic was seldom if ever mentioned in letters to family and friends in Norway. During the late 1850s several slave owners from the Lower South migrated to Bosque County, and political discussion of slavery and secession became very heated at the end of the 1850s.

Every time a Norwegian traveled to Waco to sell wheat and buy supplies, he was strongly reminded of how much more dominant slavery was in Waco and McLennan County. Norwegians were accustomed to doing their own manual labor; in Waco, a large amount of manual labor connected to city life was done by black slaves. If Bosque County was dominated by the culture of the Upper South, the slave culture of the Lower South dominated in the neighboring McLennan County and the town of Waco.

Compared with frontier Bosque County, McLennan County represented civilization. The county was organized in 1850 and experienced strong economic growth during the 1850s. Wealthy, well educated, and politically well-connected planter families migrated to Waco and

McLennan County from the Lower South. McLennan County had 6,206 inhabitants in 1860; 2,395 slaves made up 38.6 percent of the population. According to the slave schedules for McLennan County in 1860, three men had more than seventy slaves, eleven owned more than forty slaves, more than twenty-five men owned more than twenty slaves, and forty men owned between eleven and twenty slaves.[27]

The *Handbook of Waco and McLennan County* was published in 1972. The introduction stated that it was intended to be a "convenient reference volume with concise, accurate information on the widest range of subjects relevant to the Waco-McLennan area." It was further proclaimed that the handbook would deal "chiefly with people, places, notable events, organisations, institutions, geographical locations etc."[28] The institution of slavery was obviously of little interest; there is no article on the topic of "slavery." The reader will find a surprising number of articles in the handbook about well-known slaveholders, politicians, and McLennan County officers in the Confederate Army. It is hardly ever mentioned in any of these articles that the men in question were known as large slave owners.

The Gurley, Downs, and Fort families, who migrated to McLennan County from Alabama, were all leading plantation and slave owners. Members of these families moved to Waco and McLennan County in the first half of the 1850s.[29] They were strongly connected to each other by kinship, marriage, and friendship. In 1851, Richard Furman Blocker (1824–1861) came to Waco to practice law. He persuaded his brother-in-law Edward J. Gurley to join him, and the two men established the first law firm in Waco that year. Gurley was born in Franklin County, Alabama, in 1824. During the Civil War he commanded the Thirtieth Texas Cavalry.[30]

His younger brother James Henry Gurley, born in 1829, moved to Waco in 1851 and entered the general mercantile business, but also invested in farming and stock raising on his plantation along the Brazos River bottoms. Edward J. Gurley wrote his father in Alabama about the favorable prospects for cotton in McLennan County. Why not sell the plantation in Alabama and take the slaves with him to Waco? In 1853 his younger brother Davis Robert Gurley, born on October 17, 1836, led a wagon train from Alabama to Waco, which included six family members, more than fifty slaves, and livestock.[31] The journey took six weeks. Gurley bought 1,600 acres of land south of Waco.

In January 1854 28-year-old William A. Fort, born in 1826 in La Grange, Alabama, "made a trip to Waco to see if the area would be suitable for farming."[32] He was a plantation owner who persuaded Downs and several other families to move to Waco and McLennan County with him. In March 1855, a group of 500 people made the six-week trek. His own family, the family of William Woods Downs, and three others brought all their slaves, stock, and possessions with them.[33] According to oral tradition, W. W. Downs and Fort brought about 250 slaves to Waco. Fort bought land 4 miles south of Waco on the Brazos River. After the Civil War he gave up his plantation and went into business with William Trice in Waco. He later became the president of the Waco National Bank.

W. W. Downs was born in North Carolina in 1804, grew up in Georgia, and established a general mercantile business in La Grange Alabama with Fort in 1850. He later built up a wholesale and mercantile business in Waco.[34] His son John Wesley Downs, born in Georgia on November 15, 1838, came with his family in 1854 and established a plantation in Downsville in east McLennan County. He served in the Civil War with his friend L. S. Ross, and after the war he founded the first newspaper in Waco, the *Daily Examiner*.

The Harrison, Earle, and Thompson families were another important cluster of planter families settling in McLennan County in the 1850s. Several of them were related. In 1857, James Edward Harrison, born on April 24, 1815, in Greenville District, South Carolina, moved with his family to McLennan County and purchased land 10 miles south of Waco along Tehuacana Creek. According to several sources he bought 6,000 acres of land in McLennan County, but according to the tax records he owned 3,000 acres.[35] Soon after his birth, his family, one of the most wealthy and influential families in the South, had moved to a plantation in Jefferson County, Alabama, and around 1830 moved again to Monroe County, Mississippi. Harrison had been a trader among the Indians in the Indian Territory and had also served as a Mississippi state senator.

It was James's brother, Thomas Harrison, who urged the family to move from Mississippi to Texas. After many years as a lawyer in Texas he had settled in Waco. He was the uncle of John Baylis Earle and Isham Earle, who joined him in Waco in the early 1850s. In their letters home they praised the area's climate and soil. The extended Harrison family held a meeting where it was decided that the families of James E. Harrison, Louisa Jane and her husband Dr. Wells Thompson, and Dr. Earle would move to Texas. Dr. Earle sold his holdings in Monroe County, Mississippi, to Dr. Richard Harrison, who also bought James E. Harrison's Greenwood plantation. The Earle, Harrison, and Thompson families took their slaves and possessions and left Aberdeen, Mississippi, by boat, traveled down the Tombigbee River to Mobile, Alabama, sailed across the Gulf of Mexico to Galveston, and then traveled over land. According to the slave schedule for McLennan County in 1860, Dr. E. A. Earle owned 61 slaves, and B. J. Thompson 72 slaves.[36]

Bosqueville, a settlement in McLennan County located 7 miles northwest of Waco, had a population of 531 in 1860.[37] Among the first families to settle in Bosqueville were the Wortham, Scott, Cobb, McNamara, and Stuart families. N. J. W. Wortham was listed in the slave schedule for 1860 with 18 slaves. He was 43 years old, born in Virginia, and was classified as farmer, stock raiser, and medical doctor, with assets of $7,106 in real estate and $18,000 in personal estate. J. A. Cobb owned twenty slaves.[38]

By 1860, Waco had become the western outpost of the Lower South, with 949 inhabitants. Slave labor was used to clear bottomlands, plow the soil, and plant and pick cotton. Slaves were also used to build houses and work in businesses. "In 1860, plantation owners along the Brazos produced 2,320 bales of cotton and were already clearing more land for cultivation."[39] David Gurley owned more slaves than he could work. In 1854, he began renting them out to others. "Slave hiring in Texas began almost as soon as slaveholders arrived in the region," Campbell commented. "Hired slaves were expensive, but they cost less and were more dependable than free labor."[40]

The wealthiest men in McLennan County were also the leading politicians, as they had both the money and time to participate in politics. Neil McLennan and George Erath, previously introduced, were two such men; in 1860 McLennan owned eight slaves and Erath owned eleven.[41] A remarkable number of lawyers lived in McLennan County in 1860. Several of them were active in politics, were slave owners, and were active members of Baptist and Methodist churches. William A. Fort practiced law in Waco and owned seventeen slaves. James F. Davis was born in Tennessee, practiced law in Waco, and owned fifteen slaves. J. W. Nowlin came to McLennan County in 1850. He was the first attorney licensed in Waco and was associated with Richard Coke, a lawyer, politician, and later governor of Texas. Nowlin owned seven slaves. Richard Coke owned 15 slaves; another powerful lawyer, planter, and politician was Joseph W. Speight, who owned 32 slaves.

The brickyard owner Bentley B. Arnold had settled in East Waco in 1851. He was elected county commissioner in 1858, owned twenty-nine slaves in 1860, and played a leading role in the establishment of the First Baptist Church. William C. Coates (1804–63) was born in Virginia, had lived in Tennessee, and moved to McLennan County in 1856. He was a planter, with land along the Brazos River south of Waco, and owned sixteen slaves. He was county commissioner from 1858 to 1862. It can be no surprise that the leading men in McLennan County were wholeheartedly in favor of secession and the confederacy. James E. Harrison was involved in the secession movement in Texas from the beginning; he was also a delegate to the state secession convention.

INCREASING SLAVERY IN BOSQUE COUNTY IN THE 1850S

Compared with McLennan County, slavery in Bosque County was modest. Between 1856 and 1860, the number of slaves in Bosque County increased from 213 to 293. To be a plantation owner, you had to own more than twenty slaves; only two men owned more than twenty slaves in 1860. Four men owned between ten and twenty slaves (see Appendix 3).

Slavery equaled wealth. The richest man in Bosque County in 1860 was also the largest slaveholder. The total wealth of plantation owner John Jackson Smith, 61 years old, was $54,091. He owned sixty-nine slaves according to the slave schedule for Bosque County in 1860. His plantation was located at Smith's Bend on the Brazos River in the northern corner of Bosque County. Compared with Norwegian immigrants in the county he was filthy rich; compared with the richest men in Texas, Smith was only modestly wealthy. The largest slaveholder in Texas in 1860 was David Mills of Brazoria County, with real and personal property of $614,234, including more than 300 slaves.[42]

John Jackson Smith was born in Edgefield, South Carolina, on August 25, 1799, and died in Bosque County in 1867. After having purchased 7,000 acres of land inside a bend on the Brazos River in northeastern Bosque County in 1854, he migrated with his family from Scott County, Mississippi. The Smith plantation in Mississippi had approximately 200 slaves. His son Burton Smith was sent ahead of the rest of the family with about a hundred slaves, horses, cattle, and so on to begin building up the new plantation. The first cotton crop at Smith's Bend was harvested in 1855.[43]

Burton Smith suddenly died on July 5, 1856 after a long day of hunting predators on the plantation. When John Jackson Smith learned of the death of his son, he was still in Mississippi and was not yet ready to leave for Texas. He asked his daughter Margaret Ann and her husband Silas I. McCabe, who lived with their three children in De Witt County, Texas, to travel to Bosque County as soon as possible to look after the slaves and other Smith property. Silas I. McCabe was born on July 28, 1817, in Smith County, Tennessee, and died of pneumonia on April 19, 1871, at Coon Creek, Bosque County. On April 3, 1836 he bought 120 acres of land in Scott County, Mississippi, for 1,000 dollars in cash and established himself as a farmer. He married Margaret Ann Smith (1824–98) on December 23, 1841. As requested, the McCabe family traveled to Bosque County and took care of the Smith property until the rest of the family arrived. McCabe liked the land and bought several thousand acres of land west of Smith's land, some in Hill County and some in Bosque County. In 1860 McCabe owned property in Bosque County valued at $13,925.[44]

The largest slave owners in Bosque County lived on the western side of the Brazos River, on land close to the McLennan County border. Temple Spivey, who had settled in Cherokee County, Texas, in 1849, settled on the west bend of the Brazos River, one half mile upstream of Cedar Creek in 1853. The wagon crossing on the Brazos close to his home was named Spivey Crossing. Spivey owned thirteen slaves in 1860. Ephraim Snell and his family settled in Bosque County in 1858. Snell was born August 1802 in Ohio and lived in Macon and Randolph counties in Missouri from 1830 to 1858. Snell owned ten slaves in 1860.

Brooks Moon Willingham (1814–71) was born in Georgia and moved to Bosque County with his wife, Mary Louisa, eight children, and slaves in 1859. Their sons John, Isaac, Thomas, and Wilson, together with the young slave Julia Ann, moved ahead of the rest of the family, arriving in Meridian in November 1859. Willingham, his wife, and their oldest son Cashwell Augustus, who was a doctor, traveled from Charleston by steamship to Galveston in 1860. Willingham and his sons established mercantile stores in Morgan, Kopperl, and Kimball.[45]

Joel Martin Stinnett arrived with his family in 1859. He was born in Virginia in 1806 to a well-to-do family who owned land and slaves. Around 1830 the family moved to Breckinridge County, Kentucky, joining their uncles Alexander Marshall and Samuel Marshall. Samuel Marshall migrated to Lewis County, Missouri, in 1835, and Stinnett and his family followed in his footsteps the following year, settling west of the town of La Grange on the Missouri River.

Stinnett had received a good education and showed mechanical skills. He became involved in the business of building and running gins and mills, both ox tread mills and water mills. In 1847, the affluent family moved to Fannin County, Texas. Several young men helped drive the six wagons in exchange for free transportation. The women rode in carriages drawn by fine Kentucky horses, and the slaves walked. After about five weeks, the group arrived in Fannin County and settled near Orangeville, where Stinnett and his sons erected a gristmill that they operated until 1858. Stinnett also operated a cotton gin and a farm.

Stinnett sold his interests in the Fannin County mill to his oldest son Ellis in 1858, moved to Pilot Point, and set up another mill there. He continued to Clifton and Bosque County in 1859, where he bought the Clifton mill on the Bosque River. In the 1860 census, 53-year-old Joel Martin Stinnett was listed as having fifteen slaves in Bosque County and was registered with a wealth of $18,000; his 17-year old son G. M. Stinnett was registered with $9,000, and his 20-year-old son M. H. Stinnett with $6,000. All three gave merchant as their profession.

In Clifton, the Stinnetts soon established good relations with Joseph Alexander Kemp (born in Tennessee on April 2, 1840; died on January 24, 1885 in Wichita Falls). Kemp had settled in McLennan County in 1856, but moved to Clifton some months later, where he established a store.[46] The connection between the families was strengthened when Kemp married Emma Francis Stinnett on October 17, 1860.

Stinnett operated the Clifton mill until 1867, when he sold it to William A. Kemp. The new owner tore down the old mill and built a new stone mill. Kemp's mill played a major role for some years as a supplier of flour to Texas frontier settlements. Ox wagon trains hauled wheat and corn flour to places as far west as Brownwood and Fort Concho, Fort McKavett, Fort Mason, Fort San Saba, Fort Griffin, and Chadbourne.

FEW SLAVES IN THE NORSE AREA

The Norwegians in Norse lived almost 30 miles west of the Smith plantation at Smith's Bend. They hardly ever traveled in that area of the county.

The richest slave owner in their neighborhood was William R. Sedberry, county judge from 1858,[47] whose total wealth in 1860 was $18,500. He was born in North Carolina on July 13, 1823, but his parents soon moved to Tennessee. On January 26, 1843, Sedberry married Caroline Huntley Alexander in Tennessee. The couple had five sons and four daughters over the years. The Sedberry family arrived in Bosque County at the same time the first Norwegians arrived, in 1854, bringing six slaves with them. Sedberry acquired a large tract of land along the Bosque River near the Meridian Creek confluence, where he farmed and raised stock.[48]

Leroy Parks and his family lived closer to Norse. In 1860, the total value of his real and personal estate was $7,068. Parks was born in Marshall County, Mississippi, in October 1826. He migrated to Texas with his parents when he was 12 years old. "They made the long trip by wagon train with John Parks and his two oldest sons Felix and Leroy, riding horses. Elisabeth and the younger children rode in the wagons, and the slaves walked. This was in the year 1839."[49] The Parks family lived in Nacogdoches County until 1856 and then moved to Bosque County. On May 7, 1856, Leroy Parks married Susan Wheeler. The couple settled on 320 acres of land Parks had bought for 1 dollar an acre, located in the broad valley near the confluence of Meridian and Spring Creeks. This was the beginning of the Parks Ranch, which in 1895 covered 4,600 acres.

TEXAS AND SECESSION

"By 1861 Texas was so like the other Southern states economically, socially, and politically," observed Campbell, "that it joined them in secession and war. Antebellum Texans cast their lot with the Old South and in the process gave their state an indelibly Southern heritage."[50] Texas seceded from the Union in early 1861 and joined the Confederate States of America. Politicians with close ties to the plantation culture of the Lower South were strongly in favor of secession, and sooner rather than later.

In his discussion of the membership of the Texas Secession Convention, which met on January 28, 1861, Ralph A. Wooster concluded that more than 70 percent of the 177 members at the convention belonged to the age group 20 to 40, and nine out of ten members were born in slave states. Forty-three members were born in Tennessee. Lawyers constituted slightly more than 40 percent of the members. More than 70 percent were slave owners, and the median holding for these slave owners was nine slaves. "Except for the large percentage of lawyers and slaveholders present, the Texas convention seems to have been a rather typical cross-section of Texas society in 1860," Wooster commented.[51]

Only three foreign immigrants were members of the convention and all three were born in Germany. The 43-year-old physician T. Koester represented Comal County, and John Muller was a 38-year-old merchant from Galveston. Neither of them were slave owners. The 48-year-old lawyer Charles de Montel from Medina County owned nine slaves. Many German immigrants in Texas and people who had grown up in the Upper South were in favor of staying in the Union.

Of the members of the January convention, 166 voted for secession and 8 against. The most influential among the Unionists was Sam Houston, the governor of Texas between 1859 and 1861. Houston had managed to delay the secession convention until January 1861. He also played an important role in the decision to hold a public referendum on secession on February 23, 1861. Texans had held a referendum on joining the Union in 1845, and it was agreed that another referendum should be held to ratify leaving the Union.[52]

On February 23, 1861, an overwhelming number of Texans with voting rights voted to secede from the Union; a majority of 46,153 voted for and 14,747 against.[53] The secession of Texas from the Union became official on March 2, Texas Independence Day. Of the 122 counties casting votes, only 18 cast majorities against secession. The largest opposition to secession was found along the border in northern Texas—in Jack, Montague, Cooke, Grayson, Fannin, Lamar, and Collin counties, and in Angelina County in East Texas. In West Texas (the Hill Country) a majority in Mason, Burnet, Williamson, Gillespie, Blanco, Travis, Bastrop, Fayette, Uvalde, and Medina counties voted to remain in the Union. In Gillespie County 96 percent of the voters were in favor of staying in the Union. Some of the northern Texas counties had majorities in favor of staying in the Union of 60–70 percent.[54]

Several of the counties with a majority in favor of staying in the Union had strong pockets of German immigrants. For a long time the dominant view among historians was that Germans were against slavery. They did not own slaves, were sympathetic to the abolition of slavery, and were strongly in favor of staying in the Union. In *German Seed in Texas Soil. Immigrant Farmers in Nineteenth-Century Texas*, Terry G. Jordan argued that these views were too general and did not hold water when measured against the empirical evidence. "Out of 130 German farmers in Gillespie, Comal and Guadalope counties in 1850, not a single one owned slaves. Ten years later, the same was true of all 377 Germans in Gillespie, Mason, and Llano counties."[55] This might mean, Jordan argued, that Germans in the western counties on the frontier preferred to stay with their "heritage of family labor." But it could not be concluded from this that Germans generally were "antislavery and harbored abolitionist sentiments."[56] According to Jordan, few among the German farmers concerned themselves with the right or wrong of slavery. The German "peasant of the nineteenth century was not a politically oriented being, and nothing could distort the picture more than to depict the average Texas German farmer in the western settlements as an active abolitionist."[57] The US Census showed that many Germans who could afford it owned slaves in 1860. Most of the German slave owners lived in eastern counties between the lower Brazos and Colorado rivers. "Those Germans who did own slaves were, for the most part, the immigrants of the 1830s, who had longer to accumulate the necessary capital."[58]

In an article discussing Texas Germans and the Confederacy, Walter D. Kamphoefner argued that Jordan exaggerated "the degree to which Germans agreed with Anglo Texans on issues such as slavery, race, secession, and Civil War."[59] It was true that Germans owned slaves, Kamphoefner agreed, but at every level of wealth, a higher proportion of Anglos than Germans owned slaves. For example, among persons with wealth between $3,000 and $6,000, more than half of the Anglos but barely two percent of the Germans were slave owners.[60] Jordan had argued that Germans in the eastern counties had lived longer in Texas than German immigrants in the western counties and had assimilated the dominant Lower South culture to a larger degree. Kamphoefner disagreed with Jordan on this point. Wherever voting results were available on precinct level, "they show the German communities of a county to be most hostile." This was the case even in older Texas German settlements farther east. The overall vote for secession in Colorado County was 64 percent, but these numbers masked an internal polarization. The three German precincts Frelsburg, Weimar, and Mainz "voted 86 percent against secession, while five Anglo precincts cast all but six votes in favor."[61] In Bastrop County, Wendish Lutherans and German Methodists were "nearly unanimous in the opposition to secession." The political situation was very charged in late 1860 and early 1861. A lot of courage was needed to argue in favor of staying in the Union. The editor of the *New Braunfelser Zeitung* in the heavily German populated Comal County gave his readers the following advice: "When in Texas, do as the Texans do. Anything else is suicide and brings tragedy to all our Texas-Germans."[62] Both Kamphoefner and Jordan agreed that urban Germans were "more acculturated to Southern society and more subject to intimidation," and therefore voted in favor of secession.

DID NORWEGIANS HAVE A "DEEP-ROOTED HOSTILITY" TOWARD SLAVERY?

The above discussion about the attitudes and practices of German immigrants in relation to slavery and secession is very relevant to discussion of the attitudes of the Norwegian immigrants in Texas. For a long time, it has been assumed among Norwegian-American historians that Norwegians had a "deep-rooted hostility" toward slavery. If Norwegians fought in the Civil War, they fought heroically on the Union side.

Maybe the strongest Norwegian patriotism for the Union and President Lincoln was expressed by Odd S. Lovoll in *The Promise of America. A History of the Norwegian-American People*, published in Norway and the United States in 1984. The Civil War, wrote Lovoll, "created a new patriotism, a sense of having earned a legitimate place in America, for Norwegian blood had been spilled in the defense of the nation. From every settlement Norwegian immigrants went to war, willing to make the greatest sacrifice that could be asked of a citizen. For them, as for other people in the northern states, the war was an affirmation of the American ideals of freedom and equality. Norwegian immigrants, and other Scandinavians answered the call and stood behind Lincoln and the Republican Party."[63] Lovoll obviously found it self-evident that Norwegians were against slavery and would fight in the Union Army. His claims, however, were not documented by footnotes or references.

Lovoll's view was shared by Theodore C. Blegen. The great mass of Norwegian immigrants, Blegen argued, took a clear stand in the slavery controversy many years before the Civil War broke out. They were against slavery, and the slavery question was a deciding factor in why so many Norwegians chose to settle in the north and northwest rather than the south and southwest.[64] Blegen speculated that one of the reasons why the Norwegian settlement in Shelby County, Missouri, failed, was "the distaste of the majority of the Norwegian immigrants for a slave state," combined with the strong population flow into Wisconsin and Iowa.[65]

Blegen's crown witness against slavery was Ole Rynning, who in 1839 published the small booklet *Sandfærdig Beretning om Amerika til Oplysning og Nytte for Bonde og Menigmand*. Rynning commented on religion, on the many religious sects in America, and on the American form of government. The laws, government, and authorities were designed to maintain "the natural freedom and equality of men." The "infamous slave traffic," however, represented an ugly contrast to this freedom and equality. The slave traffic was still tolerated, and it continued to flourish in the southern states, he warned his readers.[66]

Blegen claimed that Norwegians had "a natural repugnancy for slavery." Immigrants in the 1850s might also have been influenced by the stir created over the translation and publication of Harriet Beecher Stowe's book *Uncle Tom's Cabin*, published in Norway in 1853. The book became a bestseller in Norway and parts of it were reprinted in Norwegian newspapers, followed by discussions on slavery and life in the United States.[67] According to Blegen, most of the Norwegian immigrants "were antislavery in view, favored a liberal public land policy, and shared the hostile frontier attitude toward the land speculator."

In 1972, C. A. Clausen and Derwood Johnson edited and published a number of Civil War letters from soldiers in the Confederate Army in the journal of the Norwegian-American Historical Association under the title "Norwegian Soldiers in Confederate Forces." In their introduction, they repeated Blegen's argument. The Norwegian settlements in the Southwest never gained any great headway because of "a deep-rooted hostility toward slavery."[68] In an article on early Norwegians in East Texas in the *Southwestern Historical Quarterly* in 1961–62, Darwin Payne presented the same argument; there could be no doubt that "the Norwegians' antagonism toward this institution (slavery) greatly hindered the growth of the Texas colonies in favor of the northern settlements."[69]

To back up their view, Clausen and Johnson used a citation from William C. Pool's 1954 *History of Bosque County*. According to Pool, "the Bosque Norwegians were inherently opposed to slavery and subsequently to secession and the Civil War. During the war years, the Bosque contingent remained aloof from the war effort and emigration from Norway ceased altogether."[70] Nowhere among the Norwegians, Clausen and Johnson concluded, could be found "any expressions of patriotism or intimations that they felt they were engaged in a noble or holy cause." Some of the Norwegian-Texans were extremely critical of the Confederates and "only hoped that the war would come to an end so that they might rejoin their friends back home and resume the pursuits of peace."[71] To support their view, they retold a story from Pool about the young Norwegian Otto Swenson, who was sent east of the Mississippi River on active duty. Since he knew "very little about the nature of the conflict, Swenson simply had no interest in the fighting. Since he had not been issued a uniform, Swenson solved his problem by walking away from his company, drifting around in enemy and neutral territory until the war officially ended, and then leisurely walking back to Bosque County."[72] One problem with this story is that Otto Swenson never served east of the Mississippi. He enlisted in McCord's Frontier Regiment and patrolled the western border with the Indians. Swenson never even served outside the borders of Texas.

Elise Wærenskjold is usually presented as the most outspoken voice against slavery among Norwegians in Texas. According to her biographer, Charles H. Russel, she reflected more on the institution of slavery than any other subject. In her "Confession of Faith," written in her diary in Van Zandt County in 1858, she was convinced "to the fullest degree that human beings are born with equal rights. Consequently it is repulsive to me to hear people read their Declaration of Independence and deliver bloated Fourth of July orations in honor of liberty while there are millions of slaves among them." Wærenskjold was further convinced, wrote Russel in a citation from her diary, that slavery "is absolutely contrary to the law of God, because the law commands us to love God and our neighbors as ourselves, and, further, that whatsoever we want others to do unto us, that we should do toward them. These rules are as simple and easily understood as they are true, and if we only accept them as the guiding lines in accordance with which we regulated our behavior toward our fellow men then we would not so easily go astray." Would white men be satisfied with being slaves, "with being sold like animals, with being separated from our mates and our children whenever it might suit our masters, with seeing our children brought up in thralldom and ignorance without the slightest possibility of rising above the miserable state into which we were born, despite the fact that we might have the highest abilities and the greatest eagerness to learn?" She concluded that slavery must be "contrary to the will of God, must be an abomination."[73]

None of the Norwegians in Bosque County owned slaves. Some of the Norwegians in East Texas, however, were slave owners. The brother of Johan R. Reiersen, George Reiersen, owned nine slaves. In the slave schedule for Kaufman County in 1860 the Norwegian Erick Bache was listed as the owner of a female slave, 33 years old. Thomas Fasting Grøgaard also owned slaves. At least one of them, a little girl named Hannah, remained with the family after the Civil War and worked for the Grøgaard family when she grew up. She followed the practice of other slaves when they became free; she took the surname of her former master. She was buried in the Negro section of a graveyard in Atlanta, Texas, and the name on her tombstone reads "Hannah Grogard, 1862–1946."[74] The Norwegian immigrants mentioned above lived in East Texas and were much better assimilated to the dominant slave culture than were the Norwegians in Bosque County.

Elise Wærenskjold never published her views on slavery; they remained in her private diary. In his biography, Russel admitted that the tax rolls for Van Zandt County in 1861 and 1862 showed that the Wærenskjolds owned one slave. In a footnote, he goes to great pains to rationalize this unfortunate fact since it clashed with her "Confession of Faith." His excuses seem farfetched, which he readily admits.[75]

As late as spring 1865 Elise Wærenskjold rented a slave to do manual agricultural work on the farm. In a letter to Carl Questad, Bosque County, her husband Wilhelm Wærenskjold wrote that spring in Van Zandt County was in full swing. While he had been away, his wife "had been fortunate enough to rehire for a year the Negro whom I had with me last fall. To be sure, I have to pay the high wages of $1,600 in the new issue, but then he is the most reliable Negro I know of. My wife has also hired a Negro woman and a little boy—the former to cook and do related jobs, the latter to tend the sheep."[76]

The slavery question proved to be complicated, for Elise Wærenskjold and the other Norwegian immigrants living in Texas, but also for immigrants in the Midwest. Before the Civil War, the orthodox Norwegian-American Lutheran Church in the Midwest defended slavery on theological grounds, and for more than a decade after the Civil War leading members held the view that slavery was not a sin.

Norwegians at Four Mile Prairie had established a Norwegian Lutheran Church, *Den Norske Lutherske Menighet i og ved Prairieville*, Texas, at a meeting in November 1848. Among the first members were

"Simon Aanenson og Kone, Knud Hansen & Marthe, O. J. Orbeck, Oline Reierson, Peder Pierson & Helene, Ole Anderson & Berget, Susanne Mjaaland, Berthe Halvorsen, Marthe Anderson, Aanen Knudtsen & Mathea, J, P. Hallin, Andreas Nilsen, E. Wærenskjold, Stian Aanensen & Kirsten, Berthe M. Foss, Gunhild Knudsen, og Ingeborg Bache." The church was organized along orthodox Lutheran Norwegian state church lines and would be based on the canonical texts in the Old and New Testament.[77] These texts were God's revealed word, and would be their only rule and guide for their faith and Christian life. The first Four Mile Lutheran Church was built in 1853, improved in 1854, and dedicated in 1855. Pastor A. E. Friedrichsen visited the Norwegian communities in Brownsboro and Four Mile between 1854 and 1857. The next pastor to visit was Elling Eielsen in 1860, the founder of the Evangelical Lutheran Church of America. During the Civil War, the Norwegian communities in Texas had to fend for themselves as best they could in both temporal and spiritual matters. Not until 1867 did Norwegian-American church authorities send someone from the north to try to do something for the Norwegian flock in the south.

THE NORWEGIAN-TEXANS AND THE SECESSION REFERENDUM

The Norwegian immigrants in East Texas lived in counties overwhelmingly in favor of secession. Only one of the thirty-five counties in East Texas "cast a majority of their ballots against secession," wrote Ralph A. Wooster and Robert Wooster.[78] In Henderson County 400 voted for secession and 49 against. "The citizens of Kaufman County voted in favor 461 ayes to 155 nayes, a three to one margin," wrote Jack Stolz.[79] In Van Zandt County 181 voted for secession and 127 against. Van Zandt County was one of the counties in East Texas where more than 40 percent of the voters cast their votes against secession. It is interesting to note that the number of slaves in this county in 1860 was only 322 persons out of a total population of 6,494.

The highest number of Norwegian immigrants in East Texas lived in Van Zandt County. According to an article on Van Zandt County in the *Handbook of Texas Online*, written by Gerald F. Kozlovsky, "many of the Norwegian settlers, who were opposed to slavery on moral grounds, and a number of small farmers, who resented the power and influence of the state's large plantation owners, spoke out against the war. In 1864 three of the Unionists were lynched, and some of the Norwegian settlers were arrested, effectively quelling the opposition."[80] As Kamphoefner has shown, even if counties voted in favor of secession, there were precincts with strong pockets of Germans who voted to stay in the Union. The precinct-level voting results in Van Zandt County might also show that the Norwegians voted against secession.

This was certainly the case in Bosque County. At the election "held in and for the County of Bosque," 198 voted for secession and sixty-eight against. The voting box for Meridian, covering the county seat Meridian and surrounding area, contained seventy-eight votes for and one against, and the "Wylies School House Box" contained twenty-eight votes, all for secession. The voters using the Clifton box, however, voted thirty for and twenty-nine against. In Norman Hill, the heart of the Norwegian colony, the voting box contained only four votes for secession and twenty-two against. Based on the election results, the county commissioners could be in no doubt about the attitudes of the Norwegian immigrants—they were in favor of remaining in the Union. What happened when the Civil Was broke out? According to the dominant view among Norwegian-American historians, the Norwegian immigrants could be expected to be unwilling to serve in the Confederate Army.

CHAPTER 8

Right or Wrong—My Texas! Norwegians in the Confederate Army during the Civil War

◊ ◊ ◊

THE CONFEDERATE ARMY

In 1981 Ralph A. Wooster and Robert Wooster wrote an article in the *Southwestern Historical Quarterly* in which they analyzed Texans in the Confederate Army. After the battle of Fort Sumter in April 1861, "thousands of young men volunteered for military service." Young Texans were "Rarin' for a fight" and could hardly wait to enlist.[1] "By late spring companies were being formed in almost every community. Often these units were organized by local political leaders or by professional men with little military knowledge or background. The lack of weapons, ammunition, and other equipment often bewildered even those with previous military experience," the Woosters wrote.[2] Many of the recruits received military instruction in their local communities or in some of the new military camps—Camp Berlin near Brenham in Washington County, Camp Honey Springs near Dallas, Camp Roberts in Smith County, or Camp Bosque, 7 miles from Waco in McLennan County, to mention just a few. The Texans who marched off to war in 1861 were dressed in a variety of uniforms. Even though the regulations "called for sabers and carbines, most cavalry units were equipped with shotguns, rifles, Bowie knives, and Colt revolvers."[3] At the end of 1861 approximately 25,000 Texans had enrolled in the Confederate Army. Most of the privates were in their early 20s, while noncommissioned and commissioned officers were somewhat older. The majority of the soldiers were farmers or sons of farmers, but a number of other occupations were represented. Many physicians enlisted, but they were dwarfed by the large number of Texas lawyers, who were strongly overrepresented among the officers.[4]

It has been estimated that around 80 percent of the Confederate soldiers from Texas were of English, Welsh, and Scottish ancestry. The majority were born in the United States, but young men from other nationalities also joined the Confederate Army. Between 4 and 5 percent were Irish and German, and less than 2 percent were French. A sprinkle of Dutch, Jews, and Mexicans were represented. Many German-Texans opposed secession, but a number also served in the Confederate Army.

At the beginning of the Civil War Anglo-Texans feared that the Germans would support the Unionist cause.[5] Their suspicions seemed to be confirmed in 1862, when sixty-five Union sympathizers among the Germans in west Texas left Kerr County, bound for Mexico. Martial law was declared, and Lieutenant C. D. McRae was sent to control the situation. He pursued the Union sympathizers commanded by Fritz Tegener, overtook them, and wiped them out in the battle of Nueces on August 10, 1862.[6]

Colorado County, located between Houston, Austin, and San Antonio, had almost 8,000 inhabitants in 1860, and about 45 percent of them were slaves. Of the remaining 4,000 almost a thousand were foreign immigrants who lived in ethnic communities, places like Frelsburg and New Mainz in the northern and northeastern part of the county. Many of the foreigners did not speak English, only German or Czech. The slave owners lived in the western and southern part of the county. Nearly half of the American-born males in Colorado County enlisted, but only one of every six Germans.[7] Nevertheless, a German infantry company was formed in Colorado County under Captain Emil Mathias Jordt.

In Fayette County, Captain Edmund Creuzbaur organized a company of artillery. Comal County organized three companies of Germans for the Confederate Army.[8] Many of the Germans fought rather reluctantly. In early 1863, after the introduction of conscription laws, Germans in Colorado, Fayette, and Austin counties voiced strong opposition. The counties were placed under martial law, and their leaders were arrested and put in jail. A number of men fled into the thickets, emerging only after the hostilities had ended.[9]

Men from the small Polish colony in Panna Maria also enlisted in the Confederate Army, Wooster and Wooster commented.[10] In a chapter on the Civil War in *The First Polish Americans. Silesian Settlements in Texas*, T. Lindsay Baker maintained that peasants who had immigrated from the Polish region Silesia, or Schlesien, appeared to never have "supported the Confederate cause enthusiastically" in Texas.[11] None of them were slaveholders. A major reason for emigration for many had been to avoid conscription in the Prussian Army. Why should they fight in someone else's war in their new land? Many Silesians "tried to avoid all Confederate officials, conscription officers in particular."[12] In the end, a number of Silesians living in Karnes and Bandera counties enlisted in the Confederate Army, mainly in the Sixth Texas Infantry and the Twenty-fourth Texas Cavalry. They were sent to Arkansas to defend the Arkansas Post. After Union forces took over this fort in early 1863, most of the Poles in the two regiments ended up as prisoners of war. Some Poles in Bandera County signed up to fight in the Texas State Troops.[13]

The Civil War also caused problems for Wendish immigrants (originally from the Lusatia region of Germany). Many Wends were unwilling to take up arms for the Confederacy, but draft evasion also proved dangerous. Some tried to escape being drafted by dressing in women's clothes when they were plowing. "Nevertheless, some young men were drafted into the Confederate forces and died fighting for the cause for which they had no sympathy. Other Wends slipped north to join the Union troops."[14]

Some of the most prominent Swedish immigrants in Texas openly defended slavery before the Civil War. In a letter from Svante Palm to the Swedish newspaper *Hemlandet* in 1855 he argued that since Swedish immigrants lived in a slave state, many had personal experience of how slaves were fed, treated, housed, and cared for. They were treated better than the "working classes" in Sweden. Some of the Swedes in Texas were slave owners. According to Palm, most Swedish immigrants would buy slaves as soon as they could afford it. His views caused angry replies among Swedes in the Midwest, but also among fellow Swedes in Texas.[15]

At the outbreak of the Civil War not more than 150 Swedish immigrants lived in Texas. Most of them grew cotton, and some owned slaves. The Swedes had divided loyalties, according to Larry E. Scott, and "only a very few Texas Swedes are known to have actually enlisted in the army of the Confederacy."[16]

Did many Norwegian-Texans enlist in the Confederate Army? In the past, historians in Texas, the Midwest, and Norway have struggled with this question. Information about Norwegians in the Confederate Army has been collected from several sources, combined, and compared. The conclusion is that a surprising number of Norwegians in Texas served in the Confederate Army.

SOURCES ON NORWEGIANS IN THE CONFEDERATE ARMY

The important genealogical work done by Derwood Johnson over many years, published in *Norge i Texas*, constitutes the backbone of the empirical evidence. Another important source has been the county history, *Bosque County: Land and People*. This book contains hundreds of biographies of families living in Bosque County since the county was organized in 1854. The book's editorial committee asked descendants of the families to write family biographies, with all the authors' subjectivity included. The Civil War can still evoke strong emotions, and if any forefathers participated in the Civil War, it was usually mentioned. A number of Norwegian immigrants in East Texas who served in the Confederate Army later migrated to Bosque County, and many of the Norwegians who joined the Confederate Army in East Texas are mentioned in that book.

The most useful electronic database on Norwegians fighting in the Civil War has been the Civil War Database at Vesterheim, the National Norwegian-American Museum and Heritage Center in Decorah, Iowa. In many cases it seems obvious that the primary source for the information in the Civil War Database is based on Confederate Army archives. Another intriguing but very dependable source for information on Norwegians in the Confederate Army is the Internet page www.borgerkrigen. info.[17] Generally, the databases have more information on Norwegians who survived the war than on Norwegians who died during the war.

A number of letters written by Norwegians who served in the Confederate Army have survived and been published. Some of the young men from Bosque County who had established a close and warm relationship with the much older Norwegian pioneer Carl Questad wrote to him after they left for the war. David Lund, who was born in Bergen or the Bergen area, was listed as 36 years old in the 1860 census. He asked Questad in March 1863 if he could invest 100 dollars on his behalf, in "livestock or something else, whatever you may think will be most advantageous for me. Act as if it was your own money and I will be entirely satisfied."[18] He knew his request would mean extra work for Questad, "but above all else I know your goodness of heart and your *willingness* to help your friends and countrymen." In July 1864 the young Norwegian Elias R. Skimland, known as more fearless than most, wrote to Questad that during periods of deep loneliness, "which

I often experience, a desire grips me to see all of you again. But it may be my fate not to have this good fortune. I wish to send you my most heartfelt thanks for all the friendliness and goodness which you have shown me on occasions too numerous to mention here, but which I nevertheless keep in fond remembrance. I can truly say that I have not found any other people like you since I left my dear parents."[19]

Most of the letters to Questad were written in the later phases of the Civil War. Sadie Hoel, Questad's granddaughter, together with Derwood Johnson, played a crucial role in saving the Civil War letters. Large parts of Questad's private archive were given to the Norwegian-American Historical Association at St. Olaf's College, Northfield, Minnesota. C. A. Clausen and Derwood Johnson edited and translated the Civil War letters from Norwegian to English and published them under the title "Norwegian Soldiers in the Confederate Forces," in *Norwegian-American Studies*. Copies of the original letters can also be found at the Cleng Peerson Research Library at the Bosque Museum in Clifton, Texas.

The quality of the Norwegian prose in the letters is impressive, not only considering the little schooling these men had had in Norway before their emigration, but also compared with the quality of prose written by Norwegian university students today. None of the letter writers had any problems communicating their inner thoughts and feelings to someone they trusted totally.[20] In the following, all citations are from the translated letters as published by Clausen and Johnson.

The letter writers all wanted to do their duty in the Confederate Army, but they did not much believe in the "Southern cause." A recurring theme in the letters is the young men asking Questad to do something for them concerning money matters. In one of the first letters to Questad, written from Camp Salmon by J. S. Jenson on December 13, 1862, he wrote that many of the Norwegian boys wanted to send money home to Bosque to pay off their debts. Would Questad be willing to make sure their money got into the hands of the right creditors? Osmund Nystøl wanted to pay $20 to Jens Ringness and have this sum subtracted from the total he owed. Nystøl sent $5 with W. Anderson and asked if Questad could be good enough to pass them on to "old man Knudson." Elias R. Skimland sent $25 with W. Anderson to be paid to Ole Arneberg, and for his brother Knud Skimland sent $20 to be paid to Ole Arneberg.[21]

Two very important Civil War letters written by Johan R. Reiersen to his oldest son Oscar in Charlottesville, Virginia, are found in the

Reierson papers at the Briscoe Center, University of Texas, Austin.[22] The letters have been published on www.borgerkrigen.info. Reiersen, the founder of the first Norwegian colony in Brownsboro, Henderson County, wholeheartedly supported the Confederate cause. If the South should lose, he wrote, he would take his family with him to Mexico rather than accept the Union.

In 1964, Alwyn Barr's "An Essay on Civil War Historiography" listed nineteen pages of titles written about Texas.[23] "Every year thousands of Americans dress up in blue or gray uniforms and take up their replica Springfield muskets to re-enact Civil War battles," wrote James McPherson in the preface to *Battle Cry of Freedom: The Civil War Era*, published in 1988. "A half-dozen popular and professional history magazines continue to chronicle every conceivable aspect of the war. Hundreds of books about the conflict pour off the press every year, adding the more than 50,000 titles on the subject that make the Civil War by a large margin the most written about event in American history."[24] This chapter will not focus on any of the big questions. It is a humble attempt to focus on the life of Norwegian soldiers in the Confederate Army and how they experienced the war. According to Reid Mitchell, the first scholarly study of the life of Civil War soldiers was *The Life of Johnny Reb: The Common Soldier of the Confederacy*, by Bell Irvin Wiley, published in 1943.[25] Wiley's book has been an important inspiration to me and has helped me understand the life of Norwegians in the Confederate Army in context.

DID THE NORWEGIANS GET ADVICE FROM THEIR RELIGIOUS LEADERS?

Decades before the Civil War broke out, the dominant Protestant denominations in the South had pronounced that owning slaves was not a sin. The three largest religious groups in the United States in the census of 1860 were the Methodists (6,259,798 members), Baptists (4,044,220 members), and Presbyterians (2,565,949). The Methodists and Baptists had split on the slavery question between North and South in 1844. In 1860 there were 2,888,338 Methodists and 2,413,818 Baptists in slave states. The Presbyterians had split on dogmatic questions in 1837, and in slave states the Cumberland Presbyterian Church was dominant, with 214,758 members.[26]

The slavery question divided Lutherans in the North and South in the same way that it divided Presbyterians, Methodists, and Baptists. During the first half of the eighteenth century some of the German Lutherans who migrated down the Shenandoah Valley into Virginia and the Carolinas became slave owners.[27] When the Civil War broke out in 1861, the Virginia Synod expressed the view of most Lutheran synods in the south. It was "fully persuaded" that the Confederate cause "is just and righteous."[28]

According to Abdel Wentz, most Norwegian and Swedish immigrants in the Midwest viewed slavery as unrighteous, but their clergy extemporized.[29] Since slavery was the law of the land and part of the Constitution, Lutherans followed the classic pattern formulated by Martin Luther during the Knights' Revolt and the Peasants' War in Germany in the 1520s. Luther had preached absolute allegiance to the temporal governing authorities.

The Norwegians in antebellum Texas had grown up with the teachings of the Norwegian Lutheran State Church. Did Norwegian Lutheran pastors preach differently on the question of slavery than Presbyterian, Methodist, and Baptist pastors? Did Norwegian immigrants in Texas or the Midwest get a clear message from their Lutheran pastors that slavery was a sin and had to be abolished? They did not.

The majority of Norwegian immigrants in the United States lived in Illinois, Wisconsin, and Iowa. The Norwegian Lutheran State Church or the Norwegian Synod was not organized in the United States until 1853. In the 1840s, many Norwegian immigrants in the United States were influenced by strong revivalist movements and American protestant sects. A number of Norwegian immigrants joined the Mormons in the 1840s and 1850s. At the outbreak of the Civil War, two different Norwegian Lutheran congregations dominated—the pietistic followers of Elling Eielsen and the Eielsen Synod, and Den Norske Synode, the orthodox Norwegian Lutheran Church in the United States. The Eielsen Synod had taken a clear stand on the question of slavery in its constitution as early as 1846; it "condemned slavery in scathing terms," wrote Blegen.[30]

A heated theological debate about the slavery question raged among Norwegian American theologians in the orthodox synod before and during the Civil War, and continued even more heatedly into the 1870s, years after the Civil War was over.[31] The synod had established a close working relationship with the German Missouri Synod in St. Louis, which dogmatically practiced Luther's teaching about worldly government and God's will. The exceptionally strong leader of the Missouri Synod was Carl Ferdinand Wilhelm Walther. He held very conservative views on a number of issues. The Missouri Synod was opposed to humanism, religious syncretism, and freemasonry. Furthermore, Walther and other pastors argued that opposition to slavery could not be based on Biblical passages. The Bible neither approved of nor condemned slavery. According to Walther, the New Testament did not support any form of earthly government, be it democracy, republic, dictatorship, monarchy, or the American Constitution. He strongly denounced the concept of so-called God-given human rights as nothing but humanism seeking an earthly utopia in place of an eternal relationship with God. A favorite quote for Walther and many of the Norwegian Lutheran ministers was Paul's first letter to the Corinthians, chapter 7, verse 1: "Everyone should remain in the condition in which he was called. Were you a slave when you were called? Do not let that trouble you, but if a chance for liberty should come, take it." Lutherans should not advise, counsel, petition, or instruct the state in any way. Walther strictly adhered to the principle "Render unto Caesar the things that are Caesar's, and unto God the things that are God's." (Matthew 22:15–22, Mark 12:13–17, Luke 20:20–26). The Norwegian Synod strictly followed the teachings of Walther.[32]

EAST TEXAS NORWEGIANS MORE EAGER TO FIGHT FOR THE CONFEDERACY THAN BOSQUE COUNTY NORWEGIANS

Many German immigrants in Texas actively opposed slavery, secession, and enlistment in the Confederate Army, and some paid for their opposition with their lives. The Norwegians never showed the same kind of public resistance. They had no strong leaders, religious or otherwise, who repeatedly upheld the sin of slave ownership.

The Norwegian farmers in Texas before the Civil War lived by the moral values they had been taught when growing up; these values did not include a positive view of slavery. The Norwegians treasured their new political and religious freedom, the chance to begin from scratch and to build up their lives without class barriers. Why should they fight for the continued life of an institution that they, with their cultural background, found morally wrong? On the other hand, they had chosen to settle in Texas, and most of them planned for a future for themselves and their children in Texas.

As long as enlistment in the Confederate Army was voluntary, many Norwegians abstained from joining. When it became apparent that the Confederates would not "whip" the Union forces within a year, the Norwegians faced strong community pressure to enlist and show their patriotism. Right or wrong—the Norwegians went to war for Texas. They joined the Confederate Army, albeit with no great enthusiasm, and hoped God, fate, or luck would help them through the war. They were not "rarin' for a fight" to the same degree as their American neighbors.

A few Norwegians in East Texas, however, strongly supported the "Southern cause." The East Texas Norwegians lived in counties overwhelmingly in favor of secession. About 1,000 men from Henderson County enlisted in the Confederate Army. In one Henderson detachment, only 13 of 150 men returned alive. There were training camps at Caldwell's farm, 3 miles northeast, and at Fincastle, 19 miles southeast of the county seat Athens. On May 8, 1861, Captain John R. Briscoe organized the first military unit in Kaufman County. From 1861 to the end of 1864 at least eighteen companies were recruited in Kaufman County, according to Jack Stolz.[33] Many in Van Zandt County also eagerly volunteered for service in the Confederate Army.

The conscription law of April 1862 can be used as a watershed in the discussion of Norwegians in the Confederate Army. It can be differentiated between Norwegians who enlisted voluntarily before the conscription law was put into practice and those who enlisted after the law was enacted. If Norwegian-American historians in the Midwest are right, hardly any Norwegians could be expected to join the Confederate Army before the conscription law. According to this view, it might be understandable that some Norwegians joined because of community pressure after April 1862. After having studied countless primary sources it can be concluded that a remarkable number of young Norwegians in East Texas enlisted voluntarily in 1861. The majority of Norwegians in Bosque County, however, enlisted in spring 1862 or later. There was a larger reluctance to fight for the institution of slavery among Norwegian immigrants in Bosque County than among those in East Texas.

NORWEGIANS IN THE THIRD TEXAS CAVALRY

"The men of the Third Texas Cavalry blundered into their first battle fifty-seven days after they enlisted in the Confederate army. As volunteers in Brigadier General Ben McCulloch's ten-thousand-man army, they rolled out of their blankets before daybreak beside a sparkling Ozark stream, called Wilson's Creek, on the morning of August 10, 1861," wrote Douglas Hale in his book on the *Third Texas Cavalry in the Civil War*.³⁴ The Third Texas Cavalry was organized by the 35-year-old plantation owner Elkanah Greer of Marshall, Harrison County. The Confederate War Department gave him a colonel's commission and ordered him to raise a thousand volunteers. "Most of the men who eagerly responded came from northeast Texas counties," and in early June, "the dusty roads to Dallas were crowded by enthusiastic volunteers." Six of ten soldiers came from the Lower South: "Alabama led in this regard with 211 natives. A third came from the Upper South. Tennessee, the birthplace of 155 recruits, was ahead in this category. Only forty-one men were natives of Northern states, and but twenty of the volunteers had been born abroad. Seven of these were native Norwegians from their settlements in Kaufmann and Henderson counties."³⁵

The Norwegian immigrants Hans Jenson, Oley (Ole) Jenson, Christian A. Reiersen, and Otto F. Reiersen enlisted as privates in Company G of the Third Texas Cavalry in Jefferson, Marion County, on June 3, 1861.

Several of the first immigrants between 1846 and 1850 came from Holt in Southern Norway, the place the members of the Reiersen family had left, where the large ironworks Næs Jernverk was located, and where many of them had worked. The family of Jens Jenson (born 1807, died Texas 1847) and his wife Ingeborg came from Holt to Texas in 1847. Four of their sons served in the Confederate Army. Their oldest son, Jens Jenson (James S. Jenson), born on March 2, 1835, served as a private in Company E, McCord's Frontier Regiment. His younger brother Hans Jenson, born Hans Jensen Møglebustad on November 13, 1836, enlisted as a private in Jefferson on June 3, 1861, for 12 months in the Third Texas Cavalry.³⁶ Jenson was listed as "absent sick" from July to October 1862. He was wounded on March 5, 1863, at Spring Hill, Tennessee, and was registered as sick in camp between July and August 1863. Jenson was sick again from May to June 1864. His 5 years younger brother, Oley Jenson, born Ole Jensen Møglebustad on January 30, 1841 at Holt, enlisted together with him in Jefferson. He served with the Twenty-First Georgia Cavalry Battalion during March and April 1863. All of the above Jenson brothers survived the Civil War, but their younger brother Peder Jenson, born on March 9, 1843, at Holt was killed in action in 1864.³⁷

Christian August Reiersen was born on September 17, 1842 in Holt, Aust-Agder. He was the son of Johan R. Reiersen in Kaufmann County. He traveled from his home in Prairieville to Palestine, Anderson County, to enlist as a private in A. T. Rainey's company on May 11, 1861, and joined the Third Texas Cavalry as a private in Company G. on June 3, 1861, in Jefferson. He was detached to ride with the 21st Georgia Cavalry Battalion during March and April 1863.[38]

His cousin Otto Theodor Reiersen enlisted in the Third Texas Cavalry, Company G, on the same day. Otto was born on December 7, 1842, in Strengereid, Holt, Norway. Before his enlistment he had been a private in Captain J. Wharton's "Texas Wide Awakes" in Kaufman County, which was part of the 13th Brigade of the Texas State Troops. Otto Reiersen was wounded at Spring Hill, Tennessee, on March 5, 1863, and was absent because of his wound through June 1863. In September and October 1863, he performed extra duty taking care of Confederate mules. During the battle at Lost Mountain, Georgia, on June 14, 1864, he was taken prisoner of war. He was interned at Camp Morton, Indiana, until he was exchanged in March 1865.[39]

The Norwegians had a different social background compared with most of the men in the Third Texas Cavalry. Each of the ten companies that met in Dallas in June 1861 had been raised as local defense forces in various East Texas counties. The companies consisted of 95 to 115 officers and privates. All commanders were wealthy men in their communities. According to Hale, mean wealth among officers of the original command of the Third Texas Cavalry Regiment was $18,464, and the mean number of slaves among the officers was twelve. "In a state where only one in every four white households owned slaves, all these original captains were slaveholders."[40] The 107 men in Company G, who enlisted in Jefferson, Marion County, elected as their captain Hinche P. Mabry, a local lawyer and legislator. Like other volunteers, the men had to supply most of their own arms, uniforms, mounts, and equipment, often at considerable personal sacrifice. Marion County pledged $10,000 to the cause, while the less affluent Van Zandt County promised to pay $200 per month. Most of the company captains had absolutely no experience as leaders of men at war.

All the men in Greer's regiment came from East Texas counties. In the 1850s a number of wealthy families from the Lower South had settled along the rich bottomlands of the Sabine, Neches, and Red rivers. The population of East Texas doubled between 1850 and 1860, and cotton production grew sevenfold. Plantation owners with more than twenty slaves, lawyers, physicians, and merchants constituted the leading social class. Even though a minority, the planter class was disproportionately represented in the political life of the region. The officers in Greer's regiment came from the wealthier classes, and they had a personal stake in the future of slavery.[41] More than half of the men in the regiment came from slave-owning households. In the view of Hale, the Norwegian "Gragard brothers, Nicholas and John, and the Reiersen brothers, Christian and Otto, were heirs of prominent leaders in the movements for Norwegian immigration to Texas."[42] Actually, the Grøgaard brothers had been orphans since 1848, and as already mentioned, Otto Reiersen was a cousin of Christian August Reiersen, not his brother. Nevertheless, Hale is right in the sense that the Grøgaard brothers as well as the Reiersens were better assimilated to the East Texas slave culture than most other Norwegian immigrants.

THE GRØGAARD BROTHERS AND THE IMPORTANCE OF ASSIMILATION

Very few of the Norwegians at Four Mile Prairie could speak or understand the English language in 1860. Several Norwegian immigrants in East Texas, however, grew up in towns, went to American schools, and practiced a trade. They were much more familiar with the dominant East Texas social and political culture than their fellow Norwegians in Van Zandt and Kaufman counties. Several of the Norwegians who joined the Confederate Army wanted to prove to their Texas neighbors that they were Texans now. Norwegians who lived in towns and were part of town culture enlisted more eagerly than their fellow countrymen living in a Norwegian colony.

The Grøgaard brothers were more assimilated at the outbreak of the Civil War than most other Norwegians and thus clear candidates for enlisting in the Confederate Army at an early stage. That in fact is what they did. Twenty-five-year-old Thomas Fasting Grøgaard enlisted in the First Texas Infantry, Company D, as a private for one year on May 26, 1861, in Linden, Cass County. He was discharged for disability on September 9, 1861, at Richmond, Virginia, but enlisted again on May 10, 1862, "for the war," at Marshall, Harrison County, and was assigned to the Twenty-eighth Texas Cavalry, Company F. He was captured at Gaines Landing, Arkansas, on July 20, 1862, and exchanged near Vicksburg, Mississippi, in early November 1862. On May 25, 1863, he was assigned to special duty in the Quartermaster Department but soon became sick and remained on the sick list until August 1863.[43]

His twin brother, Jake Grøgaard, enlisted in the Third Texas Cavalry, Company E, as a private for one year at Shelbyville, Shelby County, on May 23, 1861. He was promoted to First Sergeant on August 2, 1861, but became a private again when the regiment reorganized in May 1862. He was absent on special duty with the Regiment's Commissary department from May until late 1862. On June 1, 1863, he was assigned to the Regimental Staff as Acting Commissary and Subsistence Officer and promoted to captain on the same day. He was captured at Rolling Fork, Mississippi, on June 23, 1864, and exchanged near Vicksburg, Mississippi, on July 29, 1864.[44]

Their younger brother Nicholas C. Grøgaard enlisted as a private eleven days after Jake in the same company in the Third Texas Cavalry, at Shelbyville, Shelby County, on June 3, 1861. According to Hale, fifteen men in Company C. of the Third Texas Cavalry had studied at Larissa College, "an oasis of genteel cultivation amid the piney hills of Cherokee county."[45] Nicholas Grøgaard was one of them. During May and June 1863, he was on extra duty to procure horses. He was promoted to sergeant and transferred to Company F after September 1863. He was wounded in the battle of Lost Mountain, Georgia, on June 14, 1864. Sometime during fall 1864 he rejoined his regiment at Canton, Mississippi, was assigned as Commissary Sergeant, and was later promoted to First Sergeant. He was paroled on May 13, 1865, at Jackson, Mississippi.[46]

NORWEGIANS IN THE SEVENTH TEXAS INFANTRY
AND THE SIXTH TEXAS CAVALRY

Two Norwegians who had emigrated from Hedmark County enlisted in the Seventh Texas Infantry at Marshall in 1861. Helge J. Grann enlisted as a private in Company G on July 25, 1861, and Christian O. Strand as a private in Company A on October 1, 1861. Helge J. Grann was born Helge Jacobsen Gran on April 5, 1830, in Vang, Hedmark, and emigrated to Texas in 1851. Christian Olsen Strand was born on June

30, 1831, in Elverum, Hedmark County, and belonged to the batch of emigrants who came to Texas in 1853.

Both of them fought with the Seventh Texas Infantry at Fort Donelson, Tennessee. The fort, built in 1861, was located 100 feet above the Cumberland River, and artillery batteries could easily fire against attacking gunboats. Nevertheless, Union forces captured Fort Donelson between February 12 and 16, 1862. The Confederate forces surrendered unconditionally to General Ulysses S. Grant on the morning of February 16. Just some days earlier, on February 6, Union forces had captured Fort Henry, a key point in the defense line of Tennessee.

The captures of Fort Henry and Fort Donelson were among the first significant Union victories in the war and opened up the Tennessee River for Union troops. Nashville, the largest city in the state, had to be evacuated before the end of February 1862. Casualties at Fort Donelson were heavy. Union losses were 2,691 (507 killed, 1,976 wounded, 208 captured/missing), and Confederate losses 13,846 (327 killed, 1,127 wounded, 12,392 captured/missing). More than 7,000 Confederate prisoners of war were transported to prisoner of war camps such as Camp Douglas in Chicago and Camp Morton in Indianapolis.[47]

Both Grann and Strand were taken prisoner and ended up in Camp Douglas. Seven months later they were exchanged near Vicksburg, Mississippi, on September 20, 1862. Grann was discharged because of the illness he had acquired at Camp Douglas. He later joined the Second Texas Cavalry, Company F. In January and February 1863 Grann was assigned extra duty as a butcher. He ended up in a hospital at Yazoo City, Mississippi, in June and July 1863. Strand was taken prisoner again on July 13, 1863, and transported to Cairo, Illinois. A month later, on August 17, 1863, he enlisted in the Union Army in the Fifty-eighth Illinois Infantry. He served in the Union Army until he mustered out on April 1, 1866, at Montgomery, Alabama, and found his way home to Bosque County, Texas.[48]

Grann and Strand spent considerable time in prisoner of war camps but survived the war. Several Norwegians who enlisted lost their lives in the war. At the time of the US Census for Kaufman County in 1860, Otto Anderson, born Otto Francke Torgersen Noreødegaard in February 1839 in Moss, Norway, lived in the household of Johan R. Reiersen. He had emigrated with his parents to East Texas in 1848. In spring 1861 he enlisted as a private in the company of captain H. F. Bridges in Henderson County, and on September 12, 1861, he joined the Sixth Texas Cavalry, Company I, in Dallas. He was absent sick at Fort Gibson in the Cherokee Nation (Indian Territory) from December 22, 1861, to February 1862. He was sick again from May to August 1862. On December 21, 1862, Anderson was killed in action at Davis Mill, Mississippi.

His comrade in arms Ole W. Olson, born Ole Olsen Ramse on January 4, 1839, in Tovdal, Aust-Agder, was only 7 years old when he came to Henderson County in 1846 with his parents in the first large group of Norwegians. He enlisted with Otto Anderson as a private in the Sixth Texas Cavalry, Company I, in Dallas on September 12, 1861. He died before the year was out, on December 10, 1861, in Fayetteville, Arkansas.[49]

STRONG SUPPORTERS OF THE CONFEDERATE CAUSE: JOHAN R. REIERSEN AND HIS SONS

The most outspoken and strong supporter of the "Southern cause" among Norwegians in East Texas was Johan R. Reiersen. His views are well documented in two letters he wrote to his oldest son, Oscar, born in Copenhagen on October 29, 1836. Oscar was studying law at the University of Virginia at Charlottesville at the outbreak of the war. He enlisted in 1861 in the Thirty-fifth Virginia Cavalry, Company F, but was discharged before March 17, 1863.[50]

Oscar's younger brother Johann Henrich Reiersen, born on December 17, 1840, in Kristiansand, Norway, lived in Prairieville, Kaufman County. He enlisted for one year on March 10, 1862, as a private in Company C of the Twentieth Texas Cavalry. He had been a private in Captain J. R. Johnson's "Prairieville Company," Kaufman County, before his enlistment. He was promoted to Sergeant Major and assigned to the Regimental Staff sometime before the Battle of Honey Springs in Indian Territory on July 17, 1863. In this battle he was singled out for praise by his colonel and cited for bravery by Brigadier General Douglas Cooper. He was appointed Second Lieutenant on July 22, 1864 at Fort Washita, Chickasaw Nation, Indian Territory.[51] The third Reiersen brother, Christian August, born on September 17, 1842 in Holt, Aust-Agder, joined the Third Texas Cavalry and has already been mentioned.

In a letter from Johan R. Reiersen to his son Oscar in Virginia, dated Prairieville, December 22, 1862, written in English, he wrote that the family had gathered plenty of supplies for Christmas and the coming winter—flour, cornmeal, corn, sweet potatoes and salt, and "when I shall have killed 8 hogs that I am fattening, the want shall not knock at our doors for the meat season, unless the Yankees should come in and plunder us." They had picked up news about a battle not far from Fort Smith, Arkansas, "where our troops after a three days' fight at last succeeded in whipping the Federals, and driving them 15 miles back. I have seen no official account of the battle, only hearsay from some letters arrived here from men in Bass' Regiment." Oscar's brother Christian in the Third Texas Cavalry had "got unharmed out of both the last two bloody battles in Iuka and Corinth - and he writes me he is getting along finely since he has got his horse back and the winter clothing we sent him."[52]

Reiersen hoped Oscar was safe and would soon be "out of the turmoils of the war, and engaged in some safer business, and that at least one of my grown boys might be saved for me and for a useful and honorable after life, when the war should be ended, if ever it is to end. The expectation of the gigantic struggle that is near at hand on all sides, and the calling out of still more men" had created gloom and despondency. Nevertheless, Reiersen still had a strong hope "that we will be able to withstand the shock and drive the enemy back." Reiersen hoped the European powers would intervene in favor of the Confederacy.

The situation on the home front was "tolerably hard, and prices have risen to an unknown height. But on account of having a large supply of wheat from last year, it opened with sale of wheat at one dollar a bushel and from 4 to 5 $ pr cwb. [?] of flour, but having entirely failed in the eastern part of the state and in the western from the Brazos, wagons came in in such a number that prices rose directly up to 7 to 10 and from 10 to 12 and at last 15$ pr cw. of flour, and cannot now be had for money. Corn right here on Cedar Creek could at first be bought for 60 cents to 1 $ a bushel." So many wagons had arrived to buy from the North and West that the price of corn had increased to $1.25. The price increased further after "a fearful inundation of negro droves from Louisiana, Arkansas and Mississippi, - sometimes thousands in a drove with mules and horses." This brought the price up to $1.50 and $2.00 a bushel. The Reiersen family still had an abundance of hogs. Since they had no winter feed for them, they had been sold and driven off to Red River, "where corn seems to be inexhaustible. Thousands have in this way been sold in this neighborhood, and money distributed all around."

Union units had driven off cattle by the thousands. Salt was becoming a scarce commodity, and the price was rising to $25 a sack. New salt furnaces were being built, and new wells dug— "the whole saline prairie both at Sabine and Nueces looks like an enormous large city, composed of small loghouses, furnaces and wells, in one confused mass. Every boiler from old sawmills round about are converted into evaporating kettles, and the Jordans Saline has now a population of 4 to 5000 men - very few women."[53]

Food and clothing were bartered for salt: sweet potatoes at the exchange rate of $2 to $3 per bushel, butter at 40 cents to $1 per pound, corn at $3 per bushel, and flour at 20 cents per pound. "An old coat or a pair of old breeches easily commands a sack of salt, as also shoes or boots. Leather is almost impossible to be had since every tanyard had government contracts to deliver a certain quantity of shoes, and they work up every hide before it is more than half tanned."

Ten months later, in a letter to Oscar dated September 20, 1863, the "Southern cause" was Johan Reiersen's main theme. The news that Vicksburg had been taken by Union forces had led to a universal depression of spirit, and "a universal depreciation of our currency was the result, so that wheat went up from 2 to 6$ per bushel, and corn from 1 to 3$, and hard to be bought at that." The call for 10,000 more militiamen and the subsequent call for all men up to 45 years as conscripts, Reiersen reflected, would "pretty near clear out the country, and then besides old men are organizing companies of Minute Men to go out in case of emergency - here in Kaufman county we have a company of 60 men, to which I belong." Reiersen had truly enjoyed the Confederate exploit down at Sabine Pass; "that we took two ironclads, the Clifton of 10 and the Sachem of 5 guns, besides about 300 prisoners and killed and wounded 50 more, without the loss of a man; but it is said, that this was only the feeler for a larger expedition of 10,000 men, that will try to invade us on some point of the coast; in fact all the troops, that before were ordered to Bonham, are now counter ordered down toward Galveston, together with the greatest part of the Militia; and although such an invasion would be a calamity, I think it would have a beneficial influence in arousing the common spirit of the people till more healthy energy, and in case of our success in driving them back, effectually stop the mouth of the croakers, that are now so noisy as frogs in a pool after a summer shower."[54]

He still waited for a report on the fall of Charleston, "though I have a faint hope, that the South Carolinians will be spunky enough to fight them even if Charleston should be turned into an ash heap; but if that place is given up, I am afraid that Richmond will be untenable, and that our government will have to emigrate." He was very confused about what would happen in the end, but he still hoped that "France and Mexico, united with Spain and eventually England, should step in and form an alliance with us." If that did not happen, he was afraid that "the spirit of the people will give way."

Reiersen was very proud that his son Johann had distinguished himself at the battle at Honey Springs together with a flag bearer. When the others retreated and ran, Johann had "coolly held his ground and continued loading and firing different rounds till he was almost surrounded, then kept retreating firing all the way." The chances seemed slim that he would see "my brave boys at home safe and sound; they have indeed had a wonderful good luck, but the very luck is in the natural course of circumstance a chance for a turn in their disfavor, and I always feel under great anxiety till I recognize their next handwriting."[55]

He would never reconcile himself to "the idea of subjugation, after all we have fought so valiantly and lost such streams of the best blood in the battles." If the Confederates lost, "which God forbid, I am determined if I live to leave this disgraced country, and between the Mexican mountains and valleys to seek a hiding place for my old age." He would emigrate to Mexico, "far from the bristle of accursed Yankee race, - and with Ouline and Charley and what remains of my other boys make us a happy home under a genial climate in one of the most favored countries of the globe. I wish you had studied Spanish instead of German for such an event." He felt pessimistic, but hoped the Confederates "will conquer an honorable peace with our independence."

Johan R. Reiersen did not live to experience the end of the Civil War. He died in Prairieville on September 6, 1864, and was buried in the Four Mile Lutheran Cemetery over the county line to the east in Van Zandt County. It will never be known if he changed his thinking on the Confederate cause or if he would have joined the hardliners among the Confederates who decided to emigrate to Mexico and continue the fight for slavery from there.

CHAPTER 9

Norwegians in the Confederate Army after the Conscription Law

◊ ◊ ◊

"WITH THE PASSAGE OF TIME, recruitment of soldiers became more difficult as the early enthusiasm for military service waned," observed Wooster and Wooster.[1] By spring 1862 the Confederate Army had suffered major setbacks. The capture of Fort Henry and Fort Donelson was followed by the occupation of New Orleans. "The Confederacy was in dire need of more troops, yet volunteerism waned," observed Francelle Pruitt.[2] Almost half of the soldiers in the Confederate Army had volunteered for 12 months, as the war was not expected to last that long. The Confederacy could ill afford to lose veteran soldiers, and in April 1862 the Confederate Congress passed "An Act to provide further for the public defense." This act allowed the government to conscript all able-bodied men between ages 18 and 35 for three years. Soldiers currently in the army were required to remain for three years or the duration of the war.

The mere rumor that a conscription law would be introduced helped to pressure the population to enlist voluntarily. In her discussion of how Texas reacted to the first Confederate conscription law, Francelle Pruitt concluded that most Texans "were willing to make the sacrifice they believed was necessary by submitting to the conscription law."[3] In November 1863, Governor Lubbock reported to the Texas legislature that about 90,000 Texans had joined the Confederate Army.

There had been no great enthusiasm about joining the Confederate Army in Bosque County in 1861, but in spring 1862 the community pressure increased considerably. The Election Book for Bosque County recorded, on June 17, 1861, the need to fill vacancies for three justices of the peace and two constables who had refused to take the oath of support for the Confederacy.[4] Two months later, on August 19, 1861, the Bosque County commissioners appointed four men to "inspect and examine all the firearms" in the respective precincts fit for service, and determine which of them could be repaired at county expense.[5] A week later, on August 26, the County Court decided to appropriate 100 dollars for the purchase of clothing, camp equipment, and so forth for soldiers who were "now about to volunteer and engage in the service of the Confederate Army."[6] In February 1862 the commissioners decided that volunteers deemed poor should be given a $15 voucher "after having volunteered and been sworn in."[7] To pay for the increased expenses, the county taxes would be raised.

By spring 1862 the most eager Confederate patriots in Bosque County had lost patience with the lack of patriotism shown by their fellow citizens, not least the Norwegians. On March 17, 1862, citizens of Bosque County met at the county seat in Meridian to organize a committee on "safety." William R. Sedberry took the chair with W. T. Kemp as secretary. The Reverend John Abney stated the agenda of the meeting, and with Sam Barnes, J. K. Helton, Jerry Odle, Jack Smith, and James Lane, he was appointed to the committee on resolutions.

The committee was convinced that there were men in Bosque County "hostile to our institutions, consequently enemies to our country and its cause; besides characters who have no love of country, and who know not the meaning of patriotism."[8] The resolution concluded with the following broadside: "should any man now living in our midst, who has been or may hereafter be guilty of using language derogatory of the Southern Confederacy or its cause, or by any act giving evidence that he is unfriendly to the Confederate Government, that the committee test him by his being required to show his fidelity to our government and cause, by enlisting in the army of the Southern Confederacy; and if he refuses, give him a free pass to leave for the Lincoln Government; and if he refuses to do either, he shall be regarded and treated as an alien enemy and as a spy for which he shall be executed by the committee."[9]

NORWEGIANS IN THE ELEVENTH TEXAS INFANTRY

The community pressure on Norwegians to enlist in spring 1862 increased not only in Bosque County, but even more so in East Texas. Not every Norwegian immigrant who served in the Confederate Army is listed in this chapter. Nevertheless, the individual stories contribute to the understanding of the diversity of individual choices, the importance of the varying different contextual settings, and how the assimilation of Norwegians into Texas subcultures came to play a part in their choices.

The research has unearthed several "forgotten" Norwegians in the Confederate Army, especially in the Norwegian communities in East Texas. At the time of the Norwegians' immigration to Texas, between 1846 and 1853, many of them were children. More sons of Norwegian families in East Texas lost their lives in the war than Norwegians in Bosque County. The families in East Texas, however, have not been as diligent in promoting their participation in the Civil War as the families in Bosque County.

On February 8, 1862, a group of Norwegians in East Texas joined the Eleventh Texas Infantry in Canton. The regiment was organized by the leading Texas politician Oran Milo Roberts, commissioner of the Texas Secession Convention in 1861 and chief justice of the Texas Supreme Court. The regiment was mustered near Houston in late 1861 and early 1862 and consisted of ten companies from East Texas counties such as Nacogdoches, Rusk, Cherokee, Greg, Franklin, Harrison, Titus, Panola, Shelby, San Augustine, Kaufman, Van Zandt, and Hopkins. In May 1862 the Eleventh Texas was stationed at Camp Lubbock outside Houston. Sickness was a problem; 408 out of 1,338 soldiers were sick. The regiment marched to Camp Clough near Tyler and arrived there on June 20. The winter of 1862–63 was spent near Little Rock, Arkansas. From April 9 to May 14, 1863, the regiment was involved in operations in the Teche country of Western Louisiana. In spring 1864 the Eleventh Texas Infantry participated in the Red River Campaign in Louisiana. It fought in the battles of Mansfield and Pleasant Hill on April 8 and 9, 1864, when General Nathaniel Banks and his Union forces were turned back and prevented from capturing Shreveport. In summer 1864 the regiment was ordered to Shreveport, Louisiana, and later to Marshall and Hempstead, Texas, and ended up guarding prisoners near Tyler in early 1865.[10]

Among the Norwegians who enlisted in Canton in February, several had grown up in the municipality of Åmli, Aust-Agder County. Nels Grimland (age 22), John Nystøl (27), Thomas Targarson (20), and Knud Nelson (24), all enlisted for 12 months. Knud Anderson, born Knud Andersen Nordsveen on April 22, 1821, in Løten, Hedmark County, enlisted with them. He had earlier been a private in Capt. J. R. Johnson's Prairieville Company in Kaufman County. He was discharged at Camp Clough, Texas, on June 24, 1862, for being over 35 years of age.

The best known among the Norwegians in general was Wilhelm Wærenskjold, the husband of Elise Wærenskjold. He was born Johan Mathæus Christian Wilhelm Wærenskjold on August 24, 1823, in

Fredrikshald (Halden). He was elected second lieutenant on February 22, 1862, but was not reelected after the regiment was reorganized. He was almost 40, and was relieved from duty on June 25, 1862, by order of Brigadier General Henry E. McCulloch. Later he was a private in Company F, First Texas Cavalry regiment of Texas State Troops.

Three Grimland brothers also served in the Eleventh Infantry regiment. They had emigrated with their parents from Åmli, Aust-Agder County, to East Texas in 1850, and all of them worked as gunsmiths at Tyler and Mound Prairie in Andersen County. Gunsten Kittelsen Grimland, born on April 29, 1836, enlisted as a private in 1863 and was detailed as a gunsmith to the armory at Tyler. Ole Kittelsen Grimland, born on December 14, 1841, in Åmli, enlisted as a private in 1863, but before that he had served 18 months in the armory at Mound Prairie in Anderson County, north of Palestine. He was detailed as a gunsmith to the armory at Tyler throughout the war. His older brother Yern Grimland (born 1833) also worked as a gunsmith at the armory at Mound Prairie. The small gun- and blacksmith shop at Mound Prairie expanded considerably during the Civil War.

During a period in 1862 several Norwegians did service at the General Hospital in Tyler. A fourth Grimland brother, Nels Grimland, born Niels Kittelsen Grimeland on October 4, 1839, worked as a cook at the Tyler hospital from May 4 to June 4, 1862. Thomas Targarson worked as a nurse at the hospital from July 19 to November 1862.

One reason why so many Norwegians in the Eleventh Texas Infantry were connected to the Tyler hospital in fall 1862 might have been that a Norwegian doctor, born in Holt, Aust-Agder County, did service there. Gregory Torgason, born Gregers Torjesen Møglebustad on January 29, 1830, had studied medicine in New York and practiced his profession in Anderson, Grimes County, Texas, starting in 1856. When Torgason enlisted at Camp Lubbock on April 30, 1862, he lived in Kaufman County. Between May 17 and October 1862, he was detailed as hospital steward at the General Hospital in Tyler. On October 17 he was discharged from the regiment and commissioned as assistant surgeon. He served at the Tyler hospital until the end of the war.

Between June 25 and July 14, 1862, Nels Grimland, Knud Nelson (from Froland), John Nystøl, and Thomas Targarson were assigned to travel to Henderson and Van Zandt County to procure weapons. Grimland was promoted to Fourth Sergeant on February 12, 1863, but became a private again in February 1864. He was absent on furlough sometime between March and December 1864 and received extra duty as regimental teamster from December 19, 1864, to April 1865. Knud Nelson, born Knud Andersen Reierselmoen, got the measles in summer 1862, was hospitalized in Tyler, and died there on August 13, 1862. John Nystøl, born Johan Olsen Nystøl on November 5, 1835, in Seljaas, Åmli, was absent sick at Camp Clough, Texas, between August and October 1862, and again in Monroe, Louisiana from May to June 1863.

Two Albertson brothers, from Holt, Aust-Agder, enlisted in the Eleventh Texas Infantry in Tyler. Tom Albertson, born Terje Andersen Berge on June 15, 1832, enlisted as a private in Company G on February 10, 1862. He was transferred to Mechling's-Haldeman's Texas Battery on May 22, 1863. His younger brother Elling Albertson, born Elling Andersen Berge on December 24, 1841, in Holt, Aust-Agder County, enlisted in Company I on July 3, 1862. He was discharged on August 15, 1862, but later enlisted as a private in Company F, First Texas Cavalry on January 27, 1863, in Smith County. Elling Albertson died on August 24, 1865, in Tyler, Texas.

NORWEGIANS IN OTHER EAST TEXAS REGIMENTS

During spring 1862 some Norwegians enlisted in other regiments in East Texas. Two joined the Nineteenth Texas Infantry, one joined the Eighteenth Texas Cavalry, and four enlisted on the same day in the Twentieth Texas Cavalry. Tollef Hiljeson, born in Norway around 1830, enlisted as a private in the Nineteenth Texas Infantry, Company B, in Mount Vernon on February 15, 1862. He was absent sick at Little Rock, Arkansas, in November and December 1862, and was absent on detached duty at Little Rock in January and February 1863. Hiljeson was wounded at Milliken's Bend, Louisiana, on June 7, 1863. He was absent with leave from August 12 to September 24, 1863, and then absent without leave. In January 1864, however, Hiljeson was in the General Hospital in Shreveport, Louisiana, with acute bronchitis and died there.

Antone Shefter, born Anton Arnesen Skjefstad on February 26, 1840, in Elverum, Hedmark County, enlisted as a private in the Eighteenth Texas Cavalry, Company K, on March 1, 1862. He was captured at Arkansas Post, Arkansas, on January 11, 1863, and was imprisoned at

Camp Douglas, Illinois. Shefter was exchanged on April 10, 1863, at City Point, Virginia. He was sent to the Confederate General Hospital in Danville, Virginia, on May 14, 1863, and died in the hospital.

Eight young Norwegians enlisted in spring 1862. They had different motivations for signing up, and the consequences for their lives and health also differed. Before enlisting in Company H, First Texas Infantry, Olston O. Lee, born Ole Olsen Lia, on April 3, 1829, in Nes Verk, Holt, Norway, had been a private in Captain A. T. Rainey's Company of Texas Volunteers, also known as the Anderson County Invincibles. Lee enlisted at Mound Prairie on February 22, 1862. He was absent sick at Camp Nelson near Austin from November 23 to December 1862.

Antonio Anderson, born Anton Torgersen Noreødegaard on October 21, 1836, in Moss, Norway, also enlisted as a private in Company H of the First Texas Infantry, but at Kickapoo on March 20, 1862. In early July he was absent sick with diarrhea in General Hospital No. 18 in Richmond, Virginia. Antonio was killed in action at the Battle of Sharpsburg (Antietam), Maryland, on September 17, 1862.

On March 10, 1862, four Norwegians joined the ranks of the Twentieth Texas Cavalry. One of them was another Albertson brother from Holt, Aust-Agder County. Eilif Albertson, born Ellev Andersen Berge on December 18, 1845, enlisted in Company C. Andrew Anderson, born Andreas Svendsen Holte on July 23, 1837, in Høvåg, Aust-Agder County, had been second sergeant in Captain O. Vanpool's Trinity Guards in Kaufman County. Ole Olson, born Ole Olsen Torp on June 2, 1842, in Løten, Hedmark County, also enlisted in that company, as did Aleksander Brown, born Aleksander Brun Abrahamsen, in March 1839, in Stavanger, Rogaland County. Brown was discharged at Camp McCulloch on June 22, 1862, because of "inability to perform the duties of a soldier" (Surgeon's Certificate). All of them survived the war.

LOYAL CONFEDERATE SOLDIERS OR DESERTERS?

Simon A. Aanonsen (Simen Andreas Stiansen), born in Frolands Verk, Aust-Agder County, on September 13, 1835, enlisted in Van Zandt County on September 20, 1862, in the Second Texas Infantry, Company E. He was wounded during the siege of Vicksburg, Mississippi, and was taken prisoner of war at Vicksburg on July 4, 1863, but was paroled on July 7. He was absent on furlough from July 14 to November 8, 1863.

He did not return to service, however, and was registered as a deserter on March 26, 1864.

Much more is known about another Norwegian private in the same company. Knud Salve Knudson joined Company E in the Second Texas Infantry in Van Zandt County on August 29, 1862. He was born on December 3, 1830, on the farm Øvre Ramse in Tovdal, Aust-Agder County. In January 1863 he and a number of other soldiers were transferred from Tyler in Smith County to Camp Maury in the neighborhood of Yazoo City, Mississippi.[11] According to their colonel the unit consisted of second-hand recruits. Many should have been exempted because of health. But the largest problem, the colonel maintained, was their lack of faith in the Confederacy. About twenty of the most able-bodied soldiers had deserted.

In March 1863, Knudson was sick in the hospital. The regiment marched to Camp Timmons in the vicinity of Vicksburg in April, while the Union general Ulysses S. Grant was on his way south with his troops. Geographically and strategically, Vicksburg was very important for both sides, and the battle for control of Vicksburg would be hard. The Second Texas Infantry became integrated into the massive defense of Vicksburg in May. When the siege began, 30,000 Confederate soldiers were ready for Grant and the Union troops. President Jefferson Davis had ordered Lieutenant General John Clifford Pemberton, in command of the Confederate Army of Mississippi, Tennessee, and eastern Louisiana since October 1862, to hold Vicksburg at all costs. Pemberton conducted a stubborn defense despite the lack of adequate food, ammunition, and manpower. Grant laid siege on both land and water for forty days. On July 3 Pemberton sent a white flag through the ranks, and on July 4 he accepted Grant's terms for unconditional surrender. The capture of Vicksburg has been characterized as one of the most important strategic Union Army victories during the war.[12]

General Grant realized that it would not be possible to find transport and room in prisoner of war camps in the north for so many soldiers within a short time, so the Confederate soldiers were given an offer of amnesty. Every soldier had to sign a document before being allowed to travel back to their homes in the south, issued in two exemplars, one for the Union Army and one for the Confederate soldier. Knud Salve Knudson surrendered with his regiment on July 4, 1863. On July 7 he signed the following oath: "I will not take up arms again against the

United States, nor serve in any military police or constabulatory force in any fort, garrison or field work, held by the confederate States of America, against the United States of America, nor as guard of prisons, depots or stores, nor discharge any duties usually performed by Officers or soldiers, against the United States of America, until duly exchanged by the proper authorities."[13]

The soldiers in Knudson's regiment returned to Texas on their own or in groups. The Confederate Army granted the soldiers forty days furlough from July 17 with the hope that they would return to the army again. On July 29, Knudson was back in Tyler, where he was paid $23,08 in salary. In October he got a letter ordering him to sign up again in Houston, but he did not turn up, and on March 26, 1864, he was registered as deserter.

It is still unknown whether Knudson was a deserter or signed up with a different Confederate unit. Many years after the war, Knudson applied for a soldier's pension. In his application he wrote that he had enlisted in "Johnson's Battalion" after his return from Vicksburg. A soldier who served with him there witnessed that he had met Knudson at Fort Washita in the Indian Territory in 1864. They had served together around Houston before the war ended in May 1865. According to official documents of the Confederate Army, however, he is still registered as a deserter.[14]

The history of Knud Salve Knudson during the Civil War is meticulously reconstructed by Betty Knudson Edgar in a family history of the Knudson family. During the century after the war there was a strong tendency to romanticize the Civil War, she observed. If families could document that one of their relatives had served in the Confederate Army, they would tell their story with pride to all who would listen. Their accounts of the soldier's valor would probably have surprised the person in question, she commented. Later members of the Knudson family were no exception, recounting several stories about how Knud Salve was wounded but survived. According to one version, Knudson was saved at the battle of Vicksburg by his psalm book, which he always carried in his breast pocket: After the battle he found a bullet inside his hymnal. Another version recounted that he avoided being seriously wounded because of a thick handmade leather wallet that contained a small ring he had bought for his daughter Tomena. After the battle, he found a hole in the pocketbook and the ring was only a small piece of gold; it had saved his life. A third story was about how Knudson avoided being killed by a bullet to his head at the battle of Shiloh. The problem with this story was that he had not yet enlisted when the battle at Shiloh took place, on April 6 and 7, 1862. An even more far-fetched family story was that Knudson was sent to a Union war camp in the North after the battle of Vicksburg and survived on horsemeat, moldy bread, and large amounts of sugar.

Oral histories are often embellished. These stories were nicer for Knudson's offspring to tell than the fact that he was still registered as a deserter in the archives of the Confederate Army. If Knudson did enlist again, it probably happened after he and some other Norwegian immigrants were arrested as traitors against the Confederacy on Pentecost in 1864.

Many people in North Texas and the northern part of East Texas opposed leaving the Union. Accordingly, they were defined by the majority in the counties where they lived as opponents of the Confederate cause and as traitors. On Pentecost in 1864, a group of men from the county militia rode up to a house on a neighboring farm of the Wærenskjold farm at Four Mile Prairie. They arrested Sheriff Read, Judge McReynolds, the brother-in-law of the sheriff, Jo Holcomb, and Jeff Davis, as well as four Norwegian immigrants—Wilhelm Wærenskjold, Knud Salve Knudson, A. Knudson, and G. Sheafstad. The prisoners were brought to Canton, the county seat of Van Zandt County. It was a Sunday, and no authorities were present. After camping out that night, the militia proceeded with their prisoners to Tyler in Smith County. A sergeant registered that twelve prisoners had been brought in. Major James M. Taylor, the head of the recruitment office, defended the prisoners. All of them had their military papers in order, none of them were deserters, and Taylor had no authority outside Smith County. His message was not well received by the mob who had gathered outside the prison. Their rage was directed mainly against Read, McReynolds, Holcomb, and Davis. To protect them, Major Taylor locked them in a room on the first floor. The mob broke down the doors, took the four men, marched them out of town in the direction of Canton, and hanged them in the woods.[15]

The four Norwegians were transported from Tyler to Shreveport, Louisiana. Here it was soon documented that Knud Salve Knudson had signed the oath at Vicksburg and had been allowed to return home.

Wilhelm Wærenskjold was forced to enlist but was soon discharged. After this serious incident both understood that they were no longer safe at home in Van Zandt County. Knudson signed up as a member of Johnson's Battalion, and Wærenskjold escaped to Bosque County, where he stayed until spring 1865.[16] Did it really happen this way? Probably, but good primary sources that could prove it one way or the other were not available at the time of writing.

MANY NORWEGIANS FROM BOSQUE COUNTY ENLISTED IN REGIMENTS IN WACO

In terms of patriotism, Bosque County could not compete at all with McLennan County and Waco. Numerous wealthy planter families with many slaves lived in McLennan County. These leading families strongly believed in the culture and politics of the Lower South. During the war, seventeen military units were raised for the Confederacy in Waco. Out of a population of 2,500, McLennan County supplied the Confederate Army with nine hundred men and the state militia with another four hundred. Lawyers dominated politics in McLennan County, and lawyers organized many of the first military units. Supporters of the Southern cause in Bosque County chose to follow the leaders in McLennan County.

Norwegians enlisted in the Thirtieth Texas Cavalry (Gurley's Regiment), the Fifteenth Texas Infantry (Speight's Regiment), and the Thirty-first Texas Cavalry (Hawpe's Regiment). Soldiers were recruited to Speight's and Hawpe's regiments in Waco in March and April, and in July 1862 a small group of Norwegians enlisted in Gurley's Regiment. The letters a few Norwegians sent home to Bosque, combined with other sources, give a good picture of the life of the Norwegian immigrants in the Confederate Army.

JOINING SPEIGHT'S REGIMENT IN WACO

Waco lawyer Joseph Warren Speight organized the Fifteenth Texas Infantry Regiment and was commissioned a colonel.[17] He was born on May 31, 1825, in Green County, North Carolina. Speight had migrated to Waco in January 1854 because of declining health and took up farming and surveying. He owned thirty-two slaves in 1860. Speight was also a leading Baptist. In spring 1864 he resigned from his position as colonel of the regiment, with "ill health" given as the reason.

Waco lawyer and plantation owner Richard Coke raised Company K in the Fifteenth Texas Infantry and was elected captain of the company. Coke was born on March 18, 1829, close to Williamsburg, Virginia, and graduated from the College of William and Mary in 1848 with a law degree. He moved to Waco in 1850, where he opened a law practice. The slave schedules for McLennan County in 1860 show that Coke owned fifteen slaves. He was a delegate to the Secession Convention in Austin in 1861 and voted for secession. During the war he was wounded at Bayou Bourbeau on November 3, 1863, near Opelousas, Louisiana. Richard Coke was elected the fifteenth governor of Texas in 1873, and represented Texas in the US Senate from 1877 to 1895.[18]

William Sedberry from Bosque County was elected second lieutenant in Company K under Richard Coke. Sedberry, the county judge since 1858, played a central role at the March meeting of the Bosque County Court, where the message was join the Confederate Army, move from the county, or be treated as a traitor. Sedberry was a slave owner and one of the richest men in Bosque County in 1860, with a total wealth of $18,500.[19] He chose to lead by example and enrolled as a private in Speight's Regiment at Bosqueville on March 29, 1862. His two sons John Summerfield Sedberry and Merritt Abdon Sedberry also enlisted in this regiment.

In a letter to his wife, Caroline Sedberry, in June 1862 at Camp Speight, 105 miles southeast of Waco in Brazos County, William Sedberry wrote about measles, fever, and death among the recruits. Their son John Summerfield wrote his mother in July that in the regiment of 7,000 men, only two hundred were ready for duty. Later in the fall William wrote from Camp Nelson near Austin, Arkansas, that "venereal disease was rampant." There was also a great deal of pneumonia, and for a while twenty to thirty men died every day.[20] At Christmas-time William R. Sedberry became very ill, and on January 1, 1863, he died in Little Rock, Arkansas.

A couple of Norwegian immigrants enlisted in Speight's Regiment. One of them was George H. Poulson, born Halvor Georg Paulsen at Vik, Stange, Hedmark County, on March 31, 1843. He enlisted as a private in Company K at Bosqueville on the same day Sedberry did. Poulson was absent sick at Marksville, Louisiana, from May to June

1863. He was captured near Yellow Bayou, Louisiana on May 18, 1864, and imprisoned in New Orleans. He was sick with intermittent fever in the St. Louis General Hospital, New Orleans, from June 20 to 29, 1864, and was exchanged at Red River Landing, Louisiana on July 22, 1864.

On July 17, 1862 Andrew Thompson, born Anders Smeland in Norway around 1832, enlisted as a private in the Fifteenth Texas Infantry, Company I, at Corsicana. He was absent without leave from March to April 1863 and was sick in camp between November and December 1863. Thompson was detailed to the Quartermaster Department from January 30 to April 1864 but was sick in camp in May and June 1864. From August 15 to December 20, 1864, Thompson was on extra duty as regimental blacksmith. He was absent on furlough from December 20, 1864, to February 10, 1865, but did not return until the end of that month. The rest of spring 1865 he was detailed to the Quartermaster Department.

NORWEGIANS IN THE THIRTIETH TEXAS CAVALRY, OR GURLEY'S REGIMENT

The 35-year-old successful Waco lawyer Edward Jeremiah Gurley organized Gurley's Regiment on August 18, 1862. He was born in Franklin County, Alabama, on June 7, 1824, and moved to Waco in 1852 where he practiced law with his brother-in-law, Richard F. Blocker. Gurley bought land in McLennan, Falls, and Williamson counties and had slaves working his land.[21] Most of the men in his regiment came from Waco and the surrounding area as well as the counties of Bastrop, Johnson, Bosque, Comanche, Chambers, Erath, Hill, and Ellis. The Thirtieth Texas Cavalry remained in Texas for nearly a year before it entered battle in Arkansas and the Indian Territory with Gurley as colonel.[22]

Several Norwegians enlisted in the Thirtieth Texas Cavalry. One of them was the 34-year-old Hendrick Dahl. He and his wife Christine had settled in Bosque County in 1854. Hendrick Dahl was as solid and dependable a Norwegian as could be found. He was not much interested in fighting for the Confederates but in spring 1862 felt the pressure from his neighbors. He made no fuss about it, then nor later, and signed up as a private in Company C of the Thirtieth Texas Cavalry at Camp McCulloch, Waco, on July 1, 1862.

Three other Norwegians enlisted with him on that day—Ole Andreas

Canuteson and the brothers Eric and Niels Swenson. Canuteson had migrated from La Salle County, Illinois, to Norse, Bosque County, in 1859, and had settled on 160 acres of land adjacent to that of his uncle Canute Canuteson.[23] He was born on April 29, 1834 in Kendall County, New York. He was sick in hospital in Houston from May 26 to 31, 1863, but survived the war. During the first year of the Civil War, 20-year-old Eric Swenson, born on February 5, 1841, at Lillehogstad, Tynset, Hedmark County, carried the mail between Fort Worth and Gatesville. He rode a beautiful black horse, trained to lie down and get up on command. Eric Swenson carried the best firearms to be had at the time— two Colt cap and barrel six shooters and a Spencer repeating rifle. As a private in Company C, Thirtieth Texas Cavalry, he was often used as a message rider.[24] His brother, Niels Swenson, five years older, born on September 22, 1836, was a bugler in the same company. Niels Swenson was a fast rider who had won thousands of Confederate dollars in horse races. "The laying of wagers had a considerable vogue" in cavalry outfits, according to Bell I. Wiley. Even on duty, horse racing for bets took place to a considerable degree; in McCord's Frontier Regiment in June 1863, McCord issued an order to his officers to put a stop to horse racing and betting.[25] Niels Swenson's horsemanship did not help him much in fighting pneumonia. He died on October 20, 1863, at Fort Washita, Chickasaw Nation (Indian Territory).

NORWEGIANS IN THE THIRTY-FIRST TEXAS CAVALRY (HAWPE'S REGIMENT)

Hawpe's Regiment Texas Cavalry was organized in Waco and Dallas with eight companies. Four of the companies were recruited in Waco. Several Norwegians enlisted in the Thirty-first Texas Cavalry on April 2, 1862, in Bosque County, including Canute E. Canuteson from Norse. He was born in 1836 in Mission Township, La Salle County, Illinois. Canuteson was killed in action at Prairie Grove, Arkansas, on December 7, 1862.[26]

Evan Jenson, or Even Jensen Oddersland, was born April 2, 1843, at Tromøy, Arendal, Aust-Agder County. He enlisted in Bosque County on his 19th birthday. In late 1863 he was promoted to Second Corporal. Christopher Pederson, born Christopher Pedersen Holtet on June 14, 1841, at Stange, Hedmark County, enlisted as a private in Company B. He was promoted to Fourth Sergeant between July and December 1863.

He was absent with leave from August 2, 1863, until February 6, 1864. He was sick in camp in March and April 1865 and was in hospital in Houston in April 1865.

Casper A. Poulson, born Kasper Andreas Poulsen Vik on March 20, 1841, Stange, enlisted in Bosque County on April 2, 1862 as a private in Company B, Thirty-first Texas Cavalry. Poulson was absent on special duty for the brigade commissary from May 1 to August 1863. He was absent with leave from December 1864 to early 1865.

Alexander M. Lindburgh, born Erik Mathisen Lindberget on June 26, 1824, in Grue, Hedmark County, emigrated to Texas in 1853. He was a rock mason by profession, and in 1857 he began building a house and barn in Bosque County, which was completed in 1861. He enlisted in Bosque County on April 2, 1862, in Company B, Thirty-first Texas Cavalry. He was absent with leave from March to May 1863, but then apparently was away without leave until October 1864. A medical examination at Pass Hospital, Tyler, Texas, on April 9, 1864, found him incapable of service because of varicose veins, whereupon he was detailed to serve as shoemaker in Tyler. He was arrested for being AWOL in late October 1864. He was imprisoned for two months but was then brought back to his regiment near Minden, Louisiana, in late December 1864, and was freed of the charges against him. Lindburgh was on extra duty as a harness maker in March 1865, which probably suited him well.

Charles N. Reierson, born Carl Nicolai Reiersen on June 24, 1817, in Holt, Norway, enlisted as a private in Company A, Thirty-first Texas Cavalry, at Newtonia, Missouri, on October 3, 1862. Previously he was a private in Captain H. W. Kyser's Kaufman Guards in Kaufman County. Only a few days after he joined Company A, the 45-year-old Reierson was captured on October 12, 1862, near Bentonville, Arkansas. He was imprisoned at St. Louis, Missouri, but was exchanged on April 23, 1863, at Allen's Point, Virginia.

Sickness and epidemics were as dangerous as battles. In August 1862, Hawpe's Regiment experienced a severe measles epidemic. During September and October the regiment was in Newtonia, Missouri, but had to retreat in October because of a large Union force. In November, the regiment was ordered to fight dismounted and all cavalry horses were brought back to Texas. Colonel Hawpe found the decision unacceptable, resigned, and returned to Texas.

On December 7, 1862, a bloody battle took place at Prairie Grove,

Arkansas, not far from Fort Smith. Around 2,700 men from both armies were killed, wounded, or missing. The Confederate Army withdrew, and Missouri and northwest Arkansas came under Union control. In the weeks after this battle, the morale in the Thirty-first fell to a low, followed by desertions and a near mutiny. A number of men left their units at Christmas and New Year 1862 without formal leave and did not return before winter was over. That winter was extremely cold, and many of the soldiers who remained with their companies did not have any huts, tents, or other forms of shelter. For a while soldiers were dying every day.

In January 1863 Colonel Joseph Warren Speight took over the command of the Thirty-first Texas Cavalry. The brigade was sent to Louisiana to assist in the protection of Shreveport, under the command of Major General Richard Taylor. After the battle at Stirling's Plantation near the Mississippi River at the end of September 1863, the Thirty-first was reorganized under the command of Colonel Camille Armand Jules Marie, Prince de Polignac of France. He was put in charge of several units deemed to have a low fighting spirit: the Fifteenth Texas Infantry and the Seventeenth, Twenty-second, Thirty-first, and Thirty-fourth Texas Cavalry Regiments, all dismounted. The units endured a lot of drilling during the next months, and then took part in the battle of Mansfield, Louisiana, on April 8, 1864, where the Confederates defeated the advancing Union Army under General Nathaniel Banks. In January 1865, the brigade was reorganized under the command of Colonel James E. Harrison and ordered back to Texas.

A CLOSER LOOK AT THE LIFE OF NORWEGIANS IN THE FIFTEENTH TEXAS INFANTRY AND THIRTY-FIRST TEXAS CAVALRY

Because of surviving letters to Carl Questad from David Lund and Lars Olson, more is known about the lives of the Norwegians in the Fifteenth Infantry and the Thirty-first Cavalry than those in any other Confederate unit during the Civil War. David Lund was born in 1824, near Bergen, Norway, and arrived in Bosque County in 1859. At the time of the agricultural census of 1860 he was registered as owning 160 acres of unimproved land. Lund enlisted in the Fifteenth Texas regiment as a private in Company I at Waco on March 22, 1862. He was killed in action somewhere in Louisiana in 1864. Little is known about his death, but his reflections on the life of a soldier have survived.

In a letter to Questad from Alexandria, Louisiana, on May 28, 1863, David Lund wrote that he was tired and weary after a long march with little to eat. Moreover, he had been sick for a while with the ague. His unit had crossed the Red River and gone back to Mount Pleasant, Texas, after leaving the Indian Territory, and from there marched to Jefferson. A steamboat had ferried them to Shreveport, where they stayed two days. The soldiers had been marching 16–20 miles per day. The Yankees had been in the vicinity three weeks earlier and had taken 4,000 negroes and lots of grain. A bridge had been destroyed and the Yankees had burned all the grain they could not use, turned the horses into the fields, and generally done a lot of damage.

It seemed their unit would soon get in touch with the enemy, and "even though I am not anxious to get into a fight, it is not going to be said that D. Lund was afraid and behaved like a coward." His fate rested in the hands of God. "Unless He wills it, not even a hair on my head can be crinkled. Not every bullet hits its mark. However, if God has determined that my time has come, then I cannot do anything about it, but will accept whatever He may decree. This faith, my dear friend, has sustained me, comforted and strengthened me so many a time. And I have the trust in Providence that we two will meet again, both of us chastened by trials and adversity."[27]

Lund saw little hope for the Confederate forces. There were rumors that Vicksburg and Port Hudson had been taken. "If this be true, then we may as well give up, because Vicksburg is the key to the Mississippi. Without it, the Federal forces can do little, but if they can take Vicksburg, then our communications will be cut and we will be unable to receive money, orders, or anything else from Richmond. We know so little, but it is certain that a battle has recently been fought because we heard the thunder of cannons."[28]

Lund hoped Questad and the other Norwegians had "harvested your grain and gotten it under cover so that you can provide for your livelihood and thus be free of worries in that respect. May God hold his protecting hand over you and permit you to enjoy the fruits of your labor in peace and quiet. The hope that you, my dear friend, may remain far from the horrors of war gives me great joy. When I see abandoned houses along the roadway, open to wind and weather, I remember our quiet valleys and thank God because you are still living in peace. Would that God might shield us and permit us to live peacefully, not only among ourselves but with all our fellow beings—then come what may during the brief time we live here below. If we have a clear conscience and a kind disposition, death will not appear as an enemy but as a good friend who will lighten our burden and bring us to a better home."[29] Fate was not on his side. David Lund was presumed killed in action somewhere in Louisiana in 1864.

Even when letters from Lund ceased to arrive, Carl Questad continued to be well informed about the life of Norwegians in the Thirty-first Texas Cavalry through letters from Lars Olson, born in Holstuen, Løten, Hedmark County, on October 30, 1830. Before his emigration in 1851, Olson had been a schoolteacher in Løten. He first settled among Norwegians in Kaufman County, where he worked as a farmer and plasterer. Before his enlistment as a private in Company B, Thirty-first Texas Cavalry in Bosque County on April 2, 1862, he had been a private in Captain J. R. Johnson's Prairieville Company in Kaufman County.[30] Olson was absent sick from May through June 1863 and was absent with leave from the fall of 1863 until February 1864. After that, he was absent without leave but returned before June 1864. Olson was given extra duty as a blacksmith from December 1864 through May 1865.

Lars Olson and Questad came from the same municipality in eastern Norway and thus had the same cultural background. Both were better educated than most Norwegian immigrants, and they wrote good Norwegian prose. Lund tended to dwell on fate and being afraid of not being brave when the moment arrived; Olson cultivated the art of irony.

The Olson correspondence with Questad began after he returned to his regiment in June 1864 and continued to the end of the war. In a letter from Marksville, Louisiana, dated June 26, 1864, Olson reported that he and "the other Norwegians here are brisk and active and we live about as usual—moderately well or somewhere between average and swell. We are fed corn meal, beef, a bit of sugar and molasses, some blackberries, and good spring water."[31] Presently they had nothing to do "beyond mounting guard off and on and drilling four hours five days a week." It was getting very warm, and the "mosquitoes pester us, especially at night. Health conditions among the soldiers are good at present, but I presume the fever will arrive with the heat."

He had recently seen Andrew Thompson. "He is in good health but kicks for dear life because he is not permitted to go home and attend to his wife. There are rumors to the effect that Peder Pederson and

Aasmund, together with several Americans, have deserted—presumably a bright idea." He asked Questad to send him "a few lines and let me know what's new in Bosque." Letters should be sent to Alexandria, La., Company B, Hawpe's Regiment, Polignac's Brigade, Texas Dismounted Cavalry.[32]

On October 16, 1864, Olson sent a letter to Questad from a camp at Camden, Arkansas. His unit had walked through places like Harrisonburg, Monroe, and Monticello, and they were now building fortifications at Camden. "We have not been in contact with the enemy for a long time, but this is not necessary as we are killed off rapidly enough anyway. About fifty percent of us are sick and the rest are worn out with marching and poor food. Furthermore, most of us are barefooted and half naked. When we are on the march, there are not enough wagons to carry the sick. They must trudge along as far as they can and then remain lying by the wayside. If we then arrive at a place where we are to stay several days, wagons are sent back to pick them up. I have been sick for two weeks but am improving, at least for the time being. The other Norwegians are hale and hearty."[33] The Yankees were now at Little Rock and Pine Bluff, and "both our cavalry and theirs are somewhere between those places and Camden." Olson dreamed about spending the winter at home as he had done last year, "but no such luck, I am certain."

The next letter to Questad was written from Minden, Louisiana, on December 18, 1864. "Good news and rumors of peace are plentiful, but—like the fool I am—I tend to be dubious about all such things. Every day we hear cannonading from Shreveport." Since Casper Poulsen would be going home to Bosque for Christmas, Olson used the opportunity to "write you a few meager lines this Sunday morning while God-fearing people are attending services. I can inform you that at present I am, thank God, hale and hearty, which is a good thing. The other Norwegian boys are also in good health, and considering circumstances, we are living very well. We have established a regular camp here, and each mess has put up a house. In ours are four Norwegians and one American."[34] They had enough food and little to do.

A recurring theme in Olson's letters during fall 1864 was desertions. In the letter from Marksville, Louisiana, of June 26, Olson wrote about a "soldier by the name of McBase from Clifton, who had deserted from our company at Pleasant Hill and tried to join the Union forces." He had been captured and was now under arrest at Franklin, Louisiana.

"What punishment he will receive, I do not know."³⁵ At the end of August 1864, his regiment was 8 miles from Harrisonburg and 25 miles from the Mississippi River. Two days earlier his company had received orders to prepare for a march and cross the river. "This caused much excitement in the camp, and in the night 123 men deserted from our brigade. They marched out of the camp in regular order with their rifles and forty rounds of ammunition, and apparently no hindrances were put in their way. But they have now sent a cavalry detachment after them and some mounted infantry, our Captain Jacks being one of them. I have also heard that several hundred men absconded from Walker's division, but how true this is I do not know."³⁶

In the October 16 Camden letter, Olson told about two men from their company who had deserted to the Yankees at Monticello. "Yesterday a captain in Walker's division was shot. He is said to have encouraged his men to abscond when we were given orders to cross the Mississippi."³⁷ In a letter from Minden, Louisiana, dated December 23, 1864, Olson informed Questad that his neighbor from Norse, Lindberg, had "arrived here yesterday evening under arrest. He says that he was taken out of the shop at Tyler, two months ago, and they have held him in Rusk since then. The reason seems to be that he was absent without leave when he was detailed, and that the detail therefore was illegal. He is now in the guardhouse and I do not know as yet what they will do with him. He is brisk and active and looks like his old self. He says he has no relish for being under arrest. I can also relate that a couple of weeks ago three men of Speight's regiment were shot for attempting to desert last summer when we received orders to cross the Mississippi. In our homeland we shot old horses when they could work no more. Here they shoot soldiers when they are worn out or no longer retain the right faith and act according to it. Ross Cranfield was one of the executioners. I was also detailed but was excused because I do not like that kind of work."

DEFENDING THE FRONTIER AGAINST INDIANS

Several Norwegians living in Bosque County chose to serve in the Texas Frontier Regiment. It was established by the Ninth Legislature of Texas on December 21, 1861, to protect people living on the northern and western frontier against Indian raids. Since so many young men had left their frontier homes to fight in the Civil War, the frontier counties had become very exposed to Indian raids. Texas Governor Francis R. Lubbock appointed Colonel James M. Norris to command the Frontier Regiment on January 29, 1862. After a long and exhausting journey along the frontier in March and April 1862 with his immediate subordinates, Lieutenant Colonel Alfred T. Obenchain and Major James E. McCord, he selected eighteen frontier defense locations along a winding, serpentine line extending almost 500 miles from the Red River in North Texas to the Rio Grande in South Texas.

Nine companies of approximately 115 to 125 men were organized in March and April 1862. It was left to the enlisted men to elect their officers. Each captain was given command of two camps. Scouting patrols left one camp each day and traveled south to the next camp. The next day, the patrols returned north to the camp they had left the day before. The entire line from the Red River to the Rio Grande was traversed each day. The Indians, however, soon learned the routine of regular patrols between posts and easily slipped through the line.

Many rangers were badly mounted. There was considerable sickness in the camps, and the officers had problems keeping discipline. In February 1863, the men elected Major McCord to take over the command from Norris. James Ebenezer McCord (1834–1914) was born in South Carolina. His family moved to Henderson, Rusk County, in 1853. McCord learned the surveying business, and in 1856 he was in charge of a surveying party for the organization of a new chain of counties on the western Texas frontier. He had also served as a First Lieutenant Captain in the Texas Rangers.

McCord established his headquarters at Camp Colorado. "When McCord assumed command in February 1863, the regiment finally had someone in charge who unlike Norris, had practical experience on the Indian frontier," observed David Paul Smith in his book *Frontier Defense in the Civil War, Texas Rangers and Rebels*.[38] McCord abandoned the passive patrol system favored by Norris. Instead, he sent unexpected scouting expeditions deep into Indian country in the hope of surprising raiding parties; he also attempted to put the Indians on the defensive by attacking their home camps. He gave Major W. J. Alexander the command over four companies along the southern border from Camp Colorado to the Rio Grande, and Lieutenant Colonel James Buck Barry from Bosque County, the most experienced Indian fighter in the regiment, was given the command along the most threatened northern border.

The new type of scouting expedition was successful. "Within months the Frontier Regiment fought more engagements with Indians and captured more horses than at any time previously."[39] Several Norwegians from Bosque County enlisted in the Frontier Regiment in December 1862 and early 1863 at Fort Salmon, where the principal mission was to protect the Ledbetter Salt Works in Shackelford County against Indian attacks.[40] James S. Jenson had enlisted in Company E for one year and reenlisted for three years on December 20, 1862, at Camp Salmon. He was born Jens Jensen Møglebustad, or Jens Jensen Oddersland, on March 28, 1835, in Brovold, east of the city of Arendal in southern Norway. He emigrated to Texas with his parents and six siblings in 1853. His parents belonged to the first pioneer group of Norwegian settlers who traveled from East Texas to Bosque County in early 1854. Jenson was on wagon guard to Bosque County from April 17 through May 1863, and was discharged on June 10, 1863.[41] His substitute was Elias R. Skimland.

Otto Johan Martinsen (Johnson), born on May 28, 1840 at Aasetqværnen, Åmot, Hedmark County, had enlisted for one year at Stephenville as a private in Company E. He had previously been a private in Captain S. Fossett's Company of Bosque County Minute Men. He reenlisted "for three years or the war" on December 20, 1862, at Camp Salmon. He was on wagon guard in Bosque County from April 17 through May 1863, and was discharged on June 10, 1863, with Aadne Holverson his substitute.

Ole Ween, born Ole Larsen Grønsveen on October 23, 1822, Løten, Hedmark County, enlisted in Stephenville for twelve months and reenlisted for three years or the war on December 20, 1862, in Company E at Camp Salmon. Ween was on extra duty as a blacksmith from April 23 to May 1863. He was discharged on June 10, 1863, and his substitute was John Johnson, born Johan Johansen Vatne, on May 1, 1817, in Jæren, Rogaland County. He had arrived with his family in Bosque County in the 1850s, and in the 1860 census he was listed as a rock mason. According to his son John Bartinus Johnson, born in 1858, the day his father rode to enlist in the Frontier Regiment was etched in his mind forever. Together with his mother and sisters he accompanied their father to the edge of their farm on upper Gary Creek. "That was the last time they saw or heard from him."[42] According to official documents, Johnson was on wagon guard to Bosque County from April 17 to May 1863, and on a scouting expedition from December 17, 1863. Nobody knew what happened to him after that.

Carl Questad's son, Even Carlsen Questad, born on October 17, 1842, Løten, Hedmark County, enlisted at Stephenville on February 1, 1862 for twelve months. Before that, he had been a private in Captain S. Fossett's Company, Bosque County Minute Men. While on a scouting expedition on the frontier he contracted pneumonia. He died on August 31, 1862, at the home of his parents in Bosque County. On December 7 of that year, First Lieutenant W. C. Alexander wrote Questad a letter from Camp Salmon describing what Questad had to do to get the salary due to his son. His "Final Statement" and account would soon be ready for collection.[43]

HUNTING DESERTERS AND BECOMING DESERTERS

The main aim of the Frontier Regiment was to secure the borders against Indians, an aim that all Norwegians in the regiment agreed with. In May 1864, however, the Texas governor and legislature had to accept that the Frontier Regiment would come under Confederate control. It was renamed the Forty-sixth Texas Cavalry. Because of the threat of Union invasion from Louisiana and Arkansas, General Magruder ordered Colonel McCord and his six companies to move east toward the coast. Camp Salmon was abandoned in March 1864. On several occasions McCord asked his superiors to be allowed to return to frontier service, but this was refused by the commanding general every time. For the rest of the war the companies in the Frontier Regiment were chiefly stationed at various points closer to the coast, such as Columbus, Colorado County, and Bastrop, Bastrop County.[44] Their main job was to hunt down deserters and men trying to evade the conscription laws; fighting Indians on the frontier was very low on their list of priorities.

The Norwegians respected McCord's leadership, but were fully aware that McCord was overruled by his superiors. In a letter to Carl Questad, dated July 15, 1864, west-Norwegian E. R. Skimland described the growing tensions within the company. The company was located 14 miles from La Grange, the county seat in Fayette County, and 18 miles from Columbus, Colorado County. Colonel Bankhead was their brigadier general, and he was very much disliked by the soldiers because he was related to General Magruder. "One day when we were to drill before him, practically all the soldiers began to act up and shout as soon as they saw him coming. Then he delivered a speech, but we could hardly hear him because of the yelling and the insulting words the soldiers hurled at him." A couple of nights later thirteen men had stolen by the guards, "bent on hanging the rascal. But they did not succeed. Our Major Alexander somehow sensed that some deviltry was brewing. He got up and ran over to the general's quarters, where he found the rope in readiness and the general alone in his tent, quivering with fear for his life. His wife had taken to the woods. Then he sounded the alarm in all parts of the camp, and the people had to get out and arrest the culprits. A trial was held the next day, and they confessed that their idea had been to scare him because he had initiated too severe a regimen."[45] Some of the men's demands were met. The daily drill was reduced by a third, and "the offenders were freed because the officers found that the prisoners had more friends than they had."[46]

Two Norwegians in Company E deserted in spring 1864 after it became known that the McCord regiment would be moved to the coast. One of them was Osman Neystel, born Osmund Olsen Nystøl on November 8, 1841, in Åmli in southern Norway. He had enlisted for twelve months at Stephenville on March 7, 1862, and reenlisted for three years on December 20, 1862, at Camp Salmon. From August 21 to September 1863 Nystøl was on patrol. He was absent sick from October 15, 1863, to early 1864. After his return in spring 1864 he deserted and arrived at San Elizario, near El Paso on the Rio Grande, between May 24 and June 13. In early July he was taken to Santa Fe in New Mexico Territory, where he took the oath of allegiance to the United States and stated his willingness to take up arms in defense of the Union.

His friend Peter Pierson, born Peder Olsen Songe on February 27, 1843, on the island Tromøy south of the city of Arendal, Aust-Agder County, deserted with him. Peter was the only son of the pioneer Ole Pierson in Bosque County and had enlisted at Stephenville on May 7, 1862, for twelve months. Before that he had served with Captain S. Fossett's Company of Bosque County Minute Men. Pierson was hornblower in Company E, and he had reenlisted for three years on December 20, 1862, at Camp Salmon. With Nystøl, Pierson took the oath of allegiance to the United States in Santa Fe, New Mexico Territory, in July 1864. Pierson returned to Bosque County after the war and took over his father's farm. During the next decades he came to play a central role in the community at Norse. Pierson was never afraid of voicing his own opinions, neither in the Lutheran Church nor as a member of the Farmers' Alliance in the 1880s.

Many Norwegians in Company E shared the deserters' views on the war but did not desert just then. Two Skimland brothers, from the island Bømlo, Hordaland County, on the Norwegian west coast between Stavanger and Bergen, first immigrated to the Midwest but came to Bosque County in the late 1850s. They were as skeptical about the Southern cause as the two Company E deserters. The Skimland brothers were usually in the thick of things, especially the older brother, Elias R. Skimland. He had enlisted for one year on March 8, 1862, at Stephenville and reenlisted for three years on June 10, 1863, at Camp Salmon, as a substitute for James S. Jenson. He was on scout duty from June 27 to July 1863, and again on August 13 to September 1863. Elias was on extra duty to procure clothing from December 12, 1863 to January 1864. His brother Canute R. Skimland also enlisted in the Frontier Regiment as a private in Stephenville on March 8, 1862 and reenlisted for three years at Camp Salmon on June 10, 1863. He substituted for Ole Ween, who had previously been discharged. Before that he had served in the "Home Guard." He was on scout duty on three separate occasions between June 27, 1863, and January 1864.[47]

Five Norwegians deserted from Eagle Lake in Colorado County on August 2, 1864: the two Skimland brothers, Otto Swenson, Neil Canuteson, and Aadne Holverson. Otto Swenson was the youngest of the Swenson brothers of military age. He was born in Lillehogstad, Tynset, Hedmark County, on April 21, 1844. Swenson enlisted on March 8, 1862, at Stephenville but had been a private in Captain S. Fossett's Company before then. He reenlisted at Camp Salmon on December 20, 1862, and became a corporal in Company E. He was on wagon guard to Austin from April 20 to May 1863, and on an express assignment to Colorado from June 30 to October 1863. Neil Canuteson, born Nils Knutsen Underbakke on October 17, 1829, at Suldal, Rogaland County, enlisted as a private for one year on February 1, 1862, at Stephenville, and reenlisted for three years on December 20, 1862, at Camp Salmon. Neil Canuteson was on furlough from June 16 to August 1863. He was assigned to shell corn for his company from October 10 to December 1863. From December 21, 1863, through January 1864 he was assigned to obtain pack mules. The fifth Norwegian deserter was Aadne Holverson, born Ommund Halvorsen Egenes on March 25, 1832, in Skjold, Rogaland County. He had enlisted at Stephenville on February 1, 1862, for twelve months, and reenlisted on June 10, 1863, at Camp Salmon, as a substitute for Otto J. Johnson. All five returned to their company at Camp Felder on October 4, 1864.

Many soldiers in Company E became sick at the end of October 1864 and had to stay in the hospital in Brenham, Washington County. From October 30 to the middle of November 1864, Elias R. Skimland was detached to guard Union prisoners at Hempstead but ended up in the Brenham hospital from November 5 to mid-December 1864. On December 30, 1864, Skimland wrote Questad from their camp near La Grange, Texas. He had picked up his horse on December 14. His brother was still very feeble, but since he had two horses, "he wished to try to go along to the camp."[48] He had lived well while traveling through Brenham. Wherever he stayed he had met people who still strongly believed that the South would win the war. "And I was not beneath damning both Lincoln and those whom he sent out to fight against us. So they believed that I was staunchly on their side, but if they had known what abode in my bosom, I would have had to pay." Generally, soldiers were afraid of expressing their opinions to each other, but "we in our regiment speak what's on our minds. There are very few among us who do not have one and the same idea, so they are not afraid to tell the officers what they think." McCord had tried very hard to hold his regiment together, Skimland wrote, "but it appears that this is beyond his powers."[49]

By March 1865 several Norwegians in Company E were sick and tired of the war. Near Bastrop on March 18, 1865, five Norwegians made themselves scarce, and this time they did not return. According to official papers they had deserted to Mexico. They returned home to Bosque when the Civil War was officially over. Elias R. Skimland and his brother were again the leaders, and Otto Swenson and Aadne Holverson followed. But the fifth man this time was Christopher Johnson, born Christopher Jensen Oddersland, on April 5, 1840, in Tromøy, Aust-Agder County. He had enlisted at Stephenville on March 8, 1862, for twelve months and reenlisted for three years or the war on December 20, 1862, at Camp Salmon. He was promoted to sergeant in July 1863 but became a private again in October 1864. He was sick at the Brenham hospital from November 4, 1864, and was absent with leave from December 10, 1864, to early 1865. Neil Canuteson was not with them this time. He was absent on sick leave from March 15 to April 13, 1865 and remained absent until the war was over.

DISILLUSIONMENT AND INCREASING DESERTION

In a letter to Carl Questad from Eagle Pass on the Rio Grande on October 17, 1864, Helge J. Grann wrote critically about the behavior of his officers. His unit was in a camp on the border with Mexico. It was impossible for him to believe that "anyone who wants to serve his country can be satisfied with what he sees here every day. It appears to me that the officers do not think of anything but robbing the government as quickly as they can, and I get the impression that we are here merely to protect them."[50]

The enthusiasm for the war had been high among Texans in 1861 and 1862. But even Wooster and Wooster had to admit that "various factors, including dissatisfaction with military discipline, inadequate pay and rations, concern for families at home, increasing disillusionment over military failures, and sometimes cowardice, led to a steady increase in the number of soldiers who left the army."[51] The leaders of the Confederate Army strongly believed that the execution of captured deserters before their comrades in arms would be an effective warning against desertion. But most of the privates found executions cruel, unjust, and unnecessary butchery. The best medicine against desertion, the lower ranks argued, would be to give them better officers.

The problem of desertion was widespread not only in the Confederate Army, but even more so in the Union Army. It affected all ranks and all parts of the country but was most common among privates and low-ranking officers. In her book *Desertion during the Civil War*, published in 1928, Ella Lonn discussed the causes and consequences of desertion in both the Northern and Southern armies. After having studied heaps of official war records, she concluded that one in seven enlisted Union soldiers and one in nine Confederate soldiers deserted. There were reasons for desertion on both sides—lack of sufficient food, clothing, and weapons, weariness, hopelessness. Soldiers might be overcome by longing for family and home. Many deserted in order to help women and children back home avoid destitution. Patriotism was much stronger at the outbreak of war than toward its close, especially in the South.

Initially there was no agreement among officers about the length of time that should elapse before absence without leave was to be classified as desertion. Captured deserters were generally not shot or hanged because manpower was so precious. According to Lonn, approximately 103,400 enlisted men had deserted from the Confederate Army when the war ended, and 200,000 from the Northern armies.[52] Desertion grew worse in the Confederate Army the longer the war lasted, and in spring 1865 large parts of the army just melted away. The number of desertions in the Northern armies declined because of harsh measures against it, and because the Confederate Army was so clearly on the defensive.

Officers trained in the West Point tradition found desertion to be an evil crime. Laymen, on the other hand, condoned it and found it excusable as the hardships of war became harder than the men could be expected to bear. Both sides were lenient in dealing with defection in the beginning; they tried to persuade deserters to come back to their companies, but with little success. President Lincoln put forth two offers of general pardon, in 1863 and 1865; the Confederates offered three general amnesties: the executive offer of August 1863, the offer of the two chief commanders in the field in 1864, and finally Lee's amnesty upon assuming the post of general-in-chief of the armies. In the end both the South and the North ended up using the death penalty for deserters.

Nowhere in the North, Lonn wrote, was there such a "deserter-country" as existed in Alabama, Mississippi, and North Carolina. Toward the end of the war and after, lawless bands in the South threatened the very foundations of society. About one hundred thousand in a population of nine million were allowed to get away with robbery, arson, and murder for nearly two years. Furthermore, nearly 3,000 deserters might have lived in the woods and brush of northern Texas—in Wood, Van Zandt, Henderson, Dallas, Denton, Cooke, Grayson, and Jefferson counties.

THE END OF THE CIVIL WAR IN TEXAS AS SEEN THROUGH THE EYES OF NORWEGIANS

Despite the many battles the Confederates lost in 1863 and 1864, wrote Wooster and Wooster, "Texans continued to believe that the South would ultimately be victorious. The fall of Vicksburg and Lee's failure at Gettysburg dampened but did not destroy the confidence of Texas Confederates." Texans got a positive boost from the defeat of the Union forces led by General Banks in the Red River campaign. After the fall of Vicksburg in July 1863, the Confederate states west of the Mississippi River were cut off from the rest of the Confederacy. From then on the town of Marshall in East Texas came to play a crucial role. Toward the end of 1863 General Edmund Kirby Smith, the Confederate commander of the Trans-Mississippi Department, moved his operational headquarters to Marshall.

Most Texans found it hard to believe that Sherman would be able to take Atlanta. Some still expressed defiance, even after General Lee surrendered at Appomattox in April 1865. Diehard Texas Confederates wanted to carry on the struggle. General Kirby Smith was intent on continuing the war and urged the soldiers to remain on their posts.

In a letter to Ole Ween in Bosque County on April 23, 1865, Lars Olson commented on the depressing rumors for the Confederates from beyond the Mississippi. "It seems that Old Abe is beginning to raise real hell with our people over there. Richmond has undoubtedly fallen, and the Federal forces are advancing everywhere." In early May he wrote Questad that the soldiers knew about the fall of Richmond, the surrender of General Lee's army to General Grant, and the assassination of President Lincoln. "General Magruder has strictly forbidden the publishing of Federal news unfavorable to us." Neither Lars Olson nor his fellow comrades cared much for General John Bankhead Magruder.[53] Magruder had tried to convince the soldiers that "the fall of Richmond and the surrender of Lee's army are no serious defeats for us because they were no longer of any importance." Olson thought there would soon be peace, "because—to ordinary people—the outcome of the war can no longer be in doubt. We might as well surrender now as later, but it does not seem as if our leaders are of this opinion."

Houston newspapers reported from so-called "mass meetings" held by the different brigades and divisions of the army. According to the newspapers, "the soldiers in general want to sacrifice everything, their very dearest, and accept death rather than surrender or lose what our leaders call "liberty.""[54] Olson had attended one of the so-called mass meetings of Harrison's Brigade. "It consisted of General Harrison as president and Adjutant Tally as secretary. We had nothing to say in the matter. Then he appointed two officers from each regiment—men whom he knew to have the right faith—and these should then, in the name of God, make known the spirit of the men. It reminded me of monks in the Catholic church selling indulgences in the name of the pope. You can yourself guess what the result was. I forgot to tell that on this occasion Harrison delivered a fanatical speech in which, you can depend upon it, he gave us our instructions. The following night, however, he did not dare retire without stationing guards around his quarters, as he feared that some of the infidels might wish to get at him."[55]

High patriotism and lucrative business seemed to fit well together, Olson commented ironically. Cotton was exported from Galveston as in normal times. "All morning we have heard the roar of cannons from Galveston." During May 1865, regiments and companies collapsed. In the end General Smith had to follow orders from the Union Army at Galveston, and on June 2, he signed the terms of surrender.

How did Texans look on their fate in summer 1865? "Relatives and loved ones had died or been killed in war, slaves were now free, money was scarce, and a Union army of occupation was moving into the state." Despite all this, Wooster and Wooster concluded, most Confederate Texans "felt no bitterness at their sacrifice, but pride that they had fought gallantly for a cause in which they deeply believed."[56]

The Norwegians who served in the Confederate Army and survived would probably not have agreed. They felt that a lot of lives had been wasted for nothing; they were disillusioned by the behavior of their Confederate leaders and officers. But they knew in their bones that they were just immigrants and would not be listened to anyway. The survivors went back to their farms, bowed their heads and tilled the soil, and tried to avoid speaking their minds about how they really felt. Some of the unmarried Norwegians soon left Texas in favor of Norwegian communities in the Midwest.

◊ The ford at Brazos Point, the northwestern corner of Bosque County, Texas. The Brazos River is the longest river in Texas; it begins in New Mexico and ends in the Gulf of Mexico. After strong rainfalls, the river would swell, and people and cattle could wait for days before they could ford it. Cleng Peerson crossed the river at this point in spring 1850, on his way west from Fort Worth into Comanche County. He followed the Brazos River south to Waco.

◊ View of the Bosque Valley. Cleng Peerson was the first Norwegian to see this area in spring 1850. He walked from Brazos Point to Waco and continued back to East Texas.

◊ The Cleng Peerson monument in the churchyard of Our Savior's Lutheran Church in the Norse district. The monument was unveiled just before Christmas in 1886. It contains the following text: "Cleng Peerson, the Pioneer of Norse Emigration to America. Born in Norway, Europe May 17, 1782. Landed in America in 1821. Died in Texas December 16, 1865." Below it reads: "GRATEFUL COUNTRYMEN IN TEXAS ERECTED THIS TO HIS MEMORY."

◊ The road to the Norse district from Clifton, Bosque County, climbs in a southwesterly direction to Cranfills Gap. The two signs are placed on the right side of Cleng Peerson Memorial Highway, just before entering Norse.

◊ The Norwegian-Texans became familiar with a mighty tree that they had never seen before: the live oak. Some, like the one to the right, were several hundred years old. Limestone dominated in the Bosque County landscape, and the Norwegian immigrants continued their tradition of building stone fences. Many of the immigrants' houses were built of stone.

◊ Even Questad, the only son of Carl and Sedsel Questad from Løten, Hedmark County, served in the Civil War on the frontier against the Indians. Several Norwegians preferred to fight Indians instead of Union forces. During a long and cold patrol in the west he contracted pneumonia and died at the age of 20. He was buried on his parents' farm in Norse. The next person to be buried there was Cleng Peerson in 1865. Later, Questad gave the land where the graves were located to Our Savior's Lutheran Church

◊ Parts of a stone house still standing in the Meridian Valley, close to Meridian Creek. The house was built by Knud Salve Knudson and his father Salve Knudson in 1873. Salve Knudson, born on the farm Øvre Ramse in Aust-Agder County in 1803, emigrated to Henderson County in 1846. The Knudson family, along with several other Norwegian immigrant families, moved from East Texas to Bosque County in summer 1868, mainly due to recurrent malaria and deaths in East Texas. The Knudson families settled farthest to the west of the Norwegians in the wagon train. Father and son each registered 160 acres of land.

◊ Remains of the large Norway Mills buildings in 2019, built after the Civil War. Ole Canuteson, who emigrated from the island of Karmøy in western Norway and became a homesteader in Bosque County in spring 1854, was the driving force behind the establishment of the steam-driven mill. Once the mill began operating, local farmers were able to reduce their transportation costs considerably. The mill's flour was known for its high quality. A small village grew up around Norway Mills, which had a store, stable, blacksmith, and cotton gin, as well as its own medical doctor.

◊ This residence, located at Norway Mills, is still one of the most imposing and well kept old houses there. The Norwegian immigrant Ommund Omenson, from Skjold, Rogaland County, western Norway, built the house and lived there. He married Eline Thingstad, from Romedal, Hedmark County, eastern Norway, in 1869. During the 1870s, Eline recruited many relatives and acquaintances from Hedmark County to come to Texas. Ommund paid for their tickets, and the ticket holders paid him back by working on the Omenson farm for one year.

◊ There are many landscapes of this type in both Bosque County and the Norse district. In the valleys between the hills there is often a creek, like here at Gary Creek. The Norwegian immigrants located their houses and barns on the sides of the valleys. After heavy rain and swollen creeks, the buildings were usually secure. The Norwegians also appreciated a good view. Some of them felt that many places in Bosque were similar to places in Norway where they had grown up. The heat, however, took some time to get used to.

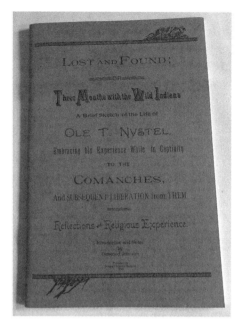

◊ Ole Tergerson Nystøl was born in East Texas. His parents had immigrated from Åmli, Aust-Agder County, in 1848. On March 20, 1867, the 14-year-old was captured by Comanche Indians. Many white prisoners lived among the Indians until they died. Nystøl was held in captivity for three months, but was bought from the Indians by an Indian trader in Kansas, and the Indian agent Colonel Leavenworth sent him home. In 1888 Nystøl published a small pamphlet in Dallas describing his experiences among the Comanche Indians.

◊ The Santa Fe Railway Station, Clifton. The Gulf, Colorado, and Santa Fe Railroad began to run through the Bosque Valley in 1881. From then on the area was well connected, to Galveston in the south and Fort Worth and Dallas in the north. Norwegian immigrants no longer needed to be picked up in Waco. Photo: The Bosque County Collection; Bosque County Historical Commission, Meridian, Texas

◊ The business quarter, Clifton, 1898. After the railroad came through, Clifton became the most important town in Bosque County. The businesses pictured had American, German, and Norwegian names, such as the C. L. Witte hardware store (German), Hill Brothers (American), and the Schouw Brothers furniture shop (Norwegian). Saturday was the most important day for shopping. In the early morning, American, Norwegian, and German families would travel to Clifton by horse and wagon. A round trip might take the whole day, both for the Norwegians from the Norse district and the Meridian Valley in the southwest, and for the German immigrants living southeast of Clifton. Photo: The Bosque County Collection; Bosque County Historical Commission, Meridian, Texas

◊ The ginning of cotton was an important yearly occurrence, and the sale of two or three bales of cotton was the most important source of family cash. But first, they had to pay their bill at the store. The Norwegians began growing cotton after the Civil War. Around 1900 cotton prices decreased, and farmers had to work increasingly harder to maintain their standard of living. Photo: The Bosque County Collection; Bosque County Historical Commission, Meridian, Texas

◊ A number of Norwegian immigrants settled in the Meridian Valley in the 1870s. Several families found it too far to travel to attend church in Norse, so they built their own church, to the right in the picture.

◊ St. Olaf's Church in the Meridian Valley, today best known as the Rock Church. Members of Our Savior's Lutheran Church who lived in the Meridian Valley were allowed to build their own church in 1885, as long as they paid for the building themselves. The stone church was used from 1892, and it only cost $1,000. It was many years before the dirt floor was covered by wooden planks.

◊ It was a mighty sight when the threshing team arrived. In the foreground are the huge steam engine and the long belt powering the threshing machine. In the background are a mountain of straw and the wagons feeding the threshing machine with wheat bundles. The farmers hired professional threshing crews or joined forces with their neighbors to form threshing teams. Photo: The Bosque County Collection; Bosque County Historical Commission, Meridian, Texas

◊ Working in the field. The most important agricultural products were maize, wheat, and cotton. Rye, oats, barley, sweet potatoes, and potatoes were also grown. An abundance of vegetables and fruits were grown closer to the houses for daily consumption or preservation. Photo: The Bosque County Collection; Bosque County Historical Commission, Meridian, Texas

◊ The wedding of Oliver Jenson and Ola Brooks in 1905. The bridegroom was Norwegian and the bride American. Before 1900, Norwegians usually married other Norwegians, but in the following years it became more common for ethnic Norwegians to marry Germans and Americans. In the back row from the left: Hans Jenson, Martin Olson, Alf Price, Gus Grimland, Lena Townsend, Lottie Snell, Alma Pederson, Herman F. Meier, Kate McSpadden, Oscar Bronstad, Ollie Carpenter, together with unknown. In the front row from the left: Will Jenson, Oliver Jenson, J. K. McSpadden, Ola Brooks, Leah Townsend, Leah Gibbs. In the front right: Atys Brooks. Photo: The Bosque County Collection; Bosque County Historical Commission, Meridian, Texas

◊ In the school yard in Cranfills Gap, 1914, nineteen girls play "Bro, bro brille." The game is well-known in Europe and was registered in Løten, Hedmark County, Norway, in 1851 as a "Christmas game." In England and the United States, it is known as "London Bridge is falling down." More than half of these girls had Norwegian or Danish parents. From right to left: Jewel Reesing, Jennie Christenson, Ada Grimland, Vivian Gardner, Agnes Arneson, Minnie Cox, Nadine Spenser, Thelma Perry, Orelia Grimland, Lorene Bronstad, Maggie Perry, Willie Mae Hanson, Veta Cox, Beatrice Bertelsen, Leona Parsons, Gladys Tindall, and Addie Cox. Photo: The Bosque County Collection; Bosque County Historical Commission, Meridian, Texas

◊ On July 1, 1900, the 90-year birthday of Liv Grimland was celebrated. She was born in Ufsvatn, Vegårdshei, southern Norway, in 1810. She emigrated with her husband Kittel Grimland and their children in 1850. The Grimland family lived among other Norwegian immigrants in Henderson County, East Texas, for 17 years before they moved to Bosque County in 1867. In the row behind the children, from left to right: Her sons J. K., Niels, Gunsten, Yern or Jørgen. To the right of Liv Grimland: her daughter Kirsti Grimland Solberg, Inger Halvorsen Grimland (the wife of Yern). Behind Kerste and Inger, standing: her daughters-in-law Mary Brown Grimland, Annie Olson Grimland, and Mary Johnson Grimland. Everybody attending this birthday party spoke Norwegian, children, in-laws, and grandchildren. Liv Grimland never learned to speak English. Photo: The Bosque County Collection; Bosque County Historical Commission, Meridian, Texas

◊ The house on the Conrad Knudson farm in the Mustang area, southwest of Meridian. The house was built in 1905 by Geof and Gunner Shefsted. In front of the picket fence stands a man with a horse, another man, and six women and girls. To the right of the house and trees is a windmill; farther to the right can be seen the large barn, built in 1909, as well as other outbuildings. Photo: The Bosque County Collection; Bosque County Historical Commission, Meridian, Texas

◊ The golden anniversary celebration (*right*) of Jørgen and Inger Grimland, July 1, 1912. Jørgen, or Yern, was born in 1833 on the Grimland farm in Åmli, Aust-Agder County. He emigrated with his parents and siblings to East-Texas in 1850, and the family moved to Bosque County in 1868. During the next decades Yern was the most politically active of the Norwegian immigrants in Bosque County. On July 1, 1862, he married the ten-years-younger Inger Halvorsen. She had immigrated with her parents from Løten, Hedmark County in 1853. Note the American flag to the left and the Norwegian flag to the right. All persons on this picture were Norwegian immigrants or the children of immigrants. In the row behind the children, from left to right from no. 3: Tellef Grimland, Maren Grimland, Clara Swenson Grimland, J. K. Grimland, Annie Olson Grimland, Ole Solberg, Kerste Grimland Solberg, Niels Grimland, Mary Johnson Grimland, Gunsten Grimland, Sophie Brown Grimland, Yern Grimland, Inger Halvorsen Grimland, Mrs. Thorson, Pastor Thorson. Behind Yern, and dressed in white, are some of Inger's sisters. To the right with a child on the arm: Hilma and Oscar J. Rea. Photo: The Bosque County Collection; Bosque County Historical Commission, Meridian, Texas

◊ The Norwegian Clifton College, under construction in 1897. The opening of this college was a watershed in the history of Norwegian immigrants in Bosque County. On the one hand, the college represented the peak of Norwegianness. On the other hand, it signified a growing transition to stronger assimilation in and adaptation to American society. Many fathers and sons of fathers were involved in the establishment of the college. From left to right: Ole Knudson, Frederick Hogevold, Aslak O. Seljos, Torger T. Hogevoll, Dan Nelson, Nils Jacob Nelson, and Hendrick Dahl. Photo: Bosque County Historical Collection

◊ The John and Martha Pederson house, Boggy community. There are many such photographs of Norwegian families from the decades before 1900 with a similar appearance. These photos were proof that the family was doing well in Texas, and could be proudly sent home to relatives in Norway. Photo: The Bosque County Collection; Bosque County Historical Commission, Meridian, Texas

CHAPTER 10

A Growing Norwegian Colony in Bosque County, despite Indian Threats

◊ ◊ ◊

RECONSTRUCTION IN TEXAS began in June 1865. Union occupation troops arrived, Confederate troops were paroled, and the Emancipation Proclamation was read on June 19, 1865. Slaves were proclaimed free. In most cases, plantation owners released their slaves as the news of Appomattox reached them, but some refused to accept it and tried to conceal knowledge of emancipation from their slaves.[1] In some cases, reality did not sink in until federal troops were literally standing on the doorstep of their plantation. In the view of James C. Kearny, the surrender of the Confederacy did not end the bloodshed of the Civil War. "Rather it ushered in another phase where the "lost cause" became a pretext and cover for terror and lawlessness, rendering Texas the most violent state in the Union for a decade after the war ended."[2]

As James L. Roark observed, "Planters' basic ideas about slavery, blacks, agriculture, and Southern civilization revealed a remarkable resistance to change," not only during the Civil War but also during Reconstruction.[3] According to Harold D. Woodman, "Northerners expected that the emancipation of the slaves would turn the plantation South into a land of small farms and shops similar to the North."[4] This would require the breakup of large landholdings and distribution of small farms and equipment to freed blacks. But the argument in favor of radical economic change did not prevail. The majority view was that the abolition of slavery would create a free economy. Freed slaves without land or other property should work for the people who owned the resources, but now in freedom. Former slaves who managed to save part of their income, it was hoped, would later be able to invest in their own land.

Union authorities expected freedmen to sign up for work. "Federal officials early observed that most freedmen desired an alternative system to that wanted by landowners," commented Carl H. Moneyhon. "The preferred arrangement, usually called sharecropping, offered the worker an opportunity to cultivate a particular piece of land in return for a share of the crop."⁵ Freedmen wanted to control their labor conditions, be free from white authority, and have economic autonomy, argued Eric Foner. "These aims led them to prefer tenancy to wage labor, and leasing land for a fixed rent to sharecropping."⁶

Agriculture in Texas experienced setbacks during the Civil War, but soon recovered and reached unprecedented heights with respect to value and productivity. Many former slaves in Texas ended up as tenants, either share tenants or sharecroppers. Share tenants might provide their own seed, work animals, and equipment. In this case they "paid the landlord one-fourth of the cotton and one-third of the corn as rent," wrote Randolph B. Campbell. Sharecroppers provided only their own labor and paid the landlord 50 percent of the cotton crop. "Share tenants had more control of their farming operations than did sharecroppers, and enjoyed a higher status. Still, one basic characteristic united all tenants—they owned no land."⁷

AGRICULTURAL GROWTH IN BOSQUE COUNTY

Bosque County was still on the Indian frontier in Texas after the Civil War. Between 1860 and 1870, the population of the county more than doubled, from 2,005 to 4,951, an increase of 147 percent; it doubled again in the 1870s and grew to 10,219 in 1880. The county had eighty-five farms in 1860, with 4,953 improved acres of land and 42,546 acres of unimproved land. In 1887, 77,050 acres of land were under cultivation in the county, and 96,882 acres were used as pasture.

The climate in Bosque County was "dry and very healthy," wrote T. C. Alexander in the *Texas Almanac* in 1867. "The soil on the Brazos and Bosque river valleys is rich sandy alluvial, and on the higher, black, stiff soil, not so productive of corn or cotton, but perhaps better for small grains." So far cotton was "but little planted, but it promises a fine yield." It was expected that the cotton crop would be in the range of 200 bales in 1867. "Wheat is the heaviest crop ever raised, and is now selling at $1 to $1.25 per bushel." The county had plenty of wood for fuel but lacked timber for fencing. "There is, however, an abundant supply of limestone rock for building houses or fences, and it is our best building material." Unimproved land could be bought for $1 to $3 per acre. "A hand can cultivate 25 acres in corn, or 40 acres in corn, wheat, and barley, except during harvest, when he will require help. The majority of negroes have done tolerably well. They do about two thirds of the work they formerly did; but most of them are very unreliable." They had decreased in numbers since emancipation. White laborers were paid 25 to 50 percent more than black laborers, except during harvest, when the great scarcity of labor and machinery brought wages for both white and black up to $1.50 and $2 per day.[8]

Hardly any Norwegian immigrants arrived in Bosque County during the Civil War. Some of the immigrants who stayed in the county during the war moved to the Midwest afterward. The number of foreign born in the county in 1870 was 277 (5.6 percent of the population). Most of them were Norwegians.

From their arrival in 1854 and until the outbreak of the Civil War, Norwegian farmers had improved only a small percentage of their land. During the war, many of the men were away in the Confederate Army. Older men, women, and children had their hands full with sowing and harvesting as much of their land as they were able to manage. A long and bloody civil war had finally ended, wrote Berger T. Rogstad in a letter to Romedal in Hedmark County in summer 1867. The war had hurt the whole of the United States. "But Texas was the least hurt of all United States so we didn't lack anything," Rogstad maintained. Wheat was the most important crop among the Norwegians. The harvest would soon begin, and it looked as if 1867 would become a very good year for wheat. Rogstad was fond of horse breeding, and he owned more than "thirty fine mares and horses."[9] The Rogstad family owned 320 acres of land. But even as late as 1867, only 40 of these acres had been cultivated. Nevertheless, in 1866 Rogstad had harvested "450 bushels of wheat, rye, oats and corn that we feed to our work horses. I'm now going to the market with 200 bushels I have left of last year's harvest. Wheat is sold for one dollar and forty cents a bushel. I've bought a reaper, and four horses, and two men cut from twelve to sixteen acres a day." The reaper had cost him more than $200, but "this money is soon returned since you have to pay a dollar an acre to have your wheat cut."[10]

The cost of labor was high during the first years after the Civil War. In a letter to her friend Karen Poppe in Lillesand, Aust-Agder County, in September 1868, Elise Wærenskjold at Four Mile Prairie wrote that it was almost impossible to get cheap workers to work the land anymore. She rented her own farm "to negroes and get half of the harvest. I would have preferred to rent it to a white man for the year if only I could have had a Norwegian, but this is no longer possible after the war." On the other hand, "the negroes are very well behaved, much better than the white Americans."[11]

People in Texas had little cash, and "prices are low on all things except for wheat, for which we've had a good price," wrote Hendrick Dahl in a letter to his father-in-law, Peder Eriksen, at the farm Furuset Nordre in Romedal, Hedmark County. Dahl had traveled to the Texas coast and bought a threshing machine in 1867. The wheat harvest in 1868 seemed to be failing all over Texas. "The grasshoppers came in such numbers the last fall that we couldn't sow the wheat until just before Christmas. They laid eggs last fall and in early March the young insects came out and destroyed almost all the wheat. We have now planted corn instead. And it looks good and we may still expect to have a good harvest. But corn doesn't give as much cash as wheat."[12]

In June 1869, after Dahl had harvested his own grain, he harvested for others. "This year I had 500 bushels of wheat, 170 of barley, and 100 bushels of rye. The corn looks as good as I've ever seen it." The price of wheat was high; many farmers had not risked growing wheat because of the grasshopper threat. Dahl still farmed on a small scale. Wheat sold for one and a half to one and three-quarters dollars, and rye and barley for 1 dollar. The prices for grain were low compared to what he had to pay in wages, and it was hardly profitable to hire workers. "We can plow all year and we don't need fertilizer. I now farm sixty acres and in good years I've harvested 1,200 bushels or even a little more, and all this has been done by myself and my two little boys, except that I had a man to plow for me for six days at seventy-five cents a day last fall when I sowed the wheat. We cut wheat for three and a half days and had six men and paid them one and a half dollar a day. So my farming expenses for this year were seventy-four dollars. But then I had my own mowing machine and threshing machine and without them my expenses would have been greater."[13]

FINALLY, THEIR OWN LUTHERAN PASTOR

Norwegian immigrants in Bosque County had repeatedly written to the Norwegian Lutheran church in the Midwest begging for a Lutheran pastor to be sent to Texas to tend to the Norwegian flock in the state. No pastor was available, was the answer, and no pastor visited them between 1854 and 1867. After the Civil War, the Norwegian colony in Bosque County became the largest in Texas. According to Hendrick Dahl, in May 1868 around 200 Norwegian immigrants including children lived in the Norse colony, and "we expect many families from Four Mile Prairie and some from Norway."[14] Elise Wærenskjold at Four Mile Prairie agreed with him. Norwegians arriving in Texas after the Civil War "go to Bosque, which now is the largest Norwegian settlement," she observed in September 1868.[15]

In early 1867, the Norwegian Lutheran Church in Wisconsin sent Pastor S. R. Reque to Texas to help the settlements establish their own churches. Aslak Nelson at Four Mile Prairie covered his visit extensively in a letter to his parents, brothers, and sisters back in Norway in April 1867. Reque had stayed at Four Mile Prairie for two weeks.[16] He held worship one day and catechetical classes the next. The reverend suggested that the Norwegians at Four Mile Prairie should elect three councilmen for their Norwegian Church. This was done, and the members were Stian Aanonsen from Frolands Verk (ironworks), Ole Evensen Mjaaland, and Aslak Nelson.[17] All of them came from the same region in Aust-Agder in southern Norway.

Pastor Reque also visited the Norwegian colony in Bosque County. Knud Salve Knudson from Four Mile Prairie accompanied him on the 180-mile journey on horseback. The visit was a big event at Norse. He stayed for two weeks, wrote Berger Rogstad in a letter to Norway. He held Communion and baptized fourteen children. "He baptized our two youngest, both girls, Anna Marthea, four and a half years, and Perine Pouline, two and a half years. We also have three boys, Johan, twelve years, Thomas Renhart, ten years, Bernt Adolf, seven years."[18]

THE ARRIVAL OF PASTOR ESTREM AND HIS FAMILY

In a letter to his parents-in-law in Romedal in June 1869, Hendrick Dahl was happy to report that Bosque County finally had their own permanent minister: Pastor Ole Estrem had arrived with his family a few weeks earlier. Hendrick felt Estrem would become important for the community, "in particular for the young people who have had little religious education. Our children have gone to English school but don't know much Norwegian, so now we'll have a Norwegian school so the children can be prepared for confirmation."[19]

On June 3, 1869, Ole Olson Estrem (Østrem), his wife Josephine, and their three children stepped down from "a covered ox wagon after a long and toilsome journey of two weeks and five days from Illinois." The family had traveled down the Mississippi River to New Orleans, then by boat to Galveston, Texas, and by train north to Bryan, where the railroad ended. The last stretch to Bosque County was done by stage and ox team "through the trackless wilderness." The Estrem family stayed in Bosque County for eight years, laying the foundation for the Norwegian Lutheran Church in Texas.

Ole Olson Østrem was born on December 18, 1835, on a farm with the same name west of Haugesund, Rogaland County, and died on July 17, 1910.[20] He emigrated in spring 1857 and arrived at Leland, Illinois, on May 28. He had worked as a teacher in Norway, but his dream in America was to become a pastor. He studied at Beloit College, Beloit, Wisconsin, from 1858 to 1861 and completed his theological studies at Augustana Lutheran Theological Seminar in 1862, followed by ordination in the Augustana Synod. Between 1862 and 1864 he worked as a pastor near Calmar, Iowa, and for a while he was a civilian chaplain in Union Army camps and on the battlefields at Missionary Ridge, Chickamaugua, and Chattanooga. He married Josephine Emilie Amundson in White Water, Wisconsin, in November 1863. The following year the newlyweds lived in Paxton, Illinois, where Estrem taught Norwegian at the Theological Seminary of the Augustana Synod. The synod financed two more years of study for him, and he and his wife moved to Philadelphia. He studied at the Mt. Erie Seminary of the General Council and finished his education at Concordia Lutheran Seminary in St. Louis, Missouri. When he left Missouri, Estrem was a dyed-in-the-wool conservative, orthodox Lutheran in the mold of the Missouri Synod.

Our Savior's Lutheran Congregation was formally established at Norse on June 14, 1869, fifteen years after the first Norwegian pioneers arrived. Pastor Estrem opened the meeting with prayer and read from the letters to the Romans, chapter 7. The men present elected Estrem president and Peder Pederson secretary. Members of the church had to be 18 years of age to have voting rights, and they had to be confirmed. Third on the agenda was a contract for the salary of the pastor. It was decided to follow the Norwegian church tradition of having an offering for the minister on holidays like Christmas, Easter, and Pentecost. Pederson was also chosen as treasurer and sexton of the congregation.[21] He was known as an excellent reader. Bergsvend Erickson, Aslak Nelson, and Hendrick Dahl were elected deacons.[22] It was decided to establish a Norwegian religious school, and from mid-July over the next four months the pupils would prepare for confirmation, taught by Pederson, who had been a schoolteacher before his emigration in 1867. The parents of boys and girls to be confirmed would pay him 20 dollars a month.[23]

TIME TO LEAVE FOUR MILE PRAIRIE FOR BOSQUE COUNTY?

Knud Salve Knudson from Four Mile Prairie had accompanied Pastor Reque during his visit in Bosque County in 1867. Knudson was very impressed by the good wheat land and the excellent drinking water in Bosque County. "With its rocky mountains and running streams of clear water, it reminded him of Norway, and there were no epidemics."[24] Some Norwegian neighbors and relatives from Four Mile Prairie had already moved to Bosque County. Upon his return home, Knudson argued very strongly in favor of leaving Four Mile Prairie and moving to Bosque County. Several of his Norwegian neighbors agreed, and it was decided they would move to Bosque County after the harvest in 1867.

Many of the Norwegian immigrants still living at Four Mile Prairie found the place unhealthy. In summer 1866 the colony had again been struck by severe illness and many deaths. In a letter to Norway in April 1867, Aslak Nelson gave an enthusiastic description of Bosque County without ever having been there. The land was excellent for wheat; some of the Norwegians had harvested 800 bushels last year! They had also harvested a great deal of corn. The place had "a very healthy climate because it is high and the Norwegians there have no or very little sickness since they came there with the exception of a few colds."[25] Bosque County had few trees, but plenty of rocks; "yes, very useful rocks, they are soft so that one can cut them, and many have built rock houses. They have also started to build rock fences and they will last forever." The valleys in Bosque County were half a mile to 3 miles wide, there were small rivers and streams, small mountains, and fields in between, and "a great deal of small trees called Cedar." Land was even cultivated on top of small hills.[26]

Nelson personally had nothing to complain about regarding where he lived. He had more work than he could handle making shoes and boots. "I make more money doing this than anything else that I am capable of doing." His oldest son, Nils Jacob, was now 17. He was big and strong and earned more than his father.²⁷ About thirty Norwegian families lived at Four Mile Prairie, altogether 150 to 160 souls. The Norwegian population was decreasing. The grass at Four Mile was good for cattle, but their cattle had to be fed in winter. The land was best suited for growing cotton; it was not good for wheat.

All plans to move to Bosque County were canceled when disturbing news was received from Bosque in early April 1867. Ole Nystøl, the 15-year-old son of the Nystøl family from Åmli, Aust-Agder County, had been captured by Comanches, and Carl Questad had been wounded but escaped. Because of the Indian threat, Nelson wrote, he and many other Norwegians had put on hold any plans to move to Bosque County; "the Indians come around and steal horses, and they don't think twice if they decide to kill people."²⁸ The Indians had also recently killed one white and three blacks and taken with them three teenagers. Elise Wærenskjold expressed the prevailing attitude toward Indians among the Norwegian immigrants most succinctly. "To tell the truth," she wrote in a letter to Norway in April 1867, "the Bosque settlement is more prosperous than ours and I also think it may be more salutary, but I wouldn't have lived so close to the Indians for anything. I grieved to bury my little Thorvald but it would have been far worse to know that he was among wild and heathen people who would torture him everyday and bring him up as a heathen."²⁹

THE INDIAN THREAT IN WESTERN TEXAS

It is very rare, almost exceptional, to find the Indian threat on the frontier mentioned in letters from Texas to Norway. Indian raids and capture of children were recurring events, and 1867 was the worst year for it in Texas history. But this time it was Ole Nystøl, one of their own, who had been captured.

The number of Indian raids was still considerable the first years after the Civil War in counties west of Fort Worth. In his book *The Settlers' War: The Struggle for the Texas Frontier in the 1860s*, Gregory Michno argued that the most violent clashes with Indians in Texas involved Indians and settlers (both Anglos and Hispanics) rather than Indians and soldiers. The 1860s was "the bloodiest decade of the Western Indian wars" and 1867 was the peak year.³⁰ Michno used Indian depredation claims in the National Archives as his primary source. Of more than

2,200 depredation claims, about 1,200 dated from the 1860s. The number of Indian raids on Texas soil declined in the 1870s, when the settlers' war also became the soldiers' war. The Red River Indian campaign of the US Army in 1874–75 more or less removed the Indian threat.[31]

"Much of the folklore that goes into creating Texas and Texans rests on the image of nineteenth century settlers fending off Plains Indian raiders," wrote David La Vere in his prizewinning book from 2004, *The Texas Indians*.[32] The classical narrative on Texas Plains Indians was still present in the third edition of Robert A. Calvert, Arnoldo de León, and Gregg Cantrell's *The History of Texas*, published in 2002. Before, during, and some years after the Civil War, they wrote, "Comanches, and their Kiowa allies, still lorded over the plains." Warfare was traditionally valued as an important source of prestige and honor in their Indian culture. Most white settlers did not dare to enter Indian territory because of the "deep enmity the Plains Indians held for white intruders, hate displayed in the sadism both tribes extended to new interlopers." Comanches and Kiowas were known to torture prisoners and to mutilate corpses. "Comanches and Kiowas, further, often abducted white women and children."[33] During the Civil War and the first years after, the Comanches controlled the territory "from southern Kansas, west to the Pecos River in New Mexico, south to the German-descent populated town of Fredericksburg, then north up the 98th meridian." The Indians swept into northern and western lands to "devastate settler's livestock and farmsteads."

Randolph B. Campbell's language was somewhat milder in *Gone to Texas: A History of the Lone Star State*. "Defenders sent to the frontier by confederate and state authorities during the war went home in 1865, and the United States, busy with numerous postwar responsibilities, did not rush to reoccupy the forts evacuated in 1861." The Indians used the opportunity and "struck settlements along a line from Gainesville to Waco with murderous effect."[34]

In an article on "The Emergence of a New Texas Indian History," Pekka Hämäläinen characterized David La Vere's book *The Texas Indians* as the "first comprehensive overview of Texas Indian history in two generations."[35] La Vere tried to give Texas Indians "center stage." Texas Indians were neither "noble savages" nor "red devils." They were humans with their own religious beliefs, political structures, kinship networks, economic strategies, and obligations of reciprocity. La Vere used a different language than many of his predecessors, observed Hämäläinen. He abandoned "dated, racially laced classifications and sets out to demystify the Indians by peeling off stereotypes. His discussions of Native cultural traditions, social arrangements, and political actions are sprinkled with adjectives like "dynamic, "complex," and "resilient"—staple epithets in the New Indian History."[36]

Between 1840 and 1880 just about every Texas community west of Waco experienced conflicts with Indians. "The raiding of Strangers, people who were not of one's own people or kin, was an ancient economic activity among the Indians."[37] The other side of the coin, argued La Vere, was that Texans "made their own raids, and they were never hesitant about attacking Indian camps, burning tepees, destroying supplies, stealing Indian horses and hides, taking captives, or gunning down warriors. Naturally both sides believed they could justify their own actions."[38]

Some raids were just simple horse thievery, where the horse owners did not even know they had been raided before the next morning. Sometimes the raids were unforgettable shows of "daring bravado, with a small raiding party swooping down and making off with the horses right in front of the wide eyes of their astonished victims," observed La Vere. "Other times, raids could be supremely violent, with men, women, and babies killed, captives taken, homes or equipment looted, or burned, and horses driven off. The question was not whether raids took place, but why so many occurred."[39]

Why was bravery in raids such an important part of status in the culture of Southern Plains Indians? "A man who continually acquitted himself well on raids eventually began leading raiding parties," La Vere explained. "Success gained him a following, which translated into political power. He acquired it through bravery and owning many horses." Because the Indians worked their horses very hard, they did not last long. A continuous supply of horses became very important. According to La Vere, the average Plains Indian household needed ten horses—five for hauling the tepee and household goods and another five for buffalo hunting and raiding. Since Plains Indians did not breed horses, they had to get them through raiding, which was easier than breaking wild mustangs.

Taking captives during raids was also part of the Plains Indian culture. Capturing women and children past infancy had first priority. White captives were needed as slaves. "Taking captives and adopting them into one's society was a time-honored method of increasing a band's population since the 17th century ... Both sides certainly had issues with the other. Texans could not understand why a peace treaty with one Comanche chief did not carry over to other Comanches."

THREE MONTHS AMONG THE INDIANS

On Saturday, March 20, 1867, Carl Questad and 14-year old Ole Nystøl were traveling by wagon to a cedar brake northwest of their farms to chop and haul some poles for fences. Soon after their arrival they were attacked by Comanche Indians, who took the teenager with them.[40] The capture of Ole Nystøl must be understood within the above context. Neither Carl Questad nor Nystøl were killed, and Nystøl was a captive among the Comanches for only three months, while many others lived as captives with the Indians for years or even the rest of their lives.[41]

Twenty years after the incident, in 1888, Nystøl published his version of what had happened to him. In the preface, dated Meridian, November 14, 1888, Nystøl asked for the indulgence of his readers with regard to its defects concerning literary style. What he had tried to write was "the naked truth, though unadorned, nevertheless the truth, which can be verified by many living witnesses in and around Clifton, Norse, and Meridian, Bosque County, Texas."

Ole Tergerson Nystøl was born in Henderson County, Texas, on January 4, 1853. His parents had arrived from Norway in 1848. When he was eight years old, his mother died. His father then moved with his family to Four Mile Prairie, Van Zandt County, but continued on to Bosque County in 1866.[42] Ole mourned the death of his mother strongly. On her deathbed she had asked her husband to look especially after Ole since he was so willful, and she prayed that "God might guide me in the way of righteousness."

On Saturday, March 20, 1867, their neighbor to the east, Carl Questad, came by the Nystøl farm with a wagon and asked if Ole would like to go with him to a cedar brake to find suitable cedar poles.[43] The two of them started working as soon as they reached their destination, five miles distant. Questad had gone off about fifty steps to commence work when Nystøl, still at the wagon, heard a noise. Looking up, he saw two Indians, "made hideous with war paint. At about the same time they saw me, and giving a few blood-curdling yells, started toward me. It appeared to my excited imagination that they were devils who had come for me and I really thought I could see a great stream of fire issuing from their mouths. Having been taught that the "devil would get me" if I was not good, it is not very strange after all that I felt as I did, under the circumstances."[44] Nystøl started to run, and "had got about forty yards when an arrow pierced my right leg, passing entirely through the flesh part, just above the knee." He fell; one of the Indians leveled a pistol at him, and the flight was over. Questad started to run at the same time he did. "In his course there was a bluff about twenty feet high, but on reaching it he never stopped to measure the distance." Questad leaped at once "down the precipice, landing safely below, none the worse save a few scratches and bruises. During the chase he was fired at several times, one shot taking effect in his right arm. I never knew his fate until I was liberated, but I supposed from their broken English and gestures that he escaped."[45] Questad later told Nystøl that a third Indian had tried to intercept him, but that fear had given him additional speed. "His bloody condition and wild, excited appearance greatly alarmed those to whom he returned." A party of men was soon organized to rescue Ole Nystøl, but the group lost the track and had to turn back.

Nystøl was brought to the Indian camp, about 40 yards away. The party consisted of six Indians; three had been engaged in cooking a meal of broiled horseflesh. They had just killed a horse and cut some choice steaks. "Comanches! Dreaded name. Synonym of all that is cruel and barbarous. What terror that name inspires along the defensive frontier. And it was amongst these monsters in human shape that I had fallen a helpless victim. By this time I was suffering severely from my wound."[46] To increase his already excruciating pain, the Indians from time to time would twist and wrench the wound area. "They would kick and knock me about just for pastime it seemed, whip my bare back until it was perfectly bloody, with frequent repetitions fire their pistols held so close to my head that the caps and powder would fly in my face, producing powder burns and bruises, until I was very much disfigured. In fact, I was used so roughly that when I got loose from them my head was a solid sore, and the scab had risen above my hair."[47]

Before the Indians left the campground, Nystøl had to take off his overshirt. He was placed on a poor, bony horse without a saddle. The Indians chose a northwesterly course. "We had traveled about three miles when we came upon a man and his son by the name of Fine who were hunting a horse that had been stolen by these Indians. When they saw us they ran to a live oak thicket near by, tying their mules and concealing themselves in the dense brush." They both escaped, the man with a wound in his hand. The Indians took the mules, however. "Just before sundown we came upon a negro man with a wagon. He saw us when a half mile off and came running toward us begging for his life. At first they seemed disposed to heed his petition, but the thirst for blood triumphed over their better nature, and amidst his cries for mercy they stabbed him to the heart. He sank to the ground without a groan, save the death-rattle in the throat. One of them then pierced him through with his spear, it coming out at his breast. They left him unscalped and showed by signs and grunts their disgust for such a scalp." They pointed to Nystøl's scalp, and indicated that his scalp was much finer. "They commanded me to laugh at this horrible deed." Nystøl thought it was his turn next.

The party traveled continuously for five days and nights. During this time Nystøl was never off a horse, except when he was lifted from one to another. In the afternoon of the fifth day, they came to a mountain in Stonewall County, where the Indians had hidden firearms, a tent, blankets, and so on. They erected the tent and slept inside while Nystøl was outside freezing. He could not escape, because his wound was still too severe, and he could not walk but had to crawl. "It became very cold during the night having commenced to sleet and snow." He found a small cave in a nearby embankment and was now more comfortable. The Indians thought he had escaped. Nystøl went into their tent and sat there until they returned. The Indians were mystified that he was in the tent, and asked through sign language where he had been. "I pointed up and indicated I had been in heaven. At that they showed signs of wonder and amazement," wrote Nystøl. "As they are very superstitious and ready to relegate anything not easily accounted for to the supernatural, this may have been of unmeasured advantage to me, as it doubtless led them to believe that I was under the protection of the 'Great Spirit.'"[48]

On the sixth day after capture, the Indians had traveled so far to the north they felt they were out of danger. When horses and mules were watered late in the afternoon, the mule Nystøl was riding on was slow to get to the water. One of the Indians became so angry that he just shot it. Nystøl landed in the water on his head. "They dragged me out and threw me on another horse with no more ado than if I had been a sack of corn. In a little while my coat was frozen to me but I dared not complain." The day after, Nystøl was forced to dig for roots to eat with his bare hands. He finally "grew desperate and jumped up from my work." An Indian knocked him into the water; upon arising, he ran away as fast as he could. "The Indian jumped to his feet and started after me, but soon found that I was too fast for him, so he got a pony and I was soon overtaken and carried back. This again seemed to awe them, to think I could outrun them."

"While here surrounded by them I fell upon my knees and prayed to God that if it was his will I might be free, but if not that I might have a saved soul. A shadow seemed to pass over me and a voice just as audible as any I have heard in my life said to me: 'Be in peace, you shall be free.' Immediately peace and quietude came to my heart, and all my fears and anxieties vanished. From that time I never despaired of my final escape." The Indians also felt the change, Nystøl wrote. They regarded him as a "spirit" or something. On the ninth day after his capture, he ate his first food again.

The Indians continued to travel in a northwesterly course over the Staked Plains. After three weeks on horseback they came upon Indian wigwams. "The duties assigned me in my new home were herding horses, carrying water, getting wood and running horse races and occasionally joining in a buffalo hunt." The Indians moved their camp from time to time. "The idea of escaping whenever a favorable opportunity presented itself, never left my mind." At last the opportunity occurred. "It was one evening just after night fall, and a very favorable night too, dark and threatening rain. I proceeded cautiously, procuring one of their fleetest horses and started, directing my course eastward and I had traveled but a short distance when it commenced raining and continued to do so all night." A little after daybreak, however, he could see the Indians pursuing him. "When I found that they would certainly overtake me, my horse having given completely out, I dismounted and lay down in the grass. I was laughing when they came up. I tried to treat it as a huge joke so as to disarm them of anger. They asked me if I was trying to run away, and the only reply I made was a laugh, having learned that that was the best way to get out of trouble."[49]

After this incident, it seemed the Indians made up their mind that it would be best to get rid of Nystøl. "They could get a ransom for me by taking me to a trading post or Indian agent and that would be better than to let me get free. We were now at the Big Bend of Arkansas River,

Kansas, which we crossed and found a trading Post kept by one Mr. Eli Bewell, and his family, together with two other men for protection." The post was located near Smoky Hills, Kansas. "After some parleying it was agreed that $250 should be the price, to be paid in brown paper, blankets, tobacco, flour and sugar and perhaps some money." A week after Bewell bought him, the family moved to Council Grove, Kansas. Bewell and his wife wanted to keep him and tried to persuade him to stay with them for good. Nystøl, however, was anxious to get back to his own people. "Mr. B became so attached to me and was so anxious to keep me that for two months he moved from place to place to evade the search of the Indian Agent, knowing that if he found me he would send me home if I wished to go."

Finally, the Indian agent, Colonel Leavenworth, heard of Nystøl's release. Colonel Jesse Henry Leavenworth had been appointed Indian agent by the Union Army in 1864 for the Southern Cheyenne, Kiowa, and Comanche tribes, with the mission of trying to keep peace with the tribes. He was instrumental in the release of a number of captive hostages taken by the tribes.[50] When they met, Nystøl told him he wanted to return home as soon as possible. After Colonel Leavenworth had a talk with Bewell, it was decided. Nystøl bid "a sad adieu to my kind friends who had released me from bondage."[51] From the mouth of the Little Arkansas River he went to Fort Washita in the Indian Territory with a government train, and from there to Sherman, Texas, with an ox wagon. He then walked most of the way to Milford, Texas, where he became very sick. "I had been there about a week and was recovering from my sickness when some of my old acquaintances, Messrs. K. Hanson, Y. and K. Grimland, called by to inquire the way to Hillsboro, Hill County." On seeing them he jumped out onto the street.

His bills were taken care of and he joined them in the direction of Hillsboro. "Just before reaching home we met my father going after me, he having received word from the people at Milford that I was there." His father did not recognize him until he jumped in front of him. The joy at the reunion of father and son was indescribable.

THE DELAYED EXODUS FROM FOUR MILE PRAIRIE BEGINS

In summer 1867, a new and very severe malaria epidemic hit the Norwegian colony at Four Mile Prairie. Twelve persons in thirty-four Norwegian families died, including Asborg, the 18-year-old daughter of Knud Salve Knudson. After the sick had recuperated and the dead been buried and mourned, the question of leaving Four Mile Prairie in favor of Bosque County came up again. The Indian threat in Bosque County might be bad, but was it worse than all the sickness and death in Van Zandt County? Despite the risks of Indian attacks in Bosque County, several families decided to leave everything they had built up at Four Mile Prairie.

Knud Salve Knudson was the driving force behind the decision, but he was supported by many others. Most of the families had originally emigrated from Aust-Agder County in Norway and had been among the first to arrive in the 1840s and early 1850s. They had stubbornly continued to live in East Texas, but enough was enough. Knudson's 65-year-old father, Salve Knudson, did not get much for his land when he left, but he "found a man with whom he could swap his three room house, his land, and an apple orchard for a wagon and three yoke of oxen. He loaded his household goods on the wagon and along with eleven other families and his cattle, headed west for Bosque County."[52]

On arrival in Bosque County, the Salve and Knud Knudson families chose land on Meridian Creek, several miles west of the existing Norwegian homesteads at Norse. This was still the frontier, and they deliberately chose to settle on 160 acres of vacant public land, where Indians might raid them from the west. The land was registered in July 1868, and the requirements necessary to get title to their land were fulfilled at the end of 1872. On January 7, 1873, they received a document, signed by the governor of Texas, which granted to them and their heirs forever 160 acres, measured and marked by cedar trees, post oaks, and stumps.[53]

Neighbors helped them build log cabins. Before the cabins were ready, the Knudson families camped under a large live oak tree at the entrance to their claim. A wide circular swath of tall grass was cut around the cabins. It would serve as a firebreak and would also be of help if they were attacked by Indians.[54] Salve Knudson built his log cabin to the east of that of his son, and they built rock houses, about one and a half mile apart, at the same time in 1873. Limestone was dug from nearby mountains, hauled to the building site, sawed into blocks, put together, and plastered inside and out with mortar produced in a large kiln nearby. Shingles for the roofs were made by squaring logs and splitting them into thin sections with a froe. Additional rooms were added in the 1890s.

Salve Knudson died in 1889, at the age of 86. Three years after his death, his wife Kari, who could not write, signed an X on a contract with her son Jacob. The agreement specified that he owed his mother 700 dollars for the homestead. If Kari chose to live with him and his family, the 700 dollars were to be considered remuneration for her support and maintenance. If Kari chose to live elsewhere or Jacob failed to support his mother, the loan would become payable. The provisions for Kari included all necessities for life, a warm and separate room, and "a friendly treatment as well in health as in sickness."[55] Jacob further agreed to pay each of the heirs of the Salve Knudson estate a specific amount upon the death of Kari. In 1896, Jacob Knudson moved with his family west to Hamilton County. His mother moved with them and lived in their new home until her death in 1913.

Knud Salve Knudson acquired close to 1,500 acres of land in Bosque County between his arrival in 1868 and 1890. When each of the children married, the parents gave the sons some land. Knud Salve and Gunhild Knudson also chose to practice the Norwegian tradition of *føderåd*. They transferred their 160-acre homestead and an additional 250 acres to their son Albert in 1899 for 1 dollar, on the condition that he maintain and support them for the rest of their lives.[56]

Aslak Nelson and his family were members of the wagon train from Four Mile Prairie to Bosque in 1868. After their arrival on June 28, 1868, the Nelson family settled on a homestead a mile west of Clifton. Until their rock house was ready, the family lived in a tent under a big live oak tree near Neil's Creek. Their youngest child, Anton Olaus, was born in this tent on September 24, 1868. At Four Mile Prairie, Nelson had worked as a tailor and shoemaker, and he continued in this line of work in Clifton. Nelson was known as a very religious man, and at times, when no pastor was in attendance, he held services and baptized children. "Many of the entries in the church book at Our Saviors Lutheran church at Norse are in his characteristic handwriting."[57]

The Kittel Grimland family from Åmli Parish in Aust-Agder County had emigrated from Norway in 1850 and settled in East Texas. Kittel Grimland was born on January 12, 1812, and his wife, Liv Grimland, was born on June 14, 1810. They married in 1835. Liv had previously been married to Kittel's brother Jørgen, who had died in 1832. The Grimland family left Arendal for New Orleans on board the ship *Arendal* with their children—Jørgen, from Liv's first marriage, was born on March 11, 1833, Gunsten on April 29, 1836, Niels or Nels on October 4, 1839, Ole on December 14, 1841, Gjeruld October 12, 1844, little Jørgen (Yern) on May 30, 1847, and Kirsti on May 7, 1849. Their youngest son, Reinart, was born in Henderson County on October 20, 1851.[58] Four of the Grimland brothers served in the Confederate Army—the elder Jørgen, Gunsten, Nels, and Ole. Yern Grimland married Gurine Tergerson in Henderson County, and they had three children when Gurine died in 1861. In 1862 he married Inger Halvorsen, and his brother Gunsten married Sophia Brown.

In Bosque County, the Grimland family acquired land in the Neil's Creek valley south of Norse, later named the Boggy community. Kittel and Liv bought 190 acres of land for $1,500 and built a stone house with 18-inch-thick stone walls. Yern and Inger Grimland purchased 360 acres from John McLennan for $320 on Gary Creek, a few miles east of his parents, on November 16, 1868. In later years Yern Grimland became well known in Bosque County as County Commissioner and Justice of the Peace.[59] Gunsten and Sophia Grimland purchased 320 acres from Jackson H. Randall on April 28, 1871, for $650. Nels Grimland had married Mary Johnson in 1867. They settled on a homestead of 160 acres between the farms of Kittel and Yern, which was deeded to them in 1871. Mary Johnson had come to Texas with her parents in 1854. She gave birth to nine children, and developed a reputation in the community as a fearless woman. According to family lore, she once drove some Indians from her kitchen door with a pot of scalding hot water. Another time, on a cold winter day, she discovered a hungry bobcat in the milk pen, which leaped onto the crib holding her baby Neal. She jerked off her apron, threw it over the cat's head, forced it to the ground, and put her knee on its throat until the cat died.[60]

Around 1870, it was still easy to find good, cheap land in this part of Bosque County. The youngest and the unmarried Grimland children continued to live at the home of their parents. Gjeruld died in 1877 when a horse fell with him in the saddle. Kirsti Grimland married Ole Solberg, while J. K. (Yern) Grimland took over the home farm after his father died in 1883 and married Annie Olson in 1888.

Ole Grimland was the only Grimland child who did not marry a Norwegian. He was born on December 14, 1841 in Norway and died on December 17, 1920 in Lampasas County. In the early 1870s he traveled south to work in Lampasas County, where he met and married Annary Utisha Eddy, born on November 21, 1856 in Virginia. The family later moved to Hamilton County, where their children were born. From the time the Grimland family arrived in Texas in 1850 until her death, Ole's mother Liv Grimland never learned to speak English. When Liv visited Ole and his family in Hamilton County, her son had to act as translator between his mother and his wife and children.[61]

Several of the best-known early Norwegian immigrants in Texas came from Holt, Aust-Agder County, not least the extended Reiersen family. Christian and Johanna Canuteson had emigrated from Arendal in fall 1851 and arrived in Brownsboro, Henderson County, in December of that year. Seven years later, the family moved to Four Mile Prairie.[62] Christian Canuteson was born in 1829 at Nes Verk. His wife, Johanna, was born in Holt in 1830. During their married life, they became the parents of thirteen children. The parents and their many children were members of the wagon train from Four Mile Prairie in 1868 and "arrived in Bosque County on July 4, 1868, and got farmland in the Turkey Creek community," where they built a rock home.[63]

The newcomers from Four Mile Prairie knew most of the Norwegians who already lived in Bosque County. Many were related; many had emigrated from the same communities in Norway; some had been on the same ships for many weeks during the Atlantic crossing. Christian and Johanna Canuteson, for instance, had sailed on the same ship as the families of Carl Questad, Jens Ringness, and Hendrick Dahl.[64] The Four Mile group did not arrive in a foreign land. They were already Norwegian-Texans; their families had endured sickness and war, and several sons had served in the Confederate Army.

On May 28, 1868, the Norwegian brig *Atalanta* arrived in Galveston with 114 Norwegian immigrants on board. Bosque County was their final destination. The newcomers represented a new era of immigration from Norway to Texas. Many of them had signed contracts to become indentured servants for Norwegians in Bosque County. The Norwegians had never used slave labor to work their land before the Civil War and did not hire freed slaves to work their land after the war. Their solution to the labor problem was the establishment of a system of indentured servitude, which continued into the twentieth century. Hedmark County was the region in Norway with the sharpest class differences between landowners and cotters. During the next decades, several hundred young men and women left their home communities in Hedmark to become indentured servants in Bosque County.

CHAPTER 11

A System for Indentured Servants from Norway

◊ ◊ ◊

IN SPRING 1867, while the Norwegian community in Texas was in uproar over Ole Nystøl's capture by the Indians, the Norwegian-Texans Poul Poulsen, his son Casper, and Ole Ween sailed from New Orleans to Norway. Their main aim was to recruit and sign indenture contracts with young Norwegians in the communities in Hedmark they had emigrated from. Poul Poulson had gone into a joint venture with the lawyer William McKerral in Waco. Laborers hired through Poulson and McKerral, it was announced, would cost farmers 65 dollars for one year. The money would cover travel expenses from Norway to Texas as well as the fee of the agent. Every laborer would sign a twelve-month contract. On arrival, the farmer would be responsible for the laborer's shelter, clothing, and provisions. At the end of one year, the employer would pay the laborer $30.[1]

The scheme was announced just two years after Appomattox and caught the interest of several newspapers. The New Orleans paper *Picayune* reported that Poulson, "a Norwegian gentleman of good standing in Northern Texas," was ready to leave New Orleans for his native land to recruit the "services of a large number of his countrymen as farm laborers in Texas." The plan to recruit Norwegian laborers reeked of indentured slave labor, some argued. Poulson's partner McKerral in Waco defended the scheme in the Louisiana newspaper *The Planter's Banner*, and his letter was later printed in the *Picayune*, as well as in the *Evening Post* in New York on March 13, 1867, under the heading "Norwegian Laborers in Texas." According to McKerral, Norwegian laborers were "strong, large boned, men and women, honest, sober and industrious. They have no disposition to run about, they are always at home, humble and tractable. It is true they do not speak English, but we do not hire men and women to talk, but to work." Prospective customers could rest assured that his Norwegian partner Poulson would only "take men who have good recommendations, as required by the laws of Norway and Sweden." If Louisiana planters should be interested in hiring Norwegians, the partners could "supply them with two and three thousand just as easily as two and three." Norwegians were raised in humble circumstances, according to McKerral. They knew nothing but work and obeyed orders promptly. "They work in Norway sixteen hours in a day, but I do not suppose we should require that much from them."[2]

Was McKerral describing Norwegian laborers as a new type of southern slave? Did Poulson and McKerral plan to revive the old American system of indenture—in this case using Norwegians to work the land? Indentured servitude had been used on a large scale in the United States from the seventeenth to the nineteenth century and was one of the most common ways for poor Europeans to finance their emigration to the American colonies. Farmers, merchants, and shopkeepers in the British colonies found it difficult to hire free workers because free immigrants found it more profitable to establish their own farms and businesses. To solve the problem of scarcity of labor, employers signed indenture contracts with young workers from Britain or one of the German states. Indentured servants had to work several years to repay their travel costs.[3] Terms of indenture ranged from one to seven years, with typical terms of four or five years.[4] Most indentured servants were younger than 25 and worked as farm laborers or domestic servants, although some were also apprenticed to craftsmen. They were not allowed to marry while under contract. It was not unusual for the father of a teenager in England to sign legal papers with a ship's captain. When the ship arrived in the American colonies, the captain would sell the legal paper for the indentured teenager to someone who needed workers. Indentured servants were not paid wages, but were provided with food, accommodation, clothing, and training. At the end of the indenture, they were free to leave.

The practice declined in the nineteenth century.[5] Many employers found it cheaper to buy and use African slaves. Rising per capita income in Europe and lower ticket prices for Atlantic crossings reduced the supply of European indentured servants on the labor market in the United States. According to David Galenson, the cost of immigration from Britain to the United States over the course of the eighteenth century dropped from 50 percent of per capita income to less than 10 percent. Consequently, it became easier for emigrants to finance their own Atlantic crossing.[6]

To American ears, the recruitment scheme for Norwegian laborers sounded almost like indenture and black slavery in Texas and Louisiana. In the Norwegian context, the sales pitch did not sound anything like slavery. The legal traditions in Norway and the mindset of farmers, tenants, and laborers were different. It sounded like an excellent opportunity to get a fresh start in another country, especially among young men and women from the cotter class.

THE COTTER AND TENANT INSTITUTION IN NORWAY

The Norwegian social institution "cotter" has been mentioned in several chapters, but without any explicit explanation. A cotter was an agricultural worker with his own household, but on land owned by the farmer. It was a permanent holding with a simple dwelling that was rented from a farmer. The cotter's acreage holdings was very small compared with the main farm. The average cotter might have a couple of cows and five to ten sheep or goats. He often reaped too little small grains to feed his family. A cotter's contract specified the duties of the cotter—for instance, how many days he was expected to work on the main farm during the spring, summer, and autumn seasons. Such contracts might specify that the cotter could live and work on his holding for life; and some even specified that his children could inherit the place after the death of the parents. Most contracts, however, specified a date when they had to be renegotiated.[7]

Cotters found themselves on the lowest rung of the agricultural ladder in Norway. They were above day laborers, but below tenant farmers and independent farmers (yeoman farmers). The number of tenant farmers in nineteenth-century Norway declined, while the number of cotters increased. The Norwegian census of 1801 enumerated 77,000 farmers and 55,000 cotters (*husmenn*).[8] The number of cotters reached its peak in the Norwegian census of 1855 with a total of 65,060 *husmenn*. During the next decades, the numbers declined because of emigration to the United States, migration to growing cities in Norway, and legal changes. More and more cotter's places became matriculated—former cotters got title to their land. The number of cotters declined from 60,296 in 1865 to 29,653 in 1890.[9]

A cotter usually needed additional income to feed his family. He could work extra days for the farmer for extra pay. In forest districts, he could help fell trees, haul timber to the river, and float the timber downriver in the spring. Cotters in mining districts could find work connected to mining operations, and in coastal areas they would participate in the fisheries or work as seamen.

If a cotter signed a contract with a farmer for a piece of land large enough to grow food for himself and his family, he was classified as a cotter with land.

Cotters without land rented the lot for a house, but had very little land around the house, maybe a garden, some pigs, and hens. Cotters without land were most prevalent in western Norway and along the coast. Historians often distinguish between east and west Norwegian cotters.[10] The largest numbers of cotters in Norway lived in eastern Norway and the Trøndelag region.[11] In western Norway, it was more common to divide farms between heirs than to partition off cotter's places. This meant that more farmers owned their own land, but the size of the land they owned became smaller over time, and they needed to have additional income from other sources such as participation in the rich spring herring fisheries. Since western Norway had a higher percentage of farmer-owners than other regions in Norway, this led to smaller class differences in the countryside in western Norway. In Hedmark County, the main county for recruitment of Norwegian immigrants to Texas, the class differences between yeoman farmers and cotters were considerable, and they remained so for a long time. This was one of the reasons why socialism had a much stronger hold in the countryside in eastern Norway than in western Norway.

The historian Ståle Dyrvik has repeatedly emphasized the complexity of the cotter institution in Norway. It represented not one type, but several types of economic adaptations. It kept the poorest part of the population surplus under control, and it continued to be an effective system for farmers to recruit cheap and stable labor. On the other hand, Dyrvik argued, the growing number of cotters could be interpreted as a sign of a growing division of labor, specialization, and economic development in Norwegian society.[12] The dominant interpretation among Norwegian historians, however, has been that the growing number of cotters was a sign of social decline in the agricultural society. In 1851, the Norwegian government appointed a commission to analyze the situation of cotters. The commission concluded that many cotters in eastern Norway were worse off than in earlier times. Nevertheless, the situation of agricultural day laborers was even less favorable because of a general decline in wages.[13]

THE BRIG *ATALANTA* BROUGHT A SHIPLOAD OF NORWEGIAN IMMIGRANTS TO GALVESTON

The Norwegian-Texans who visited Hedmark County to recruit indentured servants in 1867 had grown up with the cotter institution before they emigrated. Hedmark was the county in Norway with the highest number of cotters. Competition for jobs in the countryside increased as more and more children reached adulthood. Between 1815 and 1855 the population in the municipality of Stange, where Poulson had grown up, increased from 4,057 to 6,203, more than 50 percent. The sons of yeoman farmers had problems acquiring their own farm, and the pressure to seek work outside the municipality was considerable. Most of the migrants from Stange between 1814 and 1850 (70 percent) were sons and daughters of cotters. Sons of cotters who were married moved to neighboring municipalities, while 46 percent of unmarried migrants moved to Oslo.[14] The deal Poulson offered young Norwegians was regarded as a good offer, and many were tempted. But they could not just drop everything at home and sign an indenture contract. Some were already bound by existing work contracts.

On April 4, 1868, more than a hundred persons left Christiania on board the Norwegian brig *Atalanta*, which was chartered to sail directly to Galveston, Texas. The members of the group were a mixture of indentured servants and emigrants who paid for their own tickets. Some even brought with them a little capital to invest in a farm in Texas. The *Atalanta* had been built in 1852 in Norrköping, Sweden and was employed in the Norwegian emigration trade between Oslo and Quebec in 1864, 1865, 1866, and 1867. On May 27 or 28, 1868, the *Atalanta* arrived in Galveston. In his report to the customs authorities in Galveston the captain wrote that he had brought 114 passengers to the "Norwegian colony in the city of Bosque," 48 males and 31 females over 15 years, 18 boys and 12 girls between 15 and 1, and 3 male and 2 female infants. The passengers brought with them money totaling 1,156 Norwegian Spd. All passengers were well after fifty days at sea.

Never before had so many Norwegian immigrants arrived in Texas. A number of those who paid for their own tickets were related to Poul

Poulson. One of them was his sister Pauline Durie (born February 22, 1814), her second husband, Charles Durie, and their children. Poulson's cousin Ole Hansen and his family were also on board. Ole Hansen had owned his own farm, but economic circumstances had forced him to leave it. Gambling on a future as a farmer in Texas seemed to be as good an option as any other available to him. Hansen was born in Stange Parish, Hedmark, the son of Hans Poulson (1794–1852) and Marthe Fredriksdatter (1796–1890). At the time of the Norwegian census in 1865 he and his wife Mathea Hansdatter, born in 1839, lived on the Kjøstad farm in Løten municipality. They had two small children, Hans (4 years) and Eli (2). In 1865, they had several servants working for them, some of them around 20, while two of the maids were around 50. Ole Hansen was listed as "Sergeant" when he bought the Kjøstad farm from Ole Olsen for 4,000 Spd and *føderåd* on May 7, 1863.[15] In 1867, Ole Hansen went bankrupt, and the Kjøstad farm was sold to Lars Johansen for 2,000 Spd.[16] The farm was large in the Norwegian context; five horses, one colt, twenty cows, twelve sheep, and two pigs were registered on it in 1867.

Other Poulson relatives on board the *Atalanta* were Fredrik Hanson, the brother of Ole Hanson, and his cousins Bernt Godager, Adolf Godager, Christian Godager, Even Jensen Gillund, and Arne Poulson.[17]

BETTER TO EMIGRATE THAN TO DESCEND
THE SOCIAL LADDER IN NORWAY

The passengers on board the *Atalanta* came from divergent social backgrounds, and their individual motives for emigration differed. Some were avoiding descent on the social ladder in their communities by emigrating. The family of Ole Hansen has already been mentioned. The single girl Eline Larsdatter Thingstad also belonged in this category. She was born on September 15, 1845, in Romedal, Hedmark County, and died on December 11, 1929, in Bosque County. Eline was the oldest daughter of Lars Engebretsen Sørli from Odalen, born in 1821, and his wife, Petronelle Pedersdatter Herdal, born on May 23, 1822, Løten. Her parents, who had thirteen children (although eight died as infants or small children) bought the Thingstad farm in Romedal in April 1848 for 600 Spd. To get cheap workers to help work the farm, Lars Thingstad established three cotter's places. A cotter's contract was signed with Peder Christensen in

1853, another with Jens Jensen in 1855, and a third with Ole Johannesen in 1860. At the time of the Norwegian census of 1865, Lars Thingstad sowed a half barrel of rye, one and one-quarter barrels of barley, four barrels of mixed small grains, two barrels of oats, and six barrels of potatoes, and the farm had three cows, twelve sheep, and one pig, but no horse. A year later, in September 1866, the farm was sold at public auction for 875 Spd.[18] To have to sell a farm at auction represented a loss of face in the community. But Martin Hals, the new owner, let the Thingstad family stay on the farm for a couple of years after he had bought it.

Their daughter Eline took charge of her own future. She signed an indenture contract for one year and got her ticket paid (36 Spd) to emigrate to Texas. She had nothing to lose and much to gain. Eline was related to Sedsel Questad, who was her first cousin, and she worked for the Questad family after she arrived in Bosque County.

On November 1869 Eline Thingstad married the west Norwegian Omen Omenson, who had arrived in Bosque County as early as 1859. But not until 1868 had Omenson saved "enough money to buy one hundred and twenty-five acres from the Ole Canuteson family on Neil's Creek."[19] According to a family history, Omenson had "lived as a shepherd for ten years, saved his wages, and bought his first piece of land." The wedding took place at the Questad home. The newlyweds moved into the original log cabin on Omenson's land and lived in that cabin for several years.

Eline worried about the future of her parents, brothers, and sisters back in Romedal, who after leaving the farm lived in much poorer housing. Before Eline and Omenson were married, she had persuaded him to pay for her parents to travel to Bosque County. In fall 1869, her father, Lars Thingstad (48 years), and her mother, Petronelle (46), left Oslo with their daughter Emma, born April on 13, 1864.[20]

Omen and Eline Omenson did well over the years, not least because of the strong will and drive of Eline Omenson, who actively recruited indentured servants from relatives and acquaintances in her home community in Hedmark County. "Omen paid for the passages. The persons would work for the Omenson's for an agreed amount of time to pay off the passage fee. At the end of the agreement, the persons were free to leave. Their debt had been paid through their labor."[21] Her brother Peder Edvard Larsen Thingstad was one of them; he emigrated from Norway to Bosque in 1871. The indentured servants helped the Omensons build up their farm at Norway Mills to a level the couple would never have dared dream of before their marriage.

The *Atalanta* passenger Ole Jermstad, born on August 10, 1850, at Ljøstaddammen, Romedal, Hedmark County, belonged to the same category as Eline Thingstad. His parents, Ole Jørgensen Musstuen and Inge Andersdatter Sørli, had moved from the Musstuen farm in Stange parish to the Ljøstaddammen farm in Romedal on October 6, 1848; his brother Mikkel had bought the farm for them. Ole worked the farm until he bought it on May 7, 1863. But four years later, on September 21, 1867, the farm was sold from the bankruptcy estate of Ole Jørgensen Musstuen to Martin Hermansen Finstad for 600 Spd.[22] The 18-year-old Ole Jermstad joined the passengers on the *Atalanta*, having signed an indenture contract with Canute Canuteson. After fulfilling his contract, he worked for others, and when he married Siri Tergerson on September 23, 1875, he had acquired his own land. There was little future left for the Jermstad family in Norway, so Ole's parents emigrated to Minnesota in 1869. Ole Jermstad then persuaded his older brother Karl and his sister Mary Olsdatter to emigrate to Bosque County. They left Oslo for Bosque County on the *Oder* on April 28, 1871.

THE CHALLENGE OF HIRING INDENTURED SERVANTS

Many of the immigrants on board the *Atalanta* had done fairly well, wrote Hendrick Dahl to his parents-in-law in June 1869. Most of the persons who had signed indenture contracts had fulfilled them long ago, but not all. "Many have just done a little and others nothing, and if they decide to pay when they can, this would be good, but they can do as they wish." There was no debtors' prison in the United States, Dahl explained, "and a man has to be the owner of a certain amount of land before a debt can be collected with the law in hand. He has a right to keep 160 acres with buildings and chattels as well as the horses or oxen necessary to work his land, produce to feed his family for one year, and a few cows and pigs. And the same rules apply to other kinds of property in towns."[23]

Dahl agreed with his Norwegian neighbors in Bosque that it was a good idea to recruit young Norwegians as cheap laborers from Hedmark. The challenge was to find a way to hold laborers responsible for fulfilling the one-year indenture contract they had signed. An employer who had paid for the ticket from Norway to Texas had no legal means to recover the money a laborer might owe him. Maybe it would be a good solution to recruit laborers who were related to the Texas family who hired them, or to the people in Norway who recruited them?

Dahl tested this idea in 1870. He involved his parents-in-law in all transactions concerning the hiring and travel of Jens Larsen Fjæstad from Romedal to Bosque County. Jens Fjæstad was born on October 10, 1848, at the farm Fjæstad, created out of the farm Brunstad in 1822. He left Christiania on the *Argo* for Hull on November 18, 1870, and continued on a steamship from Liverpool.[24] Both the Brunstad and Furuset families in Texas knew his parents. In December 1871, Dahl reported to his parents-in-law that Jens had arrived safely in Bosque County and asked them to bring greetings from Jens to his parents on the Fjæstad farm. After having fulfilled his contract and worked for others as well, Jens Fjæstad ended up as a merchant in Austin.

In 1871, Dahl's parents-in-law helped him sign an indenture contract with Even Paulsen Røhne, or Evan Paulsen Rohne, from Tomteri-eie, Romedal, who was born on September 22, 1848, and died on May 11, 1901 in the Boggy community, Bosque County. Well-researched and well-documented family histories based on primary sources and footnotes have made it possible to gain a comprehensive understanding of Even Røhne's background and life in Texas.[25] He grew up in a very poor family. According to the Norwegian Census for Romedal Parish in 1865, his mother, Marthe Eriksdatter, had been a widow for about nine years. She lived with her children at the Romedal rectory farm in the house of the tenant, Mathias Christiansen Sund. Marthe, 45 years of age, worked as a *budeie* (milk maid) with her daughter Pauline (13 years), and her sons Johan Magnus (22) and Even (18) as general farmhands, *tjenestedreng*. They lived on a large farm but belonged to the lowest group of servants on the farm.

Evan Rohne left Oslo on October 14, 1871, on the *Ganger Rolf* for Hamburg. He paid 48 Spd for his ticket from Christiania to Galveston and brought with him 14 Spd in cash.[26] After some waiting in Hamburg, he boarded *Hammonia*, built for the Hamburg-America Line by the shipyard Caird & Co. in Greenock, Scotland, in August 1866. The *Hammonia* arrived in New Orleans on December 4, 1871, and Rohne continued by ship to Galveston.

Dahl paid all his traveling expenses, and after having repaid his debt through his work, Rohne worked for several other employers during the following years.[27] His dream was to own his own land, and after seven years in Texas he bought his first land in the Boggy community in Bosque County in 1878. He did not settle on this land until a couple of years later, after his marriage.

Letters from Hendrick Dahl to his parents-in-law give us insight into Bosque County's thoughts on indentured servants. Carl Questad was as interested in recruiting cheap Norwegian laborers to work his land as his neighbors. Surviving correspondence in the Carl Questad archive at the Norwegian-American Historical Association (NAHA) at St. Olaf's College, Minnesota, sheds light on how the recruitment was done on the Norwegian side.

Carl Questad continued to develop his farm during the 1870s. He bought additional land and ended up with around 1,000 acres. Much of the work on the farm was done by indentured servants recruited from Hedmark. One of the first to work for the Questad family was their relative Eline Thingstad. The brothers Lars Mikaelsen (Mickelson), who was born on August 25, 1847, in Løten, and died on October 29, 1925 in Bosque County, and Lukas or Lucas Mikaelsen (Mickelson), born on December 1851, in Løten, and died on May 29, 1903, Bosque County, signed an indenture contract with Questad in fall 1871. They crossed the Atlantic on the same ship as Evan Rohne.

The two brothers were the sons of Questad's former cotter on the Ødekvæstad farm in Løten, Hedmark. When Carl Questad emigrated, he sold the farm to his neighbor Johan Questad, and the two corresponded from time to time during the next years. In 1871, Carl wrote Johan and asked if he knew some conscientious and hardworking young men in the neighborhood that he might hire. Johan responded by suggesting that Questad hire the sons of Mikael and Lisabeth Larsen, the cotters at Ødekvæstadbekken. Johan had asked the "boys" if they would be interested in going to Texas as indentured servants for Carl Questad. They were interested, and their parents told them that they could find no better employer in the new country than Carl Questad.

Lars and Lucas Mikaelsen were to leave for Texas as soon as travel money arrived from Carl Questad, Johan Questad informed him in a letter dated May 19, 1871. The two brothers asked that Questad not send too little money, since they would also need some money to prepare for the journey.[28] On October 14, 1871, the Mikaelsens left Oslo on board the *Ganger Rolf* together with a fairly large group of Norwegians from Løten and Romedal in Hedmark County.

In a letter from Mikael Ødekvæstadbekken to Carl Questad, dated April 25, 1873, Mikael wrote that he and his wife were very sincerely grateful that Questad had been the "*Ledsager*" (mentor) for their sons after their arrival in Texas. He thanked Questad for the kindness he and his wife had shown their sons, which they had reported to their parents. The time might soon be coming when the sons would choose to leave the Questad farm to work for someone else. Could Questad still keep an eye on them?[29]

It is evident that neither the Mikaelsen sons nor their father, Mikael, regarded the indenture contract they had signed with Carl Questad as a slave contract. On the contrary, the contract was a golden opportunity to get a better start in life than would have been possible back in Løten.

HENDRICK DAHL RETURNED TO NORWAY TO RECRUIT WORKERS

Wheat and corn were the main crops for the Dahl family and their Norwegian neighbors. There was no market for rye, barley, or oats; most people fed it to the animals, wrote Hendrick Dahl. The Dahl family began growing cotton after the Civil War. They had done well in 1869, and the 1870 season also looked good.[30] The Texas economy seemed to be picking up, but the main problem was finding and hiring cheap labor.

Dahl prospered during the five years after the Civil War. In 1870, he bought a tract of 1,280 acres of land for 1 dollar per acre from the Frederick Lundt Survey. Most of the houses in the city of Clifton today are built on this land.[31] Dahl was by far the wealthiest Norwegian listed in the 1870 census. Almost all Norwegian names were spelled wrong by the census enumerator that year, and the name Hendrick Dahl is not easy to find. He is hidden behind "Dahl H Snell," 41 years of age, married to Christine Snell (36 years). His real estate was assessed to be worth 7,200 dollars and his personal estate 2,500 dollars.

In a postscript to a letter to his parents-in-law in December 1871, Dahl wrote that he planned a journey to Norway the next summer.[32] According to "The Family History of Hendrik and Christine Dahl: From the year 1854 to 1954," "after an absence of twenty-five years, Hendrick Dahl returned to his native land to visit his aged mother. On his return to Norway many relatives and friends gathered to hear about the wonderful land of America." The whole countryside was curious, and many "rallied around Hendrik to join him on his return to Texas."³³

Visiting his mother was certainly high on his agenda, and people might have become excited about the "wonderful land of America," but Dahl's main purpose was to recruit young emigrants willing to work for him and others in Bosque County. As already discussed, the scheme Poulson had tried in 1868 had not worked as planned, so Dahl and others involved relatives and friends back in Norway in their recruitment of young men and women. An important requirement was a good reputation and a strong code of honor. If they were related to a family in Texas, or their relatives in Norway had a strong connection to a family in Texas, the thinking was that they would feel morally bound to fulfill their indenture contracts before starting out on their own.

DAHL'S JOURNEY FROM HEDMARK TO TEXAS IN WINTER 1872

According to different versions of Dahl family histories, a group of "forty people left Oslo in the early morning of November 2, 1872." The passenger list of the *Louisiana*, however, filled out after arrival in New Orleans, documented that the group included eighty Norwegians, including Hendrick Dahl—fifty-three adults ages 15 years and older, and twenty-seven children and infants.³⁴ The passenger contract for one of the passengers, Martinus Pedersen Torphagen (Martinus P. Rierson), born on April 20, 1845, in Løten, has survived and gives us an excellent illustration of the conditions for their travel. His brother Peder Pederson Torphagen (Peder Pierson), born on April 19, 1851, was on the same ship. Their older brother Hans Pedersen Torphagen (Hans P. Pierson), born on July 8, 1842, had emigrated to Bosque County a year earlier with his wife, Berte Christiansdatter, born in May 1847.³⁵ The contract specified that the agent in Christiania (Oslo) had an obligation to transport the emigrant from "this place to Galveston for Sum Spd 40." The journey would be from "Christiania per steamer to Hull thence pr. railway to Liverpool and thence in the sterage (sic) with steamer Louisiana." If a passenger got sick and could not travel on the time specified, the passenger could travel on another steamship later, and if the police stopped a passenger from emigrating, the passenger would get his money back. In case of delays, it was the responsibility of the transport company to supply the passenger with food.³⁶

The Dahl group boarded the Wilson Line steamship *Hero* in Christiania, which took them across the North Sea to the East England city of Hull. The voyage across the North Sea took approximately 52 hours.

Transmigration by way of Britain had not been common among Norwegians. Around 1865 most Norwegian emigrants made the Atlantic crossing directly from a Norwegian harbor to New York or Quebec on a Norwegian sailing ship. This changed after 1870. Large trans-Atlantic steamship companies leaving British ports offered faster journeys. The crossing took less than half the time of a sailing ship, and consequently the passengers were less exposed to suffering and sickness. The number of deaths on board was also reduced. Raymond L. Cohn concluded that the change from sail to steamships in the migration trade took about fifteen years.[37] By 1873, steamships had become totally dominant in the Atlantic immigration trade to New York.

The number of deaths on sailing ships had been relatively high, although most emigrants did not know that. When the passenger lines began to compete for the migration trade, ticket prices fell. Emigrants might not understand risk, but they clearly understood lower prices combined with better accommodations and food. An increasing number of Norwegian emigrants began using British ports for their Atlantic crossing, and Norwegian shipowners could no longer compete. Foreign-owned transatlantic shipping companies such as the Allan Line, Cunard Line, American Line, National Line, Inman Line, Guion Line, Dominion Line, State Line, and White Star Line established emigration agencies in Norway. The big steamship companies, however, depended on smaller feeder ship companies to transport the emigrants from European ports to Britain. According to Nicholas J. Evans, more than 2.2 million trans-migrants from Denmark, Sweden, Norway, Finland, Germany, and Russia came through Hull between 1836 and 1914.[38]

The Wilson Line became dominant in the emigration feeder trade from Scandinavia to Hull. By 1867, the Wilson Line operated a weekly service from Christiania to Hull, usually departing from Christiania every Friday and also calling at Kristiansand. For all practical purposes the Wilson Line had a monopoly on this trade, so the company had no incentive to improve the standard of accommodation on board their ships. Complaints about conditions for steerage passengers were numerous. Most of the passengers only used the Wilson Line ships once in their life, so the shipping company did not care too much.

The Dahl group boarded the screw steamer *Louisiana* in Liverpool. The steamship was owned by the State Line Ship Co. Ltd., Glasgow. On arrival in New Orleans on December 16, 1872, Captain James Stewart filled out the "Report and List of Passengers." The ship had eight cabin passengers, one English and seven French, followed by a long list of Norwegian steerage passengers. Their ticket was from Norway to Galveston, Texas. Only one of the steerage passengers had been in the United States before, no. 45, the "laborer" O. H. Dahl, 45 years old. The Atlantic crossing was done in the winter storm season, and even though the *Louisiana* was a steamship, the crossing from Liverpool to New Orleans took eight weeks. Many of the steerage passengers became seriously ill, and three Norwegians died and were buried at sea—5-year-old Christian Christoffersen, 28-year-old John Magnus Poulsen, and a 4-year-old girl, B. M. Pedersen.

Hendrick Dahl became sick on the Atlantic crossing but managed to lead his group to Galveston and from there to Bosque County. On December 22, twenty-two people arrived at the Dahl farm and made their "headquarters with the Dahls, some camping in the barn, others in tents, until they secured permanent residence with relatives, friends, or farmers who could use an extra farm hand," to cite a family history. Dahl did not recover his health; his condition worsened after he got home. "Realizing the seriousness of his health he had his will made January 12, 1873," and on January 13 he died.[39]

THE ENTREPRENEURIAL WIDOW

Just when their life had been improving, Christine Dahl lost her husband and the children lost their father. His death was hard to live with. Why had Dahl gone back to Norway? Why had he not heeded Christine's advice to use his money to make sure he did not put his health in danger? "Our safety is in the hands of the almighty God," Christine had written to him in the middle of September 1872, "but I also see that you wish you had an overcoat and this is entirely up to you to do something about. So, my dear husband, I will advise you not to expose your health to any danger or suffer any need that you can take care of with money."[40] Why had he chosen to travel in steerage, the most dangerous place on board an emigrant ship?

Christine had been trained to take responsibility from a young age, and she was not unfamiliar with death during her years in Texas. She soon turned practical and took charge of life, work, and decisions on the farm. She proved as skilled at this work as her husband had been. The oldest children were already a great help. Their oldest son, Ole, was almost 19 years old when Hendrick died. With pride, she had written her husband in Norway, "Ole has taken over the big mule and broken it in. It is now gentle for riding and easy to handle."[41] With the help of her children and a steady stream of indentured servants from Hedmark, Christine Dahl succeeded as a farmer.

EVEN CLOSE RELATIVES WERE HIRED AS INDENTURED SERVANTS

When Evan Rohne crossed the Atlantic on the *Hammonia* from Hamburg in 1871, he probably felt he had everything to win and nothing to lose. The feelings of the Ole Andersen Halstenshov (Finstad) family might have been considerably more mixed. They had been yeoman farmers in Løten, but had now signed indenture contracts with close relatives. Ole A. Halstenshov was born on February 7, 1821, in Løten and died in Bosque County on July 18, 1903. His wife, Elisabeth Jensdatter, was born at Halstenshov, Løten, on March 14, 1819, and died in Norse on April 13, 1891. She was the 3 years older sister of Kari Ringness, the wife of Jens Ringness, who had emigrated to Bosque County as early as 1854. Their father, Jens Larsen, born in 1768, had married Gunnor Halstenshov Jensdatter in 1817 and bought the farm Halstenhov in Løten. Eight years after his death in 1846, the widow transferred the farm to her son-in-law Ole Andersen Finstad. However, he struggled to make the farm profitable, and had to declare bankruptcy in 1863; it was sold for 1,150 Spd on March 19, 1863.[42] The family remained poor until their emigration in 1871. They had no money to pay for their tickets from Hedmark to Texas, so Ole Finstad signed an indenture contract with his brother-in-law for himself, his wife, and two of their five children, daughter Juliane (20 years) and son Anders (18), who were both of working age. Their three other children stayed behind in Norway, including their 9-year old son Ole Olsen, born on April 13, 1862.

According to the stories Ole O. Finstad told his children later in life, the Ringness family was only willing to pay for adults who could work.[43] A year later, however, their 19-year old daughter Agnethe, born on May 23, 1853, brought her little brother Ole with her to Texas. "He and his sister, Agnette," as he told his children, "left Norway when life was almost unbearable, due to lack of food. He often told the story of how he was hungry and there was no food in the house. His mother scraped the old wooden bowl she used to make bread in, and they ate the crumbs. Their potato crop had failed that year." This story in *Bosque County: Land and People* contain several inconsistencies. According to this source, Agnethe and Ole Finstad crossed the Atlantic on one of England's finest steamships to New Orleans and then to Galveston. This would certainly have been nice, but actually Agnethe Olsen (19) was passenger no. 77 and Ole Olsen (10) passenger 78 in the large group of Norwegians who crossed the Atlantic with Hendrick Dahl in fall and early winter 1872 on the *Louisiana*.

Some years earlier a younger sister of Kari Ringness, Berthe (born June 27, 1829), had arrived in Bosque County with her family. She was married to Peder Olsen Vesterengen from Halstenhov, Løten. Husband and wife with eight children left Oslo on the ship *Emerald* on May 8, 1867, and arrived in Quebec on July 17, after first having waited a month in Moray, Scotland. The family had relatives in La Crosse, Wisconsin, and stayed about a year in La Crosse before continuing to Texas via Galveston. A sister and brother of Peder remained in Wisconsin. The family traveled by train from Galveston to Bryan, and from Bryan they contracted with teamsters to take them to their relatives at the Ringness farm in Bosque County. The family arrived on Christmas Day in 1868.[44]

This family had some economic means, and Peder Hoff settled on a farm in the Norse community. Why did Peder change his name to Hoff? Officers at the county seat Meridian advised Peder Olsen Vesterengen to choose a different name since so many already had Olson as their last name. He chose Peder Olsen Halstenhof but was advised to shorten it to Hoff. His younger brother Ole Olsen Vesterengen, born March 1831 in Løten, immigrated with his family to Bosque County in 1870. He changed his name to Ole O. Hagen since he had been born in Lundshagen.

A STEADY SUPPLY OF INDENTURED SERVANTS THROUGH FURUSET NORDRE

The system for hiring indentured servants that Hendrick Dahl had set up with his parents-in-law was working very well. Peder Eriksen handled all the transactions involved in bringing indentured servants to the Dahl farm: finding the right person, buying the ticket, exchanging Norwegian Spd into dollars, getting the money from Texas to pay for it all, and, not least, finding a Norwegian bank willing to honor the American money order. In a letter to her parents in January 1876, Christine Dahl commented on all the paperwork they had done for her related to the workers who had arrived at the Dahl farm on January 15, 1876. She really appreciated what her father and mother were doing for her. "I don't think I'll ever be able to do you something in return even though this would be a great pleasure."[45]

At the end of April 1877, Peder Eriksen's farm Furuset Nordre in Romedal was transferred to his son Erik Pedersen. Ten days later Peder Eriksen died. From then on, Christine wrote her letters to her mother and her brother Erik and his family. Erik took over the responsibility for finding indentured servants for the Dahl farm in Texas and the office work involved. The economic transactions were simpler now since Christine and her sisters in Texas had inherited money. Instead of transferring the inheritance money to Texas, Christine and her sister Karen kept it in an account in Norway. Every time a new indentured servant was sent from Norway to Texas, the costs for tickets and so forth were drawn from the inheritance account.

The letters to Erik from Christine were usually filled with news about her children and the harvest or the prospects for the harvest. They were in the middle of the cotton harvest, Christine informed him in September 1878, and the harvest would be good; between five and six cotton bales. "The price is low because of the yellow fever that rages in the eastern states; there is a quarantine in the ports where the cotton is shipped. Our corn crop was quite good; we harvested about 700 bushels on twenty acres. The wheat also did fairly well so we may say that we've had a fairly good year."[46]

In 1878, her brother Erik sent her the indentured servant Ole Evensen Lund, born on September 6, 1853, at Lundsbakken, a small farm in Romedal, followed in 1879 by his brother Kristoffer Evensen (Christopher Evan Lund), born on November 24, 1858. They were the sons of Even Christophersen and Lisbeth Olsdatter. Even Christophersen had taken over Lundsbakken in 1847. As of the 1865 Norwegian census, only one-sixteenth of a barrel of barley, one barrel of mixed small grains, and two barrels of potatoes had been planted. There was no livestock on the farm. On November 20, 1873, the farm had to be sold at auction for 70 *speciedaler*.[47] When the sons had the opportunity to sign indenture contracts with the Dahl farm in Texas, they were happy to do so.

In early 1879, Erik Pedersen wrote her that the parents of the two young men had asked if their sons could send them some money. Christine discussed the matter with Ole and Christopher and wrote back to Norway that "they would be glad to help them." She suggested that Erik should draw 30 Spd from her inheritance account, and the men would repay her "as soon as they can."[48]

After the Lund brothers had repaid all costs connected with their journey from Romedal to Texas, they began pursuing their own fortunes. At the time of the census in 1880, Ole was working as a sheepherder in Bell County. Christopher began to farm on his own in Bosque County, but in 1891 he bought a sheep farm in Hamilton County, midway between Hamilton and Pottsville. He married Eline Christianson on April 30, 1893, and they became the parents of three children.[49] The valley in Hamilton County where they settled is still named Lund Valley, and the school established there was named Lund School. It was classified as a high school in 1911 with nine grades.

Christine Dahl also helped relatives find a job outside agriculture. Did her brother Erik know, she wrote in 1880, if Ole Rossing had finally decided to emigrate? Ole Rossing was almost on her doorstep when she wrote the letter. In a letter dated November 25, 1880, to his uncle Erik Pedersen and aunt Beate Kristoffersdatter Furuset, the sister of Ole Rossing's father, he proudly reported that he had arrived safe and sound at the house of the "widow Dahl." After a journey of twenty-two days, he had arrived in Waco on November 20, 1880. Her son Peder Dahl had met him in Waco with "a wagon with four horses so you may imagine I had a fine ride." He was deeply impressed by what he saw on the Dahl farm. "She has thirty horses, eighty cows, twelve pigs, six of them for fattening, and a tame wolf. And they have a lot of wheat, so much that it is wonderful for a Norwegian to see it all. And they have quite a few chickens and a tame squirrel. And the fields she owns look as if we were to stand right at the center of Romedal and look over all of it."[50] Ole was trained as a baker and knew well that wheat was still a great luxury in Norway. The Dahl family had even got him a job before his arrival—at a bakery in Waco!

DOUBLE WEDDINGS AND A TREBLE WEDDING

Hendrick Dahl had done well between 1865 and his death in early 1873, and his wife Christine continued to prosper. The couple had nine children, four boys and five girls, all born in Texas. Their oldest child, Ole, was born on February 28, 1854, and their youngest, Caroline, was born on November 14, 1871.

The Dahl children were growing up, and began to marry and establish their own households. A series of marriages in and around the Dahl family from 1875 to 1887 tied together the old families in the Norse district and the newcomers from East Texas and Hedmark County in Norway. A double wedding took place on the Dahl farm on September 30, 1875. Christine's oldest son, Ole, married Gunde Hoff, born on August 15, 1851, the daughter of Peder and Berte Hoff. On the same day the oldest daughter, Syverine, married Nils Jacob Nelson, the oldest son of Aslak Nelson and his first wife Margit Hastvedt, born in East Texas in 1850. On this occasion Christine Dahl began her tradition of giving each of her children a farm or the equivalent in money when they married. Ole and Syverine were given farm land north of the Dahl farm.

Another double wedding took place on the Dahl farm in February 1880: Christine's daughter Marthe married Mikkel Hoff, the brother of Ole's wife, and her son Peder married Anna Emilia Christine Nelson, the half-sister of Nils Jacob Nelson, born in Van Zandt County on April 10, 1862. Three hundred guests attended the wedding dinner. "Some of them went home that night but most stayed over for the second day."[51] The festivities had been dampened somewhat by the death of Christine's brother Even Andreas Nelson two days before the wedding. Christine sent greetings to her family in Norway from her Aunt Anne Rogstad and her family. "Aunt looks good and is very fat."[52] She also sent greetings from Dilrud, Sinrud, Thingstad, Aunt and Wilhelm, and Ole and Christopher Lund. Her son-in-law Nils Jacob Nelson had just hired Christopher for a year.

In her letters to her brother there was a steady stream of news about grandchildren. Double weddings did not only take place on the Dahl farm. In February 1885, she reported that her Aunt Rogstad had married off two of her children at a double wedding. Her son Berendt had married Emma Andersen, and Marthea had married Johan Hoff, the brother-in-law of her son Ole.[53]

A grand wedding took place in the Norse community on Tuesday, September 27, 1887, on the Dahl farm[54] in which two of Christine Dahl's daughters and a son were married. Christian Dahl, who was born on the same day in 1864, married Olivia Adelie Pederson, born on May 13, 1864 in East Texas. His sister Helene Dahl, born on June 16, 1862 in Norse, married Emil S. Bekkelund, born on September 18, 1860 in Vang, Norway. Finally, Gina Dahl, born on January 29, 1867, married Cristian Olsen Holen.

In all these weddings, Norwegian immigrants married other Norwegian immigrants. Even though many of them were born in Texas, they grew up in Norwegian colonies, spoke Norwegian at home, learned to write Norwegian, and read Luther's Small Catechism and the prayer book to be confirmed. Two of the Dahl children married Peder Hoff's children, two married Aslak Nelson's children, and two married the children of Christopher Pederson, who had emigrated from Elverum in 1853.

Almost all the Dahl children married children of Norwegian-Texan pioneers. The exceptions were Helene Dahl, who married the former indentured servant Emil S. Bekkelund, and Gina Dahl, who married Christian O. Holen, who was born on March 20, 1859 at Løten, Hedmark County, and died on August 2, 1943. Christian had immigrated with his parents to Bosque County in 1874. His parents settled on a farm southwest of Norse, and Christian acquired a farm of his own on the other side of the road from his parents. The Holen couple also received a farm from Christine Dahl but continued to live on the farm Christian already owned.[55]

Emil S. Bekkelund and his brother Ole S. Bekkelund emigrated from Vang, Hedmark County, in 1878, having signed indenture contracts with the Ringness farm. The Bekkelunds were related to the Hoff family. Their mother, Inge O. Bekkelund (born July 25, 1835), was the younger sister of Peder Hoff, and his wife, Marte, was the sister of Kari Ringness. After fulfilling their contract with the Ringness family, the Bekkelund brothers worked for other farmers in the neighborhood and on the railroad. While working for the Terrell family, south of Clifton, "their English improved since they were living and working with English speaking people."[56]

A few weeks after Emil married Helene Dahl in September 1887, his mother and three younger brothers, Helge, Martin, and John, arrived in Bosque County. It was a great surprise to his mother that Emil had married—and married so well. As a wedding present, Christine Dahl had given Emil and Helene Bekkelund "a tract of land, ten cows with calves and one span of horses," the equivalent of 2,000 dollars.[57]

THE EAST NORWEGIAN MIGRATION CHAIN TO BOSQUE COUNTY

Because of the system of indentured servitude, the Norwegian colony received a steady infusion of new immigrants from Hedmark County. To a large degree it was this migration chain that made possible the continuation of a strong transplanted Norwegian community "Deep in the Heart of Texas." The number of sons and daughters of cotters rather than yeoman farmers increased in the 1880s and 1890s. During the same period land prices in Bosque County increased. Since the immigrants in the last decades of the nineteenth century generally were poorer on arrival than the pioneers in the 1850s and 1860s they found it harder to reach the goal of owning their own farm.

How many Norwegians immigrated to Texas in the last three decades of the nineteenth century? The table is based on the listings of Norwegian immigrants arriving in Texas between 1865 and 1900 in Syversen and Johnson's *Norge i Texas*. Hardly any Norwegian immigrants arrived in Texas during the Civil War; one person in 1862 and one in 1863. None arrived in 1866, but twenty-eight immigrants came in 1867. A record year for immigration was 1868, when the ship *Atalanta* was hired to transport emigrants directly from Oslo to Galveston, followed by more moderate migration in 1869 and 1870. The eighty-one Norwegians who arrived in Texas in 1871 are proof that migration to Texas had taken hold in eastern Norway, and in 1872, the year Hendrick Dahl went back to recruit immigrants, eighty-nine immigrants arrived.

Norwegian emigration in general declined after the world economic downturn in 1873, and emigration to Texas was also affected. When the American economy picked up again, the volume of Norwegian emigration also picked up and reached new heights. In 1881, one hundred immigrants came to Texas, and the peak year was 1883 with 113 immigrants. Between 1886 and 1899, the number of Norwegians who came to Texas leveled off and declined. Still, a number of Norwegian immigrants arrived throughout the 1890s and after 1900. The overwhelming majority came along the East Norwegian migration chain.

The Norwegian colony "retained its old world traditions and cus-

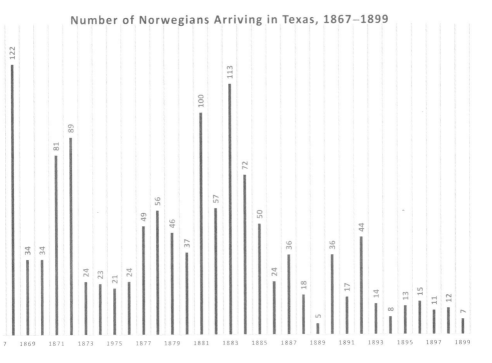

Number of Norwegians Arriving in Texas, 1867–1899

toms and remained aloof from other Bosque residents," wrote William C. Pool in 1954.[58] So many Norwegians were living in Bosque County in the two last decades of the nineteenth century that many chose to continue to speak the mother tongue. It became feasible for Norwegians to reproduce and uphold their Norwegian ethnicity. There were many marriages between recent immigrants and persons of Norwegian heritage. The tradition of Norwegians marrying Norwegians continued into the twentieth century. Norwegians with higher levels of education or living in urban environments were often the first to marry across ethnic and religious lines.

The driving force in the Norwegian migration to Texas between 1870 and 1900 was the continued recruitment of indentured servants from Hedmark County. Some immigrants continued to come from other regions in Norway, such as from western Norway through the migration chain established by Cleng Peerson and from southern Norway through the Reiersen chain. If La Salle County, Illinois, was the mother colony for Norwegian immigrants to the United States between 1835 and 1860, Bosque County was the mother colony for Norwegian-Texans between 1870 and 1900. Far from all Norwegian immigrants who came to Texas settled in Bosque County, but they usually connected with relatives or friends who lived there first before seeking work and fortune elsewhere. At the end of the nineteenth century, the life and culture in the Norwegian community in Bosque County was dominated by East Norwegians.

CHAPTER 12

Harder to Climb the Agricultural Ladder

◊ ◊ ◊

AGRICULTURE IN TEXAS experienced setbacks during the Civil War, but soon recovered and reached unprecedented heights with respect to value and productivity. The number of bales of cotton produced in Texas increased by 300 percent between 1870 and 1890. The number of improved acres increased sevenfold. "Whole areas, barely settled in 1865, were covered with productive farms in 1890." Despite the growth in production many farmers felt their situation was getting worse. "Prices fell, debts rose, and farm tenancy increased dramatically," observed Randolph B. Campbell.[1]

Very few farmers in antebellum Texas were tenants, defined as a person who resided on land owned by a landlord. According to Cecil Harper, Jr., landless farmers were "a relatively insignificant factor in the agricultural economy before the Civil War."[2] He found that the number of tenant farmers was higher in Texas frontier counties than in eastern and southern counties.

After the Civil War, former slaves usually ended up as tenants, either share tenants or sharecroppers. They might provide their own seed, work animals, and equipment and "paid the landlord one-fourth of the cotton and one-third of the corn as rent, whereas sharecroppers provided only their own labor and received the proceeds of one-half of the cotton crop. Share tenants had more control of their farming operations than did sharecroppers and enjoyed a higher status. Still, one basic characteristic united all tenants—they owned no land," noted Campbell.[3]

The highest percentage of tenant farmers lived in counties with black majorities. Randolph B. Campbell found that 80 percent of all black families in Harrison County lived as tenant farmers in 1880. Most of them were sharecroppers. No important crop in Harrison County had reached prewar levels, and farmers placed greater emphasis on cash crop production than ever before.[4] "Small farmers, especially renters and sharecroppers, generally needed credit to begin a new year, and they could obtain it only by giving a lien on their crops for that year. Those who extended credit, generally merchants in town, insisted on cotton production since it was the crop most readily converted into cash. Many farmers abandoned self-sufficiency in food production, not because they were foolishly trying to get rich and deliberately over-produced cotton, but because they actually had no choice."[5]

Agricultural Texas in the last two decades of the nineteenth century was characterized by a "great increase in the number of tenant farmers relative to the number of owners-operators," according to John Stricklin Spratt. The number of farms between 1880 and 1900 doubled; farm tenancy trebled.[6] Approximately 38 percent of all farmers in Texas were tenants in 1880, while half of all Texas farmers were tenants in 1900. The highest percentage of tenants in Texas ever recorded was almost 61 percent in 1930.[7] "The notion of an 'agricultural ladder' by which young men moved from sharecropper to share tenant to landowner, became more and more of a myth than a reality, and Texas farmers by the tens of thousands seemed doomed to live endlessly in near-poverty—working someone else's land," Campbell observed.[8] "Tenancy cast its economic blight upon blacks and whites impartially," wrote Robert A. Calvert. In the "age of agrarian discontent, white folks and black folks were becoming impoverished together."[9]

There was a pronounced shift from subsistence farming to commercialized agriculture in the closing decades of the nineteenth century.[10] The growing and land-hungry population led to increased demand for land, which in turn led to increased land prices. During the same period considerable mechanization took place. Farmers with little or no capital could not afford to buy land and had to be content with renting small farms. Farmers with large farms were in a better position to finance the purchase of agricultural machinery than farmers with smaller farms.

A bale of cotton was the most important source of cash for the farmer and his family during the year. It allowed farming families the luxury of buying clothes, coffee, or canned foods to break the monotony of the corn and pork diet. Before arriving in Texas, Norwegian immigrants had never seen cotton plants. They soon learned to grow and pick it. In 1882 12-year-old Martin Rasmussen Rohne, born on December 11, 1870, emigrated from Romedal, Hedmark County, to live with his uncle Evan Rohne and his family in Bosque County. When he arrived on the

Rohne farm on October 29, 1882, he was immediately put to picking cotton, and was able to pick 50 pounds on his first day.[11] "Cotton was our money crop," wrote the Norwegian immigrant Carl Johnson about his childhood in Hamilton County after 1900. "Sometimes it would be too cold to pick cotton, so we would pull the burrs and carry them to the barn or shed, and when it would get warm, we would pick the cotton out of the burrs." Making a 500-pound bale required 1,500 pounds of cotton. "Dad would haul the cotton and have it ginned. The owner of the gin would keep the seed to pay for having it ginned. Then we would go downtown and sell the bale of cotton and pay the grocery bill for a whole year."[12] The main problem with cotton was overproduction, and the price of cotton fell from an average of 15.5 cents per pound in 1869–71 to 8.1 cents in 1889–91.

WHAT DID IT MEAN TO CLIMB THE AGRICULTURAL LADDER?

All children who grew up on farms have stood on the first rung of the agricultural ladder. Farm children learned the skills of farming through learning by doing; sons and daughters walked literally in the footsteps of parents, grandparents, or other "role models" on the farm. They observed how their peers were doing things—plowing the field, sowing the grain, hoeing the corn, helping the cows calve and the sheep lamb, and deciding when to harvest. Farm children tried to learn as best they could what their parents tried to teach them. In agricultural communities in the nineteenth century this kind of learning had a much higher status than formal schooling.

The Bosque County immigrant Johan (John) Dahl was born at Løten, Hedmark County, on March 15, 1881, and died on December 2, 1978, in Bosque County. He was the ninth child of fourteen. Starting at the age of 10 he worked with an older woman in a summer dairy in the mountains in eastern Norway, where the livestock was grazed in summer. The two of them cared for and milked seventeen cows and made cheese and butter from the milk. He did this work for three summers, and his pay for a whole summer's work was a suit of clothes, stockings, mittens, and shoes, all homemade. He left school at the age of 14 in 1895 to emigrate to Texas with two of his older sisters.[13]

In the early 1900s his brother-in-law, immigrant Julius Johnson in Hamilton County, usually kept four or five milk cows, which his children had to milk before going to school.¹⁴ The children also played an active part in plowing and preparing the land for next year's crop, which "would consist of potatoes first crop, oats, wheat, barley, corn, maize, sweet potatoes, turnips, peas, sugar cane, millet, rye, carrots, and cabbage."¹⁵

Hired agricultural workers stood on the second rung of the agricultural ladder and the tenant farmer on the third rung. The highest rung on the ladder, and the most coveted position, was occupied by farmers who owned their own land. Landowners might need a mortgage on the farm to pay for the land, build a house, and acquire agricultural machinery as well as oxen, horses, or mules to power the equipment. Credit might also be needed to buy farm animals and seed for sowing. It could take several years to pay off debts. If the harvest was good, loans could be repaid, or land could be bought and either added to the home farm or given to young sons who wanted to follow in their father's footsteps.

In a study of farm owners covering the states of Illinois, Iowa, Kansas, Nebraska, and Minnesota published in 1919, W. J. Spillman found that 20 percent of 2,112 farm owners climbed all the rungs on the agricultural ladder. Thirteen percent skipped the tenant stage, 32 percent reached the second rung of agricultural worker, and 34 percent achieved the fourth rung of ownership by inheriting a farm from their father, marrying a girl who inherited a farm, or buying a farm from relatives for a low price or on easy terms.¹⁶ Most of the men who climbed all the rungs on the ladder had left home at a young age, usually younger than 20. It took them on average nineteen years to become owners of their own land.¹⁷

WERE THE NORWEGIANS AS INDUSTRIOUS AS THE GERMANS?

Heated debates have been raging for decades among historians regarding the history of agricultural tenancy in the United States. Some have maintained that the growth of tenancy was a natural result of the retreat of the American frontier and the disappearance of unoccupied land. Others hold the view that a misguided and ineptly practiced federal land policy made it possible for land speculators to acquire large chunks of land that were offered for rent to landless farmers. Several economic historians have argued that tenancy was a rational and socially desirable adjustment to changing conditions in a maturing economy. Tenancy served as a rung on the agricultural ladder leading to farm ownership. These discussions have focused primarily on the South and the Midwest.¹⁸

Terry G. Jordan argued in his book *German Seed in Texas Soil: Immigrant Farmers in Nineteenth-Century Texas* that antebellum Germans usually acquired their own land a few years after their arrival. Even though the policy of free land grants had been discontinued, buying land and avoiding tenancy were high on the agenda of German immigrants. Land was bought on credit, or with the money agricultural laborers had saved, or by a combination of the two. Jordan recounts a story told by a German immigrant in Austin County in 1849. He estimated that it took only three years for a hired farmhand to save enough money to buy a farm. Land could be bought for 1 to 5 dollars per acre. Jordan found that Germans had a much lower rate of tenancy than Anglo-Americans even though on arrival they had less capital than most Americans. At the time of the 1850 census, "77 percent of the Germans were landowners as opposed to 69 percent for all Anglo-Americans."[19] After the Civil War, "the Germans continued to have a lower rate of Tenancy than the Anglo-Americans." German tenants were less likely to be sharecroppers "than were native whites and Negroes." They often remained on the farms they had bought for years and "impregnated the cultural landscape with an appearance of permanency."[20] Did the Norwegians in Bosque County behave like the Germans during the last decades of the nineteenth century, or did they have more in common with their Anglo-American neighbors, who were ready to pull up the stakes and depart when opportunity knocked? Looking at the numbers, the conclusion is that the Norwegians behaved very much like the Germans.

According to the 1880 US agricultural census there were 1,228 farms in Bosque County with an average size of 226 acres. The owner of the farm and the farmer cultivating it were the same person on 929 farms (75.7 percent). Sharecroppers cultivated the land on 239 farms, and thirty farmers rented their land for a fixed sum of money. Ten years later, the number of farms in Bosque County had increased to 1,808 farms, with the average size of farms around 239 acres. Owner and farmer were the same person on 1,098 farms (57.85 percent), while seventy farmers paid a fixed yearly cash rent (3.7 percent) and 730 (38.46 percent) were sharecroppers.[21] By 1900, 1,185 farms in Bosque County were operated by owners (52.8 percent) and 1,060 farms were operated by tenants (47.2 percent), out of a total of 2,245 farms. Not until 1910 was the number of tenant farmers in Bosque County higher than 50 percent. That year there were 1,294 tenant farms (50.1 percent), while 1,289 farms were still owned and operated by farmers (49.9 percent).[22]

The agricultural census of 1910 included ownership among "foreign-born whites." Bosque County had 199 foreign-born white farmers who owned their own land (77.1 percent) and fifty-nine tenant farmers (22.9 percent), a total of 258. Since Norwegian immigrants were totally dominant among foreign-born whites in the county, it can be concluded that a much higher number of Norwegian immigrants owned their own land in Bosque County in 1910 than the average for the county and Texas.

Was Bosque County exceptional or normal with respect to tenancy rates compared with neighboring counties? In 1890, 47 percent of all farms in Bell County were owned by farmers who also operated them, while only 34.3 percent did so in 1900. In Coryell County 61.27 percent owned their farms in 1890, but that dropped to 47.6 percent in 1900. In Erath County, 61.1 percent of the farmers in 1890 owned their own land, and 47.0 percent in 1900. In Hamilton County, to the southwest, 64.9 percent owned their own land in 1890 and 47.7 in 1900. To the east, 48.1 percent of the farmers in McLennan County owned their farms in 1890, but only 33.7 in 1900. Only in Bosque County did more than 50 percent of all farmers still own their own farm in 1900.

WELL PAID AS A MILLER, BUT HIS DREAM WAS TO OWN HIS OWN FARM

Christian O. Bronstad is an example of a Norwegian immigrant who arrived in the early 1870s and was able to climb to the top of the agricultural ladder within a few years. Christian O. Bronstad was born in Romedal on August 20, 1853, and died in Bosque County in 1934. He emigrated as a member of the Dahl group in 1872. Bronstad arrived in Waco on December 19, 1872, on a severely cold day when "the Brazos River was frozen over solid." The Norwegian immigrant Peder Thingstad, the younger brother of Eline Omenson, who had immigrated from Romedal in 1871, helped Bronstad find employment as a dishwasher at the Lady Gay Restaurant in Waco. Five days later he became ill with typhoid fever. Thingstad brought him to his sister Eline at Norway Mills in Bosque County. Under the "good care of Dr. Cowan, the Omensons and the Thingstads, he was brought back to good health."[23]

Bronstad was too weak to work in spring 1873 but was hired as a miller in the flour mill at Valley Mills that summer. Did he bring miller skills from Norway? He emigrated from Saugbakken in Brunstad precinct in Romedal County. In the Norwegian census of 1865, his father, Ole Andersen (born 1816), was listed as a miller. There was no livestock on the Saugbakken place, but there was a little land, and according to the census one-eighth of a barrel of barley, three barrels of mixed grains, and half a barrel of potatoes were planted.[24] Since Christian was the oldest son, he had helped his father in the mill and learned by doing. Milling skills were scarce in Bosque County and he did well at Valley Mills. His salary was good, but when he was hired as miller at Norway Mills two years later, he was offered $35 a month, "a stupendous salary at the time."[25] Even though Bronstad earned good money as a miller, his main goal in life was to own his own farm, marry, and have children. At a double wedding on March 1, 1877, Christian married Marthe Christina Colwick, the daughter of Ovee and Johanne Colwick. She was born in La Salle County, Illinois, in 1857. Her sister Margaret Colwick married Dr. O. M. Olson on the same day. During the first two years of their marriage, the Bronstad couple lived in a two-room house at the mill.

Bronstad soon bought 100 acres of land in the Meridian Creek Valley from his brother-in-law John N. Colwick for $3 per acre. He had no problems paying for his land; his yearly salary as a miller was $420. Bronstad built a small two-story house on his farm. As the need arose, he made improvements and added rooms. The same happened with barns, granaries, sheds, lots, and corrals.[26] Over the years, Bronstad bought more land, and at the time of his retirement he owned 550 acres. Twelve children were born and raised on the farm. Bronstad and his wife believed strongly in the importance of education. He served as a trustee of the Harmony School, a trustee of the Bosque County School Board, and in his later years a trustee of the Clifton Lutheran College. "All twelve of his children pursued higher education. A remarkable record in that day."[27]

His younger brother Gulbrand Bronstad left Romedal for Bosque County in 1874, and Christian helped him get his first job. In 1878, the two brothers brought the rest of the family from Hedmark to Bosque County: their parents, Ole Andersen (born January 1816, died August 31, 1891) and Christine (born January 3, 1820, died Bosque County November 5, 1906); the oldest daughter Pernille (born March 27, 1851), Andrea (born April 17, 1858), Syverine (born September 1861), Karen (born October 13, 1863), and Otto (born August 6, 1865). The parents helped Christian and Martha build up the Bronstad farm, and Christian's sisters found work in Waco as housemaids.

AN INDENTURED SERVANT WHO DID WELL IN THE BOGGY COMMUNITY

Evan Rohne had emigrated in 1871 on the same ship as Peder Thingstad. He worked as an indentured servant for Hendrick Dahl for one year. During the next years Rohne worked for others in Bosque County, but then moved to Limestone County to find work. It was while working in Limestone County that he met and married Marianne Olianne Egeberg, from Løten, Hedmark County. She was born on May 24, 1858 at Aadalen, Løten, and died on February 25, 1947, at the house of her son Paul Bernhart Rohne in Cranfills Gap.[28] Her ticket for the Atlantic crossing and further travel to Texas was paid for by her half-brother Bernt Johnson, who had immigrated in 1870 and found work in Limestone County. Marianne left Oslo on August 30, 1878, took the *Hero* of the Wilson Line across the North Sea to Hull, traveled by train to Liverpool, and boarded a steamship that took her across the Atlantic. Bernt Johnson helped her get a job as a housemaid for Colonel John Reagan Henry on his large plantation just south of Mexia.

The wedding of Evan and Marianne took place in the house of John Reagan Henry on December 15, 1880. The couple traveled to Bosque County in a wagon driven by a span of mules. Their destination was the 320 acres of land Evan had bought on October 12, 1878, about 4 miles east of Cranfills Gap in the Boggy community in Bosque County.[29] He had paid $275 in cash and signed a promissory note for $275 with 10 percent interest due on or before January 4, 1879. Boggy got its name from a spring at a branch of the creek where a mud bottom formed during rainy weather. Wagons and buggies often got stuck in the mud there. Several Norwegians settled in the Boggy community in the Neil's Creek Valley around 1880, but Jim Hoff was the first Norwegian to settle there.[30]

The Rohne couple moved into a log cabin already on the land. A new house was not built until 1897; developing the land on the farm came first. Between 1882 and 1899, they had nine children. Minnie was the first (born September 16, 1882, died October 19, 1941), followed by Oscar (born February 20, 1884, died February 21, 1930). Then came Cora (born November 8, 1886, died October 22, 1922), Pauline (born on April 9, 1888, died September 16, 1971), Emma Mathilda (born April 9, 1889, died August 1, 1968), John Magnus (born July 18, 1891, died February 19, 1958 in Phoenix, Arizona), Paul Bernhart (born January 28, 1895, died December 10, 1981), Helen Magdalena (born September 30, 1897, died August 29, 1996 in Baytown, Harris County), and Christian Ludwig (born April 28, 1899, died October 9, 1957 in Tarrant County). All the children but two married in Bosque County when they came of age. Emma married a minister and moved to Minnesota. John Magnus studied theology and got a Ph.D. from Harvard University, moved to Minnesota, and later to Arizona. In 1926, he published a comprehensive church history of Norwegian Lutheranism in the United States, *Norwegian American Lutheranism up to 1872*. His brother Christian Rohne did very well as a banker in Cranfills Gap for more than thirty years. In addition to the many children, the couple also found room in the log cabin for Marianne's elderly mother, Olea Olsdatter Egeberg, who was born on July 16, 1817, and died on August 14, 1911, at the age of 94. She had emigrated to Texas in September 1881 after the death of her husband.

Evan Rohne never forgot his poor childhood in Hedmark County. On several occasions he helped relatives to emigrate. Martin Rasmussen Rohne, the son of his sister, was born on December 11, 1870, in Romedal, Hedmark County. Evan knew that Martin's mother, his sister Helene Paulsdatter (born June 22, 1850), lived in very poor conditions. When Martin was 11 years old, Evan offered to pay for Martin to travel from Norway to Texas and promised to take care of him until he was old enough to support himself. Martin eagerly accepted the offer, and together with an uncle he left Oslo on October 5, 1882, for New York on the *Geiser*, owned by the Danish Tingvalla Line. The ship arrived in New York on October 21 with 214 passengers on board, and on October 29, 1882, Martin arrived on his uncle's farm in the Boggy community. "I had good days with my uncle, he being very kind to me and in the spring he sent me to school at Boggy," wrote Martin Rohne in a short biography of his life in 1895, when he was 25 years old. He learned fast and before school was out had finished "the Third Reader."[31]

Around 1890, Evan Rohne was bitten by a rattlesnake in a grain field. He survived, but never fully regained his health. On May 11, 1901, Evan Rohne died of bronchial pneumonia. Shortly before his death, Rohne had bought another farm, the Hinchman ranch in Hamilton County, which included cattle, agricultural equipment, and teams of mules. For six years Marianne managed both farms with the help of the older children. She then sold 312 acres of the Hinchman ranch to her daughter Cora and her husband P. L. Christensen and set up a trust fund for the younger children with oldest son Oscar as their guardian.[32] Two months later, on March 30, 1907, Oscar Rohne applied to the Bosque County Court to sell the minors' interest in the 320 acres of the home farm. Marianne Rohne bought a third undivided interest of her minor children in the homeplace for $1,706.66. Marianne also bought the minors' interest in the personal property, cattle, horses, farm implements, and so forth, for $3,431.98 .[33]

From the time of their marriage in 1878 until Evan's death in 1901, Marianne and Evan were successful measured against their own goals. After the death of her husband, Marianne continued along the same path and did well economically. Christine Dahl gave her children a farm or the equivalent of $2,000 when her children married. Evan and Marianne Rohne gave their children the choice of their own farm or higher education.

NORWEGIANS IN LIMESTONE COUNTY

A surprising number of Norwegians moved to Limestone County after they had completed their indenture contracts. The Texas census in 1887 registered one hundred Norwegians living in McLennan County, but sixty-eight Norwegians lived in Limestone County east of Waco.[34] Limestone County had a population of 4,537 in 1860, 3,464 whites and 1,072 slaves (23 percent). The population in the county increased considerably after the Houston and Texas Central Railroad was built through Limestone County in 1869. But several race-related murders occurred in the county during Reconstruction. In October 1871 the situation was so tense that the Texas governor decided to declare the county under martial law.[35] Bosque County also experienced considerable lawlessness during Reconstruction, but not to the same degree as in Limestone County.

The brothers Lars and Lucas Michelson, who worked as indentured servants on the Questad farm, had repaid all expenses related to their tickets in 1874. Both of them left Bosque County and sought work in Limestone County. In a letter to Carl Questad, dated Mexia, October 11, 1874, Lars Michelson wrote that he and his brother had ended up in

Mexia. He worked for a farmer living about a mile outside town. He was happy with his work, and intended to stay there through the winter. Lars also sent Questad greetings from his brother "Lukkas." They had both found employment while several other Norwegians they knew had had no such luck.[36]

Bernt Fredrik Johnson came to Limestone County in 1870. He was born Bernt Fredrik Johansen Egeberg at the Årstad farm in Stange parish, Hedmark County, on October 22, 1847, and died on July 23, 1912, in Bosque County.[37] At the time of the Norwegian census in 1865, Bernt worked as a *tjenestekarl*, farm worker, at the farm of Ole Hansen (36 years) and his wife Mathea (27) in Løten.[38] Ole Hansen, the cousin of Poul Poulson in Bosque County, went bankrupt in 1867, and the Hansen family emigrated with their children to Texas on the *Atalanta* in 1868.

If Ole Hanson had good economic reasons to emigrate, Bernt Johnson had even stronger reasons to do so. At the time of the US census in 1880, Bernt Johnson lived as one of three boarders in the household of Robert Sylvester Munger.[39] Another Norwegian, listed as Hawkins Jensen (27 years) also worked for Munger. In 1881, Johnson bought 100 acres of land on Neil's Creek in Bosque County, on the border with Hamilton County, for $350, but he found his work with Munger both interesting and rewarding and did not leave during the 1880s. Robert Sylvester Munger was born in Rutersville, Fayette County, on July 24, 1854. His father, Henry Martin Munger, had established himself as a successful landowner in Limestone County in the 1870s. Robert attended Trinity University at Tehuacana for a while but left before graduation. He was put in charge of his father's gin at Mexia.[40] Munger invented a pneumatic system to convey seed cotton to the gin, and he also patented saw cleaners for ginning machines in 1878 and 1879.[41] Despite his patents, Munger could not find anyone in Mexia and Limestone County willing to finance the work necessary to make his invention an innovation. In 1885 he moved to Dallas, where he founded the Munger Improved Cotton Machine Manufacturing Company.[42] The new company produced everything needed for first class cotton ginning, from the handling of cotton direct from wagon to bale without being touched by hand. At the World's Columbian Exposition in 1893 in Chicago, the Munger system won ten gold medals.[43]

Bernt Johnson followed Munger to Dallas and became overseer for the manufacturing of the new cotton ginning machines. He married Olavia

Antoinette Simpson in Dallas in 1889. They lost their first child, but on August 16, 1891, Charles Oscar Johnson was born in Dallas. His mother, however, died on December 1, 1891. Bernt found himself in a very difficult situation, alone with a newborn child. His sister Marianne and her husband Evan Rohne gave Bernt invaluable help by taking care of the baby; Marianne nursed Charles along with her own son Magnus. Charles lived in the Rohne home until the age of 4 when Bernt remarried.

Neither the Michelson brothers nor Bernt Johnson settled permanently in Limestone County. But several Norwegian immigrants settled there as landowners. One of them was Anton Jensen (Jenson), born on March 11, 1850, at Rensvollbakken, Romedal, who emigrated in 1873. He died in Limestone County on December 22, 1925. His brother Hans Jenson, born on November 6, 1844 at Sande-Eie Romedal, also ended up as a farmer in Limestone County, where he died on April 4, 1926.

BOSQUE COUNTY IN 1887

Farming and stock raising occupied the attention of the people in Bosque County, observed the author of the 1887 Texas census. The county had 11,498 inhabitants. The number of "Americans" was 9,568; the number of "colored" people was 345. Among the foreign-born, Norwegians dominated totally: 1,247 persons born in Norway lived in the county, 140 Germans, seventy-two Danes, fifty-two Swedes, thirty-seven English, fifteen Irish, eight French, six Poles, two Scots, and one Mexican. The number of white families was 2,112 and the number of colored families was sixty-four. The county seat Meridian was still the largest town with 1,000 inhabitants, followed by Iredell with 700, Morgan 600, Clifton 600, and Valley Mills 450.[44] The county had only one bank, but 110 merchants, twenty lawyers and twenty-four physicians. Two railroads passed through the county—the Gulf, Colorado, and Santa Fe Railroad and the Texas Central Railroad. They crossed at Morgan.

The total value of property in Bosque County increased from $1,797,281 in 1877 to $4,615,815 in 1887. The cost of improved land was between $7.50 and $10 per acre, while unimproved land could still be bought for $3.50 to $7.50 per acre. Around 77,000 acres were cultivated in the county; wheat was planted on 17,905 acres, corn on 27,220 acres, barley on 7,674 acres, and cotton on 23,893 acres.

There could hardly have been a more unfortunate year to conduct a census in the 1880s than 1887. People remembered the drought in Bosque County and countless other west Texas counties for decades. The harvest was miserable. "Thirty-six per cent of the cotton crop was destroyed by worms, and the injury to all crops by drouth was 74 percent," was the comment on the situation in Bosque County.[45] In the neighboring county Hamilton, to the west, 55 percent of the cotton crop was destroyed as was 73 percent of all crops, and in McLennan County, to the east, 40 percent of the cotton was injured as was 76 percent of all other crops.[46]

WEATHER, DROUGHTS, AND HARVESTS

The weather and its influence on the harvest always loom large in the lives of farmers. In letters from immigrants to relatives in Norway, news about the harvest was usually the most important topic following the health of the family. T. Theo Colwick at Norse wrote a newspaper column titled "From Norse" for several years during the 1880s. He often commented on the weather and the harvest.

In February 1884 Colwick wrote that the wheat at Norse had been injured by heavy frost because of rain, sleet, and snow. In June, the largest problem was the heat. The wheat had been harvested, and the average yield was about 18 bushels per acre, and for oats 40 bushels. The corn crop, however, would be very good, about 25 bushels per acre.[47] When cotton picking started in the second part of July, threshing was going on at full speed. New threshers had been bought by Solberg, Ringness, Hoff, and Canuteson. The Mickelson brothers had also invested in a new thresher. At the end of September cotton picking was almost done. The yield would be from one-eighth to one-fourth of a bale per acre.[48] The wheat harvest in 1884 was very good, wrote Christine Dahl to relatives in Norway in February 1885, but prices were low. She had sold about 700 bushels, some of it for only 60 cents a bushel.[49]

The last winter had been very hard on the cattle and horses of their American neighbors, Christine Dahl wrote. Many had died of hunger because "they didn't have sufficient stores of fodder and the winter has been harsher than usual."[50] The Dahl farm was lucky and "didn't lose any. But we had to use a lot of hay to keep them alive."[51] Christine Dahl, the prudent farmer, was always planning for the worst and hoping for the best.

Threshing for farmers in the Norse community was very efficient in summer 1885. In addition to the local Norwegian threshing entrepreneurs, a big thresher from Valley Mills powered by a large steam traction engine had signed threshing contracts with farmers in Norse. Despite the competition from the Valley Mills thresher, Ringness and Solberg had threshed about 15,000 bushels of grain, and Olson and Johnson some 2,000 or 2,500 bushels.[52] According to T. T. Colwich, around 50,000 bushels of grain were harvested in the Norse community in 1885.

The "cotton crop had been cut short" in 1885 because of drought.[53] The first wagons loaded with cotton did not leave Norse for Waco until early December 1885. By the middle of December, cotton picking had been finished and the cotton ginned. Johnson and Olson had ginned some 210 bales, and T. W. Anderson had ginned 180 bales.[54]

The harvest at Norse was even worse in 1886. The yield of cotton was only one-fourth of a bale per acre.[55] The bad times continued into 1887. At the end of January, the fiercest norther anyone in Norse could remember howled through the valley. "Wheat was all frozen to ground in this section." Most of it would probably recover. "But few cattle have yet died here, but it is feared many may yet succumb to the inevitable before spring arrives unless the present mild weather continues."[56] By March spring had commenced in earnest and the corn planting had been finished. "Oats are up and look well. Wheat looks well, but need rain."[57] But the drought continued from one month into the next, and in May the wheat had been injured by frost and attacked by black rust. "It is now succumbing, over half being dead or in a dying condition." Oats would be short "in every sense of the word," it was predicted.[58] The best rainfall in two years blessed the Norse community in August 1887. A month later, on September 26, they got another good rain, and water was finally flowing in the creeks again.[59]

Surviving crops had already been harvested by then. The average harvest of wheat per acre in Bosque County in 1887 was only 4.22 bushels, corn was on average 8.30 bushels, and barley 12.12 bushels. The average cotton crop was only 0.15 bale per acre. In 1887 a total of 75,702 bushels of wheat, 225,941 bushels of corn, and 93,055 bushels of oats were harvested. The total number of cotton bales in the county was 3,617.60.[60]

SIGNS OF AFFLUENCE AMONG NORWEGIAN IMMIGRANTS

Despite several seasons with crops below average, industrious Norwegians prospered in the 1880s. "Improvements were the order of the day in the 'settlement' such as fencing in new and enlarging old farms, and building new houses," wrote T. Theo Colwick in November 1885.[61] The Norwegian colony in Bosque County had changed during the last years; it was becoming more "civilized," wrote Eli Arnesen to her friend Andrea Sleppen in Norway on August 27, 1884. Eli Arnesen and her husband Peder Arnesen had arrived in Bosque County with their children in December 1872 as members of the Dahl group. The family settled on a farm southwest of Clifton, not far from their relatives, the Sinderuds, who emigrated with them. When the family arrived, Bosque was a frontier community. Ten years later, many of the Norwegians were doing quite well, and they lived far better than their American neighbors. According to Eli their American neighbors were rather lazy and indolent, while Norwegian, Swedish, and Danish immigrants were far more industrious.[62]

Several Norwegians built new houses. It was also a sign of affluence that many painted their houses, T. T. Colwick commented. The house of John Ringness had recently been given a new coat of paint, and Dr. O. M. Olson had also painted his residence.[63] The present year, Colwick joked in June 1885, had been a productive one, not only in the field of agriculture, but also with respect to newborn babies. Three pairs of twins had been born to parents of Norwegian heritage since January.[64]

BOSQUE COUNTY WAS THE LEADING SHEEP COUNTY IN CENTRAL TEXAS

The 1887 census enumerated 12,470 horses and mules, valued at $361,449, and 43,996 cattle, valued at $318,236, in Bosque County. What really loomed large, not least compared with the neighboring counties, was the exceptionally high number of sheep. Bosque County had 100,213 sheep in 1887 compared with 52,683 in Coryell County, 66,152 in Hamilton County, 59,098 in Bell County, and 48,563 in McLennan County. The sheep in Bosque County were valued at $180,955. During the year 100,000 sheep had been sheared, producing 700,000 pounds of wool. The value of the wool clip was estimated to be $140,000.[65]

The sheep and goat business had been well established in Texas after the Civil War, observed Paul H. Carlson in his now classic book *Texas Woollybacks*. Merino and Rambouillet sheep were brought into Texas in the 1830s and 1840s. Because of their ability to walk long distances in compact flocks, Merino sheep, with their fine wool, rather than British mutton sheep "became the range sheep of Texas and the southwest."[66] Scots and German immigrants brought thoroughbred sheep to the area around San Antonio, New Braunfels, Boerne, and Fredericksburg. Walking along the Guadalupe River in the mid-1850s Frederick Law Olmsted observed fine cornfields and sheep, "a small flock of the finest Saxony. They had been selected with care, had arrived safely, and had now been, for two or three years, shifting for themselves."[67] Comal and Kendall counties became centers for sheep raising before the Civil War. Ten percent of German households in the western hill counties in the 1850s had sheep, but only small flocks.[68] About 100,000 sheep grazed east of the frontier line in Texas in the mid-1850s, according to Carlson.[69] The sheep business experienced a boom in Texas between 1870 and 1890. It increased by 115 percent, from 1,985,906 to 4,264,187 sheep. In 1890 Texas had more sheep than any other state in the United States.

At the peak of the sheep business in Texas in the late nineteenth century, Bosque County was the leading wool-producing county in Central Texas. The number of sheep there grew from 22,456 in 1880 to 100,213 in 1887, a growth of almost 350 percent. Bosque County had 109,801 sheep by 1890. Only the sparsely populated counties on the Edwards Plateau in west Texas—Edwards, Kimble, McCulloch, and Sutton—and the counties Maverick and Webb along the Mexican border west of San Antonio had more sheep. With good reason, Carlson devoted a whole chapter in *Texas Woollybacks* to the establishment and development of the sheep business on the Edwards Plateau.[70] Counties like Bosque and Hamilton, however, fell outside his field of vision (see Appendix 4).

The importance of the sheep business for the agricultural economy in Bosque County at the end of the nineteenth century has so far been totally overlooked in local historical studies. Sometimes sheep are mentioned in passing in family histories in *Bosque County: Land and People*. William Tillie Poston, for example, "always raised sheep," and in 1887 he shipped sheep by the carload to Z. T. Winfree and Co. in Galveston. Edward G. P. Kellum, who moved to Valley Mills with his family in 1874, "accumulated some two thousand acres of land and became a leading stockman with 2600 head of sheep and a large herd of cattle on his ranch."[71]

The German immigrant Adolf Landgraf, born in Posen (Poznan) in 1862, who emigrated to Texas in 1881, came to northeastern Bosque County in 1882. He tended sheep on the Fitzhugh ranch for 5 dollars a month. He saved money and bought his own farm. Because of "the large amount of pasture land, the Landgraf's had much livestock—cattle, horses and a large flock of sheep."[72] The German Jo Knust (born 1854 in Westphalen) came to Bosque County in 1881. For a while he worked with Landgraf on the Looney ranch (Fitzhugh ranch) on the north side of Steele Creek.[73] Knust always raised sheep and goats.

GO TO TEXAS OR DIE! THE O. K. REA RANCH
IN HAMILTON COUNTY

The Norwegians in Norse and the Meridian Creek Valley were most fond of growing wheat, and none of them owned large flocks of sheep at this time. But several of the Norwegians who bought land in neighboring Hamilton County became sheep owners. Most of them just crossed the border and settled less than 10 miles into Hamilton County.

In the Pottsville area in southern Hamilton County, the Norwegian O. K. Rea built up a large sheep ranch in the 1880s. He hired Norwegians from Bosque County and recent immigrants to work for him. After some years in his employ, several of them established their own sheep ranches in the Pottsville area.

O. K. Rea was born Ole Knudsen Rye in 1854 in Nord-Aurdal municipality in the mountain valley of Valdres, Oppland County, in eastern Norway. He was the oldest son of Knud Olsen Hove (born 1830) and Astri Tidemandsdatter Hilmen (born 1832). The Rye family and many others in their neighborhood emigrated from Valdres to Wisconsin in April 1867.[74] The Rye family settled on a 40-acre farm in Gibson Township, Manitowoc County, and changed their surname to Rea.

Eventually, six Rea children and their parents would migrate to Texas. Their oldest son, O. K., led the way in 1877. The damp, cold climate of Wisconsin troubled him constantly and he feared he would end up with tuberculosis and die. His doctor advised him to move to a warmer climate. He knew about the Norwegian settlement in Bosque County so traveled from Wisconsin to Waco in January 1877, and continued to Norse. He was immediately hired by the Norwegian immigrant Jim Jenson. To improve his health, he wanted work that would force him to be outdoors most of the time. He left Bosque County, went south, and became a sheep herder in Williamson County. The dry air and warm climate did wonders for his lungs, and his health improved.

Rea had lived among Germans in Wisconsin and had no problems communicating with the Texas-Germans in Williamson County. At a country dance he met Anna Caroline Zschiesche, born in September 1861 in a village near Grossenhain in Saxony, not far from Dresden. On July 17, 1881, Ole K. Rea and Anna Caroline were married in the Lutheran Church in Pflugerville, 15 miles north of Austin in northeastern Travis County.

O. K. Rea bought 638 acres of land north of Pottsville in southern Hamilton County, and in 1883, Rea, Anna, and their son John Oliver brought a flock of 900 sheep on shares to their land. Rea owned 50 percent of this flock of sheep. He did well as a sheep rancher, and he usually kept a flock of around 2,000. Around 1900 he owned, in addition to his original homestead, 1,300 acres of land on the north side of Cowhouse Creek. He cultivated only 100 acres of land, using the balance of 1,800 acres as pastures for sheep and cattle.

Hamilton County, with Hamilton as county seat, had been established as early as 1858.[75] Because of frequent Indian raids during the Civil War and the first years after the war, few people had dared to settle there. Strong population growth took place after the Indian threat was reduced in the early 1870s. Hamilton County had only 733 inhabitants in 1870, but 6,365 in 1880. Cattle ranching was the major economic activity, but several sheep ranches were established. By 1900, cotton had become the leading cash crop, while sheep ranching peaked around 1890.[76]

O. K. Rea followed the German migration chain to Hamilton County. Around 1880 many recent German immigrants and second-generation German-Texans moved from southern Texas counties to Hamilton County. By 1900, more than 50,000 foreign-born Germans lived in Texas.[77] The number of Germans in Washington County increased from 70 in 1850 to 3,281 in 1900. Germans from Washington County moved to eastern Bell County, from Frelsburg in Colorado County to western Falls County, and from Fayette, Austin, and Washington counties to the area around McGregor in Coryell County. Germans also moved to Bosque and Hamilton counties.[78]

Many of the German immigrants who settled in Hamilton County in the 1880s had first lived around Zionville close to Brenham, and in the area between Greenvine and Burton in southwestern Washington County. Some of them had lived on the border with Fayette County in Carmine and in Fayette County south to Round Top. The Germans either settled south of Hamilton, following a chain of low hills to Indian Gap, or northeast of Hamilton in places like the German Valley.[79]

In 1885, most of the Rea family in Manitowoc, Wisconsin, decided to follow the example of their oldest son and migrated to Texas. His father, Knud, mother, Astri, and their youngest daughter, Emma (age 13), were invited to stay at the house of John Rogstad in Bosque County. O. K.'s brother Tidemand Knudsen Rea was 28 years old when he moved to Hamilton County with his wife Helen and their children Alma and Clarence. He built up a sheep ranch 2 miles south of the O. K. Rea ranch. In addition to sheep and goats, T. K. Rea raised cotton, corn, grains, and maize as well as Hereford cattle and horses. T. K. was also a skilled gardener and fruit grower. Over the years T. K. Rea invested in several properties, including a ranch called Live Oak near Energy, Texas, the "320" ranch just north of his brother O. K.'s ranch, as well as property in Bosque County, and acquired four sections of land in Terry County in West Texas.

CATTLE, SHEEP, AND FENCING WARS IN HAMILTON COUNTY AND THE DEATH OF A NORWEGIAN MOTHER

As sheep owners, the Rea brothers became involved in conflicts with cattlemen in the area. When flocks of sheep came onto land the cattlemen claimed they had a right to, they retaliated by terrorizing the sheepherder and killed or drove off their sheep. The herder was usually alone, and he had small chances of defending himself or his flock if a mounted band of armed cowmen swooped down on his camp in the middle of the night.

Many of the clashes between cattlemen and sheep men were triggered by the breakthrough of barbed wire.[80] Barbed wire was first introduced in Texas in 1875. It was initially rejected by both the ranchers, in favor of free grass and open range, and the "nesters," in favor of protected fields. The breakthrough for barbed wire sales in Texas came at the end of 1876, after John Warner Gates successfully staged a barbed wire show in San Antonio.[81]

The drought in Texas in summer and fall 1883 made it hard for ranchers without land to get access to grass and water for their herds. Barbed wire fences created a lot of anger. "Wrecking of fences was reported from more than half the Texas counties," Wayne Gard observed, "and was most common in a belt extending north and south through the center of the state, the ranchman's frontier of 1883."[82]

Cattlemen and sheep men clashed several times in Hamilton County that year. Sheep men were ordered to leave Hamilton County, and when they refused to do so, raiders killed or maimed many animals. The sheep rancher M. O. Gleason, who had fenced in a watering hole on his ranch near Fairy, was confronted by a mob of cattlemen one night and told to "leave the country before daylight." Upon learning that Gleason was a Mason, one of the group who was also a Mason assured him that he would have to neither leave the county nor remove the fence.[83] Others in Hamilton County were not so lucky. On August 23, 1883, a party of eight men arrived at the ranch of E. G. Plinton, 12 miles southwest of Hico, and "after blind-folding the herder, killed nearly 100 sheep by using clubs. The depredators have not yet been recognized," the Dallas Weekly Herald reported.[84] The Texas Rangers sent Corporal B. D. Lindsey and five privates to Hamilton County to "keep the peace between the cattle and sheep men."

After heated political debates in Austin on January 8, 1884, the Texas Legislature decided to make fence cutting a felony punishable by one to five years in prison.[85] The new law ended fence cutting within a few years. From then on cattlemen and sheep men had to confine their herds to land they owned. The new law made it easier to improve the quality of livestock through controlled breeding. By 1900, the open range was gone.

Despite the new law, fencing conflicts flared up again during the droughts of 1886 and 1887. No rain fell in Hamilton County in 1886, and none in 1887 until August. The Rea ranches experienced a deadly drought-related conflict in early 1886. One night, while the men were away, cattlemen attacked the O. K. Rea ranch. Anna Rea and her mother-in-law, Astri, woke up in the middle of the night to noise outside the house. They got out of their beds, looked out, and saw burning flames along the house. The two women managed to extinguish the fire. Two weeks later the men were away again, and cowboys again used the opportunity to set the house on fire. This time they succeeded, and the house burned to the ground. The two women and the children in the house managed to get out. Anna Rea, who was pregnant at the time, wanted to save some children's clothes, and Astri dashed back into the house to try to get them. She was severely burned and died the next day, February 8, 1886.

When O. K. and T. K. Rea returned, they found the house burned down and their mother dead. Her body was laid in a coffin on the floor of a wagon, and the two brothers traveled 50 miles to the Norwegian Lutheran Church at Norse. They arrived at the church in the middle of the night, left the coffin there, and continued to the John Rogstad farm to bring the sad news to their father and sister. After this tragedy, the conflicts in Hamilton came to an end.

SEEKING NEW LAND ELSEWHERE?

The period 1860–1900 was characterized by a shift from subsistence farming to commercial agriculture. The price of land in Bosque County increased. Indentured servants who came to work in Bosque and surrounding counties had a hard time finding work after they had repaid their tickets. In the 1890s an increasing number of young Norwegian men and women tried to find work in counties west of Bosque County or in Waco, as laborers and domestic servants. But since most of them had not mastered the English language, they often felt lonely and isolated. When they got the opportunity, they returned to Bosque County to work, marry, and nurture contacts with relatives and friends. By 1900, Bosque County was the most Norwegian place in the whole of Texas.

Some still tried to climb the agricultural ladder. Christian P. Carlson, born in 1857, in Vang, Hedmark County, had emigrated to Texas at the age of 22. Like so many others, he already had a brother or sister living there, in his case, his brother Ole. Christian married Gustava Gunderson in Meridian in 1882. She was born 1862 in Brandval, Solør, Hedmark County. Christian got to know her when he worked on a neighboring farm to that of her parents. After marriage the couple rented a farm at Norway Mills, but then became tenants on a farm about 4 miles north of Meridian. A daughter was born in 1902, and in 1903 they moved to the Cove Springs community, where they lived near Christian's parents and two of his brothers. Christian Carlson was finally able to buy a farm of 200 acres in 1905. Another 150 acres were added to this farm in 1910, followed by 184 acres in 1913. In addition to farming Christian Carlson worked as a blacksmith.⁸⁶ It took him twenty-six years to reach the top of the agricultural ladder.

Earlier in this chapter the story of the Norwegian boy John Dahl was used as an example of how children were learning by doing on the first rung of the agricultural ladder. After he came to Texas in 1895 at the age of 14, he joined his sister Mina in Hamilton County, who had paid his ticket. He worked for Matt Christianson for a year herding sheep and doing miscellaneous farm work, earning enough to repay his sister. John continued to work on ranches in Hamilton County for several years, tending stock and herding sheep. As with so many other Norwegians living in Hamilton County, he got lonely and traveled the 30 miles or more each way to visit with Norwegians in Bosque County. In 1900 he met Ingvaldine Ellingson and took her to a fish fry in Hamilton County. She had immigrated with her parents from the island of Karmøy in western Norway along the Cleng Peerson chain in 1883. The couple married in 1902. They rented a farm in Hamilton County and lived there until 1909, when they moved to a place in the Norse district, followed by the purchase of a farm in the Meridian Creek Valley close to the Rock Church. John Dahl worked this farm for the next thirty years; he was also a sexton at the church for twenty-seven years.⁸⁷ It took John Dahl between fourteen and twenty years of hard work to climb to the top of the agricultural ladder.

There were others who managed to climb to the top more rapidly; Iver Olson is one example. He was born on March 27, 1858, at the farm Julusmoen in Elverum, Hedmark County, the fifth child in a family of twelve. He emigrated as an indentured servant and arrived on the farm of Peter Pierson on May 11, 1882, where he worked until he had repaid his debt. Olson brought with him considerable skill as a carpenter, and Pierson soon discovered that it would be smart to put his skills to use. During his year as an indentured servant Olson built up a considerable reputation. The wealthy rancher Leroy Parks hired him to build a huge livestock barn and an implements storage building on the large Parks ranch. This work took him the better part of a year. In the following years Olson built many residential buildings in Cranfills Gap, the Norse district, and Clifton, including the Rohne and Johnson homes in Cranfills Gap, a house for H. A. Nelson in Clifton, and the Ole Hagen house 6 miles west of Norse. While building the Hagen house he became good friends with their daughter Olena. Iver and Oleana married in 1886. During the 1890s Olson acquired some 1,100 acres of land 7 miles west of Norse. He took up farming, concentrating on livestock, and in 1912 he bought the remaining 344 acres, including all buildings and improvements, on the once vast Alonzo Cooper ranch. The family moved there in 1913, and Iver Olson died on the ranch in 1929.[88]

Some Norwegian immigrants began to look for land in West Texas. In May 1888 T. Theo Colwick reported that Geo H. Poulson, O. P. Foss, and Lars Skremstad had left on a prospecting trip to the Panhandle. They visited Fisher County, west of Abilene. After their return in June, the three men spoke in "glowing terms of the country visited and crops out there."[89] In 1884 a group of Swedish immigrants had moved from Swedish settlements in Travis and Williamson counties to northeastern Fisher County, where they established the village Swedonia. A small church was built in 1886, a post office was granted the same year, and a schoolhouse was built in 1889.

The population of Fisher County increased from 135 in 1880 to 2,996 in 1890.[90] No Norwegians migrated to Fisher County in 1888. Two years later, however, a group of Norwegians traveled by wagon train 330 miles west from Bosque County to Floyd County in the Texas Panhandle, northeast of Lubbock. Poulson bought land in Floyd County; he was also a charter member of the Masonic Lodge in Floydada, the county seat. The family of Tellev William Anderson, who was born on February 27, 1858, in Van Zandt County and died in Bosque on September 7, 1937, migrated west with this group. Anderson had moved to Bosque County with his parents in the 1870s and had married Mathilda Nystøl (born August 9, 1862) on June 7, 1879. They brought with them four girls and one son in the wagon train as well as two grandfathers, Mathilda's father, Kittel O. Nystøl (born January 6, 1824), and Anderson's father, Ole A. Anderson (born December 7, 1826, died May 2, 1914). After they arrived in Floydada, Kittel O. Nystøl died, on January 5, 1891. Terje Nystøl (Terry O. Nystel) and his family were also members of the wagon train west. The population of Floyd County was 529 in 1890. Because of droughts, grasshopper invasion, and the financial panic in 1892, many of the settlers left the county again.[91] Most of the Norwegians returned to Bosque County in 1893.[92]

THE FARMERS' ALLIANCE WAS POPULAR AMONG THE NORSEMEN

The Farmers' Alliance, a national farmers organization and a strong alliance in Texas, gained a strong foothold among Norwegian immigrant farmers in Bosque County in the 1880s, who organized the local Norse Farmers' Alliance. The members of the Alliance met every other Saturday night for a while.[93] T. Theo Colwick was an outspoken and ardent supporter of the Alliance. The strongest opponent of the Alliance was the Norwegian Lutheran minister, who threatened to expel any member of the Lutheran congregation who also became a member of the Alliance. In 1886, Helge J. Grann was president of the Alliance, O. Olson was lecturer, and John N. Colwick business manager.

The Texas Alliance had around 70,000 members by 1886, a time when the economic situation for farmers in Texas was worsening. At the Texas Alliance convention in Cleburne that year the Alliance issued "a series of demands which accurately reflected farmer and labor hopes and needs in Texas," commented Alwyn Barr. The agrarian crisis was blamed on capitalists who held the confidence of the farmer in one hand while they rifled his pockets with the other hand. "Falling farm prices and the trend toward a one-crop cotton economy interacted to aggravate each problem and to increase rural indebtedness. Cotton remained a favorite crop because of higher value per acre and its greater dependability in drought years."[94] The Alliance claimed that the economic problems of farmers were a result of excessive prices for supplies. When cotton prices fell, the farmers suffered, while middlemen continued to reap most of the profit from cotton production. At the Cleburne meeting in August 1886, farmers called for political action to address their economic problems. The most controversial demands were related to monetary reform. Alliance farmers demanded that the government immediately begin using silver in addition to gold to ease the money supply.

The Alliance experienced a dramatic increase in members, growing to around 150,000 in 1888, but then ran into problems that led to a fall in membership to 142,000 in 1889. The Alliance had its own newspaper, the *Southern Mercury*. Scores of lecturers visited local Alliance meetings, including the Norse meetings. Around 1890, the Alliance became actively involved in politics and played a major role in the election of reform Democrat James Stephen Hogg as Texas governor that year. When the People's Party was established in May 1891, it built its political platform on many of the demands formulated by the Alliance. Many Texas Alliance leaders were involved in the campaign for the People's Party, and the Alliance formally endorsed the People's Party in February 1892. The Democratic Party had dominated Texas politics. When the People's Party emerged as a strong challenger, the Democrats began to fight back. The People's Party eventually folded, and since the Alliance had supported them, the Alliance lost much of its influence.[95]

The enthusiasm among many Norsemen in Bosque County for the Farmers' Alliance was a clear sign that the Norwegians actively participated in Texas politics. Their neighbors' problems were also their problems. Most Norwegian immigrants had emigrated in the hopes of eventually becoming landowners. This was their American dream. Even though many Norwegians achieved this goal, many did not. As long as free land was still available, immigrants who brought with them a modest sum of money from Norway could step onto the top rung of the agricultural ladder after a few years. Indentured servants who did not even have money for their emigration journey might spend ten years or more as agricultural workers and tenants before they became landowners. Some remained agricultural workers the rest of their lives or migrated to cities. Norwegians who grew up in agriculture had been convinced that they could realize their American dream of owning land in Texas if they only worked hard enough. Nevertheless, around 1900 many Norwegian farmers in Bosque County began to prepare their children for a future in vocations where both higher education and leaving Bosque County were necessary to find a good job.

CHAPTER 13

A Transplanted Community

◊ ◊ ◊

THE CORE GROUP OF NORWEGIANS in Bosque County lived in the Norse district. During the 1870s and 1880s several new homesteads were established around Norway Mills, the mill to the east, and westward toward Cranfills Gap. Around 1900, several second-generation Norwegians and recent immigrants moved into Hamilton County, across the county line to the south. The Norwegianness in Bosque County was at its highest.

A GROWING COMMUNITY IN NORWAY MILLS

Omen Omenson was one of the first Norwegians to develop land east of Norse and close to Norway Mills. The leading entrepreneur in Norway Mills was Alvin Young Reeder, born in January 1836 in the Indian Territory. After the Civil War, "A. Y. Reeder and a junior partner built a mill on the south side of the creek."[1] His "junior partner" was the Norwegian-Texan Ole Canuteson, the skilled mechanic who moved to Waco to establish his own foundry and mechanical workshop in 1867. They built a two-story limestone gristmill. It was a burr mill, in which the lower millstone was stationary, while the upper stone revolved with power from a steam engine.

Around 1870 there had been three gristmills in Bosque County. The first, a mill in Clifton, had been built in the 1850s by R. J. Grant on land owned by Samuel S. Locker on the Bosque River 10 miles below Meridian. The mill was sold to J. M. Stinnett in 1859 for $2,500.50. William L. Kemp bought the mill in February 1867 and built a new mill, later known as the Old Mill.[2]

Dr. E. P. Booth and his family, former slave owners in Moorhouse County, Louisiana, moved from Waco to Bosque County in 1867. Ashbury Steagall, a cousin of Booth's wife, Elizabeth Jane, was a miller's

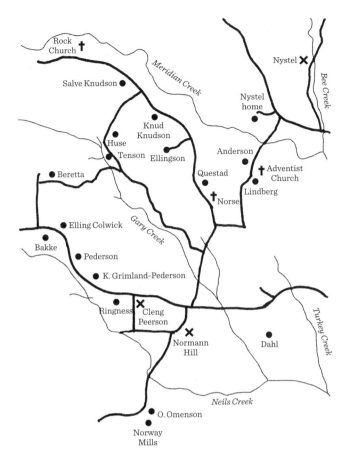

A crude map of the Norse district, undated, showing the location of some of the first Norwegian farms. Some other relevant locations have been marked as well. Note the importance of the creeks; they were a crucial factor in the Norwegian immigrants' decisions of where to establish their homesteads.

wright. Steagall thought that the crossing on the North Bosque River, 20 miles west of Waco, would be a good location for a gristmill, a lumber mill, and a town. Dr. Booth joined him in the venture. On September 22, 1867, the two men bought 300 acres from James Saddler for $15,000 and began to sell lots for the new town of Valley Mills. Steagall built a two-story gristmill powered by a water wheel, which also powered a sawmill.[3]

The three gristmills represented the first move toward industrialization in Bosque County. Texas was still characterized by lack of transportation systems in the early 1870s. Mechanization was rare, slow, and mainly connected to the processing of raw materials. According to John

Stricklin Spratt, flour mills were the most important nonagricultural industry in Texas in the 1870s.[4] The 1870 census enumerated 533 flour mills in the state with a total capital value of $1,066,893 and a production worth of $2,421,047. The average capital investment was $2,000 and the average value of production per mill was $4,542.[5]

When Norway Mills began to operate, farmers in the Norse district got easy access to a gristmill. The company charged a fee of one-sixth for grinding the grain. "Norway Mill flour was considered to be very good. It was fine and white and made fine bread," the Norwegian miller Christian O. Bronstad told his children later in life.[6] Flour was regularly delivered to Waco but was also sold farther west in counties like Hamilton, Brownwood, and Comanche, even as far west as San Angelo.

A little village grew up around Norway Mills. Reeder installed a cotton gin powered by cable from the mill shaft. He also established a merchandising store and became postmaster when Norway Mills got a post office in 1879. Dr. Barlow operated a drugstore, which Dr. O. M. Olson and Dr. Cowan later took over. For a while Norway Mills could boast of having two blacksmith's shops and a livery stable. The last investment of any size was done by Hill Brothers & Co. At the opening of a new store on November 27, 1885, they held a ball, with the Norse Brass Band supplying the music.[7]

When the roller mill became popular in the last two decades of the nineteenth century, most Texas gristmills ran into problems. Norway Mills was no exception. Ole Canuteson and his brother C. A. Canuteson sold their 50 percent ownership of the mill to Reeder at less than half of the original cost. There was a rumor in spring 1884 that Reeder planned to move his mill from Norway Mills to Jonesboro in Hamilton County.[8] In 1900, flour mills no longer held first place among Texas industries. "The small flour mill, operating as an essential part of a self-contained community, had been replaced by the large commercial establishments which marketed its products over wide areas."[9] The first roller mill in the Bosque Valley was established by Willis Sparks Helm in 1887 and named the Merchant and Exchange Flour Mill of Clifton.

Norway Mills was closed around 1890, and soon other businesses in the area dwindled away. Clifton became the preferred merchant town after the Gulf, Colorado, and Santa Fe Railroad reached Clifton in 1881. All merchants reasoned that their businesses stood to gain by moving to the vicinity of the new railway station. "Clifton experienced considerable growth as a trade and business center between the mid-1880s and 1910," wrote William C. Pool.[10]

THE UPPER SETTLEMENT AND CRANFILLS GAP

In the two decades following the Civil War, many Norwegian immigrants settled on homesteads and farms in the Meridian Creek Valley, between the Jenson and Rogstad Mountains and Meridian Creek. Bersvend Swenson and his family, who had immigrated to Texas in 1851 and settled at Four Mile Prairie, moved to Bosque County in 1857. Born Bersvend Eriksen Lillehogstad (Bersvend Swenson) on May 22, at Lillehogstad, Tynset, Hedmark County, he died in Bosque County on February 28, 1896. He had married Kari Nielsdatter, who was born on December 25, 1809, at Tynset, and died in Bosque County on September 8, 1897. They brought seven children to Texas, but the two oldest soon left Texas and moved to Wisconsin. The Swenson family chose land on the north side of Rogstad Mountain in 1857, where Swenson built a stone and wood-frame house, a stone barn, and a detached dug-out cellar.

Jens Lassen Reiersen, the brother of Johan R. Reiersen, moved to Bosque County the same year as the Swenson family. He settled on a homestead east of the Swensons. Jens Lassen Reiersen (1820–1894) had emigrated with his brother Carl to Wisconsin in 1844 but joined the rest of the family in East Texas in 1849. He had married Aase Marie (Mary) Thorbjørnsen, born in 1831, in March 1849 at Four Mile Prairie. She had immigrated with her parents from Tromøy, Aust-Agder County, to Four Mile Prairie in 1848. They had two daughters, Mathilda in December 1850 and Amelia Georgiane in 1854. Aase died between the birth of the last child in 1854 and 1857, when Jens moved to Bosque County.[11]

Berthe Marie Foss (born September 18, 1818, died at Cranfills Gap, January 22, 1886) moved from Four Mile Prairie to Bosque County with her two daughters Petra and Helen after the Civil War. She had emigrated from Toten on the western side of Lake Mjøsa in 1851 with her husband, Johan O. Foss (born 1813). He died in East Texas on September 21, 1852, after having been bitten by a rattlesnake.[12] Her daughter Petra had married the Danish immigrant Hans Jørgen Hansen at Four Mile Prairie, and they brought their firstborn daughter, Marie, with them to Bosque County.

The west-Norwegian Elias R. Skimland, who served in the Frontier

Regiment during the Civil War, settled on 160 acres near Cranfills Gap and got rights to his land on September 17, 1869. Skimland soon moved to the Midwest. Otto Swenson and his wife Eline (born Bronstad) developed a farm close to that of his parents. The three-year homestead requirements were fulfilled in 1871. Salve Knudson and his son Knud Salve got their papers the same year, while Andre and Ingeborg Johnson got their papers in 1875, A. O. Alfei in 1877, and Amund Ilseng in 1877. Peder Pederson, who settled on 160 acres, got his rights on April 3, 1877, and added another 160 acres on January 20, 1894.[13]

In the 1880s, this area in the Meridian Creek Valley became known as the Upper (Norwegian) Settlement. Growing prosperity led to the building of new houses. A schoolhouse was built in 1877, and in 1885 it was decided that Norwegian Lutherans in the Meridian Creek Valley could build their own church as long as they were able to finance it.

In the 1880s, the village of Cranfills Gap began to develop at the foot of Cranfill Mountain, on the headwaters of the North Fork of Neil's Creek. The place was named in honor of George Easton Cranfill, who came to Bosque County in 1851. The largest cattle owners in the area were Jack Pancake and Frank Hinchmann.[14] A post office was opened in Cranfills Gap in April 1879. Riley and Bud Ford established a store there in 1882, and Raz Cranfill opened a store and saloon. Doctor Whitlock, the first doctor, opened his office in the early 1880s, followed by Doctor Monroe Coston in 1885.

The Norwegian immigrant Gulbrand Olsen Bronstad became a leading citizen in Cranfills Gap. He was born on March 9, 1856 in Romedal and died in 1937 in Bosque County. Bronstad arrived in Bosque County in 1874. His brother Christian, who worked as a miller at Valley Mills and later Norway Mills, helped him get his first jobs. In the late 1870s and early 1880s he worked in Waco and Morgan, Bosque County. He was intent on learning the English language and acquiring American business skills. While working in Morgan, he met the German Morris Wiel, who worked as a bookkeeper at the Miller Merchandising Store. When the store went bankrupt during the drought in 1886, Bronstad and Wiel took over the business and moved all remaining stock to their new general merchandising store in Cranfills Gap.[15] Despite the long and hard drought that hit the region in 1886 and continued for months into 1887, the store did well from the day it opened on January 1, 1887. Wiel later sold out to Bronstad and returned to Germany.

After a while Bronstad began extending credit and loans to his customers in addition to selling general merchandise. He began to accept deposits as well as money drafts from other banks. In 1910, with the help of J. W. Butler, he secured a loan of $100,000 from an individual in Colorado to use as the capital stock for the charter of the First Guaranty State Bank of Cranfills Gap. Bronstad was on the board of directors and was elected president.

The Bronstad store at Cranfills Gap became an important meeting place for Norwegians in the Upper Settlement. In 1893 Bronstad married Laura Hansen, the daughter of Jørgen and Petra Foss Hansen. They became the parents of six children. When Bronstad went on business to Waco, Dallas, or St. Louis, he went by buggy to Clifton or McGregor, where he left horse and buggy in a livery stable before taking the train. If he went to Waco, he stayed with one of his three sisters living there—Andrea Nelson, Carrie, and Syverine Bronstad.[16] Supplies for his store came by train to Clifton and were transported to Cranfills Gap by mule-drawn wagons.

A. C. Grimland moved to Cranfills Gap in 1891. Together with Emil Sorley, Otto Reesing, and Johny Pedersen, he operated the Four Company in Cranfills Gap. Grimland was also in charge of a co-op cotton gin. At one time, there were five grocery stores in Cranfills Gap.

Alfred Anderson was the first blacksmith in Cranfills Gap. He was born in Norway in 1846 and emigrated from Oslo on the *Oder* on April 28, 1871. He had worked as a blacksmith for ten years at Hills Station (Hillsboro) in Hill County north of Bosque County.[17] In 1883 he married Bergitte Christiansdatter Enger, born on November 8, 1855, at Sandbekken, Løten, Hedmark County. She had emigrated with her parents, Christian Hansen (born February 2, 1823, died February 23, 1902) and Kari Børresdatter Enger (born 1817), from Oslo on the *Hero* on December 8, 1878. Her brother Hans Christiansen Enger (born December 9, 1851) and his wife, Eline Andersdatter Olsen (born February 13,1857), traveled with them.

UPHOLDING THE NORWEGIAN LANGUAGE

From the day he was born, December 1, 1898, in Hamilton County, Texas, and until he began school in 1906, Carl Johnson "could not speak a word of English."[18] His life story, put on tape and paper in 1979, gives rare insight into the Norwegianness among Norwegian immigrants in Texas. If Johnson had been born in the Norse community in Bosque County, it would have been no great surprise that he had not learned to speak English. Esther M. Swenson, for example, the daughter of John Belford and Alma Lund Swenson in the Harmony community, was born on a cold winter day in December 1903. "Since only Norwegian was spoken in the home, it was not surprising that Esther was unable to speak English when she started school."[19]

The Johnson family, however, lived on the O. K. Rea sheep ranch far south in Hamilton County. The Norwegian Our Savior's Lutheran Church at Norse was located 30 miles to the north. It was totally unexpected that a Norwegian family living in Hamilton County around 1900 would be able to use their Norwegian language daily. Was the family totally isolated from their neighbors? Or were their neighbors also Norwegian immigrants who communicated with each other in their mother tongue?

Carl was the second child of Julius and Mina (Dahl) Johnson. Julius was born Julius Johannesen on February 20, 1855 at Skjelset-eie, Romedal, Hedmark County, and died on August 7, 1935 in Hamilton County. Julius had emigrated from Norway on the *St. Olaf* on April 27, 1876, at the age of 21, along with two other persons from Romedal— Anne Kristensdatter (age 30) and Ludvig Pedersen (8). Tickets from Norway to Waco for the three of them cost 148 Spd and 90 skilling.[20]

Julius had signed a contract as an indentured servant with the Norwegian-Texan Otto Johnson. "He worked for him for one year for his fare to America," his son later recalled. "It was what they called the "indentured slave or labor program."[21] He arrived just in time for the grain harvest. After serving out his year, he worked for others, then worked on the railroad in Palo Pinto County for a while in the early 1880s. With his friend Jens Pederson he migrated west to Borden County and filed on a homestead. After two years they gave up the project and returned east to Hamilton County, where they became laborers again.

Julius's wife, Mina Dahl, came to Texas in 1892. She was born on December 28, 1872, at Norderhovstuen, Løten, Hedmark County, and died on December 31, 1948.[22] She emigrated with her brother Lukas H. Dahl (born 1871), who later became a farmer in Bosque County. Mina also worked as an indentured servant for a year for the people who had paid her traveling expenses from Norway to Texas. She then worked as a servant in the household of Judge Goodwin in Comanche County for four years. Soon after she moved to work in a household in Hamilton, Hamilton County, she met Julius Johnson, and on February 2, 1897, they were married.[23]

The Johnsons lived and worked on the sheep ranch of the Norwegian immigrant O. K. Rea and his German wife, Anna. T. K. Rea, O. K. Rea's brother, lived with his family on a neighboring ranch. The family of the Norwegian immigrant Martin Hagen lived nearby. Martin Hagen was born in October 1871, and his wife, Christina Carlson, was born in January 1879; both emigrated from Romedal, Hedmark County. They had four children, three boys and one girl.[24] After a while, Hagen sold his land and moved with his family to California. Christopher Evan "Chris" Lund and his family also lived in the vicinity. Lund had emigrated from Romedal, Hedmark County, in 1878 at the age of 17. He had worked as an indentured servant for Ole Dahl for one year and for his brother-in-law Nils Jacob Nelson in Bosque County for another year. He then worked as a freighter for some time between Clifton and Houston and at a cotton gin for $10 a month. He moved south and worked in Milam County for two years before moving to Williamson County, where he worked for shares on a sheep ranch. "By 1888 he had one thousand sheep of his own, so he moved to Comanche County where he lived for three years." He bought a farm in Hamilton County in 1891, later named Lund Valley, and married Eline Christianson in 1893. She had immigrated to Texas in 1887.

Eline was born in Romedal in 1867. She was the sister of Matt and John Christianson.[25] John and Karen (Dahl) Christianson also lived in the area.[26] Karen was an older sister of Mina, Carl Johnson's mother; she was born on December 12, 1867 at Løten and died in Hamilton County on October 7, 1931. She came to Texas in the 1890s and married John Christianson (born 1868, died 1954), who had emigrated to Texas in 1888. His older brother Mathias Christiansen (Matt Christianson) was born in Romedal on February 10, 1864. He arrived in Bosque County in 1885, where he had signed an indenture contract for twelve months with Otto K. Johnson. After he completed his indenture contract, he worked for both O. K. Rea and his brother T. K. Together with his brother John he then worked on the Moore ranch and in 1892 had saved enough money to buy sheep of his own. On May 22, 1900, Matt Christianson married Eline Jenson in Gatesville and brought her to the Lund place in Hamilton County, which he bought on July 28, 1900.[27]

This very intricate narrative demonstrates how a small and tight web of Norwegian immigrants, having grown up in the two neighboring municipalities of Løten and Romedal, Hedmark County, related to each other and married to each other, lived around the O. K. Rea sheep ranch in southern Hamilton County around 1900. Despite the long distance to Norse, the Norwegians in southern Hamilton County tried as best they could to stay connected with the Norwegians in Norse and the Upper Settlement in Bosque County. Some were even members of the Norwegian Lutheran church. The day Carl Johnson was baptized, the Johnson family made the 30-mile-long journey to the parsonage of Our Saviour's Lutheran Church in Bosque County and back. Carl and his older brother Harald were confirmed in the Rock Church in Bosque County on September 26, 1915.[28]

Carl Johnson was given little schooling. When they lived near Pottsville in southern Hamilton County, he reflected, "we were too far from school to walk and it was too dangerous on account of so many cattle." School was held at the Stanford Schoolhouse, not too far from where they lived. In 1906, he attended school for two months—January and February. "The next year I went four months, and the year after that I went five months. I finished the second reader, and the next year I went six months. I skipped the third reader into the fourth reader, and so on until the seventh grade." The nearest high school was 18 miles away; "that was too far to go, and we had no money to stay in town. So I stayed at home and helped work the farm."[29]

NORWEGIANS KEPT THEIR LANGUAGE IN THE MIDWEST, AND UNEXPECTEDLY DID THE SAME IN TEXAS

In the rural Midwest, Jon Gjerde observed, migrants from the eastern United States and Europe "were able not only to reestablish former ties but also to maintain patterns of language and custom. In these clustered settlements, insulated and isolated from the hostility of others, migrants were able to preserve linguistic traditions, reestablish former cultural and religious conventions, and beget new kinship ties."[30] "Many European immigrants, particularly women, lived their entire lives without learning English. These tendencies, moreover, persisted across generations." Gjerde retold a story about an old Norwegian immigrant who argued that he had nothing against the English language; he used it himself every day. "But if we don't teach our children Norwegian, what will they do when they get to heaven?"[31]

In a study of family and community in rural Wisconsin, Jane Marie Pederson argued that "nothing better illustrates the commitment to traditional cultural values than the degree to which language was retained in families and communities." She did a case study on Norwegian immigrants in Trempealeau County, Wisconsin, along the Mississippi River on the border with Minnesota. In 1880 the Norwegians were the largest ethnic group in Lincoln and Pigeon townships in the county, and people of Norwegian origin still constituted 70 percent of the population in the two townships in 1916.[32] The Polish and the Norwegian Americans proved extraordinarily resistant to efforts to replace their mother tongues with English. "Not only did the second generation of residents learn the language of their parents but so too did the third and sometimes the fourth generation."[33] A woman born in 1915 recalled later in life that she and her five younger siblings "spoke only Norwegian when they started elementary school."[34] Immigrants in rural communities clung to their native language more successfully than immigrants in urban areas.

One of the most unexpected findings of this study is how many decades their neighbors and the state of Texas allowed them to remain transplanted. The pressure to assimilate and become Texans and Americans did not increase to a significant degree until World War I. The continuous stream of indentured servants from Hedmark County to Bosque County from the 1870s and into the twentieth century was of primary importance not only for the survival of the language but also for the growth of the orthodox Norwegian Lutheran Church. The church became the main institution for conserving and upholding the Norwegian language among the immigrants.

THE NORWEGIAN LUTHERAN CHURCH IN TEXAS

According to Edwin Scott Gaustad, Lutheranism in the United States in the decades after the Civil War "revealed a pattern of development in which ethnic loyalty prevented the creation of any single Lutheran church. The large Scandinavian immigrations (Swedish, Danish, Norwegian, Finnish, Icelandic) of the second half of the nineteenth century led to separate ecclesiastical entities whose reason for being was neither doctrinal nor geographical but ethnical." At the beginning of the twentieth century there were twenty-four separate Lutheran groups in America.[35]

The Norwegian state church showed little interest in doing missionary work among immigrants in the United States until the mid-1840s. That changed with the arrival of pastor J. W. C. Dietrichson in 1844, the first Norwegian university-trained clergyman to appear among Norwegian immigrants. Dietrichson had not come to America to serve only a specific congregation. One of his main aims, according to Blegen, was to "produce ecclesiastical order out of what he considered disorder, to organize Norwegian-American Lutheranism according to the rituals of the Norwegian state church."[36] Dietrichson organized congregations along Norwegian state church lines in several Norwegian colonies in Wisconsin. Members had to accept the ritual of the Norwegian state church, pledge obedience to the minister, and agree to call no minister not regularly ordained.[37] Dietrichson returned to Norway in 1845, but his first short visit came to have a strong influence on church organization.[38] The Norwegian Evangelical Lutheran Church in America (the Norwegian Synod) was established in 1851, and the synodical constitution in 1853 reflected "faithfully the general position of the high-church Lutheran orthodoxy," Blegen commented.[39]

A pastor could not admit anyone to the congregation without satisfactory testimonial from their former minister, acceptance of the church doctrine, and submission to its order. Lay preaching was important to Elling Eielsen and his followers in the Ellingson Synod. The competing Norwegian Synod discussed lay preaching throughout the 1850s, and in 1862 the synod proclaimed that the office of public ministry was instituted by God. Teaching the word of God without a proper calling and without a university education was regarded as a sin.

The number of immigrants from Norway in the 1850s increased considerably and the demand for Norwegian orthodox Lutheran ministers was much higher than the supply. To get the right kind of pastor, the Norwegian Synod signed an agreement in 1857 with the German Missouri Synod at St. Louis and placed a Norwegian professor at St. Louis Seminary. Sixty years later, C. S. Ylvisaker of the Norwegian Synod wrote that the Missouri Synod had been "established on a truly

Lutheran foundation." By studying in St. Louis, Norwegian students became "strengthened in the knowledge of Christian doctrine and of matters pertaining to church government."[40] A total of 127 theological candidates graduated from the seminary in St. Louis before the Norwegian Synod started its own theological seminary.

ESTABLISHMENT OF THE NORWEGIAN LUTHERAN CHURCH IN BOSQUE COUNTY

"There can be little doubt that the church as an institution played a major role in the organization and development of community on nineteenth-century American frontiers, especially in the Middle West," proclaimed Robert Ostergren.[41] European immigrants' ties to the churches they had left had been severed. They felt a need for their own churches in America, and the founding of churches became an "early and widespread activity among immigrants on the frontier." Most immigration historians seem to agree with Ostergren on this point.

The history of the Norwegian immigrants in Bosque County, however, does not fit this pattern. The first immigrants arrived in 1854, but no church was formally established until 1869. The Lutheran congregation had problems recruiting and keeping members in the 1870s because of strong tensions between leading members in the community and the church. The congregation at Norse was not able to finance its own church building until 1875. By 1890, however, the church had become the central place for the Norwegian community. The continuous flow of immigrants from Hedmark County in eastern Norway to Bosque County helped the church build a more solid foundation.

It was after the Civil War that the Norwegians in East Texas and Bosque County renewed their invitation to the Norwegian Synod in the Midwest to establish a congregation among them. A constitutional meeting at Our Savior's Lutheran Congregation in Bosque County took place on June 14, 1869, soon after the arrival of Pastor Ole Estrem and his family.[42] When Estrem accepted the call to Bosque County, he had been promised a salary, that a parsonage would be built, and that land for a farm would be bought. He immediately pushed the leaders of the congregation to deliver on their promises. It was decided that members should pledge a monthly sum to cover Estrem's salary, but no fixed amount could be agreed on. The minutes from a meeting on December

27, 1869, registered that Carl Questad had donated 1 acre of land for a churchyard, and in December 1870 he donated 10 acres for a church site. In January 1871 111 acres of land were bought for the parsonage farm for $1 per acre. The parsonage, a two-story stone building, 18 × 30 feet, was built with two rooms downstairs, two rooms upstairs, and a gallery.[43] The parsonage building committee consisted of Eric Lindberg, Ole Ringness, and Peder (Peter) Pierson.

Peter Pierson came to play a central role in the church as deacon, trustee, and chairman of the financing committee for the church. Pierson always followed his own moral compass. He voiced his opinions, even if he knew the pastor would disagree. Pierson was born at Tromøy, Arendal, on February 27, 1843, the only son of Ole and Anne Helene Pierson. He came to Texas with his parents at the age of 10, and his family settled in Bosque County in early 1854. During the Civil War he served in the Confederate Army in McCord's Frontier Regiment, having enlisted in Stephenville in May 1862. When the Frontier Regiment was ordered to the Texas coast in May 1864, Pierson deserted with his friend Osmund O. Nystøl, first to El Paso and then to Santa Fe, where they joined the Union Army. He returned to Bosque County after the Civil War to take over the family farm at Norse.

To become a member of the church and get voting rights, a man had to be confirmed and 18 years of age; women had no voting rights. Pierson was certainly old enough. He was 26 in 1869, but he had never been confirmed, so he became a pupil in the first Norwegian confirmation class in summer 1869 with thirty-three others, who were confirmed in September of that year.[44] The normal confirmation age in Norway was 14, but more than half of the 1869 class was older than that.[45] The class included two married couples: Christoffer (Chris) Jenson (born April 5, 1840, at Brovold, Austre Moland) and his wife Petrine (born January 26, 1841, at Tromøy, Arendal); and Otto Swenson (born April 21, 1844, at Tynset) and his wife Eline (born Brunstad on July 9, 1846, at Romedal). Both Chris Jenson and Otto Swenson had served with Peter Pierson in the Frontier Regiment: Otto Swenson deserted two times; at the end of 1844 and in spring 1845. Chris Jenson was married to Peter Pierson's older sister. Otto Swenson's younger brother, Bersvend Bersvendson (B. B. Swenson), born on July 18, 1850, was also a pupil in the class.

FREETHINKERS IN TEXAS AND BOSQUE COUNTY

The freethinkers were ardent advocates of democracy and religious freedom. They came to the United States seeking freedom from dictatorial monarchies and clerics. They insisted strongly on the freedom to form religious opinions based on intellectual reasoning and not blind, unquestioned faith. Several of the first Norwegian settlers in Bosque County at Norse had strong sympathies with the freethinkers. Cleng Peerson was one of them. Throughout his life he seems to have been fascinated by different types of religious sects outside the official Norwegian Lutheran Church. Why he sold all he owned, joined the Swedish Bishop Hill colony, and gave them all his money is still a mystery. After his experiences at Bishop Hill and his own severe illness in 1849, he became a freethinker and showed little interest in the practice of traditional religion. In his reminiscences from Brownsboro, Texas, as a teenager, Knut Jørgensen Hastvedt commented on Peerson and religion. Peerson was "a small, rather insignificant-looking man, whom scarcely anyone would have suspected of being the pathfinder for the Norwegian people in this country." Peerson knew Norboe in Dallas County well, Hastvedt commented, and Norboe "belonged to the same religious sect as Cleng Peerson. The basis of their faith was a book published in Copenhagen called *Jesus og fornuften* (Jesus and Reason)."[46]

The best-known freethinkers in Texas were Germans who had settled in the Texas Hill Country in Kendall and Kerr counties. They were nicknamed *Die Lateiner* because so many of them had been educated at German universities and many emigrated from German states after the revolution in 1848. "The names of those who followed the early Forty-Eighters sounded like *Who's Who* in German education and popular politics," wrote Glen E. Lich.[47] They came from the intellectual core of the German states; many were highly respected noblemen, philosophers, scientists, physicians, and engineers.[48] Several had been on the losing side in the revolution in Germany in 1848 and had emigrated to the United States and Texas.[49] The town of Comfort at Cypress Creek above its confluence with the Guadalupe River became a freethinker stronghold.[50]

Peerson and Norboe were not the only Norwegians in Texas with freethinker leanings. Ovee Colwick, who took over the Peerson farm in 1860, shared Peerson's views on religion. T. Theo Colwick wrote a short biography about his father, Ovee Colwick, after his death in July 1895.

His father was "an avowed Liberal who had the moral courage to express his honest convictions upon any and all questions," he wrote. Although he was raised in the Norwegian Lutheran Church, even before he emigrated to the United States he believed in the "practical religion of good deeds rather than of theoretical creeds. He held that the Universe is governed by law, and that as certain as effects follow cause, every moral action sooner or later brings its reward, while all violations of the moral law bring consequent suffering upon the actor, just as inevitable as transgressions of the physical law."

FREEMASONS COULD NOT BE MEMBERS OF THE LUTHERAN CHURCH

Before the establishment of the Lutheran Church, several Norwegian pioneers had become Freemasons. Many men in powerful positions in the United States and Texas were Freemasons. The fraternity believed in the European Enlightenment ideals of liberty, autonomy, and God as envisioned by Deist philosophers as a Creator who largely left humanity alone. The American Constitution and the Bill of Rights were influenced by the Masonic "civil religion," which focused on freedom, free enterprise, and a limited role for the state. In a letter to Norway, dated June 1, 1867, Berger T. Rogstad wrote that there were "many freemasons here in America and almost half the Norwegians in our settlement have become freemasons. We have built a church of stone and now the second story is to be a freemason lodge."[51] The church Rogstad referred to was the Old Rock Church on Hog Creek. A. Y. Reeder and his wife had donated one acre of land in January 1866 for a combined school, meeting house, and Masonic Hall. The Masons, who would occupy the upper story, held their first meeting in this building in June 1866.[52] The Masonic Lodge was named in honor of John Armstrong, and among the other charter members were J. K. Helton, L. H. Scrutchfield, and John D. Odle, prominent men in the county. The fact that some of the Norwegian immigrants were invited to join the Masons was proof that Norwegian immigrants had done well and were respected by their American neighbors.

George Washington, the first president of the United States, had reached the top level of the Masons on August 4, 1753, in Alexandria, Virginia.[53]

Stephen F. Austin was among the first Masons in Texas in the 1820s.

The first Grand Lodge of the Republic of Texas was established in December 1837, and Sam Houston presided over the meeting. Between 1838 and 1845 the membership increased from seventy-three to 357. "Although constituting only 1½ percent of the population, Masons filled some 80 percent of the republic's higher offices. All of the presidents, vice presidents, and secretaries of state were Masons."[54] Between 1846 and 1861 five of six Texas governors were members of the fraternity. By 1860 Texas had 226 active lodges and 9,000 members, increasing to 17,000 in 1878.[55]

In addition to Berger Rogstad, it was well known in the community that Carl Questad and Ovee Colwick were Masons. When Questad died on December 13, 1886, he was buried outside the Lutheran churchyard "with Masonic honors." On the same day, within the churchyard, the dedication and unveiling of the Cleng Peerson monument took place.[56] In his biographical paper on his father Ovee Colwick in 1895, T. Theo Colwick wrote that his father had become "a Member of the Free Masons a number of years ago, he ever remained a good and consistent member of that order, until his late lamented demise." His father had passed "over the silent river which separates this breathing world from the great unknown." The funeral at his home on Neil's Creek near Norse on July 14, 1895 was conducted by the Masonic fraternity.

TENSIONS ERUPT BETWEEN NORWEGIAN BELIEVERS AND NON-BELIEVERS AT NORSE

Berger T. Rogstad had written almost enthusiastically about the visit of the Norwegian minister from Wisconsin in 1867. Their two youngest girls had been baptized. Rogstad engaged himself in the new church during the first months after the arrival of Pastor Estrem but soon learned that Estrem categorically followed the teachings of the Missouri Synod and the Norwegian Synod: Masons could not be members of the true Lutheran Church. Methodists, Baptists, and Presbyterians saw no problems with Masonry, and many of their pastors were Masons. Rogstad and other Norwegian Masons did not leave their Masonic Lodge. As a result, the most successful and affluent men in the Norwegian community did not contribute to the salary of the pastor or the building of a new Lutheran Church.

After the congregation had bought land and built a parsonage, Pastor Estrem went on the warpath regarding doctrinal questions, practicing the authoritarian orthodox theology he had learned during his education. After a meeting of the congregation on February 19, 1873, a heated doctrinal debate flared up. Anton Huse withdrew his membership because of the high expenses for the pastor's salary and the parsonage. Eric Swenson withdrew his membership because he had problems accepting that every word in the Bible was from God, as formulated in the church constitution.[57]

A new meeting was called on May 14. Some of the members found the phrasing in the draft for minutes from the last meeting too harsh. The pastor, however, was not willing to give an inch. Estrem, university educated, used all the power of his position to speak down to his "children" on what was the "right" Lutheran truth. He commented on every paragraph in the church constitution; each was important and had to be accepted. The most troublesome paragraph for several members stated that they had to accept "the Old and New Testaments as God's revealed Word and all of the Holy Scriptures as inspired by the Holy Ghost,"[58] which they felt was unnecessary. In the Lutheran Church in Norway it had been enough to be baptized and confirmed to be a member; there were no additional conditions. They thought this should also be the practice in Texas, but this suggestion was refused outright by Pastor Estrem. At the end of the long meeting, C. Pederson, O. Olson, Niels Swenson, J. Jenson, Y. Grimland, O. J. Johnson, and E. M. Lindberg demanded that it should be recorded in the minutes that they disagreed with the pastor and were in favor of practicing the tradition they knew from Norway.[59]

Estrem won in the short term, but several members left the church. The conflict between the pastor and his followers and the persons he branded "non-believers" lingered on for many years in the Norwegian settlement. New members joined the congregation, mainly Norwegian immigrants from Hedmark County, who could produce good references from their pastor in Norway.[60] Some of the new members were or had recently been indentured servants and had little money to give to the church; building their own church had to be postponed until 1875. In June of that year many members pledged $1,000 in total. A plan for the church was presented in December and bids for the building of the church were announced. One of their own, Gunerius Shefstad, was awarded the contract on March 1, 1876, and he completed the church building before the end of the year.[61] The congregation soon began using the church, but the formal inauguration did not take place until April 19–20, 1885.

It was easy to build the church compared to getting members to fund it. Peter Pierson, who was in charge of financing, spent endless hours trying to get fellow Norwegian immigrants to help with funding, but the results were disappointing. A report on the financial situation was recorded in the minutes of the congregation in January 1877. Pierson, wanting to set a good example, had pledged $100 and paid $100. The only other person who had paid $100 was "Madam Christine Dahl." Jens Ringness, Salve Knudson, Knud Salve Knudson, and Kittel Grimland had paid $50. E. B. Swenson had paid $40, and Yern Grimland, Ole Sinderud, and Pastor Estrem had each paid $30, followed by nine members who had paid $25. Smaller sums were paid by many, and when they were added up, the total was $1,365.44. Many names were still on the subscription list; $381.75, which had been pledged, had so far not been paid.[62]

The names on the subscription list reveal where in the Norse district the Lutheran Church had its strongest and weakest support. Almost all antebellum settlers were absent from the list, with the exception of Christine Dahl, Jens Ringness, B. B. Swenson, Ole Pierson, and the widow Maria Johnson. Recent immigrants from Hedmark County (those who had arrived in the the late 1860s and early 1870s) were the most eager supporters of Our Savior's Lutheran Church. Most of them had settled to the west of Norse. Several church members lived in and around Clifton. The membership was weak in the eastern part of Norse toward Norway Mills.

Estrem became very skilled in rubbing his fellow Norsemen the wrong way by always insisting on being right in doctrinal matters. The church was no democracy, and the pastor followed the teachings of the Norwegian and Missouri Synods: The pastor was the highest authority in all questions relating to faith. At a congregation meeting on January 31, 1877, after Estrem had accepted a call from a congregation in Minnesota, he poured out all the grievances with the congregation he had held since 1873; the secretary used twelve pages in the minutes to record them all. Pastor Estrem concluded with the hope that the congregation one day would turn its backs on the freethinkers.[63]

WHO OWNED THE CHURCH BELL?

J. K. Rystad, the new pastor, arrived in fall 1878 and preached his first sermon on September 29, 1878. He stayed in Norse for the rest of his professional life, until his retirement forty-seven years later. He married Bergitte Nelson in 1879, the daughter of the strong church supporter Aslak Nelson in Clifton.[64] In his reminiscences, written in Norwegian, Rystad wrote that there were still some unbelievers at the time of his arrival. "They did not believe that the Bible was the word of God." The congregation had several controversies with the "infedels" but "thanks to God" the church "always carried off the victory."[65] Despite the new church and new pastor, controversies about Lutheran beliefs and doctrine continued into the 1880s.[66] A battle broke out about the subscription for the church bell and its use. Did the church bell belong to the congregation or to the community? The freethinkers argued that the church bell should not be the property of the congregation, but the property of the whole settlement. The bell should ring for the funeral of all Norwegians, not only members of the congregation. This suggestion was rejected at a congregation meeting in April 1881. Rystad presented a long theological argument of why, according to the Bible, the congregation could not share "a church bell together with the infidels, and ring over them at funerals."[67]

Despite serious disagreements, the dedication ceremony for the church bell in September 1884 was a big event in the Norwegian colony, attended by more than 600 people.[68] Christine Dahl, always actively involved in church matters, wrote her relatives in Norway about it in April 1885: There had been "a solemn dedication of our church, but we've actually used it for several years. Our old pastor Ole Estrem and our pastor John Rystad and a pastor called A. Turmo and a dean called Koren were at the dedication. Several hundred took part and we brought dinners to the church and had a service both before and after dinner."[69] She did not mention the heated discussion about the organization of the church. At a meeting of the Lutheran Church on April 20, 1885, Bishop Koren, the pastors, and several members tried "to induce the congregation to join the Wisconsin Squad; Peter Pierson and others protested, or opposed it, and the protestors won the vote. It was decided that the congregation would remain a 'Free Congregation.'"[70]

Pastor Rystad held basically the same views on Lutheran doctrine as his predecessor, but he was more patient. The opposition to becoming a full member of the Norwegian Synod was strong, and Rystad withdrew the proposal for the time being. He used the same tactic in other instances when he met opposition from leading church members. But that did not mean that he gave up the fight. He could wait for months or even years before he got the vote he wanted.

At almost every meeting of the congregation in the late 1880s and early 1890s new members were accepted. Some of them had lived in the community for many years without becoming members. The family of Lemmick Huse, for instance, who came to Bosque County in 1872, became members of Our Savior's Lutheran Church on January 6, 1885. Lottie Huse Brown, the author of a Huse family history, could not understand "why they waited twelve years."[71] Ole Solberg had immigrated to Bosque County in 1868. According to a Solberg family history from 1979, all of "the Solbergs were active members of Our Savior's Lutheran Church at Norse."[72] That might be true, but according to congregation meeting minutes it was not until January 22, 1890, that "Ole Solberg with family wished to become members of the congregation."[73] The main point here is that it was not only recent immigrants who joined the church in the late 1880s but also many a doubting Thomas who had lived in the community for years. All new members still had to accept the church constitution on every point. The tensions between the church and the community at large had abated.

THE ROCK CHURCH IN THE UPPER SETTLEMENT

Paul Bernhart Rohne (born 1895) was the youngest son of Evan and Marianne Rohne in the Boggy community. When the family went to church, he remembered, the kids bedded down in the back of the wagon and the parents rode on spring seats. "The preacher would put in about 2 hours and then the family piled back into the wagon and drove about four miles over the rocky hills in the hot sun to their home."[74]

Our Savior's Lutheran Church had many members in the Upper Settlement. Leading church members in the Meridian Creek Valley decided they wanted their own church, closer to where they lived. In 1885 the congregation gave them permission, as long as they paid for the building without involving the Norse church in any way. It was named St. Olaf Lutheran Church, but is best known today as the Rock Church. The church was built on a rise overlooking the Meridian Creek Valley. The cornerstone of the Rock Church was laid on May 31, 1886, with pastor J. K. Rystad performing the foundation ceremony. Andrew Mickelson was architect and builder in collaboration with his brothers Christian and Ole. Most of the rock work and building was done by church members in their spare time. Many local farmers quarried the native limestone 2 miles away. At a congregation meeting in spring 1892, Pastor Rystad informed the meeting that the St. Olaf's Church was nearing completion.[75] The total cost of the church was approximately $1,000. For many years to come the churchgoers had to be content with only a dirt floor.

REVIVAL MEETINGS—THE ODDEST FORM OF CHRISTIAN WORSHIP?

In his book *Buck Barry, Texas Ranger and Frontiersman*, Barry remembered that the "church provided a welcome social opportunity and an outlet for pent-up emotionalism" in frontier society. "This was especially true for our women. Their life was hard, nerve-racking at times, and despite their hopes for a better day, something was needed to give them a chance to lift the restrictions of the daily existence." Camp meetings were "thoroughly enjoyed. We attended Methodist, Presbyterian, etc., at Meridian. One of the best attended camp meetings in our section was held in Meridian in September 1860, by a Methodist preacher, and we carried bedding, food etc., to spend several days."[76]

The first Norwegian immigrants in Texas were not attracted by the revivalist religion of their American neighbors. Most of them would probably have agreed with Elise Wærenskjold, who in 1852 characterized camp meetings as "the oddest form of Christian worship that any person can imagine." She had attended such a meeting somewhere in the woods, where five preachers were present, and they continued preaching "day and night for eight days." At times the attendants became so inspired, she wrote, "that one after another they begin to sing and clap their hands, crying out 'Glory! Glory!' as loudly as they can. They begin pounding on the ones nearest to them, throwing themselves on their knees or on their backs, laughing and crying—in short conducting themselves like perfectly insane people."[77]

In the early 1880s members of Our Savior's Lutheran Church raised the issue of arranging camp meetings at several congregation meetings. Rystad turned the promoters down again and again. The Norwegian

Synod was very much against lay preaching and had pronounced in 1862 that the office of public ministry was instituted by God. Even the conservative minister Rystad had to admit that times were changing. If he continued to prohibit camp meetings like those held by their neighbors the Methodists, Baptists, and Presbyterians, he faced the risk that some members would leave the Lutherans and become members of competing congregations. Eventually he relented, but insisted that the Lutheran Church should have full control of the content of the camp meeting. In July 1885, the first camp meeting organized by the Norwegian Lutheran Church took place. It was well attended, and there were no signs of the "unseemly excitement and boisterous kind of 'getting religion' with which it is generally accompanied, such as shouting and other modes of becoming 'happy.' The Norwegian Lutherans had behaved with 'quiet and decorum,'" wrote T. Theo Colwick.[78]

"NORSE IS UP TO THE FULL STANDARD IN PICNICS AND PRETTY GIRLS"

Church life might have constituted the backbone of social life for many Norwegian immigrants in Bosque County. However, they had immigrated with the hopes of finding a better life in this world. They worked hard and wanted to enjoy the fruits of their labor here and now.

The Norse community developed a strong tradition of organizing picnics in connection with the Norwegian National Day, May 17, and the American National Day, July 4. The 4th of July picnic in particular was attended by many people far beyond the circle of Norwegian immigrants. The Norwegians were proud of their brass band, and on May 17, 1884, the Norse Brass Band played at the Neil's Creek schoolhouse. On the same day in 1886, a large crowd met on Sugar Loaf Mountain to celebrate the Norwegian National Day. Again, the Norse Brass Band played throughout the day.[79] Because of threatening weather, attendance on May 17, 1887, was less than anticipated. In the afternoon Reverend Rystad delivered a well-written speech. Some liked it, but the always outspoken T. Theo Colwick thought Rystad spoiled it by "his laudation of Lutheranism and by his unwarranted attack upon labor organizations." It seemed to Colwick "to come with bad grace for a Lutheran minister—himself but the mouth-piece of a tyrannical church—to denounce labor organizations as being tyrannical."[80]

The 4th of July picnic became a Norse tradition. The 4th of July, 1886, was celebrated on Sugar Loaf Mountain and was very well attended. "We found the celebrated patriotic Norse Band on hand discussing its sweetest strains and lending enlivening interest to the occasion." After the speaker of the day gave his speech, an ample dinner was served, including "all the good things palatable and refreshing for which the hospitable and generous Norse people are noted, and which they delight and dispense to their friends and guests." During the afternoon candidates for various political offices gave speeches outlining their intended actions if elected. After listening to political speeches for hours, many in the audience were happy when dancing finally commenced.[81]

Two years later, at the 1888 4th of July picnic, visitors came from far beyond the Norwegian settlement. From 10 a.m. onward, buggies and wagons arrived carrying people from Meridian, Clifton, Norway Mills, Cranfills Gap, Walnut, Indian Gap, and Waco, even from as far away as Fort Worth. "About 11 o'clock the well known Norse Brass Band opened with a Quick Step March, after which speaking was announced and short speeches and announcements were made by the candidates for the various county offices, the band furnishing music at intervals. At 1 o clock the crowd scattered all round under trees to join in for dinner. But wood had to be gathered and fires built to cook the coffee with." At 7 p.m. dancing commenced. "The platform was filled to its utmost capacity with dancers, who all seemed to enjoy themselves. The light 'fantastic' toe together with the splendid music furnished by Messrs' Swenson, Nelson and Prather of Clifton, was tripped till 2 o'clock in the morning."[82]

Dancing was obviously an integral part of the May 17 and July 4 picnics. But occasional dances in homes were also very popular. In 2009 Herbert V. Schulz put on paper some reminiscences from the time he grew up as a German child among Norwegians in the Norse community. "Up to the late nineteen twenties," Schulz wrote, "it was custom in the southwest part of Bosque County to hold an occasional dance at someone's home." When the participants arrived, all furniture was moved out into the yard. "When the dance was over, they would mop the floors and put the furniture back into the house."[83] Schulz did not know when this tradition started.

T. Theo Colwick frequently reported about occasional dances among the young people in the community in the 1880s. On March 31, 1885, a party had been held at the house of T. W. Grimland "where all seemed to have a gay time," Colwick wrote.⁸⁴ In the middle of July the "young people had a picnic last Sunday on Gary Creek opposite Andrew Anderson's residence, and I am told enjoyed themselves." In early September there had been "a private sociable at Y. Grimland's residence." On Thursday night, November 14, there was a party at the house of O. M. Olson, and ten days later a party was held at G. Grimland's house.⁸⁵

In August 1887, the big drought of Bosque County finally ended, and it rained more than it had during the last two years. The night of the rain, a dance had taken place at the house of E. M. Lindberg. The next morning, "several bays were water bound on Gary's Creek, the first time ever. There had been another dance at the house of O. Colwick on the night of September 5."⁸⁶

Picnics and dances were at their peak during the summer months. Young girls who worked in Waco or went to school outside the colony came home to visit during the summer. On July 22, 1884, T. Theo Colwick brought the news that "a good many young ladies" were up from Waco and other places visiting relatives and friends. "They will make some of our young men's hearts beat faster I presume."⁸⁷

NORWEGIANS BECAME MORE INVOLVED IN AMERICAN POLITICS AND ASSOCIATIONS

Even though immigrants continued to speak Norwegian among themselves, they became more and more involved in American politics and associations. On the last Saturday in February 1884, the "Stock Club or Association" had a meeting. Farmers and stockraisers in the Norse community joined forces to pool their livestock for sale; they hoped to get a higher price than if each of them had sold on his own and to avoid all "middlemen."⁸⁸ The Norwegian foreman Otto J. Johnson reported that 227 head of young cattle had been subscribed and sold in April at a price of $12 for yearlings, $16 for two-year-olds, $23 for cows and heifers, and $30 for cow and calf.

During the 1880s a Norse Lecture Club was established with the aim of inviting scientific and liberal lecturers to give talks in the community from time to time. On the last Saturday in April 1884, no less than four public meetings were held in Norse. The first was a meeting about school districts, followed by a precinct meeting to elect delegates to the Democratic County Convention to be held at Meridian on May 24.

Otto J. Johnson, Helge Y. Grann, Calvin Holmes and Peter Pierson—three Norwegians and one American—were elected and were instructed to vote for farmers at the meeting in Meridian.

The Norse Protective Association, established in 1882, also met on that day. The association had pursued horse thieves in the community with success; Sheriff Will Terrell and Holver Canuteson had done a very good job.[89] In 1886, J. N. Colwick was president, B. B. Swenson secretary, and his brother Otto Swenson treasurer, and the association had 94 members. Every horse under the protection of the association that had been stolen had been recovered. "Thieves had been arrested and brought to justice in every instance."[90] Some horse thieves, who had stolen horses and saddles from people on Hog Creek, came through Norse in May 1887. "J. Dansby, Hill Bros, and others came in hot pursuit, tracking and overtaking them on Meridian Creek last Friday." The thieves had refused arrest and shots were exchanged. The thieves escaped into the brush into the mountain, but "two youthful accessories were arrested, two horses and saddles recovered."[91]

A COLLEGE IN CLIFTON—THE PEAK OF NORWEGIANNESS?

In the Midwest, Jon Gjerde observed, "Ethnic leaders often attempted to encourage the formation of parochial schools for their descendants. They argued that their church schools succeeded in providing ethical and religious instruction without the secular education that was by definition nonsectarian and included students and teachers with varying religious beliefs."[92] Even prior to the Civil War, Norwegian immigrants in Bosque County accepted that their children attended school with non-Norwegian children. The question of establishing parochial schools was not addressed until the Lutheran Church was established in 1869. To be confirmed in Norway it was required that children could read and write. The same tradition was practiced by the Norwegian Lutheran Synod in the United States and Texas. Instruction in the catechism, the Bible, and the hymnal was also a condition for being confirmed. All instruction beforehand, and all examinations on the confirmation day, were conducted in Norwegian. Even if children were born in Bosque County, they had to learn to read and write in Norwegian. Such learning took place in the "Norwegian summer school."[93]

In the last two decades of the nineteenth century it became harder for Norwegian immigrants to realize the dream of owning their own farm. Many families had numerous children; it was obvious that not all of them could become farmers or find other well-paying work in Bosque County. More and more parents began to send their children to schools of higher education outside the county.

In 1893 the German Lutherans proposed to establish a college in Clifton, and the Norwegian Lutherans were invited to join them. The proposal was discussed at a special congregation meeting at Our Savior's Lutheran Church on October 9, 1893. Attendees agreed that a school for secondary education was needed, and that the Norwegians could not finance such a project on their own.[94] The German Synod was going to discuss the question of establishing a college at a meeting of the synod in New Orleans in February 1894. Would the Norwegian Lutherans be willing to commit themselves financially to help realize such a project? At a congregation meeting on December 28, 1893, it was decided that the Norwegian congregation would be willing to cover half of the expenses when construction of a college building began,[95] with the condition that one of the teachers should be qualified to teach Norwegian.

At the meeting in New Orleans, however, the German Synod turned the proposal down. The Norwegians had suggested that the college be a co-educational institution and students did not need to have a calling for the ministry. The Germans were only interested in a school for young men preparing for the ministry.

Two years later, on November 19, 1895, the Norwegians decided to establish their own junior college. Pastor Rystad argued that the community needed its own Lutheran college so that the children of Norwegian immigrants and their children would be able to uphold their Lutheran faith. The Norwegian language would have a central place in the curriculum. The congregation agreed that it was time to take up the challenge but to begin on a small scale. A committee was appointed consisting of Rystad, Dr. Fronshang, Nils Jacob Nelson, and Gulbrand Bronstad.[96] The plan was well received in the Norwegian community. A trial subscription list was circulated and between 1,200 and 1,500 dollars was subscribed.

On May 6, 1896, about thirty men, members of the Norwegian Lutheran Church, met at the home of T. T. Hogevold. "They wished the boys and girls of their congregation to receive instruction upon a Christian basis of a more advanced nature than their Parochial schools gave their children during the summer."⁹⁷ A Board of Trustees was elected and a charter adopted. The name of the institution was to be the Lutheran College of Clifton, Texas. Pastor J. K. Rystad was elected president, O. T. Rikansrud vice-president, Nils Jacob Nelson secretary, and Mikkel Hoff treasurer. Y. Grimland, Dr. S. J. Fronshang, and J. J. Ringness were elected as trustees.

Nils Jacob Nelson donated 5 acres of land and T. T. Hogevold donated 3 acres for the college on Clifton Hill. The land gift was valued at $800 and it cost $6,000 to build the first college building. "This money was donated by the Norwegian people, principally the second generation."⁹⁸ The dedication of the building took place on October 14, 1896. Professor P. H. Bothne from Luther College, Iowa, was the speaker of the day. Pastor Rystad, the driving force behind the college, delivered the dedication address. For him and those present it was self-evident that he would speak in Norwegian on this occasion. The singing was also in Norwegian, commented Elif Albertson Moore. Most of those present "appreciated the Norwegian more than they did the English language."⁹⁹

The first class of the college had only eleven students, but the number increased in the spring term to thirty-two. "Most of the students were very well acquainted before they came together, as they had met each other at various church gatherings and at the picnics which were so common at the time."¹⁰⁰ English reading and English grammar as well as Norwegian and religion were taught during every one of the four years the students stayed at the college. The college had no boarding facilities until such were built in 1904. All teachers held at least a B.A. degree.

After an enthusiastic beginning, the level of operation of the college had to be scaled down because of lack of students. To survive, the college needed to be flexible. Parents belonging to other denominations had begun to express a strong interest in enrolling their children in the college. Within a few years it had become the best school in the county. The stumbling block for those parents was that their children would have to take a compulsory Bible course if they enrolled in the Norwegian college. The Board of Trustees solved this problem in 1902 by deciding that parents could write a letter requesting that their son or daughter not attend the Lutheran Bible course and it would be accepted, as would requests from non-Norwegian parents that their children be excused from attending classes in Norwegian.¹⁰¹

Initially the German Lutherans had withheld their support of the college, but in 1906 an agreement of cooperation was signed with the German Lutherans. This agreement alone did not solve the problem of sufficient enrollment to pay running expenses. Of greater importance was the decision to cooperate with the city of Clifton, and in 1907 a record number of seventy-three students were registered. The Methodists of Clifton began supporting the college financially as well. On October 14, 1907, the college passed a resolution stating that English would be the official language of Clifton College Corporation.[102]

The opening of the Lutheran College of Clifton, Texas, in 1897 can be looked upon as a watershed in the history of Norwegians in Bosque County. In one sense it marked the peak of Norwegianness in the county. In another sense it marked the transition to a new level of assimilation. The transplanted community was strongly intertwined with a preference for the agrarian community and the dominance of the Lutheran Church. It was the Lutheran Church that more than any other institution insisted on using the Norwegian language to keep Norwegian traditions intact, not least those connected to church life.

Until the turn of the century the most significant measure of success among Norwegian immigrants was to own so much land that all boys could get their own farm when they came of age and married. After 1900, having enough money to be able to pay for higher education of all children, male or female, became at least as important. In this context the practice of Evan and Marianne Rohne regarding the future of their children can be said to represent the beginning of the transition: Their children could choose either getting their own farm or getting a higher education.

Both Martin Sorenson and Christian O. Bronstad were on the the list of men who signed the Articles of Incorporation for the Lutheran College of Clifton, Texas, on May 20, 1896. They still had to find the money to pay for school every year. Martin Sorenson, born in Denmark, and his wife, Astrid, born in Norway, had come to Bosque County in 1880 and married in 1881. All of their children were students at the college: "the boys stayed in the Dormitory and the girls living with families in exchange for light housekeeping. In 1899 tuition for a year was thirty dollars a year, a handsome sum for a Norwegian farmer."[103]

All of Christian O. Bronstad's children also got a higher education. Cora Josephine, born on January 17, 1878, studied at Waco Female College; Millie Olivia, born on October 5, 1879, studied at the Lutheran

College of Clifton for two years and attended teacher Summer Normal School at Iredell and Hico; Oscar Alfred, born on March 13, 1881, enrolled at Toby's Practical Business College in Waco; Ovee Kleng, born on June 11, 1883, also enrolled at Toby's Practical Business College; while Morris Theodore, born on July 31, 1885, studied a year at the Lutheran College of Clifton before attending Toby's Practical Business College. His twin brother, Conrad Gulbrand, studied at Lutheran College of Clifton and Toby's Practical Business College. Jesse Marion, born on December 24, 1887, studied at Lutheran College of Clifton and attended teacher's training courses at Summer Normal School in Meridian and at Baylor University, Waco. May Pernella, born on February 17, 1890, studied at Lutheran College of Clifton and later took a business course at Hill Business College; Clarence Syverin, born on October 19, 1891, graduated from Lutheran College of Clifton in 1912; Alwyn Lawrence, born on March 13, 1894, graduated from Lutheran College of Clifton in 1914, and got a B.A. degree at Luther College, Iowa; Olga Belle, born on September 6, 1896, graduated from Lutheran College of Clifton; and Viola Cecelia, born on October 7, 1898, graduated from Lutheran College of Clifton. She taught at the Meridian Creek School for a while, but in 1920 together with her sister May and Anna Grimland she enrolled as a nurse student at Baptist Hospital in Waco, graduated, and became a registered nurse in 1923.[104]

Petter and Helen (Fjæstad) Olsen had a strong drive to give their eleven children higher education, which they themselves could not have dreamed of when they grew up in Norway. Petter Olsen, born on December 27, 1860, in Bjørkholen, Romedal, came to Clifton in 1881 and was first employed at Norway Mills. For many years he was an employee in the furniture department of P. E. Schouw and Brothers in Clifton. On September 14, 1887, Petter Olsen married Helen Fjæstad in Fort Worth. She was born on September 18, 1866, in Gaustadeie, Romedal, and died on November 13, 1955. She had arrived in Clifton on August 10, 1881, as an indentured servant and worked as a housekeeper in the Otto Johnsen home near Norse. Later she worked as a servant in Cleburne and Fort Worth.

The Olsen family lived on 2 acres of land bought from Jacob Nelson at 315 S. Avenue S., Clifton. Helene worked "all day and far into the night to carry out her desire that her children should receive college education. She also instilled that desire in her children, who went on to work their way through college, and to become successful in their chosen fields of endeavor."[105] Helene held a number of jobs to make ends meet during the year, including working for the Turkey Creek Thresher Crew for many years. Three Olsen sons studied at Texas A&M and became engineers. "Back home" in Norway, sons and daughters of parents with their social background did not typically get a college education until after the Second World War.

NORWEGIAN BUT ALSO AMERICAN

The Norwegian settlement in Bosque County was strongly influenced by migration networks with a common ancestry in Norway and by the Norwegian state religion. They transplanted the traditions they still considered important to their new environment, with large distances between neighbors, retaining old beliefs in a new context. On the other hand, they had gradually assimilated and became more American.

The Norwegians in Bosque County were proud of what they had achieved in Texas, but their connection to family and friends in the communities they had left in Norway was still vital to how they defined their identity. Every year new Norwegians arrived who could tell stories about family and social life in Norway.

The longer they lived in Texas, the more interwoven their lives became with the broader community and American society. Although they held on to their ethnic practices and beliefs, new ideas and beliefs coursed through the region. The Norwegians eagerly embraced new agricultural technologies and devices that lightened the workload. New steam threshers created a stir in the Norse community in the 1880s. The fruits of their hard work were converted into new and modern houses, and the houses were filled with modern household furnishings as far as the household economy allowed. At the end of October 1884, T. Theo Colwick informed the community that Ole Holen was up from Waco, selling sewing machines for the Singer Sewing Machine Co.[106] Owning a sewing machine was a goal for many a housewife; those who owned one were visited by neighbors asking to use the machine.

American traditions and the English language became increasingly important in the immigrants' lives; the Norse celebration of the 4th of July, for example, was a fusion of Norwegian and American traditions.[107] What could be more American than playing baseball? A. C.

Grimland, the son of Yern Grimland, introduced baseball to the Norse community and organized the first team in the county in 1886. He had learned to play as a child in Prairieville, Kaufman County. Twins Morris Theodore and Conrad Gulbrand Bronstad, born on July 31, 1885, played baseball on the Harmony School district team—there were not many Americans there. "They played in homemade uniforms. The baseball pants were made of white ducking, with colored streamers sewed along the side seams."[108] Teenagers who immigrated from Norway to Texas around 1900 had never seen such uniforms, and the sport was totally unknown to them.

CHAPTER 14

Conclusion

◊ ◊ ◊

ONE OF THE MOST SURPRISING observations of this study has been how long the Norwegians in Bosque County held on to or were permitted by the larger society to hold on to their Norwegian ethnicity. Did Norwegian immigrants in Texas preserve the Norwegian language, culture, and ways of life longer or shorter than other European immigrant groups? To get a better understanding of these questions we can compare with the ethnicity and assimilation of German, Polish, and Czech immigrants in Texas during the same decades.

At the Texas census in 1887 there were 129,610 German immigrants and children of German immigrants in Texas. German-Texans constituted the largest group of European immigrants in Texas, and it is a reasonable conjecture that German-Texans were able to preserve their own culture and ethnic identity for a long time. But other ethnic groups also held on to their ethnicity for decades, especially the Polish-Texans and the Czech-Texans.

GERMAN-TEXANS

The Germans were by far the largest group of Europeans in Texas. They had immigrated from several German states. More than 190,000 persons of German heritage lived in Texas in 1900, about 6 percent of the total population of Texas. The percentage of Germans in the population leveled off in the first decade of the twentieth century, and people of German background were gradually more assimilated into Texan and American culture and way of life. English expressions became part of the German vocabulary, and English advertisements were printed on the pages of German newspapers in Texas.

Two types of Germans maintained the German language and culture in Texas: *Kirchendeutsche*, Church Germans, and *Vereinsdeutsche*, members of organizations and clubs; sometimes one person was a member of both. For the Church Germans, their religious beliefs were the most important in their lives. Around two-thirds of the German immigrants were Protestants, mainly Lutherans, and three out of ten were Catholics. German Protestants and Catholics lived in peaceful coexistence with each other in both towns and the countryside in Texas,[1] as they had done in mixed-religion villages in Germany since at least the seventeenth century. Most of the time Protestants married Protestants and Catholics married Catholics, but there were exceptions. If someone were to marry outside their own denomination, it was better that they married a German immigrant than an American. Even in the late 1950s seven out of ten Germans in areas with a high density of Germans married a person of German heritage.[2]

Both Protestant and Catholic congregations had their own religious parochial schools preparing young people for confirmation. The instruction was conducted in German. Far more important for maintaining the German language were Texan laws that made it possible to use Spanish, German, or other languages as the teaching language in private schools through large parts of the nineteenth century. In certain periods even public schools in German colonies used German as the main language of instruction.[3]

Lutheran and Catholic Germans were generally very conservative—they were against liberal radicalism and socialism. They seldom participated in politics on county and state levels, except when they found that political decisions could have adverse consequences for their congregation, especially legislation that could make it more difficult to maintain German language education.

Members of German organizations and clubs were more secular, and many of them were freethinkers and Freemasons. A number of early German immigrants in Texas had a university education, and many were liberals who had escaped from German states after having been on the losing side in the March 1848 revolution in Germany. Liberals were often elected as leaders of German organizations and clubs. Several were editors of German newspapers in Texas. It was well known in Texas that the German immigrants loved their *Gesangvereins*, their singing societies, and their beer.

The German influence in Texas was considerable until the First World War. Between 75,000 and 100,000 Texans had German as their main language. The service in German-Lutheran and German-Catholic churches was conducted in German. In areas with a large percentage of German-Texans the language spoken in the public schools was German, and it was certainly German in parochial schools, which were operated by German Protestants and Catholics. Despite increasing assimilation among second- and third-generation Germans, German culture and traditions remained a crucial part of their identity.

During the First World War, German immigrants in the United States experienced strong political pressure to finally leave behind their German-ness and become true Americans. In a speech in Philadelphia in May 1915, President Woodrow Wilson urged all immigrants from all over the world to leave their past behind them. "You cannot dedicate yourself to America unless you become in every respect and with every purpose of your will thorough Americans."[4]

After the United States entered the war on the side of the Allies, German-Texans were accused of supporting Germany in the war. The pressure on German immigrants increased considerably, sometimes ending in violence.[5] In counties where between 20 and 30 percent of the population was of German heritage, they were more or less left alone. But in counties where the pockets of Germans were smaller, the majority demanded that it was high time for the Germans to assimilate into the larger society; if they were unwilling to do that, they could return to Germany. Although demand for stronger assimilation of Germans in Texas was considerable, German groups in states in the Midwest experienced much greater pressure. They also experienced more violence.

After the war ended, the political call for Americanization of German ethnic groups weakened. Patterns of behavior well known before the war were continued, and German language and culture in Texas resurged somewhat, especially in counties where German-Texans still constituted a large part of the total population. At the time of the 1910 US Census, 26.7 percent of the population in Washington County had a German background, 25.8 percent in Comal County, 24.8 percent in Gillespie County, 23.3 percent in Kendall County, and 22.4 percent in Austin County.[6] In these counties and in counties with smaller communities of German-Texans, German culture and language were preserved for a surprisingly long time, into the Second World War and at least until 1960.

THE POLISH-TEXANS

At the time of the 1887 census, 4,987 Polish immigrants lived in Texas. The first immigrants arrived from Silesia in fall 1854, and they continued to come for some decades. Before the Civil War there were three Polish colonies on the east and west side of San Antonio. The largest colony was Panna Maria in Karnes County, some 60 miles southeast of San Antonio. *Panna Maria* in Polish meant Virgin Mary, and all the Polish immigrants were Catholic.[7] In the decades following the Civil War, the major cause for population growth in the Polish colonies was that the immigrant families had many children, on average nine per family.

The Polish Catholic Church in Texas constituted the backbone for maintaining the traditional Polish agricultural society in Texas. When a new Polish colony was established, it did not take long before a new Catholic congregation was established. The second most important institution was the Polish private school, financed by parents, where instruction was given in the Polish language. The Polish schools were attended by "virtually all Silesian children in the years before World War I," Lindsay T. Baker wrote. The Polish were insulated from the influences they would have been exposed to in non-Polish schools. The schools "reinforced the Silesian culture they brought with them from home."[8] The authorities in counties with strong Polish colonies tried to get Polish children to attend the public schools with instruction in English, but without any large success.

Daily life among the Polish in Texas differed from life among their American neighbors. They spoke Polish among themselves, which was unintelligible to their American neighbors. Until the First World War the majority of Polish-Silesians were born, grew up, married, had children, and died within their own colonies. They were little known in larger Texas society, and even less understood by their closest neighbors.[9] Their churches and schools as well as their geographical concentration were of crucial importance for maintaining their distinct Silesian culture in Texas, at least until after the Second World War.[10]

According to classic assimilation theory, the deciding marker of full assimilation is the time span before children of immigrants marry outside their own ethnic group. It is common that first-generation immigrants stand out as distinct from the larger society where they live. Second- and third-generation immigrants are more willing to participate in the political, economic, and cultural processes of the larger society. A high degree of marriage across ethnic lines can be interpreted as a sign that social and ethnic separateness is becoming blurred or is on the verge of disappearing.[11] According to this definition of assimilation, there is no doubt that the assimilation of the Polish-Texans took longer than that of the German-Texans, who outnumbered them by tens and tens of thousands. The Silesians upheld the tradition of marrying within their own group for a long time. Fewer than one in ten Silesians married outside their ethnic group before the Second World War. Even in the first decade after the war, 80 percent of Polish immigrants kept up the tradition of marrying each other, especially in remote agricultural settlements.[12]

THE CZECH-TEXANS

The first group of poor Slavic agricultural workers from Bohemia and Moravia arrived in Texas in 1851 and settled in Cat Spring, Austin County. A new group of Czech immigrants arrived two years later. The Czechs in Texas tended to settle near German-Texans in Southeast Texas. When the Civil War broke out, around 700 Czechs lived in Texas. Thousands of Czechs immigrated to Texas in the decades after the Civil War. At the time of the Texas census in 1887, 2,483 Czechs were living in Austin County, 6,084 in Fayette County, 2,758 in Lavaca County, 645 in Washington County, and 975 in Burleson County.

In 1910 15,074 Czechs were enumerated in Texas, and approximately 250 Czech colonies still existed in Texas. Before the First World War relatively cheap land could still be bought. Farming was a distinct way of life among the Czechs as late as the 1920s and 1930s, wrote Robert L. Skrabanek. "We ate, drank, and slept farming."[13] The families lived on relatively small farms where everyone in the family, from toddlers to old people, participated in the different phases of the yearly farming circle. Most of the Czechs who immigrated to Texas had grown up among the land-hungry peasantry in a country where land was scarce. Success in Texas meant ownership of a farm with the necessary buildings and equipment. "Not only did the ownership of land have a high place in our system of values, but it also was to be treated with the greatest respect."[14] If hard work was good for the parents, it had to be good for the children. The Czech-Texans produced almost everything they needed for their sustenance, and even as late as the 1930s spent very little money in stores in towns or villages.

Almost 90 percent of the Czechs were Catholics. A significant minority were Protestants and belonged to the Moravian Brethren.[15] Their form of pietism had its roots with Jan Hus and the Hussitian Church in Bohemia in the fifteenth and sixteenth centuries. Organized teaching of the Czech language in Texas began before the Civil War; it was taught in both public and parochial schools. The maintenance of the Czech language was very important to Czech-Texans, as it was an important part of their identity, in relation to both the many German immigrants living in the neighborhood and the Americans.[16] The number of public schools that offered instruction in Czech declined during the last decades of the nineteenth century, but Czech Catholic congregations and Protestant Brethren congregations taught Czech in their parochial schools at least until the Second World War.

The heyday of Czech ethnicity in Texas was during the First World War and the first years after, when the state of Czechoslovakia was established. As in the case of the German-Texans and the Polish-Texans, the Czech immigrants had a strong tradition of marrying other Czechs. The Czech ethnicity, as measured against the rate of intermarriage, was still very strong in Texas before the Second World War.[17]

ETHNICITY AMONG NORWEGIAN-TEXANS

In 1887, 1,958 Norwegian immigrants or children of immigrants were registered as living in Texas, and of those 1,247, or two-thirds, lived in Bosque County. Many of them lived in neighboring counties. Compared with the European immigrant groups already discussed there were few Norwegians in Texas. From 1854 until 1880, the Norwegians were much more open to assimilation than were the German, Polish, and Czech immigrants, for several reasons. The school system in Bosque County was undeveloped in the years before and after the Civil War, and there were few school-age Norwegian children. The first Norwegian immigrants in Bosque chose to send their children to the public school in the district, which taught in English. Later they participated actively in the establishment of new schools, but always as part of a community endeavor in the school districts where they lived. No serious discussion about using Norwegian as the teaching language arose.[18] Neither did the Norwegians in Bosque demonstrate an urgent drive to establish their own Norwegian Lutheran congregation in the two first decades.

The use of the Norwegian language and culture was strengthened between 1880 and the turn of the century. Between 1870 and 1900 there was a steady increase in the number of Norwegians settling in Bosque County. The Norse colony got its own Lutheran pastor, a church was built, and the service was conducted in the Norwegian language. But not even then, in the two last decades of the nineteenth century, did the Norwegian-Texans try to establish their own school where teaching was conducted in Norwegian. If the Norwegian immigrants had been convinced of its importance, they could have followed in the footsteps of the German, Polish, and Czech immigrants and organized private schools where the mother tongue was the language of instruction.

When the Lutheran congregation got its own minister and a formal organization, a Norwegian parochial school was immediately established. The goal of this school was to instruct the students in Norwegian and prepare them for confirmation. Many children who were born in Texas and who had never visited Norway learned to read and write in Norwegian at this school. The Norwegian-Lutheran congregations in Norse, the Upper Settlement in the Meridian Creek Valley, Clifton, and Cranfills Gap became the hub for the maintenance of the Norwegian language in Texas.

Newcomers, if they had any choice, settled close to other Norwegians. The Norwegianness of the Norse colony and the Upper Settlement in Meridian Creek Valley became increasingly more distinct. To illustrate this development, let us take a closer look at the neighbors of Hendrick and Christine Dahl around 1860 and compare them with Christine's neighbors in 1900. The Dahl farm is located a few miles uphill from Clifton on the left side of the Cleng Peerson Highway, Highway 219. In 1860 their neighbors were almost exclusively of American heritage, but in 1900 almost all of Christine Dahl's neighbors were of Norwegian heritage. Down the hills toward Clifton were the farms of E. M. Colwick, her son Peder Dahl, and Ole Sinderud. Close family lived on three farms east of the Dahls—the families of her sons Andrew Dahl and Chris Dahl, and her daughter Martha, who was married to Mikkel Hoff. Further uphill was the farm of Christopher Pederson and Christine's sister Syverine Dillerud, as well as the farm of her daughter

Helene, who was married to Emil Bekkelund. The Bergman schoolhouse was located on the farm of Swedish-born John Bergman.

Turkey Creek Valley was full of Norwegian immigrants and their children, and at the end of the 1880s the new school district Harmony was established to serve the settlers toward the west in Meridian Creek and Spring Creek Valley. The majority of families with children of school age were Norwegian. Christian O. Bronstad was the driving force behind the establishment of the school. The school committee bought 2 acres of land from the Norwegian farmer O. P. Carlson. A good source of potable drinking water was found on the land, and it was close to a public road. A new schoolhouse was ready in fall 1890. The teacher, McGinn, was American, but the majority of schoolchildren had Norwegian parents; among forty-two pupils in spring 1891, only eight had American surnames. The rest of them, thirty-four children, had Norwegian surnames.[19] The language of instruction was English. Many of the pupils had been taught by their parents that it was very important to master the English language orally as well as in writing if they wanted to succeed in Texas society. Still, at recess, even the American children spoke Norwegian.

At the beginning of the twentieth century many first- and second-generation Norwegians in Bosque County might have been even less assimilated than the first wave of immigrants, due to the yearly influx of indentured servants who arrived from specific communities in municipalities in Hedmark County between 1870 and the First World War. When a new young man or woman arrived in Bosque County from Hedmark County, the Norwegians in Bosque heard news about relatives and neighbors in the communities they had left behind. Who had married and had children, who had died, who was doing well in life, and who had built new houses or moved to Oslo? The immigrants in Bosque County viewed themselves as Texans, but they still had a strong Norwegian identity. Not only was Norwegian spoken in the settlements and on the streets of Clifton and Cranfills Gap, but many spoke it with the distinct Hedmark dialect. The Norwegians in Bosque County were aware of their Norwegianness, and they were proud to live in the county in Texas with the highest number of Norwegians.

The Norwegian influence in Bosque County reached its zenith at the turn of the twentieth century. The county had a total population of 17,390 in 1900 and 19,013 in 1910, a level never again surpassed. The US Census of 1910 enumerated the "white population of foreign birth or foreign parentage": 4,445 persons were of Norwegian parentage in Texas, 0.7 percent of the foreign born in the state. Bosque County was still dominant as the Texas county with the highest number of Norwegian immigrants and their children; 407 persons were born in Norway and 568 were children of parents who both had been born in Norway, a total of 975.[20] The number of persons in the county who had one parent born in Norway was not included. But if we presume that the number with one parent born in Norway was the same as for all of Texas, the number of Norwegian immigrants in Bosque County was still around 1,200 persons, at the same level as in 1887.

Waco was the city closest to Bosque County. Many immigrants traveled through Waco on their way to Bosque, and many found work there, especially Norwegian girls working as servants. In 1910, 168 Norwegians lived in McLennan County, and 148 of them lived in Waco. Between 1887 and 1910, Hamilton County became the most important county outside Bosque for immigrants who still dreamed of their own farm. The numbers increased from seventeen persons in 1887 to almost 200 persons in 1910, while Limestone County was almost as important for Norwegian immigrants in 1910 (ninety-nine Norwegians) as it had been in 1887 (sixty-eight).

The Gulf, Colorado, and Santa Fe Railroad was built through the Bosque Valley in 1881. Its starting point was Galveston on the Gulf of Mexico, and it reached Fort Worth and Dallas in North Texas, where more and more Norwegians were seeking work, in 1881 and 1882. If young people were willing to travel some distance from Bosque County to find work in Dallas and Fort Worth or Houston and Galveston, the railway connections were excellent. But the most surprising insight concerning Norwegians in Texas in 1910 is how many had chosen to settle in counties bordering the Gulf of Mexico or close to the Gulf. In 1910, 394 Norwegians were enumerated in Galveston County, and 332 of them lived in Galveston City.[21] Another 148 Norwegians lived in Jefferson County, northwest of Galveston. Norwegian immigrants were attracted by the "new and industrial Texas" emerging after the giant oil discovery at Spindletop in 1901 and the construction connected with the Houston Ship Channel. Many oil refineries and chemical factories began operating in this area during the first decades of the twentieth century, and a number of Norwegians from Bosque County moved to the Houston area to find work. The change in settlement patterns among Norwegian immigrants in the first half of the twentieth century is certainly worthy of further research but is beyond the confines of this study.

Weddings among the Norwegians in Bosque County were discussed at some length in an earlier chapter. The main practice continuing into the twentieth century was that of Norwegian men marrying Norwegian women. On some occasions both husband and wife immigrated from Norway, or second- or third-generation immigrants married each other or a first-generation immigrant.

Where they grew up in Norway, differences in social class would have made many of these marriages difficult. In some instances, Norwegian farm owners in Bosque County found it more acceptable that a son or daughter marry someone who had grown up on a cotter's place in Norway than marry an American. Norwegian teenagers in Texas might have had a fair amount of social interaction with Americans of their own age during primary school and high school, and perhaps even dated each other. But when the question of whom they would marry, or whom the parents would prefer they marry, surfaced in earnest, the ideals of shared ethnicity and a common cultural and religious background prevailed. If one family came from Aust-Agder County or Rogaland County and the other from Romedal in Hedmark County, at least they shared the same language, culture, and religion, even though their dialect might be different.

In the late 1880s and early 1890s many first-generation, but mostly second-generation, German immigrants moved to Bosque County. They settled on the prairie on the north side of the Bosque River. The distance from Norse to Womack along Highway 219 through Clifton was around 13 miles. The Germans were not close neighbors with the Norwegians, but both groups met on Saturdays when they went to Clifton to shop. The majority of the Germans belonged to the German Orthodox Lutheran Church, and the Norwegians belonged to the Norwegian Lutheran Church, which was also orthodox in its teachings. After 1900, German parents began sending their children to the Norwegian college in Clifton. As young people attending the same school, the Norwegian and German students got to know each other, and some fell in love and even married. Neither the Norwegian nor the German mothers were happy when their children began dating. But the Norwegians knew the German farmers were very hardworking, and the Germans admitted that even though the Norwegian farmers were not quite as industrious, they were on the same level. They were certainly much more hardworking than their American neighbors, and Germans and Norwegians shared the same religion. After 1900, when the Norwegians began to marry outside their own ethnic group, marriages between Norwegians and Germans dominated.

Immigrant parents in Bosque County in the first decade of the twentieth century were aware that to find good jobs, their children would have to move out of the county and work in fields other than agriculture. The best investment they could make was to secure a higher education for their children. The American dream among Norwegians changed from owning your own farm to being able to pay for the higher education of your children. Second-generation sons and daughters became more urbanized, certainly compared with their parents. At the time of the Second World War they were more assimilated as Texans than previous generations. Socially, they played down their Norwegianness and worked hard to erase the Norwegian accent from their spoken English.

APPENDIX I

Norwegians in Texas in 1860

County	Total
Henderson	157
Bosque	105
Van Zandt	86
Kaufman	23
Travis	20
Anderson	13
Galveston	13
Rusk	11
Smith	11
Fannin	10
Cherokee	9
Matagorda	7
Caldwell	6
Collin	6
Nacogdoches	6
Titus	5
Sabine	4
Colorado	3
Cooke	3
Hays	3
Milam	3
Calhoun	1
Dallas	1
Ellis	1
Freestone	1

County	Total
Jackson	1
Karnes	1
McLennan	1
Nueces	1
Orange	1
Presidio	1
Robertson	1
Shelby	1
Total	516
Norwegian-born	306

APPENDIX 2

The Wealthiest Men in Bosque County, Texas, 1860 (in US dollars)

Name	Business	Born	Real estate	Personal estate	Total
J. J. Smith	Farmer	South Carolina		54,091	54,091
James Bentley	Farmer	Kentucky	3,600	42,713	46,313
J. W. Smith	Merchant	Kentucky	7,500	35,000	42,500
Green Powell	Farmer	South Carolina	4,912	27,460	33,372
B. Willingham	Merchant	Georgia	18,000	5,000	23,000
Temple Spivey	Farmer	North Carolina	9,070	13,270	22,340
Jackson Randall	Farmer	North Carolina	9,300	12,210	21,150
William Sedberry	Farmer	North Carolina	5,500	13,000	18,500
B. M. Willingham	Merchant	Georgia	18,000	500	18,500
H. L. White	Merchant	Vermont	4,300	14,050	18,350
J. M. Stinnett	Merchant	Virginia	3,000	15,000	18,000
R. J. Hart	Farmer	Florida	500	15,830	16,330
Allen S. Anderson	Stock Raiser	Missouri	2,250	13,000	15,250
Ephraim Snell	Farmer	Ohio	1,600	12,780	14,380
Ezra Wibber	Stock Raiser	New York	11,105	3,000	14,105
Henry Fossett	Lawyer	Maine	13,000	1,000	14,000
Silas McCabe	Farmer	Tennessee	2,100	11,825	13,925
John Abney	Preacher	South Carolina	3,000	8,555	11,555
I. B. Standifer	Farmer	Tennessee	8,000	3,155	11,155
R. L. Barnes	Farmer	Tennessee	3,690	7,225	10,915
Robert J. Singleton	Stock Raiser	South Carolina	400	10,325	10,725
Stinson Holloway	Farmer	Louisiana	7,535	3,000	10,535
R. or B. Scott	Farmer	Mississippi	4,000	6,300	10,300
F. M. Kell	Farmer	Indiana	3,768	6,485	10,253
Benjamin Weeks	Farmer	South Carolina	4,000	6,200	10,200
William B. Reed	Farmer	South Carolina	6,612	3,533	10,145

SOURCE: "Heart of Texas Records," Central Texas Genealogical Society, Inc., Waco, vol. XIX, no. 1, 2, 3, and 4.

APPENDIX 3

Slave Owners in Bosque County, Texas, 1860

Slave Owner's Name	Number of Slaves	From 1860 Bosque County Census					
		Family #	Post Office	Profession	Age	Gender	Place of Birth
ABNEY, John	9	299	Meridian	Methodist Preacher	39	M	South Carolina
ADAMS, Sylvester	1	225	Cyrus	Stock Raising	26	M	Tennessee
ALEXANDER, C. N.	1	171	Clifton	Farmer	26	M	Georgia
ALEXANDER, T. C.	3	285	Meridian	Lawyer	35	M	Tennessee
BARNES, R. S. (Robert Samuel)	1	63	Meridian	Farmer	44	M	Tennessee
BARRY, J. B. (John Buckner "Buck")	15	99	Meridian	Farmer	37	M	Arkansas
BENTLEY, James	22	241	Clifton	Farmer	43	M	Kentucky
BRIDGES, W. H. (William H.)	5	300	Meridian	Physician	35	M	Georgia
DEWEY, C. C.	2	4	Clifton	Millwright	37	M	New York
EVERETT, Thos. E. (Thomas Ewell)	3	168	Clifton	Farmer	67	M	Kentucky (?)

From 1860 Bosque County Census

Slave Owner's Name	Number of Slaves	Family #	Post Office	Profession	Age	Gender	Place of Birth
FRAZIER, Jas. C. as admin. of Ridley Robinson	5	83	Meridian	Farmer	29	M	Tennessee
GREER, T. L. (Thomas Lacy)	6	88	Meridian	Farmer	33	M	Georgia
HAMPTON, B. F.	3	121	Flag Pond	Farmer	29	M	Georgia
HAMPTON, Roenna	1	122	Flag Pond		49	F	Georgia
HART, R. J. (Richard Jefferson)	6	217	Cyrus	Farmer	34	M	Florida (?)
HILL, H. D.	2	152	Clifton	Farmer	32	M	South Carolina
HOWELL, J. C.	3	70	Meridian	Farmer	30	M	Missouri
IRVIN, A. R.	1	87	Meridian	Stock Raising	33	M	Georgia
JAMES, Benj. (Benjamin)	1	8	Clifton	Farmer	52	M	Tennessee
JAMES, Wm. (William)	2	103	Clifton	Farmer	39	M	Tennessee
LOCKER, S. S. (Samuel S.)	1	2	Clifton	Farmer	72	M	North Carolina

APPENDIX 3 ◊ 346

Slave Owner's Name	Number of Slaves	From 1860 Bosque County Census					
		Family #	Post Office	Profession	Age	Gender	Place of Birth
MABRY, J. H. (James H.)	1	166	Clifton	Farmer	41	M	Alabama
MAURY, M.	3	189	Clifton	Stock Raising	29	M	Alabama
McADOO, J. E. (James E.)	3	76	Clifton	Farmer	43	M	Tennessee
McCABE, Silas	11	220	Cyrus	Farmer	43	M	Tennessee
McKISICK, J. W.	2	235	Cyrus	Farmer	52	M	North Carolina
MOORE, Wm. B.	6	269	Meridian	Farmer	39	M	Tennessee
NELSON, Mary S.	5	117	Flag Pond		33	F	Georgia
ODLE, John	2	164	Clifton	Stock Raising	36	M	Tennessee
PARKS, Felix	1	15	Clifton	Stock Raising	35	M	Tennessee
PARKS, Leroy	1	16	Clifton	Stock Raising	33	M	Tennessee
RANDALL, Jackson	9	64	Meridian	Farmer	54	M	North Carolina

Slave Owners in Bosque County, Texas, 1860   ◊   347

Slave Owner's Name	Number of Slaves	From 1860 Bosque County Census					
		Family #	Post Office	Profession	Age	Gender	Place of Birth
REED, Pleasant	1	265	Meridian	Stock Raising	42	M	Tennessee
REED, Wm. (William B.)	2	264	Meridian	Farmer	62	M	South Carolina
REEDER, Dorcas	1	161	Norman Hill	Farmer	55	F	North Carolina
SEDBERRY, Wm. R. (William Rush)	7	238	Cyrus	Chief Justice & Farmer	37	M	North Carolina
SINGLETON, Robt. J. (Robert)	7	298	Meridian	Stock Raising	43	M	South Carolina
SMITH, J. J. (John J.)	63	210	Cyrus	Farmer	61	M	South Carolina
SMITH, Joseph	3	281	Meridian	Merchant	45	M	Kentucky
SNELL, Eph. (Ephraim)	10	40	Norman Hill	Farmer	56	M	Ohio
SPIVEY, Temple	13	206	Cyrus	Farmer	65	M	North Carolina

APPENDIX 3 ◊ 348

Slave Owner's Name	Number of Slaves	From 1860 Bosque County Census					
		Family #	Post Office	Profession	Age	Gender	Place of Birth
STEADHAM, Georgia V.	2	298	Meridian		19	F	Alabama
STINNETT, J. M.	15	1	Clifton	Merchant	55	M	Virginia
TALBOT, Wm. (William)	5	271	Meridian	Stock Raising	56	M	Georgia
WALKER, Dixon	4	111	Clifton	Farmer	56	M	Tennessee
WALTERS, Nancy	5	156	Norman Hill	Farmer	52	F	Tennessee
WEEKS, B. F. (Benjamin Franklin)	4	173	Clifton	Farmer	29	M	Mississippi
WHETSTONE, Warrick	2	204	Cyrus	Farmer	34	M	Arkansas
WHITE, J. B.	1	212	Cyrus	Overseer	60	M	Kentucky
WILLINGHAM, A. (Cashwell Augus)	7	273	Clifton	Physician	25	M	Georgia
WILLINGHAM, R. M.	4	274	Clifton	Merchant	46	M	Georgia

Total Slaves Registered in Bosque County = 293.

APPENDIX 4

Number of Sheep in Some Texas Counties, 1880–1900

County	1880	1887	1890	1900
Bandera	32,974	72,911	60,293	10,763
Bell	21,224	53,098	42,003	5,447
Bexar	24,200	21,789	35,510	7,046
Blanco	19,004	39,066	20,892	12,900
Bosque	22,456	100,213	109,801	27,923
Brown	9,976	25,591	10,566	1,741
Burnet	24,915	49,831	58,850	45,780
Coleman	28,419	65,400	48,700	44,028
Comal	4,847	19,474	18,540	4,242
Comanche	2,925	11,480	15,003	3,799
Concho	-	62,931	41,774	15,332
Coryell	4,280	52,683	74,582	24,920
De Witt	70,524	31,644	20,348	3,291
Duval	196,684	160,793	60,160	2,901
Edwards	610	30,549	114,834	111,585
Erath	9,807	7,893	13,628	6,618
Gillespie	27,158	44,407	40,007	18,173
Hamilton	11,004	66,152	81,529	55,075
Kendall	16,259	28,178	22,252	6,134
Kerr	15,504	51,168	23,301	29,314
Kimble	11,196	67,109	120,574	11,223
Kinney	55,597	113,227	97,044	17,967
Lampasas	8,814	47,294	44,569	41,503
Lasalle	69,309	70,996	50,560	8,040
Llano	13,841	28,523	24,329	3,553

County	1880	1887	1890	1900
Maverick	11,246	105,716	149,310	9,651
McCullloch	12,497	86,850	120,833	18,432
McLennan	25,424	48,563	25,908	6,896
Menard	27,586	72,000	90,363	15,164
Nolan	1,301	41,928	88,850	2,545
Nueces	44,400	94,540	-	3,548
Pecos	1,299	15,178	150	102,969
Runnels	-	42,312	28,601	4,486
San Saba	17,915	51,672	53,983	13,842
Schleicher	-	-	57,250	13,496
Starr	29,686	139,022	50,966	17,944
Sutton	-	-	136,372	69,804
Taylor	5,837	23,718	36,632	6,325
Throckmorton	7,420	41,662	59,415	23,703
Tom Green	-	158,407	10,800	43,707
Uvalde	75,068	98,457	77,309	7,328
Val Verde	-	102,220	812	101,009
Webb	181,616	198,995	184,590	114,233
Williamson	18,400	61,828	80,961	14,759

NOTES

◊ ◊ ◊

INTRODUCTION

1. Theodore C. Blegen, "Cleng Peerson and Norwegian Immigration," *Mississippi Valley Historical Review* 7, no. 4 (1921): 320.
2. Blegen, *Norwegian Migration to America*, 62.
3. *Utvandringsstatistikk, Norwegian Official Statistics.* NOS VII, 25; 5.
4. Henry J. Cadbury, "Four Immigrant Shiploads of 1836 and 1837," *Norwegian-American Studies* 2 (1927): 20–52.
5. *Utvandringsstatistikk*, 5–6.
6. Østrem, *Den store utferda*, 78.
7. *Population of the United States in 1860 compiled from the original Returns of The Eighth Census* (Washington DC: Bureau of the Census Library, 1864), xxii.
8. ibid., xxviii.
9. Harper, *Emigration from Scotland between the war*; Harper, *Adventurers & Exiles*; Harper and Constantine, eds., *Migration and Empire*; Harper, *Scotland No More?*; Harper, "Rhetoric and Reality: British Migration to Canada, 1867–1967," 160–80.
10. Harper, *Adventurers & Exiles*, 106.
11. ibid., 107.
12. ibid., 107.
13. Devine, *Scotland's Empire*; Devine, *To the Ends of the Earth*; Calder, *Scots in Canada*; Calder, *Frontier Scots*; Hunter, *Scottish Exodus*; Ray, *Highland Heritage*; Gibson, *Highland Cowboys*; Fry, *How the Scots Made America*; Kay, *The Scottish World*; Bueltmann, Hinson, and Morton, eds., *The Scottish Diaspora*; Leith and Sim, eds., *The Modern Scottish Diaspora*; Calloway, *White People, Indians, and Highlanders*.
14. Harper, *Adventurers & Exiles*, 107.
15. E. G. Ravenstein, "The Laws of Migration," *Journal of the Royal Statistical Society* 48, no. 2 (June 1885): 167–235; E. G. Ravenstein "The Laws of Migration," *Journal of the Royal Statistical Society* 52, no. 2 (June 1889): 241–301. See also D. B. Grigg, "E. G. Ravenstein and the 'Laws of Migration.'" *Journal of Historical Geography* 3, no. 1 (1974): 41–54.

16. Ravenstein, "Laws of Migration," 281.
17. ibid., 284, 286.
18. See, e.g., the discussion in Syversen and Johnson, *Norge i Texas*, 90–96.
19. Baines, *Emigration from Europe 1815–1930*, 29.
20. Kamphoefner, "German Emigration Research, North, South, and East: Findings, Methods, and Open Questions."
21. Gjerde, *The Minds of the West*, 89–90.
22. Ostergren, *A Community Transplanted*, 8.
23. ibid., xiii.
24. ibid.
25. Syversen and Johnson, *Norge i Texas*.
26. Bosque County History Book Committee, ed., *Bosque County: Land and People*.
27. ibid.,105.
28. There is no lack of books on how to do family history in a professional way. See, Anthony Adolph, *Tracing Your Family History* (London: Colling, 2005, reprint 2007); Peter Christian, *The Genealogist's Internet* (London: Bloomsbury, 2012).
29. Jack Temple Kirby, "ANCESTRYdotBOMB: Genealogy, Genomics, Mischief, Mystery, and Southern Family Stories," *Journal of Southern History* 76, no. 1 (February 2010): 3–38.
30. Keats-Rohan, ed., *Prosopography Approaches and Applications*; Bonnie H. Erickson, "Social Networks and History: A Review Essay." *Historical Methods* 30, no. 3 (1997): 149–57; Düring and Stark, "Historical Network Analysis."
31. Fischer, *Albion's Seed*.
32. ibid., 5.
33. ibid., 7.
34. Geertz, "Thick Description: Toward an Interpretive Theory of Culture," 9.

CHAPTER I

1. Johan Reinert Reiersen, "Behind the Scenes of Emigration: A Series of Letters from the 1840's," *Norwegian-American Studies* 14 (1944): 78–117.
2. Reiersen, "Behind the Scenes," letter, January 30, 1843.
3. ibid., March 12, 1843.
4. Paul Thyness, *Jacob Aall* in *Norsk biografisk leksikon*, https://nbl.snl.no/Jacob_Aall; Stubhaug, *Jacob Aall i sin tid*.
5. Nelson, "Introduction," in Johan Reinert Reiersen, *Pathfinder for Norwegian Emigrants* (Northfield, Minnesota: Norwegian-American Historical Association, 1981); Syversen and Johnson, *Norge i Texas*, 23–56; Evensen, *From Canaan to the Promised Land*, 258–84.

6. Reiersen, "Behind the Scenes," 78–117.
7. Reiersen, "Behind the Scenes," letter, April 9, 1843.
8. ibid.
9. ibid.
10. ibid.
11. ibid., May 2, 1843.
12. Evensen, *From Canaan to the Promised Land*, 271–72.
13. Davis, *Frontier Illinois*, 174.
14. Reiersen, "Behind the Scenes," letter, Iowa City, January 24, 1844.
15. ibid.
16. ibid.
17. ibid.
18. Schwieder, *Iowa: The Middle Land*, 28–33.
19. The Indian Ho-Chunk Nation (Winnebago) were moved from their traditional homeland in Wisconsin to the "Neutral Ground" in northeastern Iowa, with the promise that the military would protect them from their enemies. Fort Atkinson was built in 1840 on a bluff overlooking Rogers Creek, a tributary to the Turkey River.
20. Reiersen, "Behind the Scenes," letter, Iowa City, January 24, 1844.
21. ibid.
22. ibid. Letter, Cincinnati, March 30, 1844.
23. ibid.
24. Calvert, De Leon, Cantrell, *The History of Texas*, third ed., 94.
25. Biesele, *The History of the German Settlements in Texas*; Jordan, *German Seed in Texas Soil*; Lich, *The German Texans*; Biggers, *German Pioneers in Texas*; King, *John O. Meusebach*.
26. Reiersen, "Behind the Scenes," letter, Cincinnati, March 30, 1844.
27. ibid.
28. Blegen, *Norwegian Migration to America*, 243.
29. Blegen, "Introduction," in Reiersen, "Behind the Scenes."
30. Nelson, "Introduction," in Reiersen, *Pathfinder*, 30.
31. Johan R. Reiersen, "Norwegians in the West in 1844: A Contemporary Account by Johan R. Reiersen and translated and edited by Theodore C. Blegen," *Norwegian-American Studies* 1 (1926): 110–25.
32. Evensen, *From Canaan to the Promised Land*, 274.
33. Nelson, "Introduction" in Reiersen, *Pathfinder*, 39.
34. Reiersen, "Behind the Scenes," letter, Holt, October 29, 1844.
35. Nelson, "Introduction," in Reiersen, *Pathfinder*, 40.
36. Reiersen, *Pathfinder*, 207.
37. ibid., 215.
38. Reiersen, "Behind the Scenes," letter to Grøgaard, n.d., spring 1845.

39. Nelson, "Introduction," in Reiersen, *Pathfinder*, 42.
40. *"Angaaende Udvandring til Fremmede Verdensdele,"* *Innstilling fra Finansdepartementet til Stortinget*, no. 6 (Christiania, November 8, 1843), 5.
41. Nelson, "Introduction," in Reiersen, *Pathfinder*, 43.
42. Evensen, *From Canaan to the Promised Land*, 284–5.
43. Syversen and Johnson, *Norge i Texas*, 203–6.
44. On June 12, 1845, Ole Reiersen wrote in a letter to his wife that land had been bought. Syversen and Johnson, *Norge i Texas*, 38–9.
45. Letter from Ole Reiersen in Norwegian, dated New Orleans, June 9 (12), 1845, cited in Syversen and Johnson, *Norge i Texas*, 39, 41.
46. Parsons, "A Texas Immigration Story: The Grøgaard Family from Norway"; Syversen and Johnson, *Norge i Texas*, 42.
47. Syversen and Johnson, *Norge i Texas*, 42.
48. *Handbook of Texas Online*, Linda Sybert Hudson, "Starr, James Harper," http://www.tshaonline.org/handbook/online/articles/fst22
49. Ericson, *Nacogdoches, Gateway to Texas*; *Handbook of Texas Online*, Linda Sybert Hudson, "Amory, Nathaniel C.," http://www.tshaonline.org/handbook/online/articles/fam04
50. Syversen and Johnson, *Norge i Texas*, 42.
51. Marilyn M. Sibley, "The Texas-Cherokee War of 1839," *East Texas Historical Journal* 3, no.1 (1965): 18–33.
52. *Handbook of Texas Online*, Linda Sybert Hudson, "Henderson County," http://www.tshaonline.org/handbook/online/articles/hch13
53. Nelson, "Introduction," in Reiersen, *Pathfinder*, 44.
54. Syversen and Johnson, *Norge i Texas*, 43.
55. ibid., 42–43.
56. Nelson, "Introduction," in Reiersen, *Pathfinder*, 45.
57. Syversen and Johnson, *Norge i Texas*, 35.
58. Nelson, "Introduction," in Reiersen, *Pathfinder*, 46.
59. Syversen and Johnson, *Norge i Texas*, 35.
60. ibid., 45.
61. Reiersen, *Pathfinder*, 217.
62. McDonald, ed., *Hurrah for Texas*, 174.
63. Reiersen, *Pathfinder*, 217.
64. ibid., 224.
65. Letter in Norwegian in Syversen and Johnson, *Norge i Texas*, 208–9.
66. Evensen, *From Canaan to the Promised Land*, 294–5.

CHAPTER 2

1. Edgar, *From Generation to Generation, Part I*, 7. Compared with most other writers of family histories of Norwegian immigrants to Texas in the 19th century, Edgar was very thorough in her historical research. Her sources are documented throughout the text with footnotes.
2. "Angaaende Udvandring til Fremmede Verdensdele," 24.
3. Anders A. Svalestuen, "Emigration from the community of Tinn, 1837–1907," *Norwegian-American Studies* 29 (1983), 43–89.
4. Edgar, *From Generation to Generation*, 7.
5. ibid.
6. Syversen and Johnson, *Norge i Texas*, 220.
7. Knud Jørgensen Hastvedt, "Recollections of a Norwegian Pioneer in Texas," translated and edited by C. A. Clausen, *Norwegian-American Studies* 12 (1941), 91.
8. The classic study on the impact of steamboats on the development and settling of the west is Hunter, *Steamboats on the Western Rivers*. Steamboat travel on the Red River is covered several places in Hunter's book. For a more recent discussion, see Gudmestad, *Steamboats and the Rise of the Cotton Kingdom*, 131–33. See also Grant Foreman, "River Navigation in the Early Southwest, *The Mississippi Valley Historical Review* 15, no. 1 (July 1928): 34–55.
9. Jørgen Olsen Hastvedt was not the only one who got seriously ill from drinking river water. Salve Knudson became so ill that he had to be carried in the two-wheeled wagon they had built to transport their belongings. See "The Man with America Fever. A History of Salve Knudson," Cleng Peerson Research Library, n.d., 14.
10. Hastvedt, "Recollections of a Norwegian Pioneer," 91.
11. Syversen and Johnson, *Norge i Texas*, 25–6.
12. Mildred O. Hogstel and Alice Marie Berg, "History and Family Tree of Aslak Nelson," manuscript, June 1973.
13. "Aslak Nelson," F858, *Bosque County: Land and People*, 535.
14. Edgar, *From Generation to Generation*, 16.
15. *The History of Van Zandt County, Texas* (1984), 10.
16. Mills, *History of Van Zandt County*, 165, 167.
17. *History of Van Zandt County, Texas* (1984), 32.
18. Hall, *A History of Van Zandt County*, 18.
19. ibid., 20.
20. www.us-census.rg/pub/usgenweb/census/tx/henderson/1850/pg00249.txt
21. Kenneth Howell, "The Economic Development of the Dixie Frontier:

Henderson County, Texas 1850–1860," *East Texas Historical Journal* 37, no. 2 (1999): 36.

22. Øverland, *From Norway to America*, Vol. 1, 109.

23. Nelson, "Introduction," in Reiersen, *Pathfinder*, 51.

24. Ericson, *Nacogdoches Headrights*; Faulk, *History of Henderson County*; *Handbook of Texas Online*, Linda Sybert Hudson, "Brown, John [Red]," http://www.tshaonline.org/handbook/online/articles/fbr91

25. Hans Jacob Grøgaard's ABC was published in 1815, and *Læsebog for Børn, især i Omgangsskoledistrikterne* (*Reader for Children, Particularly in the Ambulatory School Districts*) was published in 1816. Both books came in several editions. The reader was the most frequently used textbook in the ambulatory schools of Norway until the 1860s. See Eide Johnsen, "En stat 'hvortil Udvandring fra Norge kunde ansees fordelaktig," 43–60.

26. The children were Anna Maria (Marie) (1826–78), Helene Sophie (1828–1902), Elise (1830–57), Christopher (1833–58), Hans Jacob (1836–99) and his twin brother Thomas Fasting (1836–1901), Emma Hilda (1838–1915), Nicolai Christian Keyser (1840–1919), Wilhelm and Johannes.

27. *Statistiske Tabeller for Kongeriget Norge*, III.

28. Syversen and Johnson, *Norge i Texas*, 210.

29. ibid.

30. Citation from Russel, *Undaunted*, 44.

31. Olmsted, *A Journey Through Texas*, 78.

32. McDonald, ed., *Hurrah for Texas*, ix–xi.

33. McDonald, ed., *Hurrah for Texas*, ix–xi; *Handbook of Texas Online*, Archie McDonald, "Sterne, Nicholas Adolphus," http://www.tshaonline.org/handbook/online/articles/fst45

34. Johannes Ostendorf, "Zur Geschichte der Auswanderung aus dem alten Amt Damme (Oldb.) insbesondere nach Nordamerika, in den Jahren 1830–1880," *Oldenburger Jahrbuch* 46/47 (1942–1943): 164–279.

35. Brancato, *A Glimpse of Life in Nineteenth-Century East Texas*, 52.

36. McDonald, ed., *Hurrah for Texas*, 6, 22, 214.

37. ibid., 54, 56, 97, 117.

38. ibid., 197.

39. Ericson, transc., *1847 Census, Nacogdoches County*.

40. Parsons, "A Texas Immigration Story."

41. Parsons, "A Texas Immigration Story," 34.

42. ibid., 35.

43. ibid., 37.

44. James G. Dickson, Jr., "A. A. Nelson, Sailor, Surveyor, and Citizen: A Personal Profile," *East Texas Historical Journal* 3, no. 2 (October 1965): 119–29.

45. McDonald, ed., *Hurrah for Texas*, 56.
46. Some of Nelson's papers can be found at the Dolph Briscoe Center for American History, The University of Texas at Austin, as "The Albert Aldrich Nelson Papers, 1833–1853," and include two bound volumes of typescripts of correspondence and a diary. His papers can also be found as "A. A. Nelson Papers," East Texas Collection, Steen Library, Stephen F. Austin State University.
47. Parsons, "A Texas Immigration Story," 35.
48. ibid., 22.
49. Smallwood, *The History of Smith County Texas*, 64.
50. Øverland, *From America to Norway: vol. 1*, 272.
51. Gordon, *Assimilation in American Life*.
52. Ralph A. Wooster, "East of the Trinity: Glimpses of Life in East Texas in the Early 1850s," *East Texas Historical Journal*, 13, no. 2 (1975): 6.
53. S. W. Geiser, "A Note on 'Vinzent's Texanische Pflanzen,' 1847," *Field and Laboratory* 25, (1957): 45–53.
54. Parsons, "A Texas Immigration Story," 17.
55. Øverland, *From America to Norway,* Vol. 1, 89.
56. Parsons, "A Texas Immigration Story," 17.
57. ibid.
58. ibid., 20.
59. Smallwood, *The History of Smith County Texas*, 122.

CHAPTER 3

1. "From Norse," January 10, 1887, *Uncle Theo's Scrapbook, 1883–1891, Norse, Texas. Newspaper Items from a Pioneer Texas Community*.
2. Anderson, *The First Chapter of Norwegian Immigration*, 193.
3. Flom, *A History of Norwegian Immigration to the United States*, 126.
4. Blegen, "Cleng Peerson and Norwegian Immigration," 320.
5. Nils Olav Østrem, "*Cleng Peerson—skaparen av den store forteljinga om Amerika,*" *Ætt og heim. Lokalhistorisk årbok for Rogaland 1999* (Stavanger, 2000): 9–10.
6. Østrem, "*Cleng Peerson—skaparen av den store forteljinga,*" 16–17.
7. Ropeid, "Kristenlivet i åra før 1850," 144–7.
8. Blake McKelvey, "The Genesee County Villages in Early Rochester's History," *Rochester History* 47, no. 1 and 2 (January and April 1985): 6; Conover, ed., *History of Ontario County, New York*, chapter 21.
9. Alexander M. Stewart, "Sesquicentennial of Farmington, New York 1789–1939," *Bulletin of Friends' Historical Association*, 29, no. 1 (Spring 1940): 37–43.

10. "The Society of Friends in Western New York," *The Canadian Quaker History Newsletter*, 37 (July 1985): 6–11.

11. Richard Canuteson, "A Little More Light on the Kendall Colony," *Norwegian-American Studies* 18 (1954): 82–101.

12. Cited after Blegen, *Norwegian Migration to America*, 39.

13. Blake McKelvey, "The Population of Rochester," *Rochester History* 12, no. 4 (October 1950): 4–5; Bernstein, *Wedding of the Waters*, 272; Dearinger, *The Filth of Progress*, 31, 37.

14. Blegen, *Norwegian Migration to America*, 51.

15. Gunleif Seldal, "The Sloopers" (paper presented at the Cleng Peerson Conference, Bosque Museum, Clifton, October 2015), 1.

16. Seldal, "The Sloopers," 16.

17. A short notice was published in *Den Norske Rigstidende*, Oslo, July 25, 1825. Citation from Seldal, "The Sloopers," 23.

18. Blegen, *John Quincy Adams and the Sloop Restauration*, 18.

19. Shaw, *Erie Water West. A History of the Erie Canal 1792–1854*.

20. Canuteson, "A Little More Light," 82–101. The investors in the Pulteney Association were a group of men led by Sir William Pulteney, 5th Baronet (1729–1805); Jeffrey M. Johnstone, "Sir William Johnstone Pulteney and the Scottish Origins of Western New York," *Crooked Lake Review* (Summer 2004); John H. Martin, "The Pulteney Estates in the Genesee Lands," *Crooked Lake Review* (Fall 2005); James D. Folts, "The 'Alien Proprietorship': The Pulteney Estate during the Nineteenth Century," *Crooked Lake Review* (Fall 2003).

21. Anderson, *The First Chapter of Norwegian Immigration*, 77.

22. Rynning, *Sandfærdig Beretning om Amerika til Oplysning og Nytte for Bonde og Menigmand*, 6.

23. Canuteson, "A Little More Light," 90.

24. ibid., 91.

25. ibid., 93.

26. Rynning, *Ole Rynning's True Account of America*, 73.

27. Rohrbough, *The Trans-Appalachian Frontier*, 33–4.

28. Seldal, "The Sloopers," 32. Tormod Jensen Madland died in 1826 and his wife Siri in 1829. Aanen Thoresen Brastad (Oyen Thompson) and his youngest daughter Birthe Karine died in Rochester in 1826. Cornelius Nelson Hersdal, the brother-in-law of Cleng Peerson, died in 1833. His widow Kari Pedersdatter was left with seven children aged from a few months to twenty years. Sven Jacobsen Aasen (Swaim Jacobson), who emigrated from Tysvær in 1829 with his wife Johanna Johnsdatter Hervig and three children, died in 1831.

29. Henry J. Cadbury, "The Norwegian Quakers of 1825," *The Harvard Theological Review*, 18, no. 4 (October 1925): 293–319; Anderson, *The First Chapter of Norwegian Immigration*, 185.

30. Blegen, *Norwegian Migration to America*, 61.

31. Louise Phelps Kellogg, "The Story of Wisconsin, 1634–1848," *Wisconsin Magazine of History* 3, no. 2 (December 1919): 190.

32. Wyman, *The Wisconsin Frontier*, 158.

33. Blegen, *Norwegian Migration to America*, 61–2.

34. Qualey, *Norwegian Settlement in the United States*, 22.

35. Citation from Qualey, *Norwegian Settlement*, 23.

36. Blegen, *Norwegian Migration to America*, 61.

37. Joseph Fellows Jr. was born in Redditch, Worcestershire, England, on July 2, 1782, and died April 29, 1873 in Corning, Steuben County, New York. His father, Joseph Fellows, was born in Dudley, Worcestershire, England, in 1755 and died in Scranton, Lackawanna County, Pennsylvania, on June 18, 1836. His mother, Catherine Turvey Fellows, died April 27, 1814, in Scranton, Pennsylvania.

38. Isaac L. Kip was born on April 6, 1767, in New York City and died there on January 20, 1837. Jacobus Hendrickson Kip (1631–90) built a large brick house in 1655 at Kip's Bay on the East River, Manhattan Island. See Kip and Hawley, *History of The Kip Family in America*, 354–55. During the American War of Independence, a military battle took place there, known as the "Landing at Kip's Bay." See Fischer, *Washington's Crossing*, 102–6.

39. Roe, *Rose Neighborhood Sketches, Wayne County, New York*, XII–XIII. Robert Troup was born in Elizabethtown, New Jersey, on August 19, 1756, and died on January 14, 1832, in New York City. He studied law and shared a room with Alexander Hamilton at King's College (Columbia University). He served in the American army 1776–80 and ended up a lieutenant colonel. Congress appointed him Minister of Defense in 1778, and he served as Minister of Finance between 1779 and 1780. He worked as a lawyer in New York City between 1784 and 1796.

40. Canuteson, "A Little Light," 90–1.

41. Rosdail, *The Sloopers, Their Ancestry and Posterity*.

42. Rosdail, *The Sloopers*, 62.

43. Remini, *Andrew Jackson*; Ward, *Andrew Jackson—Symbol for an Age*; Wallace, *The Long, Bitter Trail. Andrew Jackson and the Indians*; Inskeep, *Jacksonland*.

44. Ray A. Billington, "The Frontier in Illinois History," *Journal of the Illinois Historical Society* 43, no. 1 (Spring 1950): 28–45.

45. Baldwin, *History of La Salle County*, 87.

46. Davis, *Frontier Illinois*, 193–98; Wyman, *The Wisconsin Frontier*, 145–56.

47. George N. Fuller, "Settlement of Michigan Territory," *The Mississippi Valley Historical Review*, 2, no. 1 (June 1915): 39.

48. George J. Miller, "Some Geographic Influences in the Settlement of Michigan and in the distribution of its Population," *Bulletin of the American Geographical Society* 45, no. 5 (1913): 346.

49. Gray, *The Yankee West*, 1–2.
50. Fuller, "Settlement of Michigan Territory," 36.
51. Miller, "Some Geographic Influences," 332.
52. Coolidge, *A Twentieth Century History of Berrien County Michigan*.
53. Davis, *Frontier Illinois*, 207.
54. ibid., 246.
55. William Vipond Pooley, "The Settlement of Illinois from 1830 to 1850," *Bulletin of the University of Wisconsin History Series* 1 (1908): 287–595.
56. Paul W. Gates, "Tenants of the Log Cabin," *Mississippi Valley Historical Review* 49, no. 1 (June 1962): 3–31; Roy M. Robbins, "Preemption — A Frontier Triumph, *Mississippi Valley Historical Review* 18, no. 2 (December 1931): 331–49; Paul Wallace Gates, "Southern Investments in Northern Lands Before the Civil War," *Journal of Southern History* 5, no. 2 (May 1939): 155–85.
57. Baldwin, *History of La Salle County*, 163.
58. Meyer, *Making the Heartland Quilt*, 33.
59. The LaSalle Co. land records at Ottawa show that Fellows purchased 4,970 acres in the townships 34-4, 34-5, and 35-5 alone. This included 7 entire sections. Rev. J. J. Wang found that the purchases totaled 7,000 acres. Strand, ed., *A History of the Norwegians of Illinois*.
60. *Utvandringsstatistikk*, 5.
61. Brandal, *Hjelmeland*, 30.
62. *Stavanger Adresseavis*, May 27, 1836; *Stavanger Adresseavis*, June 7, 1837; *Stavanger Adresseavis*, October 13, 1837.
63. Cadbury, "Four Immigrant Shiploads of 1836 and 1837," 20.
64. *Utvandringsstatistikk*, 5–6.
65. Baldwin, *History of La Salle County*, 176.
66. Blegen, *Norwegian Migration to America*, 480–1.
67. Blegen, "Cleng Peerson and Norwegian Immigration," 321; Blegen, *Norwegian Migration to America*, 112.
68. Bingham, *General History of Shelby County Missouri*, 31.
69. Anderson, *The First Chapter of Norwegian Immigration*, 186.
70. Arne Odd Johnsen, "Johannes Nordboe and Norwegian Immigration. An 'America Letter' of 1837," *Norwegian-American Studies* 8 (1932): 23–8; Hovdhaugen, *Frå Venabygd til Texas*, 41.
71. Blegen, "Cleng Peerson and Norwegian Immigration," 321.
72. Østrem, "*Cleng Peerson — skaparen av den store forteljinga om Amerika*," 22–3.
73. Blegen, *Norwegian Migration to America*, 248–9.
74. William Mulder, "Norwegian Forerunners Among the Early Mormons," *Norwegian-American Studies* 19 (1956): 46, 47. See also Anderson, *First Chapter*

of Norwegian Immigration, 399–408; Blegen, *Norwegian Migration to America*, 248; Blegen, *Norwegian Migration to America: The American Transition*, 112–4.

75. Mulder, "Norwegian Forerunners Among the Early Mormons," 47.

76. ibid., 48.

77. Anderson, *First Chapter of Norwegian Immigration*, 400; Blegen, *Norwegian Migration to America*, 112–18.

78. Mulder, "Norwegian Forerunners Among the Early Mormons," 51.

79. Ronald E. Nelson, "The Bishop Hill Colony and Its Pioneer Economy," *Swedish Pioneer Historical Quarterly* 18, no. 1 (1987): 32–48, has argued that a few years after its establishment "the colony became the economic center of its immediate area." From the standpoint of the development of the American Midwest, "the real significance of the Bishop Hill Colony was its role in paving the way for the settlement of tens of thousands of Swedes in the region during the latter half of the nineteenth century."

80. Elmen, *Wheat Flour Messiah*.

81. Blegen, "Cleng Peerson and Norwegian Immigration," 323.

82. Elmen, *Wheat Flour Messiah*, 130.

83. Østrem, "*Cleng Peerson—skaparen av den store forteljinga om Amerika*," 22–3.

84. Elmen, *Wheat Flour Messiah*, 19.

85. ibid., 142.

86. Letter from Cleng Peerson to Knud Langeland, August 30, 1850. Translation to English, NAHA Archives, St. Olaf College.

87. Letter from Peerson to Langeland, August 30, 1850.

88. ibid.

89. ibid.

90. Hovdhaugen, *Frå Venabygd til Texas*.

91. Johnsen, "Johannes Nordboe and Norwegian Immigration," 23.

92. Letter from Johannes Norboe to *Statsborgeren*, a Norwegian newspaper, July 13, 1837, cited in Johnsen, "Johannes Nordboe and Norwegian Immigration."

93. Letter from Johannes Nordboe, Five Mile Creek, Dallas Co., to Taale Andreas Gjestvang, Løten, Hedmark, February 1852, in *Fra Amerika til Norge*, vol. 1, eds. Øverland and Kjærheim, 212–20.

94. Johnsen, "Johannes Nordboe and Norwegian Immigration."

95. *Handbook of Texas Online*, Cecil Harper, Jr., "Bryan, John Neely," http://www.tshaonline.org/handbook/online/articles/fbran

96. Enstam, *Women and the Creation of Urban Life*, 4–5; Trent, *John Neely Bryan*.

97. *Handbook of Texas Online*, Lisa C. Maxwell, "Dallas County," http://www.tshaonline.org/handbook/online/articles/hcd02

98. See US Census for Dallas County 1850, enumerated on Oct. 31-Dec. 19, 1850, by Ben Merrell, sheet 82B, Dallas County. The assets for John Norboe were valued at $900, John Jr. $320, Powel $500, and Peter Norboe $1,000.

99. Seymour V. Connor, "A Statistical Review of the Settlement of the Peters Colony, 1841–1848," *Southwestern Historical Quarterly* 57 (July 1953-April 1954): 49.

100. Connor, "A Statistical Review," 54; Connor, *The Peters Colony of Texas*, 350.

101. Connor, "A Statistical Review," 51.

102. ibid., 50.

103. *Handbook of Texas Online*, Kenneth E. Hendrickson, Jr., "Brazos River," http://www.tshaonline.org/handbook/online/articles/rnb07; Hendrickson, Jr., *The Waters of the Brazos*; Puryear and Winfield, Jr., *Sandbars and Sternwheelers*.

104. Calvert, de León, Cantrell, *The History of Texas*, 102.

105. Hoig, *Jesse Chisholm*, 60–5.

106. Campbell, *Gone to Texas*, 194–9.

107. *Handbook of Texas Online*, W. Kellon Hightower, "Tarrant County," http://www.tshaonline.org/handbook/online/articles/hct01

108. Sandra L. Myres, "Fort Graham, Listening Post on the Texas Frontier," *West Texas Historical Association Yearbook* 59 (1983): 33–52.

109. Syversen and Johnson, *Norge i Texas*, 202–3. The three oldest Johnson children migrated with their grandmother and their uncle and aunt to California in 1855 or 1856, after the death of John Norboe.

110. Letter from Johannes Nordboe, Five Mile Creek, Dallas Co. to Taale Andreas Gjestvang, Løten, Hedmark, February 1852, in *Fra Amerika til Norge*, vol. I., eds. Øverland and Kjærheim, 216.

111. Kallelid, *Stavanger bys historie*, 192.

112. Syversen and Johnson, *Norge i Texas*, 255–56.

113. "Brig Favoriten, J. A. Køhler & Co," http://www.norwayheritage.com/p_ship.asp?sh=favor

114. Brandal, *Hjelmeland*, 28–32.

115. *A Memorial and Biographical History of McLennan, Bell and Coryell Counties, Texas*, 505–9.

116. Anderson, *First Chapter of Norwegian Immigration*, 189.

117. ibid., 190.

118. *A Memorial and Biographical History of McLennan, Bell etc., Texas*, 505.4.

CHAPTER 4

1. Syversen, "*Prestedatteren, skipsrederen og bonden*," in Syversen and Johnson, *Norge i Texas*, 60–61.

2. Syversen and Johnson, *Norge i Texas*, 64–73.

3. ibid., 65.

4. ibid., 66.

5. ibid., 90.

6. ibid., 91.

7. Morthoff, *Romedalsboka. Garder og slekter*, Vol. 1, 689–93.

8. Morthoff, *Romedalsboka. Garder og slekter*, Vol. 2, 374–5; Sveen, "Introduction," to "Reisebrev fra Texas 1852," by Brunstad. In *Gammalt frå Stange og Romedal 1975*, 10.

9. *Beretning om Kongeriket Norges økonomiske Tilstand i Aarene 1846–1850*.

10. Braastad, *Slekter på garder i Vardal og Redalen*, 89.

11. Syversen and Johnson, *Norge i Texas*, 92.

12. ibid., 93.

13. Letter from T. A. Gjestvang, Hougstad, October 20, 1851, to Carl Ødeqvæstad. Carl Questad papers. NAHA Archives 1455, P 737, St. Olaf College, Northfield, Minnesota.

14. Syversen and Johnson, *Norge i Texas*, 262–3.

15. ibid., 249.

16. Brunstad, "Reisebrev fra Texas 1852," 12–15.

17. ibid., 16.

18. ibid., 16.

19. ibid., 17.

20. On the role of T. A. Gjestvang, see Syversen and Johnson, *Norge i Texas*, 64–73.

21. Syversen and Johnson, *Norge i Texas*, 257.

22. Lyder L. Unstad, "The First Norwegian Migration into Texas. Four 'America Letters.'" Translated and edited by Lyder L. Unstad, *Norwegian-American Studies*, 8 (1934): 39–57.

23. Unstad, "First Norwegian Migration into Texas."

24. *Handbook of Texas Online*, W. Kellon Hightower, "TARRANT COUNTY," accessed June 07, 2020, http://www.tshaonline.org/handbook/online/articles/hct01 Uploaded on June 15, 2010. Modified on May 6, 2019. Published by the Texas State Historical Association.

25. Selcer, *The Fort That Became a City*, 2–5; Knight, *Fort Worth*; Pate, *North of the River*.

26. *Handbook of Texas Online*, Janet Schmelzer, "Fort Worth, TX," http://www.tshaonline.org/handbook/online/articles/hdf01; Schmelzer, *Where the West Begins*.

27. Unstad, "First Norwegian Migration into Texas."

28. ibid.

29. ibid.
30. ibid.
31. ibid.
32. ibid.
33. ibid.
34. Archive 3P150. Reierson (Johan Reinert) Family Papers, 2. Personal Correspondence 1840–1853.
35. ibid.
36. Øverland, *From America to Norway*, vol. 1, 144–5.
37. ibid., 145.
38. Seok Chul Hong, "The Burden of Early Exposure to Malaria in the United States, 1850–1860: Malnutrition and Immune Disorders," *Journal of Economic History* 67, no. 4 (December 2007): 1001–35; R. H. von Ezdorf, "Malaria in the United States: Its Prevalence and Geographic Distribution," *Public Health Reports* 30, no. 22 (May 28, 1915): 1603–24.
39. Rocco, *Quinine. Malaria and the Quest for a Cure That Changed the World*, 172–3.
40. Øverland, *From America to Norway*, Vol. 1, 149.
41. ibid., 149.
42. ibid., 150.
43. Hastvedt, "Recollections of a Norwegian Pioneer in Texas," 91–104, 104.
44. "Aslak Nelson," F858, *Bosque County: Land and People*, 535.
45. Syversen and Johnson, *Norge i Texas*, 62–3.
46. ibid., 67.
47. ibid. 63.
48. *Beretning om Kongeriket Norges økonomiske Tilstand i Aarene 1851–1855*, 44.

CHAPTER 5

1. *Handbook of Texas Online*, "Bosque River," http://www.tshaonline.org/handbook/online/articles/rnb05
2. Citation from Pool, *Bosque Territory*, 57.
3. Miller, *The Public Lands of Texas 1519–1970*, 35–6.
4. Connor, *Kentucky Colonization in Texas*, 69; Connor, *The Peters Colony of Texas: A History and Biographical Sketches of the Early Settlers*.
5. *Handbook of Texas Online*, Sandra L. Myres, "Fort Graham," http://www.tshaonline.org/handbook/online/articles/qbf21
6. *Bosque County: Land and People*, 285.
7. Citation from Pool, *Bosque Territory*, 57.
8. ibid.

9. Lucy A. Erath, "Memoirs of major George Bernard Erath," *Southern Historical Quarterly* 26, no. 3 (January 1923): 207–33.

10. Pool, *Bosque Territory*, 30.

11. The following registered land: Clairborne Pool, November 20, 1839; John C. Pool, November 20, 1839; John McLennan, November 22, 1839; Anson Darniel, November 19, 1839; James Hughes, November 19, 1839. Pool, *Bosque Territory*, 34.

12. *Handbook of Texas Online*, Thomas W. Cutrer, "Erath, George Bernard," http://www.tshaonline.org/handbook/online/articles/fer01

13. Meyer, *The Highland Scots of North Carolina 1732–1776*, 69–101.

14. McKinnon, *History of Walton County*, 13.

15. Reynolds, *Waking Giant*, 23–4; Rohrbaugh, *The Trans-Appalachian Frontier*, 246–61.

16. McKinnon, *History of Walton County*, 14–9.

17. ibid., 39–43.

18. ibid., 98–100.

19. ibid., 101.

20. McLean, ed., *Papers Concerning Robertson's Colony in Texas, Volume 10*, 34. The University of Texas at Arlington purchased a part of the Robertson papers in 1976. Until then three volumes had been published. Until McLean retired in 1992, he continued compiling and editing the series, beginning with Volume 4 and ending with Volume 18.

21. Pool, *Bosque Territory*, 24; *Handbook of Texas Online*, Margaret E. Lengert, "Nashville-on-the-Brazos, Texas," http://www.tshaonline.org/handbook/online/articles/hvn04; *Handbook of Texas Online*, Evelyn Clark Longwell, "McLennan's Bluff," http://www.tshaonline.org/handbook/online/articles/rjm51

22. McLean, ed., *Papers Concerning Robertson's Colony in Texas*, 142.

23. *Handbook of Texas Online*, Evelyn Clark Longwell, "McLennan, Neil," http://www.tshaonline.org/handbook/online/articles/fmc89; *A Memorial and Biographical History of McLennan, Bell, and Coryell Counties*; Batte, *History of Milam County, Texas*; Conger, *A Pictorial History of Waco*.

24. Pool, *Bosque Territory*, 36.

25. ibid., 46–7.

26. Bracken, ed., *Historic McLennan County*, 7–8.

27. Cureton and Cureton, *Sketch of the Early History of Bosque County*, 3.

28. Texas General Land Office Land Grant Search, GLO Land Grants Records, Abstract Number 81, Milam 3rd District/Class, File No. 913, Patents Volume 13, on January 4, 1857.

29. Pierson, *Norwegian Settlements in Bosque County*, 40.

30. Boyd, *Texas Land Survey Maps for Bosque County*, 35.

31. Pierson, *Norwegian Settlements in Bosque County*, 40.
32. Penciled notes written by Jacob Olson, date unknown. Transcribed by George W. Larson, May 3, 2003.
33. Letter from Taale Andreas Gjestvang, dated Hougstad, May 11, 1854, to Carl Questad, Questad Archive, NAHA.
34. Abstract #671, patentee Carl Questad, patent date October 9, 1858, Patent #842, 286 acres. Boyd, *Texas Land Survey Maps for Bosque County*, 61.
35. Letter to Carl Questad from Sigurd Ørbæk, Prairieville, July 9, 1856. Questad Archive, NAHA.
36. Syversen and Johnson, *Norge i Texas*, 262–63.
37. US Census 1860 for Bosque County.
38. *Bosque County: Land and People*, 411.
39. Abstract #652, Ole Pederson, Patentee Ole Person, Patent #298, patent date December 15, 1859, 320 acres.
40. Pierson, *Norwegian Settlements in Bosque County*, 41.
41. "Jasper Newton Mabry," F766, *Bosque County: Land and People*, 492.
42. "An Act for the relief of Cling Pearson," Chapter 121, *Texas Legislature*, 1856, 599.
43. James C. Frazier came to Bosque County in 1849, then left for a year until he married in 1855. He had acquired a good deal of land, including in Bosque County. Frazier built a two-story house on land he owned between Kopperl and Kimball.
44. Pierson, *Norwegian Settlements in Bosque County*, 21–2.
45. Barry, *Buck Barry, Texas Ranger and Frontiersman*, 91.
46. ibid., 90.
47. ibid.
48. Pierson, *Norwegian Settlements in Bosque County*, p 21.
49. "Petition to his Excellency E. M. Pease, Governor of the State of Texas," dated Neil's Creek, Bosque County, April 14, 1854. Copy from Texas State Archives, CPRL, Bosque Museum.
50. "Petition to his Excellency E. M. Pease, Governor of the State of Texas," 1854.
51. Kenneth A. Breisch and David Moore, "The Norwegian Rock Houses of Bosque County, Texas: Some Observations on a Nineteenth-Century Vernacular Building Type," *Perspectives in Vernacular Architecture* 2 (1986): 64–71.
52. Jordan, *Texas Log Buildings*, 12.
53. "Pioneer Homebuilding," CPRL, Bosque Museum.
54. Rebecca D. Radde, "Ole Pierson Homestead" (manuscript, Meridian, July 30, 1985), 3.
55. See FamilySearch, US Census 1860, https://www.familysearch.org, Texas, Bosque County, household 27.

56. Dahl, *The Family History and Descendants of Hendrik and Christine Dahl From the Year 1854–2004*, 9.

57. See FamilySearch, US Census 1860, https://www.familysearch.org, Texas, Bosque County, household 18.

58. "The Questad House," *The Clifton Record*, October 16, 1975.

59. "The Quest of the Questad Place" (manuscript by Mary Ellen Ellingson and Martha Louise Springer 1992), 1.

60. "Strand, Christian and Regina," *Bosque County. Land and People*, 701.

61. Pool, *Bosque Territory*, 77.

62. Republished in *Haugesunds Dagblad*, June 27, 1955. The relatives in Norway got the letter in early November 1860, and they replied on November 20, 1860.

63. *Haugesunds Dagblad*, June 27, 1955.

CHAPTER 6

1. Cureton and Cureton, *Sketch of the Early History*, 5.

2. ibid., 3.

3. Pool, *Bosque Territory*, 62.

4. "F1103 Scrutchfield, Lowry Hampton," *Bosque County: Land and People*; George M. Lewis, "Lowry Hampton Scrutchfield," in *Profiles of Pioneer Families in the Valley Mills Area*, eds. Gerald and Jo Meyer, 97–100; Pool, *Bosque Territory*, 40.

5. "Barnes, Robert Samuel," F58, *Bosque County: Land and People*, 135.

6. "Barnes, Elisabeth Oakes Barton," F59, *Bosque County: Land and People*, 135.

7. Pool, *Bosque Territory*, 55.

8. For literature on the Tutt-Everett War in Marion County, Arkansas, see W. B. Flippin, "The Tutt and Everett War in Marion County," *Arkansas Historical Quarterly* 17 (Summer 1958): 155–63; Blevins, *The Tutt & Everett War*. On feuds in Texas, see Douglas, *Famous Texas Feuds*; Sonnichsen, *I'll Die Before I'll Run*; *Handbook of Texas Online*, C. L. Sonnichsen, "Feuds," http://www.tshaonline.org/handbook/online/articles/jgf01

9. "Everett, Thomas Ewell Family," F355, *Bosque County: Land and People*, 285.

10. Syversen and Johnson, *Norge i Texas*, 290. The author in *Bosque County: Land and People* wrote that "Paul Paulsen and his wife, Oline Mary Halvorsdatter emigrated from the Vik estates, Stange Parish, Norway, to Bosque County in 1857."

11. "Vik gård," in *Stange Bygdebok. II. Gards- og slektshistorien*, ed. M. Veflingstad, 856–86.

12. Syversen and Johnson, *Norge i Texas*, 94.

13. ibid., 856–63.

14. ibid. 94.

15. ibid., 95.

16. ibid., 85.

17. Cleng Peerson and Ole Canuteson witnessed the transaction on March 10, 1857.

18. "History of the Joseph Olson Family," written by Jacob Olson.

19. Blegen, "Cleng Peerson and the Norwegian Immigration," 325.

20. Eli was born at Rossedal, Tysvær, and baptized October 11, 1807; Ove was born at Rossedal and baptized on December 17, 1809; Lars was born at Rossedal and baptized January 31, 1813; Johannes was born at Rossedal June 4, 1821; Helga Carine was born at Rossedal, February 12, 1825. A sixth child, Caroline, was born in Kendall on April 1, 1829.

21. Foldøy, *Jelsa 1. Gards- og Ættesoga*, 273.

22. Bergitte Cecilie Naadland, born October 7, 1837, Jelsa, died June 17, 1936, in Bosque County. She married David Martin, born ca. 1837 in Tennessee. He was first a farmer in Erath County and later in Coryell County.

23. Larson, *Norwegian Emigration to Canada, 1850–1874*.

24. Laxton, *The Famine Ships*.

25. Colwick, *The Colwicks*, 1.

26. Anna E. Wynne with Elma White, "The Eggen Family History. Ancestors and Descendants of Thore Anstensen Eggen and His Brothers and Sisters," copyright Anna Wynne, 2001, 42.

27. Colwick, *The Colwicks*, 2.

28. ibid. 2–3.

29. *Haugesunds Dagblad*, June 27, 1955.

30. Colwick, *The Colwicks*.

31. Copy in Van Sickle, "The Reconstruction of the History of the Cleng Peerson/Colwick Farm in Bosque County, Texas," May 2017, 180.

32. On the herring fisheries in Rogaland County, western Norway, in the 19th century, see Langhelle, "*Oppsving i næringane ved sida av jordbruket,,*" 333–46; Østrem, *Karmøys historie — mot havet du deg vender*, 31–81.

33. Van Sickle, "The Reconstruction of the History of the Cleng Peerson Farm," 69f.

34. Wynne, with White, "The Eggen Family History," 41.

35. Bosque County, Texas Deed Record, Volume D, 558.

36. The deed from Ovee Rosdail to Ovee Colwick was signed in Illinois, July 17, 1860, and was recorded in Volume I, page 10, and filed for the record in Bosque County, April 20, 1869, duly recorded in Book II, pages 10 and 11, Bosque County, Volume E, 102.

37. "Indenture made the twenty sixth day of November, in the year one thousand eight hundred and sixty between Cleng Peerson of the county of Bosque and state of Texas of the first party and Ove Colwick of the state and County aforesaid of the first part," Jacob Olson papers, NAHA Archives, St. Olaf College.

38. Abstract #199, patentee Ole Andreas Canuteson, patent #50, patent date May 15, 1879, 160 acres; see Boyd, *Texas Land Survey Maps for Bosque County*, 37.

39. After his second wife died in 1891, he was married a third time, to Martha Renne. "Canuteson, Ole Andrew," F176, *Bosque County: Land and People*.

40. Syversen and Johnson, *Norge i Texas*, 297.

41. ibid., 289

42. ibid., 254.

43. Anderson, *The First Chapter of Norwegian Immigration*, 190.

44. ibid., 388.

45. Letter from T. Theo Colwick to Rasmus B. Anderson, Madison, Wisconsin, dated Norse, Bosque County, Texas, November 4, 1895. Anderson papers, University of Wisconsin.

46. Letter from T. T. Colwick to Rasmus B. Anderson, November 4, 1895.

47. Contract signed January 16, 1865, Jacob Olson papers, NAHA Archives, published in Syversen and Johnson, *Norge i Texas*, 102.

48. Anderson, *The First Chapter of Norwegian Immigration*, 190.

49. Letter from T. Theo Colwick to Rasmus B. Anderson, November 4, 1895.

50. ibid.

51. Cureton and Cureton, *Sketch of the Early History of Bosque County*, 10–11.

52. Winfrey and Day, eds., *The Indian Papers of Texas and the Southwest, 1825–1916*, 207.

53. ibid., 214.

54. ibid., 207.

55. Pool, *Bosque Territory*, 61.

56. Winfrey and Day, eds., *The Indian Papers of Texas and the Southwest*, 207.

57. Pool, *Bosque Territory*. 61.

CHAPTER 7

1. Kenneth Howell, "The Economic Development of the Dixie Frontier: Henderson County, Texas 1850–1860," *East Texas Historical Journal* 37, Issue 2 (1999): 36.

2. *Texas 1860 Agricultural Census*, vol. 1, transcribed and compiled by Green, 76.

3. On the Swenson family, see Syversen and Johnson, *Norge i Texas*, 260-2.

4. *Texas 1860 Agricultural Census*, 76.

5. Campbell, *An Empire for Slavery*.

6. Terry G. Jordan, "The Imprint of the Upper and Lower South in Mid-Nineteenth-Century Texas," *Annals of the Association of American Geographers* 57, no. 4 (Dec. 1967): 667.

7. Campbell, *An Empire for Slavery*, 36.

8. Jordan, "The Imprint of the Upper and Lower South," 688.

9. William O. Lynch, "The Westward Flow of Southern Colonists before 1861," *Journal of Southern History* 9, no. 3 (Aug. 1943): 318.

10. 1860 US Census of Bosque County, Bosque County Historical Commission—1860 US. Census, www.bosquehc.org/census-1860.shtml

11. Jordan-Bychkov, *The Upland South*.

12. Leyburn, *The Scotch-Irish*, 184f.; Kennedy, *The Scots-Irish in the Carolinas*, 105–9.

13. Lynch, "The Westward Flow of Southern Colonists before 1861," 317.

14. Rohrbough, *The Trans-Appalachian Frontier*, 272–3.

15. Ownsley, *Plain Folk of the Old South*, V.

16. John Solomon Otto, "The Migration of the Southern Plain Folk: An Interdisciplinary Synthesis," *Journal of Southern History* 51, no. 2 (May 1985): 192.

17. Terry G. Jordan, "The Texas Appalachia," *Annals of the Association of American Geographers*, September 1970 60, no. 3 (September 1970): 409–27.

18. ibid., 410.

19. ibid., 416.

20. Ralph A. Wooster, "East of the Trinity: Glimpses of Life in East Texas in the Early 1850s," *East Texas Historical Journal* 13, Issue 2 (1975): 6.

21. Campbell, *An Empire for Slavery*, 58.

22. ibid., 68.

23. ibid., 51.

24. ibid., 58.

25. Wooster, "East of the Trinity," 3.

26. ibid., 95.

27. "McLennan, TX, 1860 Federal Slave-Schedule Census," http://ftp-us-census.org/pub/usgenweb/census/tx/mclennan/1860/sl

28. Kelly, ed. "Introduction" in *The Handbook of Waco and McLennan County, Texas*.

29. Wallace, *Waco. A Sesquicentennial History*, 22

30. Kelly, ed., *Handbook of Waco and McLennan County*, 118.

31. ibid., 117.

32. ibid., 104.

33. Wallace, *Waco. A Sesquicentennial History*, 23.

34. Kelly, ed., *Handbook of Waco and McLennan County*, 89.

35. "Carter-Harrison Family Papers, 1837–1932 (bulk 1845–1895)," The Texas Collection, Baylor University, n.d. The archive contains documents from the Harrison families, one of the most well-connected families in the American South.

36. "McLennan, TX, 1860 Federal Slave-Schedule Census," http://ftp-us-census.org/pub/usgenweb/census/tx/mclennan/1860/sl

37. Bracken, "Bosqueville," in *Historic McLennan Count*, 19.

38. "McLennan, TX, 1860 Federal Slave-Schedule Census," http://ftp-us-census.org/pub/usgenweb/census/tx/mclennan/1860/sl.

39. Wallace, *Waco. A Sesquicentennial History*, 1999, 33.

40. Campbell, *An Empire for Slavery*, 82–92.

41. According to the slave schedules for McLennan County in 1860, William Blocker owned 9 slaves, and R. F. Blocker 8 slaves. W. A. Fort owned 24 slaves and A. H. Fort was listed as the owner of 18 slaves. W. H. Downs had 5 slaves, and F. C. Downs had 18 slaves, but all of them were surpassed by W. W. Downs with 72 slaves. "McLennan, TX, 1860 Federal Slave-Schedule Census," http://ftp-us-census.org/pub/usgenweb/census/tx/mclennan/1860/sl

42. Wooster, "Wealthy Texans," 184.

43. Marshall V. Bonds, "Silas McCabe: First Settler at Coon Creek Community," *The Clifton Record — Centennial Edition*, 1954, section 1, page 6.

44. "Coon Creek Community History," *Bosque County: Land and People*, 26.

45. "Willingham Family," F1309, *Bosque County: Land and People*, 765.

46. *Bosque County, Land and People*, 434.

47. Pool, *A History of Bosque County, Texas*, 25.

48. *Bosque County, Land and People*, 666; Pool, *A History of Bosque County, Texas*; *Handbook of Texas Online*, "Sedberry, William Rush," http://www.tshaonline.org/handbook/online/articles/fse35; "William Rush Sedberry, Sr." Find A Grave Memorial, http://www.findagrave.com/cgi-bin/fg.cgi?page=gr&GRid=73240229

49. *Bosque County: Land and People*, 569.

50. *Handbook of Texas Online*, Randolph B. Campbell, "Antebellum Texas," http://www.tshaonline.org/handbook/online/articles/npa01

51. Wooster, "An Analysis of the Membership of the Texas Secession Convention," *Southwestern Historical Quarterly* 62, no. 3 (January 1959): 322–35.

52. *Handbook of Texas Online*, Walter L. Buenger, "Secession," http://www.tshaonline.org/handbook/online/articles/mgs02; Robin E. Baker and Dale Baum, "The Texas Voter and the Crisis of the Union, 1859–1861," *Journal of Southern History* 53 (August 1987): 395–420; Anna Irene Sandbo, "The First Session of the Secession Convention of Texas," *Southwestern Historical Quarterly* 18 (October 1914): 162–94.

53. Joe E. Timmons, "The Referendum in Texas on the Ordinance of Secession, February 23, 1861: The Vote," *East Texas Historical Journal* 11, no. 2 (1973): 12–28, discusses the election, errors and discrepancies concerning total votes cast in Texas as a whole and in individual counties.

54. Percentages computed on the basis of Joe E. Timmons, "The Referendum in Texas," Table 1: Comparison of votes on the ordinance of secession, 15–6.

55. Jordan, *German Seed in Texas Soil*, 181.

56. ibid., 182.

57. ibid., 182.

58. ibid., 110.

59. Walter D. Kamphoefner, "New Perspectives on Texas Germans and the Confederacy," *Southwestern Historical Quarterly* 102, no. 4 (April 1999): 442.

60. ibid.

61. ibid., 445.

62. ibid., 447.

63. Lovoll, *The Promise of America*, 75.

64. Blegen, *Norwegian Migration to America*, 186.

65. ibid., 114.

66. Rynning, *Ole Rynning's True Account of America*, 48.

67. Blegen, *Norwegian Migration to America*, 346.

68. C. A. Clausen and Derwood Johnson, "Norwegian Soldiers in the Confederate Forces," *Norwegian-American Studies* 25 (1972): 108.

69. Darwin Payne, "Early Norwegians in Northeast Texas," *Southwestern Historical Quarterly* 65, no. 2 (July 1961–April 1962): 196.

70. Pool, *A History of Bosque County, Texas*, 41.

71. Clausen and Johnson, "Norwegian Soldiers in the Confederate Forces," 110.

72. Pool, *A History of Bosque County, Texas*, 42.

73. Citations from Russel, *Undaunted*, 102.

74. Clausen and Johnson, "Norwegian Soldiers in the Confederate Forces," 109.

75. Russel, *Undaunted*, 204, note 6.

76. Clausen and Johnson, "Norwegian Soldiers in the Confederate Forces," 136.

77. "Four Mile Lutheran Church," (manuscript, July 4, 1964).

78. Ralph A. Wooster and Robert Wooster, "A People at War: East Texans during the Civil War, *East Texas Historical Journal* 28, no. 1, (1990): 3–16.

79. Jack Stolz, "Kaufman County in the Civil War," *East Texas Historical Journal* 28, no. 1 (1990): 37–43.

80. *Handbook of Texas Online*, Gerald F. Kozlowski, "Van Zandt County," http://www.tshaonline.org/handbook/online/articles/hcv02

CHAPTER 8

1. Ralph A. Wooster and Robert Wooster, "Rarin' for a Fight," Texans in the Confederate Army," *Southwestern Historical Quarterly* 84 (April 1980-April 1981): 387–426.
2. ibid., 387.
3. ibid., 391.
4. ibid., 397.
5. Kamphoefner and Helbich, eds., *Germans in the Civil War*. Germans comprised nearly 10 percent of all Union troops. The book contains letters written by German immigrants to friends and family back home. See also Shook, *German Unionism in Texas during the Civil War and Reconstruction*.
6. *Handbook of Texas Online*, "Nueces, Battle of the," http://www.tshaonline.org/handbook/online/articles/qfn01
7. Stein, "Distress, Discontent, and Dissent: Colorado County, Texas, during the Civil War," 301–16.
8. *Handbook of Texas Online*, Rudolph L. Biesele, "German Attitude Toward the Civil War," http://www.tshaonline.org/handbook/online/articles/png01
9. Bill Stein, "The German Draft Revolt in Colorado, Austin, and Fayette Counties," *Journal of the German-Texas Heritage Society* 14, no. 3 (Fall 1992): 221-4.
10. Wooster and Wooster, "Rarin' for a Fight," 395–6.
11. Baker, *The First Polish Americans. Silesian Settlements in Texas*, 64.
12. ibid., 65.
13. ibid., 69.
14. Grider, *The Wendish Texans*, 38–9.
15. Scott, *The Swedish Texans*, 102.
16. Ibid.,105.
17. The bulk of the information is compiled and edited by Lars Gjertveit, who is also presented as the "Commander of 7th Texas Infantry Reenactors in Norway." Roy H. Larsen compiled and edited the data for the web, and is the webmaster of www.borgerkrigen.info.
18. Clausen and Johnson, "Norwegian Soldiers in the Confederate Forces," 117.
19. ibid., 123.
20. Civil War letters have been an important source for many writers of books and articles about the Civil War. This chapter has been inspired by Bonner, *The Soldier's Pen*; Clinton, ed., *Southern Families at War*; Faust, *Mothers of Invention*; Taylor, *The Divided Family in Civil War America*; Rable, *Civil Wars. Women and the Crisis of Southern Nationalism*; Waggoner and Nemmers, eds., *Yours in Filial Regard*.

21. Clausen and Johnson, "Norwegian Soldiers in the Confederate Forces," 114.

22. Reierson (Johan Reinert) Family Papers, Archive 3P150 3, Personal correspondence, 1860–1891, Briscoe Center, University of Texas at Austin.

23. Barr, "An Essay on Texas Civil War Historiography," 257–76; Bell, "Civil War Texas: A Review of the Historical Literature," 14–44.

24. McPherson, *Battle Cry of Freedom*, viii–ix.

25. Mitchell, "Not the General but the Soldier," 82; Wiley, *The Life of Johnny Reb*; Madden ed., *Beyond the Battlefield*.

26. Velde, *The Presbyterian Churches and the Federal Union, 1861–1869*; Smith, *Revivalism and Social Reform*, 278–303; Walter B. Posey, "The Slavery Question in the Presbyterian Church in the Old Southwest," *Journal of Southern History* 15, no. 3 (August 1949): 311–24; Lewis M. Purifoy, "The Southern Methodist Church and the Proslavery Argument," *Journal of Southern History* 32, no. 3 (August 1966): 325–41.

27. Roeber, *Palatines, Liberty, and Property*.

28. Anderson, *Lutheranism in the Southeastern States, 1860–1886*, 49.

29. Wentz, *A Basic History of Lutheranism in America*, 164–6.

30. Blegen, *Norwegian Migration to America*, 149.

31. Robert Fortenbaugh, "American Lutheran Synods and Slavery, 1830–60," *Journal of Religion* 13, no. 1 (January 1933): 72–92; Heathcote, *The Lutheran Church and the Civil War*; Haraldsø, *Slaveridebattene i Den norske synode*.

32. On the connection between the Missouri Synod and the Norwegian Synod, see Carl S. Meyer, "Lutheran Immigrants' Churches Face the Problems of the Frontier," *Church History* 29, no. 4 (December 1960): 451–42.

33. Stolz, "Kaufman County in the Civil War," 37–43.

34. Hale, *The Third Texas Cavalry in the Civil War*, 3.

35. Ibid., 44.

36. Syversen and Johnson, *Norge i Texas*, 227–8.

37. www.borgerkrigen.info, "Peder Jenson." Another Jenson, born Jens Jensen Angelstad on September 23, 1821, in Holt parish, also served in the Confederate Army. He came from the same community in Norway and traveled with his family on the same ship from Hamburg in 1847. In 1858, Jens Jenson Angelstad and his family lived in Cherokee County, but moved to Henderson County where he had a smithy. He survived the war and died on November 20, 1888, in Brownsboro, Texas. Syversen and Johnson, *Norge i Texas*, 226.

38. www.borgerkrigen.info, "Reierson, Christian."

39. ibid., "Reiersen, Otto Theodor."

40. Hale, *The Third Texas Cavalry in the Civil War*, 28.

41. ibid., 39.

42. ibid., 44.
43. www.borgerkrigen.info, "Grogaard, Thomas Fasting."
44. ibid., "Grogard, John J."
45. Hale, *The Third Texas Cavalry in the Civil War*.
46. www.borgerkrigen.info, "Grogard, Nicholas C."
47. James S. Hamilton, "The Battle of Fort Donelson," *Journal of Southern History* 35, no. 1 (1969): 99–100; Knight, *The Battle of Fort Donelson*; Engle, *Struggle for the Heartland*; Tucker, *Unconditional Surrender*.
48. www.borgerkrigen.info
49. Syversen and Johnson, *Norge i Texas*, 222.
50. www.borgerkrigen.info, "Reierson, Oscar."
51. ibid., "Reiersen, Johann Henrich."
52. Reierson (Johan Reinert) Family Papers, letter dated Prairieville, December 28, 1862, Archive 3P150 3, Briscoe Center, University of Texas at Austin.
53. ibid.
54. ibid., letter dated Prairieville, September 20, 1863.
55. ibid.

CHAPTER 9

1. Wooster and Wooster, "Rarin' for a Fight," 399.
2. Francelle Pruitt, "'We've Got to Fight or Die:' Early Texas Reaction to the Confederate Draft, 1862," *East Texas Historical Journal* 36, no. 1 (1998): 4.
3. Pruitt, "We've Got to Fight or Die," 16.
4. "Election Record Book," Bosque County, June 17, 1861, 25.
5. "Commissioner's Court Minute Book," Bosque County, volume A, 122.
6. ibid., 123.
7. ibid., 145.
8. Cureton and Cureton, *Sketch of the Early History of Bosque County*, 9.
9. ibid., 10.
10. *Handbook of Texas Online*, Brett J. Derbes, "Eleventh Texas Infantry," http://www.tshaonline.org/handbook/online/articles/qke06
11. Edgar, *From Generation to Generation*, 83.
12. McPherson, *Battle Cry of Freedom*, 626–38.
13. Edgar, "*From Generation to Generation*, 86.
14. ibid., 88.
15. Clair Hines, "A Personal Civil War: The Murder of Wilhelm Warenskjold," *Norwegian-American Studies* 35 (2000), 51.
16. ibid., 52.
17. After the Civil War, Speight withdrew from his law practice and concen-

trated on his agricultural business. He died in Waco on April 26, 1888. *Handbook of Texas Online*, Roger N. Conger, "Speight, Joseph Warren," http://www.tshaonline.org/handbook/online/articles/fsp07

18. Richard Coke was elected as an associate justice to the Texas Supreme Court in 1866. The following year military Governor General Philip Sheridan fired Coke and four other judges as "an impediment to reconstruction." Coke won the Democratic nomination for governor in 1873. In a bitter and sometimes violent election he defeated the sitting Governor Edmund J. Davis, the Republican candidate. Coke resigned as governor in December 1876 following his election to the United States Senate. He died in Waco on May 14, 1897. *Handbook of Texas Online*, John W. Payne, Jr., "Coke, Richard," 2020, http://www.tshaonline.org/handbook/online/articles/fco15

19. *Bosque County, Land and People*, 666.

20. Ewing, "Caroline Sedberry, Politician's Wife," 47.

21. *Handbook of Waco and McLennan County, Texas*, 118; *Handbook of Texas Online*, "Gurley, Edward Jeremiah," http://www.tshaonline.org/handbook/online/articles/fgu08

22. *Handbook of Texas Online*, Charles D. Grear, "Thirtieth Texas Cavalry," http://www.tshaonline.org/handbook/online/articles/qkt08

23. Abstract #199, patentee Ole Andreas Canuteson, patent #50, patent date 15 May 1879, 160 acres; see Boyd, *Texas Land Survey Maps for Bosque County*, 37; www.borgerkrigen.info, "Ole A. Knudsen."

24. Copies made available by Maurice Jenson, Stillwater, Minnesota; www.borgerkrigen.info

25. Wiley, *The Life of Johnny Reb*, 38.

26. Syversen and Johnson, *Norge i Texas*, 297.

27. Clausen and Johnson, "Norwegian Soldiers in the Confederate Forces," 117.

28. ibid., 118.

29. ibid., 119.

30. Lars Olson survived the Civil War and lived in Bosque County until his death on August 6, 1928. After the war he married Anne Marie (Mary Olson) in Bosque County, born February 13, 1850, and died February 7, 1932. She was the oldest child of Joseph and Anne Karine Olson, and the oldest sister of Jacob Olson. The couple raised a family of ten children.

31. Clausen and Johnson, "Norwegian Soldiers in the Confederate Forces," 122.

32. ibid., 126.

33. ibid., 127.

34. ibid., 130.

35. ibid., 122.

36. ibid., 126.

37. ibid., 127.
38. Smith, *Frontier Defense in the Civil War*, 53.
39. ibid.
40. *Handbook of Texas Online*, Jeanne F. Lively, "Camp Salmon," http://www.tshaonline.org/handbook/online/articles/qcc36
41. According to a family story he buried his guns on his home farm after the war; they were never found. He died on January 3, 1912 (www.borgerkrigen.info).
42. "Johnson, John Bartinus (Teen) and Louisa," F639, *Bosque County: Land and People*, 424.
43. Letter from W. C. Alexander to Mr. Questad, dated Camp Salmon, December 7, 1862. Questad Archive, NAHA.
44. Smith, *Frontier Defense in the Civil War*, p.120.
45. Clausen and Johnson, "Norwegian Soldiers in the Confederate Forces," 124.
46. ibid., 123.
47. A third Skimland brother, R. R. Skimland, served in the Home Guard in the Indian wars and in Texas units during the Civil War (www.borgerkrigen.info).
48. Clausen and Johnson, "Norwegian Soldiers in the Confederate Forces," 132.
49. ibid., 133.
50. Clausen and Johnson, "Norwegian Soldiers in the Confederate Forces," 128.
51. Wooster and Wooster, "Rarin' for a Fight," 405.
52. Lonn, *Desertion during the Civil War*. See also Weitz, *A Higher Duty*.
53. Clausen and Johnson, "Norwegian Soldiers in the Confederate Forces," 139. John Bankhead Magruder was born on May 1, 1807, at Port Royal, Virginia. He left the Military Academy at West Point in 1830. When the Civil War broke out, he resigned from the US Army on April 20, 1861, and was appointed brigadier general and soon became a major general. He took over the command of the District of Texas, New Mexico, and Arizona in October 1862. His greatest success was his recapture of Galveston on January 1, 1863. After the Civil War, Magruder escaped to Mexico and offered his skills as a mercenary to Emperor Maximilian of Mexico. *Handbook of Texas Online*, Thomas W. Cutrer, "Magruder, John Bankhead," http://www.tshaonline.org/handbook/online/articles/fma15
54. Clausen and Johnson, "Norwegian Soldiers in the Confederate Forces," 139.
55. ibid., 140.
56. Wooster and Wooster, "Rarin' for a Fight," 426.

CHAPTER 10

1. See the chapter "Free at Last," in *The Slave Narratives of Texas*, eds. Tyler and Murphy, 113–27.
2. Kearney, *Nassau Plantation*, 184.
3. Roark, *Masters Without Slaves*, 95.
4. Harold D. Woodman, "Class, Race, Politics, and the Modernization of the Postbellum South," *Journal of Southern History* 63, no. 1 (Feb. 1997): 3–22.
5. Moneyhon, *Texas after the Civil War*, 24.
6. Foner, *A Short History of Reconstruction, 1863—1877*, 46.
7. Campbell, *Gone to Texas*, 311.
8. T. C. Alexander, "Bosque County," *Texas Almanac 1867*, 80.
9. Letter from Berger Rogstad, Norman Hill, Texas, June 1, 1867, to Peder Eriksen Furuset, Romedal, in *From Norway to America. Volume One, 1838–1870*, ed. Orm Øverland, 379–80.
10. Letter from Berger Rogstad, Norman Hill, Texas, June 1,1867, to Peder Eriksen Furuset, Romedal, in *From Norway to America. Volume One, 1838–1870*, ed. Orm Øverland, 380.
11. Letter from Elise Wærenskjold, Prairieville, Sept. 29–30, 1868, to Karen Poppe, Lillesand, in *From Norway to America. Volume One, 1838–1870*, ed. Orm Øverland, 411–2.
12. Letter from Hendrik Dahl, Norman Hill, Texas, May 30, 1868, to Peder Eriksen Furuset, Romedal, in *From Norway to America. Volume One, 1838–1870*, ed. Orm Øverland, 402.
13. Letter from Hendrik Dahl, Norman Hill, Texas, June 27. 1869, to Peder Eriksen and Siri Kristiansdatter Furuset, Romedal, in *From Norway to America. Volume One, 1838–1870*, ed. Orm Øverland, 426–7.
14. Letter from Hendrik Dahl, Norman Hill, Texas, May 30, 1868, to Peder Eriksen Furuset, Romedal, in *From Norway to America. Volume One, 1838–1870*, ed. Orm Øverland, 402.
15. Letter from. Elise Wærenskjold, Prairieville, Sept. 29–30, 1868, to Karen Poppe, Lillesand, in *From Norway to America. Volume One, 1838–1870*, ed. Orm Øverland, 411–2.
16. Letter from Aslak Nilsen Smeland to "precious parents, sisters and brothers," dated Four Mile Prairie, April 26, 1867.
17. Letter from Aslak Nilsen Smeland to "precious parents, sisters and brothers."
18. Letter from Berger Rogstad, Norman Hill, Texas, June 1, 1867, to Peder Eriksen Furuset, Romedal, in *From Norway to America. Volume One, 1838–1870*, ed. Orm Øverland, 380.

19. Letter from Hendrik Dahl, Norman Hill, Texas, June 27, 1869, to Peder Eriksen and Siri Kristiansdatter Furuset, Romedal, Stange, in *From Norway to America. Volume One, 1838–1870*, ed. Orm Øverland, 426.

20. Gaard, "No. 318, Ole Olsson Østrem," *Gaardslekten: med sidegreinene Utsiraslekten og Austreimslekten*.

21. Secretary book, volume 1, June 14, 1869–December 29, 1886, "Our Savior's Lutheran Church, Norse, Bosque County."

22. "1869–1969. Our Savior's Lutheran Church Centennial Celebration, June 28–29, 1969," 1.

23. "A History of Peder Pederson from 1821–1888," family history.

24. Edgar, *From Generation to Generation. Part I*, 89.

25. Letter from Aslak Nilsen Smeland to "precious parents, sisters and brothers."

26. ibid.

27. ibid.

28. ibid.

29. Letter from Elise Wærenskjold, Four Mile Prairie, to Thorvald Dannevig, Lillesand, Norway, April 30, 1867, in *From Norway to America. Volume One, 1838–1870*, ed. Orm Øverland, 376.

30. Michno, *The Settlers' War*.

31. Utley, *Frontier Regulars*, 219–233.

32. La Vere, *The Texas Indians*, 180.

33. Calvert, de León, and Cantrell, *The History of Texas*, 180–81.

34. Campbell, *Gone to Texas*, 291.

35. La Vere, *The Texas Indians*, 180.

36. Hämäläinen, "Into the Mainstream. The emergence of a New Texas Indian History," 75.

37. La Vere, *The Texas Indians*, 180.

38. ibid., 181.

39. ibid., 181.

40. Nystel, *Lost and Found or Three Months with the Wild Indians*, 22.

41. Rimmer, *Captured by the Indian*.

42. Nystel, *Lost and Found or Three Months with the Wild Indians*, 22.

43. ibid., 23.

44. ibid., 23.

45. ibid., 23.

46. ibid., 24.

47. ibid., 25.

48. ibid., 27.

49. ibid., 28.

50. The Indian agent Jesse Henry Leavenworth was the son of Brigadier General Henry Leavenworth and was born March 29, 1807, in Danville, Vermont. He graduated at West Point in 1830. He served as a second lieutenant in the Rocky Mountain Rangers near Denver, Colorado, but resigned in 1836. In February 1862, after the outbreak of the Civil War he was commissioned as a colonel in the volunteers and was given command of the Second Colorado Infantry Regiment. He was honorably discharged in September 1863.

51. Nystel, *Lost and Found or Three Months with the Wild Indians*, 29.

52. Edgar, *From Generation to Generation*, 19.

53. ibid., 20.

54. ibid., 20.

55. ibid., 33.

56. "Knud S. Knudson Homestead," n.d.; Edgar, *From Generation to Generation*, 99.

57. "N. J. Nelson," F861, *Bosque County: Land and People*, 537.

58. Syversen and Johnson, *Norge i Texas*, 240–3.

59. ibid., 240.

60. *Bosque County: Land and People*, 334.

61. ibid.

62. "Christian and Johanna Larsdatter Canuteson," genealogy of the family.

63. "Ole and Matthilda (Canuteson) Bronstad," F139, *Bosque County: Land and People*, 179.

64. "Christian and Johanna Larsdatter Canuteson."

CHAPTER 11

1. Newspaper article reprinted in Syversen and Johnson, *Norge i Texas*, 114.

2. Syversen and Johnson, *Norge i Texas*, 114.

3. Jordan and Walsh *White Cargo*; Diffenderfer, *The German Immigration into Pennsylvania Through the Port of Philadelphia, 1700–1775*); Christopher Tomlins, "Reconsidering Indentured Servitude: European Migration and the Early American Labor Force, 1600–1775," *Labor History*, 42, no. 1 (2001): 5–43; Farley Grubb, "The Incidence of Servitude in Trans-Atlantic Migration, 1771–1804," *Explorations in Economic History* 22, no. 3 (July 1985): 316–39.

4. David Galenson, "The Rise and Fall of Indentured Servitude in the Americas: An Economic Analysis," *Journal of Economic History* 44, no. 1 (March 1984): 1–26.

5. Farley Grubb, "The Disappearance of Organized Markets for European Immigrant Servants in the United States: Five Popular Explanations Reexamined," *Social Science History* 18, no. 1 (Spring 1994): 1–30; Farley Grubb, "The

End of European Immigrant Servitude in the United States: An Economic Analysis of Market Collapse, 1772–1835," *Journal of Economic History* 54, no. 4 (December 1994): 794–824.

6. David W. Galenson, "White Servitude and the Growth of Black Slavery in Colonial America," *Journal of Economic History* 41, no. 1 (March 1981): 39–47; David W. Galenson, "The Market Evaluation of Human Capital: The Case of Indentured Servitude," *Journal of Political Economy* 89, no. 3 (June 1981), 446–67; Galenson, "The Rise and Fall of Indentured Servitude in the Americas: An Economic Analysis," 1–26; Sharon V. Salinger, "Colonial Labor in Transition: The Decline of Indentured Servitude in Late Eighteenth-Century Philadelphia," *Labor History* 22, no. 2 (1981): 165–91; Salinger, *To Serve Well and Faithfully*.

7. The classic book about the cotter institution in Norway is Skappel, *Om Husmandsvæsenet i Norge, Dets oprindelse og utvikling*.

8. Several articles in *Heimen. Lokalhistorisk Tidskrift*, no. 2 (2004), discussed the cotter institution in Southern Norway. Was it established earlier than the 18th century? An overview is found in Urkedal York, Solli, and Dyrvik, *Vi Arbeidere er ingen faatallig Hær*.

9. Nagel, "*Oversikter, årstall, tabeller*," 238.

10. Try, *Norges historie*, 339–51.

11. In 1855, the county of Oppland had the highest number of cotters with 8,871, followed by Nord-Trøndelag with 6,070, and Hedmark with 6,053. In 1890, the county of Oppland still had the highest number with 4,249, followed by Hedmark with 3,493.

12. Dyrvik in Helle, Dyrvik, Hovland, and Grønlie, *Grunnbok i norsk historie*, 138–9; Dyrvik, "Sosial differensiering," and "Sosiale endringar på 1800-talet," 236–45, 399–417; Ståle Dyrvik, "Husmannsvesenet i Noreg," *Jord og Gjerning. Årbok for Norsk landbruksmuseum* (1990): 51–61.

13. Seip, *Nasjonen bygges 1830—1870*, 90–91.

14. Stensrud, "'Til Christiania for at søge Arbejde.': 49–68; see also Stensrud, *Økonomiske og sosiale kår innen husmannsklassen i Stange med særlig vekt på de første tiår av det 19. århundre*; Melby, *Husmannsvesenet i Ringsaker hovedsogn 1875–1900*.

15. Folketelling Norge 1865, 0415P Løten prestegjeld, Digitalarkivet.

16. Morthhoff, *Løtenboka*, 13, 14, 17.

17. "Ole Hanson and Mathea," F487, *Bosque County: Land and People*, 350.

18. Morthoff, *Romedalsboka*, 70.

19. Bell, *The Omenson Family History and the Norway Mills Community*.

20. Sveen, *Utvandrerne 1845–1870*, 213.

21. Bell, *The Omenson Family History and the Norway Mills Community*.

22. Morthoff and Løland, *Romedalsboka*, 301–2.

23. Letter from Hendrik Olsen Dahl, Norse, Texas, June 27, 1869, to Peder

Eriksen and Siri Kristiansdatter Furuset, in *From Norway to America. Volume One, 1838–1870,* ed. Orm Øverland, 427.

24. "Fjæstad" in *Romedalsboka,* vol. 2, 398–400.

25. The author leans very heavily on the work done by Elroy Christensen and Wayne Rohne on the Rohne family.

26. *Gammalt fra Stange og Romedal 1976,* 104.

27. Alice Easum, "Rohne Family History: Norwegian Immigrant Started Farm Here with Yoke of Oxen," *Cranfills Gap Index* 47, no. 44 (Nov. 27, 1964).

28. Letter from Johan Questad, Kvæstad, May 19, 1871, to Carl Questad, Bosque County, Texas. Questad Papers, NAHA, St. Olafs' College, Northfield, Minnesota.

29. Letter to Karl Kvæstad from "Hengivne" Mikhal og Lisabet Ødekvæstadbekken, dated Ødekvæstadbekken, Løiten, April 25, 1873. Questad Archive, NAHA.

30. Letter from Hendrik Olsen Dahl, Norse, Texas, July 10, 1870, to Peder Eriksen and Siri Kristiansdatter Furuset, Romedal, in *From Norway to America. Volume One, 1838–1870,* ed. Orm Øverland, 454.

31. Frederick Lundt was registered as the patentee on this land on June 23, 1848. See Boyd, *Texas Land Survey Maps for Bosque County,* 53.

32. Letter from Hendrik Dahl, Norman Hill, Texas, December 10, 1871, to Peder Eriksen and Siri Kristiansdatter Furuset, Romedal, in *From Norway to America. Volume One, 1838–1870,* ed. Orm Øverland, 60.

33. Dahl, *The Family History of Hendrik and Christine Dahl from the Year 1854 to 1954*; Dahl, *The Family History and Descendants of Hendrik and Christine Dahl,* 10.

34. "Microcopy no. 259. Passenger Lists of Vessels Arriving at New Orleans, 1820–1902, Roll 57, November 5–December 28, 1872, Nos. 159–97," The National Archives, Washington 1958. Online on Family Search.

35. Syversen and Johnson, *Norge i Texas,* 356, 341, 33.

36. "Passenger Contract No. 30, Martinus Pedersen," signed Christiania, October 31, 1872.

37. For a thorough discussion on mortality rates in the Atlantic migration trade, see Raymond L. Cohn, "Mortality on Immigrant Voyages to New York, 1836–1853," *Journal of Economic History* 44, no. 2. (June 1984): 289–300; Raymond L. Cohn, "The Transition from Sail to Steam in Immigration to the United States," *Journal of Economic History* 65, no. 2. (June 2005): 469–95, 492.

38. Evans, "Indirect Passage from Europe. Transmigration via the UK, 1836–1914," *Journal of Maritime Research* 3, no. 1 (2001): 70–84.

39. Dahl, *The Family History and Descendants of Hendrik and Christine Dahl. From the Year 1854 to 2004,* 10.

40. Letter from Christine Dahl, Norse, September 15, 1872, to Hendrik Dahl, Romedal, Orm Øverland, ed., *From Norway to America. Volume One, 1838–1870*, 70.

41. Letter from Christine Dahl, Norse, September 15, 1872, to Hendrik Dahl, Romedal, 70.

42. Morthoff, *Løtenboka*, 689, 693.

43. "Ole O. Finstad and Julia Pederson," F381, *Bosque County: Land and People*, 296.

44. "Hoff, John Family," F553, *Bosque County: Land and People*, 380.

45. Letter from Christine Dahl, Norse, Texas, January 17, 1876, to Peder Eriksen Furuset, Romedal, in *From Norway to America. Volume Two, 1871–1892*, ed. Orm Øverland, 132.

46. Letter from Christine Dahl, Norse, Texas, September 21, 1878, to Erik Pedersen Furuset, Romedal, in *From Norway to America. Volume Two, 1871–1892*, ed. Orm Øverland, 183.

47. Løland, *Romedalsboka. Garder og slekter*, 565–66; "Folketeljing 1865 for 0416 Romedal prestegjeld," household 0034, krets 006 Mælum, 115.

48. Letter from Christine Dahl, Norman Hill, Texas, June 12, 1879, to Erik Pedersen Furuset, Romedal, in *From Norway to America. Volume Two, 1871–1892*, ed. Orm Øverland, 202.

49. Syversen and Johnson, *Norge i Texas*, 384.

50. Letter from Ole Rossing, Norse, Texas, November 25, 1880, to Erik Pedersen and Beate Kristoffsdatter Furuset, Romedal, in *From Norway to America. Volume Two, 1871–1892*, ed. Orm Øverland, 248.

51. Letter from Christine Dahl, Norman Hill, Texas, March 11, 1880, to Erik Pedersen Furuset, Romedal, in *From Norway to America. Volume Two, 1871–1892*, ed. Orm Øverland, 226.

52. ibid., 227.

53. Letter from Christine Dahl, Norse, Texas, February 4, 1885, to Erik Pedersen Furuset, Romedal, in *From Norway to America. Volume Two, 1871–1892*, ed. Orm Øverland, 348.

54. "From Norse," October 2, 1887, *Uncle Theo's Scrapbook*.

55. "Holen C. O. and Gina Dahl family," F561, *Bosque County: Land and People*, 384.

56. "Bekkelund family," F85, *Bosque County: Land and People*, 149.

57. ibid.

58. Pool, *Bosque Territory*, 47.

CHAPTER 12

1. Campbell, *Gone to Texas*, 311.
2. Harper, Jr., *Farming Someone Else's Land*, ii.
3. Campbell, *Gone to Texas*, 311.
4. Campbell, *A Southern Community in Crisis*, 375.
5. ibid., 376.
6. Spratt, *The Road to Spindletop, Economic Change in Texas, 1875–1901*, 56.
7. Louis Ferleger, "Sharecropping Contracts in the Late-Nineteenth-Century South," *Agricultural History* 67, no. 3 (Summer 1993): 31–46; Ferleger andMetz, *Cultivating Success in the South*; *Handbook of Texas Online*, Cecil Harper, Jr., and E. Dale Odom, "Farm Tenancy," http://www.tshaonline.org/handbook/online/articles/aefmu
8. Campbell, *Gone to Texas*, 312.
9. Robert A. Calvert, "Nineteenth-Century Farmers, Cotton, and Prosperity," *Southwestern Historical Quarterly*, 73 (April 1970): 509; Calvert, de León, Cantrell, *The History of Texas*, 222.
10. Jeremy Atack, "The Agricultural Ladder Revisited: A New Look at an Old Question with Some Data for 1860," *Agricultural History*, 63, no. 1 (1989): 1–25; Jeremy Atack, "Tenants and Yeomen in the Nineteenth Century," *Agricultural History* 62, no. 3 (1988): 6–32.
11. "Biography of M. R. Rohne," written Cranfills Gap, Texas, December 24, 1895.
12. Johnson, *I, Carl Johnson. Born of Immigrant Parents*, An autobiography as recorded on tape during the year 1979 at my home, 3416 Street, Fort Worth, Texas, 22.
13. Dahl, John, F276, *Bosque County: Land and People*, 247–8.
14. Johnson, *I, Carl Johnson*, 21.
15. ibid.
16. W. J. Spillman, "The Agricultural Ladder," *American Economic Review* 9, no. 1 (March 1919): 170–9.
17. ibid., 172.
18. B. H. Hibbard, "Farm Tenancy in 1920," *Journal of Farm Economics* 3, no. 4 (Oct. 1921): 168–75; John D. Black and R. H. Allen, "The Growth of Farm Tenancy in the United States," *Quarterly Journal of Economics* 51, no. 3 (May 1937): 393–425; Allan G. Bogue, "Farming in the Prairie Peninsula," *Journal of Economic History* 23, no. 1 (March 1963): 3–29; Robert Swieringa, "The Equity Effects of Public Land Speculation in Iowa: Large versus Small Speculators," *Journal of Economic History* 34, no. 4 (Dec. 1974): 1008–20; Gavin Wright and Howard Kunreuther, "Cotton, Corn and Risk in the Nineteenth Century," *Journal of Economic History* 35, no. 3 (Sep. 1975): 526–51; J. C. Hsiao, "The Theory of Share Tenancy

Revisited," *Journal of Political Economy* 83, no. 5 (Oct. 1975): 1023–32; Donald L. Winters, "Tenancy as an Economic Institution: The Growth and Distribution of Agricultural Tenancy in Iowa, 1850–1900," *Journal of Economic History* 37, no. 2 (June 1977): 382–408; Harry N. Scheiber, "The Economic Historian as Realist and as Keeper of Democratic Ideals: Paul Wallace Gates's Studies of American Land Policy," *Journal of Economic History* 40, no. 3 (Sep. 1980): 585–93; Nancy Virts, "The Efficiency of Southern Tenant Plantations, 1900–1945," *Journal of Economic History* 51, no. 2 (June 1991): 385–95; Mary Eschelbach Hansen, "Land Ownership, Farm Size, and Tenancy after the Civil War," *Journal of Economic History* 58, no. 3 (Sep. 1998): 822–9; Lee J. Alston and Kyle D. Kauffman, "Agricultural Chutes and Ladders: New Estimates of Sharecroppers and 'True Tenants' in the South, 1900–1920," *Journal of Economic History* 57, no. 2 (June 1997): 464–75; Lee J. Alston and Joseph H. Ferrie, "Time on the Ladder: Career Mobility in Agriculture," *Journal of Economic History* 65, no. 4 (Dec. 2005): 1058–81.

19. Jordan, *German Seed in Texas Soil*, 115, 116, 189.

20. ibid., 190.

21. Report on the Productions of Agriculture, Tenth Census (June 1, 1880)[1] (Washington, DC, Government Printing Office, 1883), 88–9; *Report on the Statistics of Agriculture in the United States at the Eleventh Census: 1890* (Washington, DC, Government Printing Office, 1895), 182–3.

22. Thirteenth Census of the United States taken in the Year 1910, Volume 5, Agriculture 1909 and 1910. General Report and Analysis. (Washington DC, Government Printing Office, 1914).

23. Alvin L. Bronstad, "The C. O. Bronstad Family History," 7.

24. Folketeljing 1865 (Census 1865) for 0416 Romedal prestegjeld, Digitalarkivet, 22.09.2014, 81.

25. ibid., 7.

26. ibid., 9.

27. "C. O. Bronstad," F132, *Bosque County: Land and People*, 175.

28. Alice Easum, "Rohne Family History: Norwegian Immigrant Started Farm Here with Yoke of Oxen," *Cranfills Gap Index* 47, no. 44 (Nov. 27, 1964).

29. Deed, recorded in Volume R, page 151, Deed Records, Bosque County, Texas, dated October 12, 1878, Hugh S. Dickson and wife Sarah Dickson of Philadelphia, Pennsylvania, conveyed 320 acres in the John Pool Survey, Bosque County, Texas to E. Rohne.

30. *Bosque County: Land and People*, 45–6.

31. "Biography of M. R. Rohne," written Cranfills Gap, Texas, December 24, 1895.

32. Deed from Oscar as Guardian to L. Christenson, dated March 23, 1907, Volume 33, page 600. Deed Records of Hamilton County, Texas.

33. Deed from Oscar Rohne as Guardian to Marianne Rohne, dated May 22,

1907, Volume 55, page 608. Deed Records, Bosque County, Texas/records of Wayne Rohne.

34. Foster *Forgotten Texas Census*, 17.

35. Jayne, *Martial Law in Reconstruction Texas*, 59–82; Steele, *A History of Limestone County, 1833–1860*; Walter, *A History of Limestone County*, 626–27.

36. Letter from Lars Mickelson to Carl Questad, dated Mexia, October 11, 1874. Questad Archive, NAHA, St. Olaf's College, Northfield.

37. His parents were not married when he was born, but six years later his mother, Olea Olsdatter, born July 16, 1817, and died August 15, 1911, at Cranfills Gap in Bosque County, married the tailor Knud Olsen Egeberg (born September 19, 1806, died June 10, 1878).

38. Folketelling Norge 1865, 0452P for Løten Prestegjeld. Digitalarkivet.

39. 1880 US Census, Mexia, Limestone, Texas. Series: T9 Roll: 1317 Enum. Dist. 97, Page 437, June 18, 1880.

40. *A Memorial and Biographical History of Navarro, Henderson, Anderson, Limestone, Freestone, and Leon Counties, Texas*, 626–7.

41. *Handbook of Texas Online*, "Munger, Robert Sylvester," http://www.tshaonline.org/handbook/online/articles/fmu04

42. "Miscellaneous. Manufacturing," *The Dallas Herald*, October 13, 1887.

43. Several cotton machine manufacturers were merged into the Continental Gin Company in October 1899, which became the largest manufacturer of cotton gins in the United States. "Continental Gin Company, Dallas Plant," National Register of Historic Places Inventory—Nomination Form, Austin, January 1983, 7.

44. Foster, *Forgotten Texas Census*, 17–18.

45. ibid., 18.

46. Roy Sylvan Dunn, "Drought in West Texas, 1890–1894," *West Texas Historical Association Yearbook* 37(1961): 121–36; W. C. Holden, "West Texas Droughts," *Southwestern Historical Quarterly* 32 (October 1928): 103–23; J. W. Williams, "A Statistical Study of the Drought of 1886," *West Texas Historical Association Yearbook* 21 (1945): 85–109.

47. "From Norse," February 8, 1884; "From Norse," February 19, 1884; "From Norse," June 24, 1884, *Uncle Theo's Scrapbook*.

48. "From Norse," July 22, 1884; "From Norse," September 30, 1884, *Uncle Theo's Scrapbook*.

49. Letter from Christine Dahl, Norse, Texas, February 4, 1885, to Erik Pedersen Furuset, Romedal, in *From America to Norway. Volume Two, 1871–1892*, ed. Orm Øverland, 349.

50. ibid.

51. Letter from Christine Dahl, Norse, Texas, April 30, 1885, to Erik Pedersen

Furuset, Romedal, in *From America to Norway. Volume Two, 1871–1892*, ed. Orm Øverland, 358.

52. "From Norse," July 18, 1885; "From Norse," August 12, 1885; "From Norse," September 9, 1885, *Uncle Theo's Scrapbook*.

53. "From Norse," August 25, 1885, *Uncle Theo's Scrapbook*.

54. "From Norse," November 14, 1885; "From Norse," December 5, 1885; "From Norse," December 19, 1885, *Uncle Theo's Scrapbook*.

55. "From Norse," June 6, 1886; "From Norse," June 29, 1886; "From Norse," September 14, 1886, *Uncle Theo's Scrapbook*.

56. "From Norse," February 3, 1887; "From Norse," February 6, 1887, *Uncle Theo's Scrapbook*.

57. "From Norse," March 20, 1887, *Uncle Theo's Scrapbook*.

58. "From Norse," May 16, 1887, *Uncle Theo's Scrapbook*.

59. "From Norse," May 3, 1887; "From Norse," September 11, 1887; "From Norse," October 2, 1887, *Uncle Theo's Scrapbook*.

60. Foster, *Forgotten Texas Census*, 17–18.

61. "From Norse," November 15, 1885, *Uncle Theo's Scrapbook*.

62. "Letters from Eli and Peder Arnesen," *Gammalt frå Stange og Romedal 1975*, 86–89. Peder Arnesen was born on April 31, 1831, at Stange, Hedmark, and died October 10, 1884 in Bosque County. He married Eli Nielsdatter (born July 21, 1837 at Stange, died December 21, 1901 in Bosque County), on January 16, 1861.

63. "From Norse," November 15, 1885, *Uncle Theo's Scrapbook*.

64. "From Norse," June 24, 1884, *Uncle Theo's Scrapbook*.

65. Foster, *Forgotten Texas Census*, 18.

66. Carlson, *Texas Woollybacks*, 23.

67. Olmsted, *A Journey through Texas*, 93.

68. Jordan, *German Seed in Texas Soil*, 148–50.

69. Carlson, *Texas Woollybacks*, 26, 28.

70. ibid., 101–20.

71. F972, *Bosque County: Land and People*, 596, 83, 589.

72. Landgraf, Adolf Family, F702, *Bosque County: Land and People*, 457.

73. *Thus Far God Has Brought Us. A History of the Jo Knust Family of Bosque County, Texas*, 9.

74. "The Hovie Family Tree, 1765–1995," copyright 1995 John & Dorothy Hovie. Bruce L. Wiland has put "The Family of Knud Olsen Rea and Astri Tidemandsdatter Hilmen" on the Internet, but it is almost verbatim the same document as "The Hovie Family Tree." The citations are from the original.

75. *A History of Hamilton County Texas*, 8.

76. *Handbook of Texas Online*, John Leffler, "Hamilton County," http://www.tshaonline.org/handbook/online/articles/hch03

77. Terry G. Jordan, "The German Settlement of Texas after 1865," *Southwestern Historical Quarterly* 73 (October 1969): 195, 198.
78. Jordan, "The German Settlement of Texas after 1865," 200.
79. *A History of Hamilton County, Texas*, 36.
80. Krell, *Devil's Rope*, 28.
81. McCallum and McCallum, *The Wire That Fenced the West*, 65–74.
82. *Handbook of Texas Online*, Wayne Gard, "Fence Cutting," http://www.tshaonline.org/handbook/online/articles/auf01
83. *A History of Hamilton County, Texas*, 34.
84. *Dallas Weekly Herald*, August 23, 1883.
85. Wayne Gard, "The Fence-Cutters," *Southwestern Historical Quarterly* 51, no. 1 (July 1947): 1–15; Roy D. Holt, "The Introduction of Barbed Wire into Texas and the Fence Cutting War," *West Texas Historical Association Yearbook* 6 (1930): 65–79.
86. "Christian Carlson family," F179, *Bosque County; Land and People*, 199.
87. "Dahl, John," F276; "Dahl, Hans B. and family," F273, *Bosque County, Land and People*, 246–48.
88. *Bosque County: Land and People*, pp 550–51.
89. "From Norse," May 6, 1888, "From Norse," June 22, 1888, *Uncle Theo's Scrapbook*.
90. *Handbook of Texas Online*, Hooper Shelton, "Fisher County," http://www.tshaonline.org/handbook/online/articles/hcf04
91. Worster, *Rivers of Empire*.
92. "Anderson T. W. and Tilda," F33, *Bosque County: Land and People*, 122.
93. "From Norse," April 5, 1886, *Uncle Theo's Scrapbook*.
94. Barr, *Reconstruction to Reform*, 94–5.
95. Barnes, *Farmers in Rebellion*:; McMath, Jr., *Populist Vanguard*; *Handbook of Texas Online*, Donna A. Barnes, "Farmers' Alliance," http://www.tshaonline.org/handbook/online/articles/aaf02

CHAPTER 13

1. *Bosque County: Land and People*, 614.
2. "Story of the Old Mill," compiled by Mary Ellen Swenson Ellingson. Clifton, 1984, 3–4.
3. "The E. Booth Family," *Profiles of Pioneer Families in the Valley Mills Area* (Meridian, Texas, n.d.), 94–6.
4. Spratt, *The Road to Spindletop*, 255–6.
5. ibid., 302.
6. Bronstad, "The C. O. Bronstad Family History," 7.

7. "From Norse," December 5, 1885, *Uncle Theo's Scrapbook.*
8. "From Norse," April 30, 1884, *Uncle Theo's Scrapbook.*
9. Spratt, *The Road to Spindletop,* 256.
10. Pool, *Bosque Territory,* 109.
11. Syversen and Johnson, *Norge i Texas,* 206, 232.
12. Syversen and Johnson, *Norge i Texas,* 266.
13. Kenneth A. Breisch and David Moore, "The Norwegian Settlement in Bosque County," National Register of Historic Places. Thematic Nomination (Austin, Texas Historical Commission, 1983), 113.
14. *Bosque Territory,* 119.
15. *Bosque County. Land and People,* 176.
16. "Gulbrand Olson Bronstad family," F134, *Bosque County: Land and People,* 176.
17. *Bosque County. Land and People,* 118.
18. Johnson, *I, Carl Johnson. Born of Immigrant Parents,* An autobiography as recorded on tape during the year 1979 at my home, 3416 Street, Fort Worth, Texas.
19. *Bosque County: Land and People,* 520.
20. Sveen, *Utvandrere 1871–1879,* 107.
21. Johnson, *I, Carl Johnson,* 5.
22. Syversen and Johnson, *Norge i Texas,* 372, 480.
23. Johnson, *I, Carl Johnson,* 7.
24. ibid.
25. The Lund family sold their farm to Matt Christianson in 1897, when they moved to the Meissner ranch west of Hamilton. Three years later they bought a ranch 12 miles west of Hamilton in Hamilton County; see *A History of Hamilton County, Texas,* 220–1.
26. Johnson, *I, Carl Johnson,* 13.
27. *A History of Hamilton County, Texas,* 133–4.
28. Johnson, *I, Carl Johnson,* 3.
29. ibid., 2.
30. Gjerde, *The Minds of the West,* 108
31. ibid., 109.
32. Pederson, *Between Memory and Reality,* 22
33. ibid., 35.
34. ibid.
35. Gaustad, *A Religious History of America,* 183–4.
36. Blegen, *Norwegian Migration to America: The American Transition,* 141.
37. ibid., 143.
38. ibid., 144.

39. ibid., 163–64.

40. Ylvisaker, "The Missouri Synod and the Norwegians," 265.

41. Robert C. Ostergren, "The Immigrant Church as a Symbol of Community and Place in the Upper Midwest," *Great Plains Quarterly* 10, no. 1 (1981), 225–38, 225.

42. "Journal for den norsk evangelisk lutherske Menighed i Bosque County, Texas," Secretary book, volume 1, June 14, 1869–December 29, 1886.

43. "1869–1969. Our Savior's Lutheran Church Centennial Celebration, June 28–29, 1969," 3.

44. ibid., 2.

45. The first confirmation class in 1869 consisted of Christoffer Jenson and wife Petrine Jenson, Otto Swenson and wife Helene Swenson, Nils Jacob Nilson, Bersvend Bersvendsen (Swenson), Peder Pierson, Martin Skjefstad, John Ringnes, Andreas Staach, Ole Nystøl, Jensine Maria Grimeland, Helene Knutsen, Ragnhild Olsen, Bergista Knutsen, Marthe Lindberg, Siri Jensen, Inger Christine Brunstad, Marthe Svendsen, Petrine Johnson, Gina Brunstad, Jørgine Knutsen, Ole Ringness, Karen Vold, Anna Maria Olsen, Betsy Randine Paulsen, Anne Olive Johnsen, Oli Johnson, Andrea Knutsen, Arthur Wolseley Durie, Petra Hanson, Siri Knutsen.

46. Hastvedt, "Recollections of a Norwegian Pioneer in Texas," 91.

47. Lich, *The German Texans*, 41.

48. Lich and Reeves, eds., *German Culture in Texas*; *Handbook of Texas Online*, Glen E. Lich, "Freethinkers," http://www.tshaonline.org/handbook/online/articles/pfflg

49. Carl Wittke, "The German Forty-Eighters in America: A Centennial Appraisal," *American Historical Review* 53, no. 4 (July 1948): 711–25.

50. Lich, *The German Texans*, 43.

51. Letter from Berger Tollevsen Rogstad, Norman Hill, Texas, June 1, 1867, to Peder Eriksen Furuset, Romedal, Stange, Hedmark, in *From America to Norway. Volume One, 1838–1870*, edited by Orm Øverland, 380.

52. "Old Rock Church on Hog Creek," C49, *Bosque County: Land and People*, 92.

53. Carter, *Masonry in Texas*, 119–54.

54. *Handbook of Texas Online*, William Preston Vaughn, "Freemasonry," http://www.tshaonline.org/handbook/online/articles/vnf01

55. Joseph W. Hale, "Masonry in the Early Days of Texas," *Southwestern Historical Quarterly* 49 (January 1946): 374–83.

56. "From Norse," December 19, 1886, *Uncle Theo's Scrapbook*.

57. Secretary book, volume 1, June 14, 1869–December 29, 1886, 11–12.

58. "1869–1969. Our Savior's Lutheran Church Centennial Celebration," 3.

59. Secretary book, volume 1, June 14, 1869–December 29, 1886, 13–17.
60. ibid., 20.
61. "1869–1969. Our Savior's Lutheran Church Centennial Celebration," 4, 6.
62. Secretary book, volume 1, June 14, 1869–December 29, 1886, 40–3.
63. ibid., 47–58, 54.
64. Rystad, "Memoirs of the Rev. John Knudson Rystad." Rystad wrote his memoirs in Norwegian.
65. ibid., 10.
66. "1869–1969. Our Savior's Lutheran Church Centennial Celebration," 6.
67. Rystad, "Memoirs," 13–14.
68. Colwick, "From Norse," September 30, 1884, *Uncle Theo's Scrapbook*.
69. Letter from Christine Dahl, Norse, April 30, 1885, to Erik Pedersen Furuset, Romedal, in *From America to Norway. Volume Two, 1871–1892*, ed. Orm Øverland, 358.
70. From Norse," May 2, 1885, *Uncle Theo's Scrapbook*.
71. Brown, "The Huse Book," n.d.
72. "The Solberg Family from Norway to Texas," several authors, Clifton, 1979, 15.
73. "*Journal for den norsk evangelisk lutherske Menighed i Bosque County, Texas*," Secretary book, volume 2, 1887—1910, July 30, 1873, 8.
74. "Rohne Family History: Norwegian Immigrants Started Farm Here with Yoke of Oxen."
75. Secretary book, volume 2, 1887–1910, July 30, 1873, 45.
76. Greer, ed., *Buck Barry Texas Ranger and Frontiersman*, 82.
77. Clausen, ed., *The Lady with the Pen*, 40–1.
78. Colwick, "From Norse," July 29, 1885, *Uncle Theo's Scrapbook*.
79. "From Norse," May 5, 1884; "From Norse," May 18, 1886, *Uncle Theo's Scrapbook*.
80. "From Norse," May 22, 1887, *Uncle Theo's Scrapbook*.
81. "Norse Celebration," July 10, 1886, *Uncle Theo's Scrapbook*.
82. "Norse Picnic," July 1888, *Uncle Theo's Scrapbook*.
83. "The Memories of Herbert Schulz."
84. "From Norse," March 31, 1885, *Uncle Theo's Scrapbook*.
85. "From Norse," July 18, 1885; "From Norse," September 9, 1885. "From Norse," November 14, 1885, "From Norse," December 5, 1885, *Uncle Theo's Scrapbook*.
86. "From Norse," September 11, 1887, *Uncle Theo's Scrapbook*.
87. "From Norse," July 22, 1884, *Uncle Theo's Scrapbook*.
88. "From Norse," February 19, 1884, *Uncle Theo's Scrapbook*.
89. "From Norse," March 26, 1884, *Uncle Theo's Scrapbook*.
90. "From Norse," April 12, 1886, *Uncle Theo's Scrapbook*.

91. "From Norse," May 16, 1887, *Uncle Theo's Scrapbook*.
92. Gjerde, *The Minds of the West*, 233.
93. Elif Albertson Moore, *The History of Clifton College* (MA thesis, University of Texas, Austin, June 1927), 22.
94. Secretary book, volume 2, 1887–1910, 56–57.
95. ibid., 63.
96. ibid., 78–79.
97. Moore, *The History of Clifton College*, 26.
98. ibid., 29, 38.
99. ibid., 38–39.
100. ibid., 46.
101. ibid., 55.
102. ibid., 68.
103. Edgar, *The Milkman and the Maid*, 35.
104. Bronstad, "The C. O. Bronstad Family History," 9–19.
105. *Bosque County: Land and People*, 549.
106. "From Norse," October 30, 1884, *Uncle Theo's Scrapbook*.
107. "From Norse," April 12, 1886, *Uncle Theo's Scrapbook*.
108. Bronstad, "The C. O. Bronstad Family History," 14.

CHAPTER 14

1. Tippens, *Turning Germans into Texans*, 47.
2. Lich, *The German Texans*, 164–5.
3. Blanton, *The Strange Career of Bilingual Education in Texas, 1836–1981*, 11–41.
4. Citation from Chrislock, *Ethnicity Challenged*, 39.
5. See the chapters "The Anti German Hysteria in Texas, 1917–1918" and "Attacking the German Language in Texas, 1917–1918" in Tippens, *Turning Germans into Texans*, 89–159.
6. Tippens, *Turning Germans into Texans*, 43.
7. Baker, *The First Polish Americans*, 21–44.
8. ibid., 125.
9. ibid., 126.
10. ibid., 125.
11. Gordon, *Assimilation in American Life*; Alba and Nee, *Remaking the American Mainstream*.
12. Baker, *The First Polish Americans*, 137.
13. Skrabanek, *We're Czech*, 16.
14. ibid.

15. ibid., 192–224.
16. Blanton, *The Strange Career of Bilingual Education*, 39–40.
17. Hudson and Maresh, *Czech Pioneers of the Southwest*; Machann, ed., *The Czechs in Texas*; Machann and Mendl, *Krásná Amerika*.
18. Bronstad, *The History of Education in Bosque County, Texas*.
19. ibid., 163–77.
20. *Thirteenth Census of the United States Taken in the Year 1910, Volume 3, Population 1910. Reports by States with Statistics for Counties, Cities and other Civil Divisions*, Washington DC: Government Printing Office, 1913, 779, 799, 806.
21. ibid., 804–52.

BIBLIOGRAPHY

◊ ◊ ◊

A History of Hamilton County Texas. Hamilton, TX: Hamilton County Historical Commission, 1979.

A Memorial and Biographical History of McLennan, Bell and Coryell Counties, Texas (Chicago: Lewis Publishing Company, 1893).

A Memorial and Biographical History of Navarro, Henderson, Anderson, Limestone, Freestone, and Leon Counties, Texas (Chicago: The Lewis Publishing Co., 1893).

Adolph, Anthony. *Tracing Your Family History* (London: Colling, 2005).

Alba, Richard, and Victor Nee. *Remaking the American Mainstream: Assimilation and Contemporary Immigration* (Cambridge, MA: Harvard University Press, 2003).

Alexander, T. C. "Bosque County." *Texas Almanac 1867*.

Alston, Lee J., and Joseph H. Ferrie. "Time on the Ladder: Career Mobility in Agriculture." *Journal of Economic History* 65, no. 4 (December 2005): 1058–81.

Alston, Lee J., and Kyle D. Kauffman. "Agricultural Chutes and Ladders: New Estimates of Sharecroppers and "True Tenants" in the South, 1900–1920." *Journal of Economic History* 57, no. 2 (June 1997): 464–75.

Anderson, Hugh George. *Lutheranism in the Southeastern States, 1860–1886: A Social History* (The Hague: Mouton, 1969.)

Anderson, Rasmus B. *The First Chapter of Norwegian Immigration (1821–1840). Its Causes and Results: With an Introduction on the Services Rendered by the Scandinavians to the World and to America* (Madison, WI: privately printed, 1895).

"Angaaende Udvandring til Fremmede Verdensdele," *Innstilling fra Finansdepartementet til Stortinget*. No. 6, Christiania, November 8, 1843.

Atack, Jeremy. "Tenants and Yeomen in the Nineteenth Century." *Agricultural History* 62, no. 3 (1988): 6–32.

———. "The Agricultural Ladder Revisited: A New Look at an Old Question with Some Data for 1860." *Agricultural History* 63, no. 1 (1989): 1–25.

Baines, Dudley. *Emigration from Europe 1815–1930* (Cambridge: Cambridge University Press, 1995).

Baker, Robin E., and Dale Baum. "The Texas Voter and the Crisis of the Union, 1859–1861." *Journal of Southern History* 53 (August 1987): 395–420.

Baker, T. Lindsay. *The First Polish Americans. Silesian Settlements in Texas* (College Station: Texas A&M University Press, 1979).

Baldwin, Elmer. *History of La Salle County. Its topography, geology, botany, natural history, history of the Mound builders, Indian tribes, French explorations, and a sketch of the pioneer settlers of each town to 1840* (Chicago: Rand McNally, 1877).

Barnes, Donna A. *Farmers in Rebellion: The Rise and Fall of the Southern Farmers' Alliance and People's Party in Texas* (Austin: University of Texas Press, 1984).

Barr, Alwyn. "An Essay on Texas Civil War Historiography." In *Texas. The Dark Corner of the Confederacy. Contemporary Accounts of the Lone Star State in the Civil War*, third edition, ed. B. P. Gallaway (Lincoln, NE: University of Nebraska Press 1994), 257–76.

———. *Reconstruction to Reform. Texas Politics, 1876–1906* (Dallas: Southern Methodist University Press, 1971).

Barry, James Buckner. *Buck Barry, Texas Ranger and Frontiersman*, ed. James K. Greer (First published 1932; new edition, Waco 1978; reprinted Lincoln: University of Nebraska Press, 1984).

Batte, Lelia M. *History of Milam County, Texas* (San Antonio: Naylor Company, 1956).

Bell, Walter F. "Civil War Texas: A Review of the Historical Literature." In *Lone Star Blue & Gray. Essays on Texas and the Civil War,*, eds. Ralph A. Wooster and Robert Wooster (Denton, TX: Texas State Historical Association, second edition 2015), 14–44.

Beretning om Kongeriket Norges økonomiske Tilstand i Aarene 1846–1850 med tilhørende Tabeller (Christiania: Norwegian Ministry of the Interior, 1853).

Beretning om Kongeriket Norges økonomiske Tilstand i Aarene 1851–1855 med tilhørende Tabeller (Christiania: Norwegian Ministry of the Interior, 1858).

Bernstein, Peter L. *Wedding of the Waters. The Erie Canal and the Making of a Great Nation* (New York: W. W. Norton & Company, 2005).

Biesele, Rudolph Leopold. *The History of the German Settlements in Texas, 1831–1861* (Austin, 1930. Reprinted German-Texan Heritage Society, 1987).

Biggers, Don Hampton. *German Pioneers in Texas* (Fredericksburg, TX: Fredericksburg Publishing Co., 1925).

Billington, Ray A. "The Frontier in Illinois History." *Journal of the Illinois Historical Society* 43, no. 1 (Spring 1950): 28–45.

Bingham, William H. *General History of Shelby County Missouri* (Chicago: H. Taylor & Co., 1911).

Black, John D., and R. H. Allen. "The Growth of Farm Tenancy in the United States." *Quarterly Journal of Economics* 51, no. 3 (May 1937): 393–425.

Blanton, Carlos Kevin. *The Strange Career of Bilingual Education in Texas, 1836–1981* (College Station: Texas A&M University Press, 2004).

Blegen, Theodore C. "Cleng Peerson and Norwegian Immigration." *Mississippi Valley Historical Review* 7, no. 4 (1921): 301–31.
———. *John Quincy Adams and the Sloop Restauration*. 1940.
———. *Norwegian Migration to America, 1825–1860* (Northfield, MN: Norwegian-American Historical Association, 1931).
———. *Norwegian Migration to America: The American Transition* (Northfield, MN: Norwegian-American Historical Association, 1940).
Blevins, Bill B. *The Tutt & Everett War. A Family Feud That Became a War in Marion County Arkansas, 1844–1850* (Mountain Home, Arkansas: privately printed, 2002).
Bonds, Marshall V. "Silas McCabe: First Settler at Coon Creek Community." *The Clifton Record – Centennial Edition* 1 (1954): 6.
Bogue, Allan G. "Farming in the Prairie Peninsula." *Journal of Economic History* 23, no. 1 (March 1963): 3–29.
Bonner, Robert E. *The Soldier's Pen. Firsthand Impressions of the Civil War* (New York: Hill & Wang, 2006).
Bosque County History Book Committee, ed. *Bosque County: Land and People (A History of Bosque County)* (Dallas: Curtis Media Corporation, 1985).
Boyd, Gregory A. *Texas Land Survey Maps for Bosque County* (Norman, OK: 2010).
Braastad, Per. *Slekter på garder i Vardal og Redalen* (Gjøvik: Gjøvik og Toten Slektshistorielag, 2008).
Bracken, Sharon, ed. *Historic McLennan County: An Illustrated History* (San Antonio: McLennan County Historical Commission, 2010).
Brancato, Jennifer Michelle. "A Glimpse of Life in Nineteenth-Century East Texas: Material Culture and the Durst-Taylor House" (MA thesis, Stephen Austin University, 2008).
Brandal, Trygve. *Hjelmeland. Bygdesoge 1800–1990* (Stavanger: Hjelmeland Kommune, 1994).
Bronstad, Alwin Lawrence. *The History of Education in Bosque County, Texas* (Clifton, TX: Bosque Memorial Museum Press, 2004).
Breisch, Kenneth A., and David Moore. "The Norwegian Rock Houses of Bosque County, TX: Some Observations on a Nineteenth-Century Vernacular Building Type." *Perspectives in Vernacular Architecture* 2 (1986): 64–71.
———. "The Norwegian Settlement in Bosque County." National Register of Historic Places. Thematic Nomination (Austin, TX: Texas Historical Commission, 1983).
Brunstad, Johan Olsen. "Reisebrev fra Texas 1852." *Gammalt frå Stange og Romedal 1975*. Hamar: Stange Historielag (1975): 12–25.
Bueltmann, Tanja, Andrew Hinson, and Graeme Morton, eds. *The Scottish Diaspora* (Edinburgh: Edinburgh University Press, 2013).

Burnet County History. A Pioneer History, 1847–1979, Vol. 1 (Burnet, TX: Burnet County Historical Commission, 1979).

Cadbury, Henry J. "Four Immigrant Shiploads of 1836 and 1837." *Norwegian-American Studies* 2 (1927): 20–52.

———. "The Norwegian Quakers of 1825." *Harvard Theological Review* 18, no. 4 (October 1925): 293–319.

Calder, Jenni. *Frontier Scots. The Scots Who Won the West* (Edinburgh: Luath Press, 2014).

———. *Scots in Canada* (Edinburgh: Luath Press, New Edition, 2013).

Calloway, Colin G. *White People, Indians, and Highlanders. Tribal Peoples and Colonial Encounters in Scotland and America* (Oxford: Oxford University Press, 2008).

Calvert, Robert A. "Nineteenth-Century Farmers, Cotton, and Prosperity." *Southwestern Historical Quarterly* 73 (April 1970): 509–21.

Calvert, Robert A., Arnoldo De León, and Gregg Cantrell. *The History of Texas*, 3rd ed (Wheeling, IL: Harlan Davidson, 2002).

Campbell, Randolph B. *An Empire for Slavery. The Peculiar Institution in Texas, 1821–1865* (Baton Rouge: Louisiana State University Press, 1989).

———. *Gone to Texas. A History of the Lone Star State* (New York: Oxford University Press, 2003).

———. *A Southern Community in Crisis. Harrison County, Texas, 1850–1880* (Originally published in 1983, reprinted by Texas State Historical Association, Austin, 2016).

Canuteson, Richard. "A Little More Light on the Kendall Colony." *Norwegian-American Studies* 18 (1954): 82–101.

Carlson, Paul H. *Texas Woollybacks. The Range Sheep & Goat Industry* (College Station: Texas A & M University Press, 1982).

Carter, James David. *Masonry in Texas: Background, History and Influence to 1846* (Waco, 1955).

Chrislock, Carl H. *Ethnicity Challenged. The Upper Midwest Norwegian-American Experience in World War I* (Northfield, MN: Norwegian-American Historical Association, 1981).

Christian, Peter. *The Genealogist's Internet* (London: Bloomsbury, 2012).

Chul Hong, Seok. "The Burden of Early Exposure to Malaria in the United States, 1850–1860: Malnutrition and Immune Disorders." *Journal of Economic History* 67, no. 4 (December 2007): 1001–35.

Clausen, C.A., ed. *The Lady with the Pen. Elise Wærenskjold in Texas* (Northfield, MN: Norwegian-American Historical Association, 1961).

Clausen, C.A., and Derwood Johnson. "Norwegian Soldiers in the Confederate Forces." *Norwegian-American Studies* 25 (1972): 104–41.

Clinton, Catherine ed. *Southern Families at War. Loyalty and Conflict in the Civil War South* (Oxford: Oxford University Press, 2000).

Cohn, Raymond L. "Mortality on Immigrant Voyages to New York, 1836–1853." *Journal of Economic History* 44, no. 2 (June 1984): 289–300.

———. "The Transition from Sail to Steam in Immigration to the United States." *Journal of Economic History* 65, no. 2 (June 2005): 469–95.

Conger, Roger Norman. *A Pictorial History of Waco* (Waco: Texian Press, 1964).

Connor, Seymour V. *Kentucky Colonization in Texas: A History of the Peters Colony* (Baltimore, Maryland: Genealogical Publishing Co., 1989).

———. *The Peters Colony of Texas: A History and Biographical Sketches of the Early Settlers* (Austin: Texas State Historical Society, 1959).

———. "A Statistical Review of the Settlement of the Peters Colony, 1841–1848." *Southwestern Historical Quarterly* 57 (July 1953-April 1954): 38–64.

Conover, George S., ed. *History of Ontario County, New York* (Syracuse, NY: Lewis Cass Aldridge, 1893).

Coolidge, Orville W. *A Twentieth Century History of Berrien County Michigan* (Chicago, IL: Lewis Publishing Company, 1906).

Cureton, H. J. and C. M. *Sketch of the Early History of Bosque County* (Meridian, TX, 1904).

Davis, James E. *Frontier Illinois* (Bloomington, IN: University of Indiana Press, 1998).

Dearinger, Ryan. *The Filth of Progress. Immigrants, Americans, and the Building of Canals and Railroads in the West* (Oakland, CA: University of California Press, 2016).

Devine, T. M. *Scotland's Empire. The Origins of the Global Diaspora* (London: Penguin Books, 2003).

———. *To the Ends of the Earth. Scotland's Global Diaspora* (London: Allen Lane, 2011).

Dickson, James G., Jr. "A. A. Nelson, Sailor, Surveyor, and Citizen: A Personal Profile." *East Texas Historical Journal* 3, Issue 2 (October 1965): 119–29.

Diffenderfer, Frank Ried. *The German Immigration into Pennsylvania Through the Port of Philadelphia, 1700–1775* (Lancaster, Pennsylvania: privately printed, 1900).

Douglas, C. L. *Famous Texas Feuds (*Dallas: Turner, 1936; Reprint Abilene: State House Press, 2007).

Dunn, Roy Sylvan, "Drought in West Texas, 1890–1894," *West Texas Historical Association Yearbook* 37(1961): 121–36.

Düring, Marten, and Martin Stark. "Historical Network Analysis." In *Encyclopedia of Social Networks*, ed. George A. Barnett (London: Sage Publishing, 2011).

Dyrvik, Ståle. "*Sosial differensiering*" and "*Sosiale endringar på 1800-talet.*" In *Fra*

Vistehola til Ekofisk. Rogaland gjennom tidene, eds. Edgar Hovland and Hans Eyvind Næss (Oslo: Universitetsforlaget, 1987), Vol. 1, 236–45, 399–417.

———. "Husmannsvesenet i Noreg." *Jord og Gjerning. Årbok for Norsk landbruksmuseum* (1990): 51–61.

Easum, Alice. "Rohne Family History: Norwegian Immigrant Started Farm Here with Yoke of Oxen." *Cranfills Gap Index* 47, no. 44 (Nov. 27, 1964).

Eide Johnsen, Berit. *"En stat 'hvortil Udvandring fra Norge kunde ansees fordelaktig'. Utvandringen fra Agder til Texas på 1840-tallet."* In *På vandring og på flukt. Migrasjon i historisk perspektiv*, ed. Berit Eide Johnsen (Oslo: Cappelen Damm Akademisk, 2017).

Elmen, Paul. *Wheat Flour Messiah. Eric Jansson of Bishop Hill* (Carbondale, IL: Southern Illinois University Press, 1976).

Engle, Stephen Douglas. *Struggle for the Heartland: The Campaigns from Fort Henry to Corinth* (Lincoln: University of Nebraska Press, 2001).

Enstam, Elizabeth York. *Women and the Creation of Urban Life. Dallas, Texas, 1843–1920* (College Station: Texas A&M University Press, 1998).

Erath, Lucy A. "Memoirs of major George Bernard Erath." *Southern Historical Quarterly* 26, no. 3 (January 1923): 207–33.

Erickson, Bonnie H. "Social Networks and History: A Review Essay." *Historical Methods* 30, no. 3 (1997): 149–57.

Ericson, Carolyn Reeves, transc. *1847 Census Nacogdoches County* (Nabu Press, 2010).

———. *Nacogdoches, Gateway to Texas: A Biographical Directory* (Fort Worth: Arrow-Curtis Printing, 1974, 1987).

———. *Nacogdoches Headrights: A Record of the Disposition of Land in East Texas and in Other Parts of that State, 1838–1848* (New Orleans: Polyanthos, 1977).

Evans, Nicholas J. "Indirect Passage from Europe. Transmigration via the UK, 1836–1914." *Journal of Maritime Research* 3, no. 1 (2001): 70–84.

Evensen, Erik Aalvik, "From Canaan to the Promised Land. Pioneer Migration from Hommedal Parish (Landvik and Eide Sub-Parishes), Southern Norway, to St. Joseph, Missouri and East Norway, Kansas" (PhD dissertation, University of Oslo, 2008).

Ewing, Dorothy. "Caroline Sedberry, Politician's Wife." In *Women in Civil War Texas. Diversity and Dissidence in the Trans-Mississippi*, eds. Debora M. Liles and Angela Boswell (Denton, TX: University of North Texas Press, 2016).

Ezdorf, R. H. von. "Malaria in the United States: Its Prevalence and Geographic Distribution." *Public Health Reports* 30, no. 22 (May 28, 1915): 1603–24.

Faulk, J. J. *History of Henderson County* (Athens, TX: Athens Review Printing, 1929).

Faust, Drew Gilpin. *Mothers of Invention. Women of the Slaveholding South in the*

American Civil War (Chapel Hill, NC: University of North Carolina Press, 1996).

Ferleger, Louis A. "Sharecropping Contracts in the Late-Nineteenth-Century South." *Agricultural History* 67, no. 3 (Summer 1993): 31–46.

Ferleger, Louis A., and John D. Metz. *Cultivating Success in the South. Farm Households in the Postbellum Era* (New York: Cambridge University Press, 2014).

Fischer, David Hackett. *Albion's Seed. Four British Folkways in America* (New York: Oxford University Press, 1989).

———. *Washington's Crossing* (New York: Oxford University Press, 2004).

Flippin, W. B. "The Tutt and Everett War in Marion County." *Arkansas Historical Quarterly* 17 (Summer 1958): 155–63.

Flom, George T. *A History of Norwegian Immigration the United States* (Iowa City, IA: privately published, 1909).

Foldøy, Ola. *Jelsa 1. Gards- og Ættesoga* (Sand: Suldal Kommune, 1967).

Folts, James D. "The 'Alien Proprietorship': The Pulteney Estate during the Nineteenth Century." *Crooked Lake Review*, Fall 2003.

Foner, Eric. *A Short History of Reconstruction, 1863–1877* (New York: Harper and Row, 1990).

Foreman, Grant. "River Navigation in the Early Southwest." *Mississippi Valley Historical Review* 15, no. 1 (July 1928): 34–55.

Fortenbaugh, Robert. "American Lutheran Synods and Slavery, 1830–60." *Journal of Religion* 13, no. 1 (January 1933): 72–92.

Foster, L. L. *Forgotten Texas Census: First Annual Report of the Agricultural Bureau of the Department of Agriculture, Insurance, Statistics, and History, 1887–88*. Introduction by Barbara J. Rozek (Austin, TX: State Printing Office, 1889; reprint with introduction, Austin, TX: Texas State Historical Association, 2001).

Fry, Michael. *How the Scots Made America* (New York: St. Martin's Press, 2003).

Fuller, George N. "Settlement of Michigan Territory." *Mississippi Valley Historical Review* 2, no. 1 (July 1915): 25–55.

Gaard, Harald. "No. 318, Ole Olsson Østrem." *Gaardslekten: med sidegreinene Utsiraslekten og Austreimslekten* (Haugesund, privately published, 1951).

Galenson, David W. "The Market Evaluation of Human Capital: The Case of Indentured Servitude." *Journal of Political Economy* 89, no. 3 (June 1981): 446–67.

———. "The Rise and Fall of Indentured Servitude in the Americas: An Economic Analysis." *Journal of Economic History* 44, no. 1 (March 1984): 1–26.

———. "White Servitude and the Growth of Black Slavery in Colonial America." *Journal of Economic History* 41, no 1 (March 1981): 39–47.

Gard, Wayne. "The Fence-Cutters." *Southwestern Historical Quarterly* 51, no. 1 (July 1947): 1–15.

Gates, Paul Wallace. "Southern Investments in Northern Lands Before the Civil War." *Journal of Southern History* 5, no 2 (May 1939): 155–85.

Gates, Paul W. "Tenants of the Log Cabin." *Mississippi Valley Historical Review* 49, no. 1 (June 1962): 3–31.

Gaustad, Edwin Scott. *A Religious History of America* (San Francisco: Harper & Row, 1990).

Geertz, Clifford. "Thick Description: Toward an Interpretive Theory of Culture." In *The Interpretation of Cultures: Selected Essays* (New York: Basic Books, 1973).

Geiser, S. W. "A Note on 'Vinzent's Texanische Pflanzen', 1847." *Field and Laboratory* 25 (1957): 45–53.

Gibson, Rob. *Highland Cowboys. From the Hills of Scotland to the American Wild West* (Edinburgh: Luath Press, 2003).

Gjerde, Jon. *The Minds of the West. Ethnocultural Evolution in the Rural Middle West, 1830–1917* (Chapel Hill: University of North Carolina Press, 1997).

Gordon, Milton M. *Assimilation in American Life: The Role of Race, Religion, and National Origins* (New York: Oxford University Press, 1964).

Gray, Susan E. *The Yankee West. Community Life on the Michigan Frontier* (Chapel Hill, NC: University of North Carolina Press, 1996).

Grider, Sylvia Ann. *The Wendish Texans* (San Antonio: Institute of Texan Cultures, 1982).

Grigg, D. B. "E. G. Ravenstein and the 'Laws of Migration.'" *Journal of Historical Geography* 3, no. 1 (1974): 41–54.

Grubb, Farley. "The Disappearance of Organized Markets for European Immigrant Servants in the United States: Five Popular Explanations Reexamined." *Social Science History* 18, no. 1 (Spring 1994): 1–30.

———. "The End of European Immigrant Servitude in the United States: An Economic Analysis of Market Collapse, 1772–1835." *Journal of Economic History* 54, no. 4 (December 1994): 794–824.

———. "The Incidence of Servitude in Trans-Atlantic Migration, 1771–1804." *Explorations in Economic History* 22, no. 3 (July 1985): 316–39.

Gudmestad, Robert. *Steamboats and the Rise of the Cotton Kingdom* (Baton Rouge, LA: Louisiana State University, 2011).

Hale, Douglas. *The Third Texas Cavalry in the Civil War* (Norman, OK: University of Oklahoma Press, 1993).

Hale, Joseph W. "Masonry in the Early Days of Texas." *Southwestern Historical Quarterly* 49 (January 1946): 374–83.

Hall, Margaret Elisabeth. *A History of Van Zandt County* (Austin: Jenkins Publishing Co., 1976).

Hamilton, James J. "The Battle of Fort Donelson." *Journal of Southern History* 35, no. 1 (1969): 99–100.

Handbook of Texas Online, Linda Sybert Hudson, "Amory, Nathaniel C.," accessed June 7, 2020, http://www.tshaonline.org/handbook/online/articles/fam04. Uploaded on June 9, 2010. Published by the Texas State Historical Association.

Handbook of Texas Online, Randolph B. Campbell, "Antebellum Texas," accessed June 7, 2020, http://www.tshaonline.org/handbook/online/articles/npa01. Uploaded on June 9, 2010. Modified on July 9, 2019. Published by the Texas State Historical Association.

Handbook of Texas Online, "Bosque River," accessed June 7, 2020, http://www.tshaonline.org/handbook/online/articles/rnb05. Uploaded on June 12, 2010. Published by the Texas State Historical Association.

Handbook of Texas Online, Kenneth E. Hendrickson, Jr., "Brazos River," accessed June 7, 2020, http://www.tshaonline.org/handbook/online/articles/rnb07. Uploaded on June 12, 2010. Modified on May 1, 2019. Published by the Texas State Historical Association.

Handbook of Texas Online, Linda Sybert Hudson, "Brown, John [Red]," accessed June 7, 2020, http://www.tshaonline.org/handbook/online/articles/fbr91. Uploaded on June 12, 2010. Published by the Texas State Historical Association.

Handbook of Texas Online, Cecil Harper, Jr., "Bryan, John Neely," accessed June 7, 2020, http://www.tshaonline.org/handbook/online/articles/fbran. Uploaded on June 12, 2010. Published by the Texas State Historical Association.

Handbook of Texas Online, Jeanne F. Lively, "Camp Salmon," accessed June 7, 2020, http://www.tshaonline.org/handbook/online/articles/qcc36. Uploaded on June 12, 2010. Published by the Texas State Historical Association.

Handbook of Texas Online, John W. Payne, Jr., "Coke, Richard," accessed June 7, 2020, http://www.tshaonline.org/handbook/online/articles/fco15. Uploaded on June 12, 2010. Modified on May 28, 2020. Published by the Texas State Historical Association.

Handbook of Texas Online, Lisa C. Maxwell, "Dallas County," accessed June 7, 2020, http://www.tshaonline.org/handbook/online/articles/hcd02. Uploaded on June 12, 2010. Modified on October 9, 2019. Published by the Texas State Historical Association.

Handbook of Texas Online, Brett J. Derbes, "Eleventh Texas Infantry," accessed June 7, 2020, http://www.tshaonline.org/handbook/online/articles/qke06. Uploaded on March 8, 2011. Modified on April 11, 2011. Published by the Texas State Historical Association.

Handbook of Texas Online, Thomas W. Cutrer, "Erath, George Bernard," accessed June 7, 2020, http://www.tshaonline.org/handbook/online/articles/fer01. Uploaded on June 12, 2010. Modified on September 18, 2019. Published by the Texas State Historical Association.

Handbook of Texas Online, Cecil Harper, Jr., and E. Dale Odom, "Farm Tenancy," accessed June 7, 2020, http://www.tshaonline.org/handbook/online

/articles/aefmu. Uploaded on June 12, 2010. Modified on September 4, 2013. Published by the Texas State Historical Association.

Handbook of Texas Online, Donna A. Barnes, "Farmers' Alliance," accessed June 7, 2020, http://www.tshaonline.org/handbook/online/articles/aaf02. Uploaded on June 12, 2010. Modified on September 4, 2013. Published by the Texas State Historical Association.

Handbook of Texas Online, Wayne Gard, "Fence Cutting," accessed June 7, 2020, http://www.tshaonline.org/handbook/online/articles/auf01. Uploaded on June 12, 2010. Modified on September 27, 2019. Published by the Texas State Historical Association.

Handbook of Texas Online, C. L. Sonnichsen, "Feuds," accessed June 7, 2020, http://www.tshaonline.org/handbook/online/articles/jgf01. Uploaded on June 12, 2010. Modified on October 2, 2019. Published by the Texas State Historical Association.

Handbook of Texas Online, Hooper Shelton, "Fisher County," accessed June 7, 2020, http://www.tshaonline.org/handbook/online/articles/hcf04. Uploaded on June 12, 2010. Modified on October 2, 2019. Published by the Texas State Historical Association.

Handbook of Texas Online, Zelma Scott, "Fort Gates," accessed June 7, 2020, http://www.tshaonline.org/handbook/online/articles/qbf20. Uploaded on June 12, 2010. Modified on October 9, 2019. Published by the Texas State Historical Association.

Handbook of Texas Online, Sandra L. Myres, "Fort Graham," accessed June 7, 2020, http://www.tshaonline.org/handbook/online/articles/qbf21. Uploaded on June 12, 2010. Modified on October 9, 2019. Published by the Texas State Historical Association.

Handbook of Texas Online, William Preston Vaughn, "Freemasonry," accessed June 7, 2020, http://www.tshaonline.org/handbook/online/articles/vnf01. Uploaded on June 12, 2010. Modified on February 16, 2017. Published by the Texas State Historical Association.

Handbook of Texas Online, Janet Schmelzer, "Fort Worth, TX," accessed June 7, 2020, http://www.tshaonline.org/handbook/online/articles/hdf01. Uploaded on June 12, 2010. Modified on June 1, 2017. Published by the Texas State Historical Association.

Handbook of Texas Online, Glen E. Lich, "Freethinkers," accessed June 7, 2020, http://www.tshaonline.org/handbook/online/articles/pflg. Uploaded on June 12, 2010. Modified on April 13, 2016. Published by the Texas State Historical Association.

Handbook of Texas Online, Rudolph L. Biesele, "German Attitude Toward the Civil War," accessed June 7, 2020, http://www.tshaonline.org/handbook/

online/articles/png01. Uploaded on June 15, 2010. Modified on March 8, 2011. Published by the Texas State Historical Association.

Handbook of Texas Online, "Gurley, Edward Jeremiah," accessed June 7, 2020, http://www.tshaonline.org/handbook/online/articles/fgu08. Uploaded on June 15, 2010. Published by the Texas State Historical Association.

Handbook of Texas Online, John Leffler, "Hamilton County," accessed June 7, 2020, http://www.tshaonline.org/handbook/online/articles/hch03. Uploaded on June 15, 2010. Modified on February 5, 2016. Published by the Texas State Historical Association.

Handbook of Texas Online, Linda Sybert Hudson, "Henderson County," accessed June 7, 2020, http://www.tshaonline.org/handbook/online/articles/hch13. Uploaded on June 15, 2010. Modified on October 30, 2019. Published by the Texas State Historical Association.

Handbook of Texas Online, Thomas W. Cutrer, "Magruder, John Bankhead," accessed June 7, 2020, http://www.tshaonline.org/handbook/online/articles/fma15. Uploaded on June 15, 2010. Modified on January 18, 2013. Published by the Texas State Historical Association.

Handbook of Texas Online, Evelyn Clark Longwell, "Mclennan, Neil," accessed June 7, 2020, http://www.tshaonline.org/handbook/online/articles/fmc89. Uploaded on June 15, 2010. Published by the Texas State Historical Association.

Handbook of Texas Online, Evelyn Clark Longwell, "Mclennan's Bluff," accessed June 7, 2020, http://www.tshaonline.org/handbook/online/articles/rjm51. Uploaded on June 15, 2010. Published by the Texas State Historical Association.

Handbook of Texas Online, "Munger, Robert Sylvester," accessed June 7, 2020, http://www.tshaonline.org/handbook/online/articles/fmu04. Uploaded on June 15, 2010. Modified on October 28, 2013. Published by the Texas State Historical Association.

Handbook of Texas Online, Margaret E. Lengert, "Nashville-on-the-Brazos, Texas," accessed June 7, 2020, http://www.tshaonline.org/handbook/online/articles/hvn04. Uploaded on June 15, 2010. Modified on February 21, 2018. Published by the Texas State Historical Association.

Handbook of Texas Online, "Nueces, Battle of the," accessed June 7, 2020, http://www.tshaonline.org/handbook/online/articles/qfn01. Uploaded on August 31, 2010. Modified on March 13, 2019. Published by the Texas State Historical Association.

Handbook of Texas Online, Walter L. Buenger, "Secession," accessed June 7, 2020, http://www.tshaonline.org/handbook/online/articles/mgs02. Uploaded on June 15, 2010. Modified on March 8, 2011. Published by the Texas State Historical Association.

Handbook of Texas Online, "Sedberry, William Rush," accessed June 7, 2020, http://www.tshaonline.org/handbook/online/articles/fse35. Uploaded on September 4, 2014. Published by the Texas State Historical Association.

Handbook of Texas Online, Roger N. Conger, "Speight, Joseph Warren," accessed June 10, 2020, http://www.tshaonline.org/handbook/online/articles/fsp07. Uploaded on June 15, 2010. Published by the Texas State Historical Association.

Handbook of Texas Online, Linda Sybert Hudson, "Starr, James Harper," accessed June 7, 2020, http://www.tshaonline.org/handbook/online/articles/fst22. Uploaded on June 15, 2010. Modified on July 25, 2016. Published by the Texas State Historical Association.

Handbook of Texas Online, Archie P. McDonald, "Sterne, Nicholas Adolphus," accessed June 7, 2020, http://www.tshaonline.org/handbook/online/articles/fst45. Uploaded on June 15, 2010. Modified on April 22, 2019. Published by the Texas State Historical Association.

Handbook of Texas Online, W. Kellon Hightower, "Tarrant County," accessed June 7, 2020, http://www.tshaonline.org/handbook/online/articles/hct01. Uploaded on June 15, 2010. Modified on May 6, 2019. Published by the Texas State Historical Association.

Handbook of Texas Online, Charles D. Grear, "Thirtieth Texas Cavalry," accessed June 7, 2020, http://www.tshaonline.org/handbook/online/articles/qkt08. Uploaded on March 31, 2011. Modified on April 11, 2011. Published by the Texas State Historical Association.

Handbook of Texas Online, Gerald F. Kozlowski, "Van Zandt County," accessed June 7, 2020, http://www.tshaonline.org/handbook/online/articles/hcv02. Uploaded on June 15, 2010. Modified on February 8, 2019. Published by the Texas State Historical Association.

Hansen, Mary Eschelbach. "Land Ownership, Farm Size, and Tenancy after the Civil War." *Journal of Economic History* 58, no. 3 (Sep. 1998): 822–9.

Haraldsø, Brynjar. *Slaveridebattene i Den norske synode. En undersøkelse av slaveridebatten i Den norske synode i USA i 1860-årene med særlig vekt på debattens kirkelig-teologiske aspekter* (Oslo: Solum Forlag, 1988).

Harper, Cecil Jr. *Farming Someone Else's Land: Farm Tenancy in the Texas Brazos River Valley, 1850–1880* (Denton, TX, December 1988).

Harper, Marjory. *Adventurers & Exiles. The Great Scottish Exodus* (London: Profile Books, 2004).

———. *Emigration from Scotland between the wars: opportunity or exile?* (Manchester: Manchester University Press, 1998).

———. "Rhetoric and Reality: British Migration to Canada, 1867–1967." In *Canada and the British Empire*, ed. Philip Buckner (Oxford: Oxford University Press, 2008).

———. *Scotland No More? The Scots Who Left Scotland in the Twentieth Century* (Edinburgh: Luath Press, 2012).

Harper, Marjory, and Stephen Constantine, eds. *Migration and Empire* (Oxford: Oxford University Press, 2010).

Hastvedt, Knud Jørgensen. "Recollections of a Norwegian Pioneer in Texas." Translated and edited by C. A. Clausen. *Norwegian-American Studies* 12 (1941): 91–104.

Heathcote, Charles William. *The Lutheran Church and the Civil War* (Burlington, IA: Lutheran Literary Board, 1919).

Helle, Knut, Ståle Dyrvik, Edgar Hovland, and Tore Grønlie. *Grunnbok i norsk historie. Fra vikingtid til våre dager* (Oslo: Universitetsforlaget, 2013).

Hämäläinen, Pekka. "Into the Mainstream. The Emergence of a New Texas Indian History." In *Beyond Texas Through Time. Breaking Away from Past Interpretations*, eds. Walter L. Buenger and Arnoldo de León (College Station: Texas A&M University Press, 2011).

Hendrickson, Kenneth E., Jr. *The Waters of the Brazos: A History of the Brazos River Authority, 1929–1979* (Waco, TX: The Texian Press, 1981).

Hibbard, B. H. "Farm Tenancy in 1920." *Journal of Farm Economics* 3, no. 4 (Oct. 1921): 168–75.

Hines, Clair. "A Personal Civil War: The Murder of Wilhelm Warenskjold." *Norwegian-American Studies* 35 (2000): 37–89.

The History of Van Zandt County, Texas (Willis Point: The Van Zandt County History Book Committee, 1984).

Hoig, Stan. *Jesse Chisholm. Ambassador of the Plains* (Norman, OK: University of Oklahoma Press, 1991).

Holden, W. C. "West Texas Droughts." *Southwestern Historical Quarterly* 32 (October 1928): 103–23.

Holt, Roy D. "The Introduction of Barbed Wire into Texas and the Fence Cutting War." *West Texas Historical Association Yearbook* 6 (1930): 65–79.

Houston, Cecil J., and William J. Smyth. *Irish Emigration and Canadian Settlement: Patterns, Links, and Letters* (Toronto: University of Toronto Press, 1990).

Hovdhaugen, Einar. *Frå Venabygd til Texas. Pioneren Jehans Nordbu* (Oslo: Det Norske Samlaget, 1975).

Howell, Kenneth. "The Economic Development of the Dixie Frontier: Henderson County, Texas 1850–1860." *East Texas Historical Journal* 37, issue 2 (1999): 36–43.

Hsiao, J. C. "The Theory of Share Tenancy Revisited." *Journal of Political Economy* 83, no. 5 (Oct. 1975): 1023–32.

Hudson, Estelle, and Henry R. Maresh. *Czech Pioneers of the Southwest* (Dallas: South-West Press, 1934).

Hunter, James. *Scottish Exodus. Travels Among a Worldwide Clan* (Edinburgh: Mainstream Publishing, third edition, 2007).

Hunter, Louis C. *Steamboats on the Western Rivers. An Economic and Technological History* (Cambridge, MA: 1949; reprint New York: Dover, 1977).

Inskeep, Steve. *Jacksonland. President Andrew Jackson, Cherokee Chief John Ross, and a Great American Land Grab* (New York: Penguin Books, 2015).

Jayne, Reginald D. "Martial Law in Reconstruction Texas" (MA thesis, Sam Houston State University, May 2005).

Johnsen, Arne Odd. "Johannes Nordboe and Norwegian Immigration. An 'America Letter' of 1837." *Norwegian-American Studies* 8 (1932): 23–8.

Johnstone, Jeffrey M. "Sir William Johnstone Pulteney and the Scottish Origins of Western New York." *Crooked Lake Review.* Summer 2004.

Jordan, Terry G. *German Seed in Texas Soil: Immigrant Farmers in Nineteenth-Century Texas* (Austin: University of Texas Press, 1966).

———. "The German Settlement of Texas after 1865." *Southwestern Historical Quarterly* 73 (October 1969): 193–212.

———. "The Imprint of the Upper and Lower South in Mid-Nineteenth-Century Texas." *Annals of the Association of American Geographers* 57, no. 4 (December 1967): 667–90.

———. "The Texas Appalachia." *Annals of the Association of American Geographers* 60, no. 3 (September 1970): 409–27.

———. *Texas Log Buildings. A Folk Architecture* (Austin: University of Texas Press, 1978).

Jordan-Bychkov, Terry G. *The Upland South: The Making of an American Folk Region and Landscape* (Harrisonburg, VA: University of Virginia Press, 2003).

Jordan, Don, and Michael Walsh. *White Cargo: The Forgotten History of Britain's White Slaves in America* (New York: New York University Press, 2008).

Kallelid, Ole. *Stavanger bys historie. Sild og seil 1815–1890*, Vol. 2 (Stavanger: Wigestrand Forlag, 2012).

Kamphoefner, Walter D. "German Emigration Research, North, South, and East: Findings, Methods, and Open Questions." In *People in Transit. German Migrations in Comparative Perspective, 1820–1930*, eds. Dirk Hoerder and Jörg Nagler (New York: Cambridge University Press, 1995).

———. "New Perspectives on Texas Germans and the Confederacy." *Southwestern Historical Quarterly* 102, no. 4 (April 1999): 441–55.

Kamphoefner, Walter D., and Wolfgang Helbich, eds. *Germans in the Civil War. The Letters They Wrote Home* (Chapel Hill, NC: University of North Carolina Press, 2006).

Kay, Billy. *The Scottish World. A Journey into the Scottish Diaspora*. E(dinburgh: Mainstream Publishing, 2006).

Kearney, James C. *Nassau Plantation. The Evolution of a Texas German Slave Plantation* (Denton, TX: University of North Texas Press, 2010).

Keats-Rohan, Katherine S. B., ed. *Prosopography Approaches and Applications: A Handbook* (Oxford: Prosopographica et Genealogica, 2007).

Kellogg, Louise Phelps. "The Story of Wisconsin, 1634 – 1848." *Wisconsin Magazine of History* 3, no. 2 (December 1919): 189–208.

Kelly, Dayton, ed. *The Handbook of Waco and McLennan County, Texas* (Waco, TX: Texian Press, 1972).

Kennedy, Billy. *The Scots-Irish in the Carolinas* (Belfast, Ireland: Emerald House Group, 1997).

King, Irene M. *John O. Meusebach, German Colonizer in Texas* (Austin, TX: University of Texas Press, 1967).

Kip, Frederic E., and Margarita L. Hawley. *History of The Kip Family in America* (Boston, 1928).

Kirby, Jack Temple. "ANCESTRYdotBOMB: Genealogy, Genomics, Mischief, Mystery, and Southern Family Stories." *Journal of Southern History* 76, no. 1 (February 2010): 3–38.

Knight, James R. *The Battle of Fort Donelson: No Terms but Unconditional Surrender* (Charleston: The History Press, 2011).

Knight, Oliver. *Fort Worth. Outpost on the Trinity* (Norman, OK: Oklahoma University Press, 1953).

Krell, Alan. *Devil's Rope. A Cultural History of Barbed Wire* (Chicago: University of Chicago Press, 2004).

La Vere, David. *The Texas Indians* (College Station: Texas A&M University Press, 2004).

Langhelle, Svein Ivar. "*Oppsving i næringane ved sida av jordbruket.*" In *Fra Vistehola til Ekofisk. Rogaland gjennom tidene,* eds. Edgar Hovland and Hans Eyvind Næss, Vol. 1 (Oslo: Universitetsforlaget, 1987), 333–46.

Larson, Lars Erik. *Norwegian Emigration to Canada, 1850–1874* (Whitewater: University of Wisconsin-Whitewater, 2010), http://www.chequamegonbay-history.com/files/NorwegianEmigrationToCanada.pdf(accessedJune11,2020).

Laxton, Edward. *The Famine Ships. The Irish Exodus to America* (New York: Henry Holt and Company, 1996).

Leith, Murray Stewart, and Duncan Sim, eds. *The Modern Scottish Diaspora. Contemporary Debates and Perspectives* (Edinburgh: Edinburgh University Press, 2014).

Leyburn, James G. *The Scotch-Irish. A Social History* (Chapel Hill, NC: North Carolina University Press, 1962).

Lich, Glen E. *The German Texans* (San Antonio: The University of Texas, Institute of Texan Cultures, 1996).

Lich, Glen E., and Dona B. Reeves, eds. *German Culture in Texas* (Boston: Twayne, 1980; *Handbook of Texas Online*).

Lonn, Ella. *Desertion during the Civil War* (New York: Century, 1928; reprint by Peter Smith, Gloucester, MA:, 1966).

Lovoll, Odd S. *The Promise of America. A History of the Norwegian American People* (Oslo: Universitetsforlaget, 1984).

Løland, Jacob Sverre. *Romedalsboka. Garder og slekter.* Vol. 5, no. 1 (Elverum: Elverum Trykk, 1992).

Lynch, William O. "The Westward Flow of Southern Colonists before 1861." *Journal of Southern History* 9, no. 3 (Aug. 1943): 303–27.

Machann, Clinton (ed.). *The Czechs in Texas* (College Station: Texas A&M University Department of English, 1979).

Machann, Clinton, and James W. Mendl. *Krásná Amerika: A Study of the Texas Czechs, 1851–1939.* Austin: Eakin Press, 1983).

Madden, David, ed. *Beyond the Battlefield. The Ordinary Life and Extraordinary Times of the Civil War Soldier* (New York: Touchstone Book, 2000).

Martin, John H. "The Pulteney Estates in the Genesee Lands." *Crooked Lake Review* Fall 2005.

McCallum, Henry D., and Frances T. McCallum. *The Wire That Fenced the West* (Norman, OK: University of Oklahoma Press, 1965).

McDonald, Archie P., ed. *Hurrah for Texas. The Diary of Adolphus Sterne, 1838–1851* (Austin: Eakin Press, 1986).

McKelvey, Blake. "The Genesee County Villages in Early Rochester's History." *Rochester History* 47, no. 1 and 2 (January and April 1985): 1–32.

———. "The Population of Rochester." *Rochester History* 12, no. 4 (October 1950): 1–24.

McKinnon, John L. *History of Walton County* (Atlanta: Byrd Printing, 1911).

McLean, Malcolm, ed. *Papers Concerning Robertson's Colony in Texas, Volume 10: March 21 through July 25, 1835: The Ranger Rendesvouz* (Arlington, Texas, 1983).

McMath, Robert C., Jr. *Populist Vanguard: A History of the Southern Farmers' Alliance* (Chapel Hill, NC: University of North Carolina Press, 1975).

McPherson, James M. *Battle Cry of Freedom. The Civil War Era* (New York: Oxford University Press, 1988).

Melby, Kjell Gudmund. *Husmannsvesenet i Ringsaker hovedsogn 1875–1900* (Hovedfagsoppgave i historie, Universitetet i Trondheim, 1990).

Meyer, Carl S. "Lutheran Immigrants' Churches Face the Problems of the Frontier." *Church History* 29, no. 4 (December 1960): 440–62.

Meyer, Douglas K. *Making the Heartland Quilt: A Geographical History of Settlement and Migration in Early-Nineteenth-Century* (Carbondale, IL: Southern Illinois University Press, 2000).

Meyer, Duane. *The Highland Scots of North Carolina 1732–1776* (Chapel Hill, NC: University of North Carolina Press, 1957).

Michno, Gregory. *The Settlers' War: The Struggle for the Texas Frontier in the 1860s* (Caldwell, ID: Caxton Press, 2011).

Miller, George J. "Some Geographic Influences in the Settlement of Michigan and in the distribution of its Population." *Bulletin of the American Geographical Society* 45, no. 5 (1913): 321–48.

Miller, Thomas Lloyd. *The Public Lands of Texas 1519 – 1970* (Norman, OK: University of Oklahoma Press, 1972).

Mills, W. S. *History of Van Zandt County* (Canton, TX: privately printed, 1950).

Mitchell, Reid. "'Not the General but the Soldier.' The Study of Civil War Soldiers." In *Writing the Civil War. The Quest to Understand*, eds. James M. McPherson and William J. Cooper (Columbia, SC: University of South Carolina Press, 1998), 81–95.

Moneyhon, Carl H. *Texas after the Civil War. The Struggle of Reconstruction* (College Station: Texas A&M University Press, 2005).

Morthoff, Bjarne. *Romedalsboka. Garder og slekter*, Vol. 1 (Elverum: Elverum Trykk, 1967).

———. *Romedalsboka. Garder og slekter*, Vol. 2 (Elverum: Elverum Trykk, 1970).

Morthoff, Bjarne, and Jacob Sverre Løland. *Romedalsboka. Garder og slekter*, Vol. 3 (Elverum: Elverum Trykk, 1979).

Morthoff, J. B. *Løtenboka: Garder og slekter* (Løten: Løten Historielag, 1955).

Mulder, William. "Norwegian Forerunners Among the Early Mormons." *Norwegian-American Studies* 19 (1956): 46–61.

Myres, Sandra L. "Fort Graham, Listening Post on the Texas Frontier." *West Texas Historical Association Yearbook* 59 (1983): 33–52.

Nagel, Anne-Hilde. "*Oversikter, årstall, tabeller.*" In *Norges historie*, ed. Knut Mykland, Vol 15 (Oslo: Cappelen, 1980).

Nelson, Frank G. "Introduction." In Johan Reinert Reiersen, *Pathfinder for Norwegians Emigrants* (Northfield, MN: Norwegian-American Historical Association, 1981).

Nelson, Ronald E. "The Bishop Hill Colony and Its Pioneer Economy." *Swedish Pioneer Historical Quarterly* 18, no. 1 (1987): 32–48.

Nystel, Ole. *Lost and Found or Three Months with the Wild Indians. A Brief Sketch of the Life of Ole T. Nystel Embracing his Experience While in Captivity to the Comanches and Subsequent Liberation from Them* (Dallas, 1888; reprinted Clifton: Bosque Memorial Museum, 1994).

Olmsted, Frederick Law. *A Journey Through Texas. Or a Saddle-trip on the Southwestern Frontier* (First published New York, 1857; Lincoln, NE: University of Nebraska Press, 2004).

Ostendorf, Johannes. "*Zur Geschichte der Auswanderung aus dem alten Amt Damme (Oldb.) insbesondere nach Nordamerika, in den Jahren 1830–1880.*" Oldenburger Jahrbuch 46/47 (1942–1943): 164–279.

Ostergren, Robert C. *A Community Transplanted. The Trans-Atlantic Experience of a Swedish Immigrant Settlement in the Upper Middle West, 1835–1915* (Madison, WI: University of Wisconsin Press, 1988).

———. "The Immigrant Church as a Symbol of Community and Place in the Upper Midwest." Great Plains Quarterly 10, no. 1 (1981): 225–38.

Otto, John Solomon. "The Migration of the Southern Plain Folk: An Interdisciplinary Synthesis." *Journal of Southern History* 51, no. 2 (May 1985): 183–200.

Ownsley, Frank Lawrence. *Plain Folk of the Old South* (Baton Rouge, LA: Louisiana State University, 1949).

Pate, J'Nell. *North of the River. A Brief History of North Fort Worth* (Fort Worth, TX: Texas Christian University Press, 1994).

Payne, Darwin. "Early Norwegians in Northeast Texas." *Southwestern Historical Quarterly* 65, no. 2 (July 1961–April 1962): 196–203.

Pederson, Jane Marie. *Between Memory and Reality. Family and Community in Rural Wisconsin, 1870–1970* (Madison, WI: University of Wisconsin Press, 1992).

Pierson, Oris E. *Norwegian Settlements in Bosque County, Texas* (Clifton, TX: Bosque Memorial Museum, 1979).

Pool, William C. *A History of Bosque County, Texas* (San Marcos, TX: Record Press, 1954).

———. *Bosque Territory: A History of an Agrarian Community* (Kyle, TX: Chapparal Press, 1964).

Pooley, William Vipond. "The Settlement of Illinois from 1830 to 1850." *Bulletin of the University of Wisconsin History Series* 1 (1908): 287–595.

Population of the United States in 1860 Compiled from the Original Returns of The Eighth Census (Washington, DC: Bureau of the Census Library, 1864).

Posey, Walter B. "The Slavery Question in the Presbyterian Church in the Old Southwest." *Journal of Southern History* 15, no. 3 (August 1949): 311–24.

Pruitt, Francelle. "'We've Got to Fight or Die:' Early Texas Reaction to the Confederate Draft, 1862)." *East Texas Historical Journal* 36, Issue 1 (1998): 3–17.

Purifoy, Lewis M. "The Southern Methodist Church and the Proslavery Argument." *Journal of Southern History* 32, no. 3 (August 1966): 325–41.

Puryear, Pamela A., and Nathan Winfield, Jr. *Sandbars and Sternwheelers: Steam Navigation on the Brazos* (College Station: Texas A&M University Press, 2005).

Qualey, Carlton C. *Norwegian Settlement in the United States* (Northfield, MN: Norwegian-American Historical Association, 1938).

Rable, George C. *Civil Wars. Women and the Crisis of Southern Nationalism* (Urbana, IL: University of Illinois Press, 1989).

Ravenstein, E. G. "The Laws of Migration." *Journal of the Royal Statistical Society* 48, no. 2 (June 1885): 167–235.

———. "The Laws of Migration." *Journal of the Royal Statistical Society* 52, no. 2 (June 1889): 241–301.

Ray, Celeste. *Highland Heritage. Scottish Americans in the American South* (Chapel Hill, NC: University of North Carolina Press, 2001).

Reiersen, Johan R. "Norwegians in the West in 1844: A Contemporary Account by Johan R. Reiersen and translated and edited by Theodore C. Blegen." *Norwegian-American Studies* 1 (1926): 110–25.

———. "Behind the Scenes of Emigration: A Series of Letters from the 1840's." *Norwegian-American Studies* 14 (1944): 78–117.

Remini, Robert V. *Andrew Jackson* (New York: Harper & Row Publishers, 1966).

Report on the Productions of Agriculture, Tenth Census (June 1, 1880) (Washington, DC: Government Printing Office, 1883).

Report on the Statistics of Agriculture in the United States at the Eleventh Census: 1890 (Washington, DC: Government Printing Office, 1895).

Reynolds, David S. *Waking Giant. America in the Age of Jackson* (New York: Harper, 2008).

Rimmer, Frederick. *Captured by the Indians. 15 Firsthand Accounts, 1750–1870* (New York: Dover, 1961).

Roark, James L. R. *Masters Without Slaves. Southern Planters in the Civil War and Reconstruction* (New York: E.W. Norton, 1977).

Robbins, Roy M. "Preemption – A Frontier Triumph." *Mississippi Valley Historical Review* 18, no. 2 (December 1931): 331–49.

Rocco, Fiammetta. *Quinine. Malaria and the Quest for a Cure That Changed the World* (New York: Perennial, 2003).

Roe, Alfred Seelye. *Rose Neighborhood Sketches, Wayne County, New York: with glimpses of the adjacent towns: Butler, Wolcott, Huron, Sodus, Lyons and Savannah* (Worcester, MA:, 1893).

Roeber, A. G. *Palatines, Liberty, and Property: German Lutherans in Colonial British America* (Baltimore: Johns Hopkins University Press, 1993).

Rohrbough, Malcolm J. *The Trans-Appalachian Frontier. People, Societies, and Institutions 1775 – 1850* (New York: Oxford University Press, 1978).

Ropeid, Andreas. "Kristenlivet i åra før 1850." In *Stavanger på 1800-tallet* (Stavanger: Stabenfeldt, 1975).

Rosdail, J. Hart. *The Sloopers, Their Ancestry and Posterity* (Broadview, IL: Photopress, 1961).

Russel, Charles H. *Undaunted. A Norwegian Woman in Frontier Texas* (College Station: Texas A&M University Press, 2006).

Rynning, Ole. *Ole Rynning's True Account of America*, translated and edited by Theodore C. Blegen (Minneapolis, MN: Norwegian-American Historical Association, 1926).

———. *Sandfærdig Beretning om Amerika til Oplysning og Nytte for Bonde og Menigmand. Forfattet af En norsk, som kom derover i Juni Maaned 1837* (Christiania, 1839).

Salinger, Sharon V. "Colonial Labor in Transition: The Decline of Indentured Servitude in Late Eighteenth-Century Philadelphia." *Labor History* 22, no. 2 (1981): 165–91.

———. *To Serve Well and Faithfully: Labor and Indentured Servants in Pennsylvania, 1682–1800* (New York: Cambridge University Press, 1987).

Sandbo, Anna Irene. "The First Session of the Secession Convention of Texas." *Southwestern Historical Quarterly* 18 (October 1914): 162–94.

Scheiber, Harry N. "The Economic Historian as Realist and as Keeper of Democratic Ideals: Paul Wallace Gates's Studies of American Land Policy." *Journal of Economic History* 40, no. 3 (September 1980): 585–93.

Schmelzer, Janet L. *Where the West Begins: Fort Worth and Tarrant County* (Northridge, CA, 1985).

Schwieder, Dorothy. *Iowa: The Middle Land* (Iowa City, IA: University of Iowa Press, 1996).

Scott, Larry E. *The Swedish Texans* (San Antonio: Institute of Texan Cultures, 1990).

Seip, Anne Lise. *Nasjonen bygges 1830–1870*, Vol. 8 (Oslo: Aschehoug, 1997).

Selcer, Richard F. *The Fort That Became a City: An Illustrated Reconstruction of Fort Worth, Texas, 1849–1853* (Fort Worth, 1995).

Shaw, Ronald E. *Erie Water West. A History of the Erie Canal 1792–1854* (Lexington, KY, 1966).

Shook, Robert W. "German Unionism in Texas during the Civil War and Reconstruction," (MA thesis, North Texas State College, 1957).

Sibley, Marilyn M. "The Texas-Cherokee War of 1839." *East Texas Historical Journal* 3, no. 1 (1965): 18–33.

Skappel, Simen. *Om Husmandsvæsenet i Norge, Dets oprindelse og utvikling*, Videnselskabets Skrifter 2. Hist. Filos. Klasse. no. 4, Kristiania 1922.

Skrabanek, Robert L. *We're Czechs* (College Station: Texas A&M University Press, 1988).

Smallwood, James. *The History of Smith County Texas, vol. 1. Born in Dixie. Smith County Origin to 1875* (Austin: Eakin Press, 1999).

Smith, David Paul. *Frontier Defense in the Civil War. Texas Rangers and Rebels* (College Station, Texas A&M University Press, 1992).

Smith, Timothy L. *Revivalism and Social Reform: American Protestantism on the Eve of the Civil War* (Eugene, OR: Abingdon Press, 1957).

"The Society of Friends in Western New York. *The Canadian Quaker History Newsletter* 37 (July 1985), 6–11.

Sonnichsen, C. L. *I'll Die Before I'll Run: The Story of the Great Feuds of Texas* (New York: Harper, 1951; 2nd ed., New York: Devin-Adair, 1962).

Spillman, W. J. "The Agricultural Ladder." *American Economic Review* 9, no. 1 (March 1919): 170–9.

Spratt, John Stricklin. *The Road to Spindletop. Economic Change in Texas, 1875–1901* (Austin: University of Texas Press, 1970).

Statistiske Tabeller for Kongeriget Norge. Åttende Række (Christiania: Departementet for det Indre, 1847).

Steele, Hampton. *A History of Limestone County, 1833–1860* (Mexia, TX: Mexia News, n.d.).

Stein, Bill. "The German Draft Revolt in Colorado, Austin, and Fayette Counties." *Journal of the German-Texas Heritage Society* 14, no. 3 (Fall 1992): 221–4.

———. "Distress, Discontent, and Dissent: Colorado County, Texas, during the Civil War." In *The Seventh Star of the Confederacy*, ed. Kenneth W. Howell. Denton, TX: University of North Texas Press, 2009: 301–16.

Stensrud, Odd. "Økonomiske og sosiale kår innen husmannsklassen i Stange med særlig vekt på de første tiår av det 19. århundre," (MA thesis, University of Oslo, 1974).

———. "'Til Christiania for at søge Arbejde.' Utflytting fra Stange før Amerika-feberen." In *Gammalt frå Stange og Romedal 1976*. Hamar: Stange Historielag, 1976: 49–68.

Stewart, Alexander M. "Sesquicentennial of Farmington, New York, 1789–1939." *Bulletin of Friends' Historical Association* 29, no. 1 (Spring 1940): 37–43.

Stolz, Jack. "Kaufman County in the Civil War." *East Texas Historical Journal* 28, no. 1 (1990): 37–43.

Strand, A. E., ed. *A History of the Norwegians of Illinois* (Chicago, IL: John Anderson Publishing, 1905).

Stubhaug, Arild. *Jacob Aall i sin tid* (Oslo: Aschehoug, 2014).

Svalestuen, Anders A. "Emigration from the Community of Tinn, 1837–1907." *Norwegian American Studies* 29 (1983): 43–89.

Sveen, Kåre. "Introduction," in "*Reisebrev fra Texas 1852*" by Johan Olsen Brunstad. In *Gammalt frå Stange og Romedal 1975* (Hamar: Stange Historielag, 1975).

———. *Utvandrerne 1845–1870. Gammalt frå Stange og Romedal 1975* (Hamar: Stange Historielag, 1975).

———. *Utvandrerne 1871–1879. Gammalt frå Stange og Romedal 1976* (Hamar: Stange Historielag, 1976).

Swieringa, Robert P. "The Equity Effects of Public Land Speculation in Iowa:

Large versus Small Speculators." *Journal of Economic History* 34, no. 4 (December 1974): 1008–20.

Syversen, Odd Magnar, and Derwood Johnson. *Norge i Texas. Et bidrag til norsk emigrasjonshistorie* (Elverum: Stange Historielag, 1982).

Taylor, Amy Murrell. *The Divided Family in Civil War America* (Chapel Hill, NC: North Carolina University Press, 2005).

Texas 1860 Agricultural Census, Vol. 1, transcribed and compiled by Linda L. Green (Westminster, Maryland, Willow Bend Books, 2008).

Thirteenth Census of the United States Taken in the Year 1910, Volume 3, Population 1910. Reports by States with Statistics for Counties, Cities and other Civil Divisions, (Washington, DC: Government Printing Office, 1913).

Thirteenth Census of the United States taken in the Year 1910, Volume 5, Agriculture 1909 and 1910. General Report and Analysis (Washington, DC: Government Printing Office, 1914).

Thyness, Paul. "Jacob Aall," in *Norsk biografisk leksikon,* https://nbl.snl.no/Jacob _Aall (accessed June 11, 2020).

Timmons, Joe E. "The Referendum in Texas on the Ordinance of Secession, February 23, 1861: The Vote." *East Texas Historical Journal* 11, no 2 (1973): 12–28.

Tippens, Matthew D. *Turning Germans into Texans. World War I and the Assimilation and Survival of German Culture in Texas, 1900–1930* (Kleingarten Press, 2000).

Tomlins, Christopher. "Reconsidering Indentured Servitude: European Migration and the Early American Labor Force, 1600–1775." *Labor History* 42, no. 1 (2001): 5–43.

Trent, Lucy C. *John Neely Bryan, Founder of Dallas* (Dallas: Tardy Publishing, 1936).

Try, Hans. *Norges historie. To kulturer – en stat 1851–1884* (Oslo: J. W. Cappelen Forlag, 1986).

Tucker, Spencer C. *Unconditional Surrender. The Capture of Forts Henry and Donelson* (Abilene, TX: Whiney Foundation Press, 2001).

Tyler, Ron, and Lawrence A. Murphy, eds. *The Slave Narratives of Texas* (Austin: State House Press, 1997).

Unstad, Lyder L. "The First Norwegian Migration into Texas. Four 'America Letters'." Translated and edited by Lyder L. Unstad. *Norwegian-American Studies* 8 (1934): 39–57.

Urkedal York, Eyvind, Arne Solli, and Ståle Dyrvik. *"Vi Arbeidere er ingen faatallig Hær": Om husmannsvesenet i Norge* (Luster: Luster Sogelag, 2011).

Utley, Robert M. *Frontier Regulars. The United States Army and the Indian, 1866–1891* (Lincoln, NE: University of Nebraska Press, 1973).

Utvandringsstatistikk. Norwegian Official Statistics. 7.25 (Oslo: Norwegian Interior Ministry, 1921).

Veflingstad, M., ed. *Stange Bygdebok. II. Gards- og slektshistorien* (Stange: Stange historielag, 1952).

Velde, Lewis G. Vander. *The Presbyterian Churches and the Federal Union, 1861–1869* (Cambridge, MA: Harvard University Press, 1932).

Virts, Nancy. "The Efficiency of Southern Tenant Plantations, 1900–1945." *Journal of Economic History* 51, no. 2 (June 1991): 385–95.

Waggoner, Kassia, and Adam Nemmers, eds. *Yours in Filial Regard. The Civil War Letters of a Texas Family* (Fort Worth: TCU Press, 2015).

Wallace, Anthony F. C. *The Long, Bitter Trail. Andrew Jackson and the Indians* (New York: Hill and Wang, 1993).

Wallace, Patricia Ward. *Waco. A Sesquicentennial History* (Virginia Beach, VA: Wallsworth Publishing, 1999).

Walter, Ray A. *A History of Limestone County* (Austin: Von Boeckmann-Jones, 1959)

Ward, John William. *Andrew Jackson – Symbol for an Age* (Oxford: Oxford University Press, 1953).

Weitz, Mark A. *A Higher Duty: Desertion among Georgia Troops during the Civil War* (Lincoln, NE: University of Nebraska Press, 2000).

Wentz, Abdel Ross. *A Basic History of Lutheranism in America* (Philadelphia: Fortress, rev. edition, 1964).

Wiley, Bell Irwin. *The Life of Johnny Reb. The Common Soldier of the Confederacy* (Baton Rouge: Louisiana State University Press, 1993).

Williams, J. W. "A Statistical Study of the Drought of 1886." *West Texas Historical Association Yearbook* 21 (1945): 85–109.

Winfrey, Dorman H., and James M. Day, eds. *The Indian Papers of Texas and the Southwest, 1825–1916* (Austin: Pemberton Press, 1966).

Winters, Donald L. "Tenancy as an Economic Institution: The Growth and Distribution of Agricultural Tenancy in Iowa, 1850–1900." *Journal of Economic History* 3, no. 2 (June 1977): 382–408.

Wittke, Carl. "The German Forty-Eighters in America: A Centennial Appraisal." *American Historical Review* 53, no. 4 (July 1948): 711–25.

Woodman, Harold D. "Class, Race, Politics, and the Modernization of the Postbellum South." *Journal of Southern History* 63, no. 1 (February 1997): 3–22.

Wooster, Ralph A. "An Analysis of the Membership of the Texas Secession Convention." *Southwestern Historical Quarterly* 62, no. 3 (January 1959): 322–35.

———. "East of the Trinity: Glimpses of Life in East Texas in the Early 1850s." *East Texas Historical Journal* 13, Issue 2 (1975): 3–10.

———. "Wealthy Texans, 1870)." In *Texas Vistas*. Edited by Ralph A. Wooster and Robert A. Calvert. Austin: Texas State Historical Association, 1987: 184–93.

Wooster, Ralph A., and Robert Wooster. "'Rarin' for a Fight'. Texans in the Confederate Army." *Southwestern Historical Quarterly* 84 (April 1980-April 1981): 387–426.

———. "A People at War: East Texans during the Civil War." *East Texas Historical Journal* 28, no. 1 (1990): 3–16.

Worster, Donald. *Rivers of Empire: Water, Aridity, and the Growth of the American West* (New York: Oxford University Press, 1985).

Wright, Gavin, and Howard Kunreuther. "Cotton, Corn and Risk in the Nineteenth Century." *Journal of Economic History* 35, no. 3 (September 1975): 526–51.

Wyman, Mark. *The Wisconsin Frontier* (Bloomington, IN: Indiana University Press, 1998).

Ylvisaker, S. C. "The Missouri Synod and the Norwegians," in *Ebenezer. Review of the Work of the Missouri Synod during three Quarters of a Century*, ed. W. H. T. Dau (St. Louis, MO: Concordia Publishing House, 1922), 264–6.

Østrem, Nils Olav. "*Cleng Peerson – skaparen av den store forteljinga om Amerika.*" Ætt og heim. Lokalhistorisk årbok for Rogaland 1999. Stavanger, 2000: 5–44.

———. *Karmøys historie – mot havet du deg vender* (Bergen: Fagbokforlaget, 2010).

Østrem, Nils Olav. *Den store utferda. Utvandring frå Skjold og Vats til Amerika 1837–1914* (Oslo: Scandinavian Academic Press, 2015).

Øverland, Orm, ed. *From Norway to America. Norwegian-American Immigrant Letters 1838–1914. Volume One, 1838–1870* (Northfield, MN: Norwegian-American Historical Association, 2012).

———. *From Norway to America. Norwegian-American Immigrant Letters 1838–1914. Volume Two, 1871–1892* (Northfield, MN: Norwegian-American Historical Association, 2015).

Øverland, Orm, and Steinar Kjærheim, eds. *Fra Amerika til Norge. Norske utvandrerbrev 1838–1857*, Vol. 1 (Oslo: Solum Forlag, 1992).

DOCUMENTS FROM THE ARCHIVES OF CLENG PEERSON RESEARCH LIBRARY (CPRL), CLIFTON, AND BOSQUE COUNTY HISTORICAL COMMISSION, MERIDIAN, BOSQUE COUNTY

"1869–1969. Our Savior's Lutheran Church Centennial Celebration, June 28–29, 1969."

Bell, Elaine Bakke. *The Omenson Family History and the Norway Mills Community* (Clifton, 2005).

"Biography of M. R. Rohne." Cranfills Gap, Texas, December 24, 1895.

Bronstad, Alvin L. "The C. O. Bronstad Family History." Clifton, Texas, March 1970.

Brown, Lottie Huse, "The Huse Book," n.d.

"Christian and Johanna Larsdatter Canuteson." Genealogy of the family, n.d.

Colwick, Mary Latimer. *The Colwicks. Ovee, John and Family 1825–1966.* August 1966.

"Commissioner's Court Minute Book." Bosque County, volume A.

Dahl, Folke. *The Family History of Hendrik and Christine Dahl from the Year 1854 to 1954.* Clifton, 1954.

———. *The Family History and Descendants of Hendrik and Christine Dahl. From the Year 1854 to 2004.* Clifton, 2004.

Edgar, Betty Knudson. *The Milkman and the Maid. A History of the Martin Sorenson Family of Bosque County, Texas.* Manuscript 1995. Bosque County Historical Commission, Meridian, Texas.

———. *From Generation to Generation, Part I. Histories of Salve Knudson, Tellef Tergerson, and Knud Salve Knudson.* Published privately, 1996. Bosque County Historical Commission, Meridian, Bosque County.

"Election Record Book." Bosque County, June 17, 1861.

"Four Mile Lutheran Church." July 4, 1964.

"A History of Peder Pederson from 1821–1888." Family history, n.d.

Hogstel, Mildred O., and Alice Marie Berg. "History and Family Tree of Aslak Nelson." Manuscript, June 1973. Cleng Peerson Research Library, Bosque Museum, Clifton, Texas.

"The Hovie Family Tree, 1765–1995." copyright 1995 John & Dorothy Hovie, 121 Briar Drive, Neenah, Wisconsin.

Johnson, Carl. *I, Carl Johnson. Born of Immigrant Parents*, An Autobiography as recorded on tape during the year 1979 at my home, 3416 Street, Fort Worth, Texas.

Knud S., Knudson Homestead. No date, archive no. 051.001.043, CPRL.

Letter from Aslak Nilsen Smeland to "precious parents, sisters and brothers," dated Four Mile Prairie, April 26, 1867.

"The Man with America Fever. A History of Salve Knudson." Cleng Peerson Research Library, n.d.

"The Memories of Herbert Schulz." CPRL, Bosque Museum, archive no. 050.001.001.

"Passenger Contract No. 30, Martinus Pedersen." Christiania, October 31, 1872, 020.001.02, CPRL.

Profiles of Pioneer Families in the Valley Mills Area, compiled and indexed by Gerald and Jo Meyer for the Bosque County Collection, Meridian, Texas, n.d.

"The Quest of the Questad Place." Manuscript by Mary Ellen Ellingson and Martha Louise Springer, 1992.
"The Questad House." *The Clifton Record*, October 16, 1975.
Radde, Rebecca D. "Ole Pierson Homestead." Manuscript July 30, 1985, Meridian, CPRL.
"Rohne Family History: Norwegian Immigrants Started Farm Here with Yoke of Oxen."
Rystad, John Knudson Rystad. "Memoirs of the Rev. John Knudson Rystad." Clifton, 1967.
"The Solberg Family from Norway to Texas." Several authors, Clifton, Texas, 1979.
"Story of the Old Mill." Compiled by Mary Ellen Swenson Ellingson, Clifton, 1984.
Thus Far God Has Brought Us. A History of the Jo Knust Family of Bosque County, Texas.
Uncle Theo's Scrapbook, 1883–1891, Norse, Texas. Newspaper Items from a Pioneer Texas Community, Cleng Peerson Research Library, Bosque Museum.
Wynne, Anna E., with Elma White. "The Eggen Family History. Ancestors and Descendants of Thore Anstensen Eggen and His Brothers and Sisters." Copyright Anna Wynne 2001. Bosque County Historical Commission, Meridian.

FAMILY BIOGRAPHIES IN BOSQUE COUNTY: LAND AND PEOPLE

"Anderson T. W. and Tilda." F33, *Bosque County: Land and People*, 122.
"Barnes, Elisabeth Oakes Barton." F59, *Bosque County: Land and People*, 135.
"Barnes, Robert Samuel." F58, *Bosque County: Land and People*, 135.
"Bekkelund family." F85, *Bosque County: Land and People*, 149.
"C. O. Bronstad." F132, *Bosque County: Land and People*, 175.
"Gulbrand Olson Bronstad family." F134, *Bosque County: Land and People*, 176.
"Ole and Matthilda (Canuteson) Bronstad." F139, *Bosque County: Land and People*, 179.
"Canuteson Ole Andrew." F176, *Bosque County: Land and People*, 198.
"Christian P. Carlson family." F179, *Bosque County; Land and People*, 199.
"Dahl, Hans B. and family." F273, *Bosque County, Land and People*, 246–248.
"Dahl, John." F276, *Bosque County: Land and People*, 247–248.
"Ole O. Finstad and Julia Pederson." F381, *Bosque County: Land and People*, 296.
"Hanson, Ole and Mathea." F487, *Bosque County: Land and People*, 350.
"Hoff, John P. Family." F553, *Bosque County: Land and People*, 380.
"Holen C.O. and Gine Dahl family." F561, *Bosque County: Land and People*, 384.

"Johnson, John Bartinus (Teen) and Louisa." F639, *Bosque County: Land and People*, 424.
"Landgraf, Adolf Family." F702, *Bosque County: Land and People*, 457.
"Jasper Newton Mabry." F766, *Bosque County: Land and People*, 492.
"Aslak Nelson." F858, *Bosque County: Land and People*, 535.
"N. J. Nelson." F861, *Bosque County: Land and People*, 537.
"Scrutchfield, Lowry Hampton." F1103, *Bosque County: Land and People*, 662–663.
"Strand, Christian and Regina." *Bosque County. Land and People*, 701.
"Willingham Family." F1309, *Bosque County: Land and People*, 765.

DATABASES, UNPUBLISHED SOURCES, AND INFORMALLY PUBLISHED MATERIAL

www.borgerkrigen.info. "Grogaard, Thomas Fasting."
www.borgerkrigen.info. "Grogard, John J."
www.borgerkrigen.info. "Grogard, Nicholas C."
www.borgerkrigen.info. "Peder Jenson."
www.borgerkrigen.info. "Reierson, Christian."
www.borgerkrigen.info. "Reiersen, Johann Henrich."
www.borgerkrigen.info. "Reierson, Oscar."
www.borgerkrigen.info. "Reiersen, Otto Theodor."
Folketeljing Norge 1865 for Løten Prestegjeld. http://xml.arkivverket.no/folketellinger/hefter/1865/04Hedmark/f8650415_hefte.pdf
Folketeljing Romedals prestegjeld 1865. http://xml.arkivverket.no/folketellinger/hefter/1865/04Hedmark/f8650416_hefte.pdf
www.us-census.rg/pub/usgenweb/census/tx/henderson/1850/pg00249.txt.
http://us-census.org/pub/usgenweb/census/tx/mclennan/1860/sl-index.txt.
"McLennan, TX 1860 Federal Slave-Schedule Census." http://ftp-us-census.org/pub/usgenweb/census/tx/mclennan/1860/sl.
http://us-census.org/pub/usgenweb/census/tx/mclennan/1860/sl-index.txt
1860 US Census of Bosque County. www.bosquehc.org/census-1860.shtml. Bosque County Historical Commission–1860 U. S. Census.
"Brig Favoriten, J. A. Køhler & Co." http://www.norwayheritage.com/p_ship.asp?sh=favor
Briscoe Center for American History. Reierson (Johan Reinert) Family Papers. University of Texas at Austin.
"Continental Gin Company, Dallas Plant." National Register of Historic Places Inventory – Nomination Form, January 1983.
Dallas Weekly Herald. August 23, 1883.

FamilySearch. "Microcopy no. 259. Passenger Lists of Vessels Arriving at New Orleans, 1820–1902, Roll 57, November 5–December 28, 1872, Nos. 159–197." The National Archives, Washington 1958.

FamilySearch. "United States Census, 1860." Database with images. Texas, Bosque County, household 27 and 18. https://www.familysearch.org/search/collection/1473181. From "1860 US Federal Census - Population." Washington, DC: National Archives and Records Administration, n.d.

"Heart of Texas Records." Vol. XIX, no. 1, 2, 3, and 4. Central Texas Genealogical Society, Waco.

"Journal for den norsk evangelisk lutherske Menighed i Bosque County, Texas," Secretary book. Volume 1, June 14, 1869–December 29, 1886; Volume 2, 1887–1910. "Our Savior's Lutheran church," Norse, Bosque County.

"Miscellaneous. Manufacturing." *The Dallas Herald*, October 13, 1887.

Parsons, Lee. "A Texas Immigration Story: The Grøgaard Family from Norway," manuscript, November 2019.

Questad Archive. NAHA archives. Northfield, MN: St. Olaf's College.

Seldal, Gunleif. "The Sloopers." Paper presented at the Cleng Peerson Conference, Bosque Museum, October 2015.

Van Sickle, Dale Orbeck, "The Reconstruction of the History of the Cleng Peerson/Colwick Farm in Bosque County, Texas." Manuscript, Austin 2019.

INDEX

◊ ◊ ◊

Aadals Brug Iron Works, Løten, Norway, 87–90
Aall, Jacob, iron works owner, 14
agricultural ladder, 252, 273–91; climbing the, 275–80, 293–97; and cotters, 252–54; and ethnicity, 277–78
agriculture: subsistence farming, 42, 120, 123, 145–49, 274, 293; cotton, 33, 53, 55, 84, 105, 109, 124, 148–49, 153, 155, 157, 159, 160, 172, 181, 216, 223, 232, 238, 260, 266, 273–75, 2838–6, 290–91, 296, 300, 303, 305; maize, 75–76, 124, 146, 276, 291; oats, 76, 123, 233, 256, 260, 276, 285–86; sweet potatoes, 75–76, 145, 185–86, 276; wheat, 76, 78, 91, 95, 123–24, 130, 149, 153–54, 160, 185–86, 232–34, 237–38, 260, 266–67, 276, 284–86, 289
Albany: capital of New York, 2, 62, 66
Albertson brothers, 56–57, 192–93
Amerika, brig, 84–85
Åmli, 36, 40
Åmli, Aust-Agder County, Norway, 36
Ancona, 34, 36, 38
Arendal, 14, 21, 37, 47, 82, 85, 89, 90, 99–101, 110, 113, 200, 208, 210, 247, 249, 310

Arendal, barque, 85, 86, 89, 90, 114, 146, 247
assimilation, 1, 2, 181, 189, 230, 325, 329, 331–34
Atalanta, brig, 249, 254–55, 257, 270, 283
Austin, Texas, 20, 29, 54, 79, 95, 121, 126, 132, 142, 149, 171, 193, 198, 211, 258, 277, 290, 292, 312, 331, 333

Baptists, 157, 158, 175, 176, 198, 313, 319
Barry, Buck, 118–19, 207, 318, 344
Bell County, Texas, 267, 278, 287, 290
Bishop Hill Colony; and Jansson, Eric 72–74, 311
Black Hawk War, 62, 64, 67–69
Bondies, George, 28, 49–52
Bosque County: Norwegians in, 19, 101, 109, 123, 125–44; Norwegian freethinkers 311, 316
Bosque River, 80, 103–15, 108, 112, 126–27, 160–61, 232, 298–99, 338
Bosque Valley, 39, 103–5, 106, 109, 111–2, 114, 125, 217, 222, 300, 337,
Brazos Point, 80, 217
Brazos River, 21, 24, 39, 79–80, 103–4, 107–9, 112, 118, 125–26, 143, 156, 158–59, 217, 278,
breakthrough for Norwegian emigration, 3, 37, 70,

Bretten, Anders, 109–13, 116
Bronstad, Christian O., 278, 300, 325–26, 336
Bronstad, Gulbrand O., 279, 302–3, 323
Brownsboro, Henderson County, 5, 34, 39–45, 57, 99, 168, 175, 249, 311
Brunstad, Anne and Johan, 86–88, 90–92, 114–15

Canton, Texas 39, 41, 91, 182, 190, 196
Canuteson, Canute, 83, 109, 111, 116, 118, 120, 126, 130, 137, 138, 146, 147, 148, 210, 257
Canuteson, Christian and Johanna, 249
Canuteson, Ole Andreas, 137–38
Canuteson, Ole, 83, 103, 109–11, 117–18, 120, 123, 127, 140–41, 146–48, 220, 256, 298–301
Cherokee Indians, see Native Americans
Chevaillier, Charles, 50–52
Chicago, 2, 64, 67–70, 183, 283
church bell controversy, 316
Cincinnati, Ohio, 19–22, 105
Civil War Deserters, 9, 193, 196, 202, 205–6, 209–14, 310
Civil War, 19, 24, 260, 262, 296, 398; and Union Army, 24, 319, 350, 353; letters from the Civil War, 334–36; see Lund, David; see Lars Olson; see Reierson, Johan R.
Clifton College, 322–25
Clifton, Bosque County, 104, 112, 114–15, 125, 131, 160, 169, 223, 241, 247, 260, 269, 284, 294, 298, 300, 303
Colwick, Ovee, 132–37, 313

Comanche Indians, see Native Americans
Confederate army, 9, 131, 138, 155, 165, 169, 170–75, 177–79, 181, 188–90, 194–98, 202, 213–14, 216, 233, 248, 249, 310,
Congregationalists, 55
Conscription Law 1862, 171, 178, 188, 209
Coryell County, 108, 278, 287, 290, 349
Cotters 59, 82, 127, 249–54, 259, 270, 338
Cranfills Gap, 301–3. *See* Upper Settlement

Dahl, Hendrick O. and Christine, 113, 315
Dallas, 75, 77–78
dancing in Bosque County, 319–20. *See* picnics
drought 1887, 124, 152, 285, 291, 292, 295–96, 302, 321

Eighteenth Texas cavalry, 182, 192
Eleventh Texas infantry, 189–92
emancipation of slaves: Texas, June 19th, 231, 233
emigration from Telemark County, 36–37
Erath County, 103, 106, 142, 278
Erath, George, surveyor, Texas Ranger, 104–6, 110
Erie Canal, 36, 62
Estrem (Østrem), Ole, pastor, 235–36, 309–10, 314–16
Evangelical, 151, 168, 308
Everett, Thomas Ewell, 126–27, 205, 256

Farmers' Alliance, 211, 296–97
Farmington, New York, 59–60
Favoriten, brig, 82–83, 200–202
Fellows, Joseph, 40, 44–47, 62, 65–67,70
Fifteenth Texas infantry, 197–99, 202–4
financial crisis 1837, 71
Finstad, Ole O., 264–65
Fond du Lac, Wisconsin, 103
Fort Gates, 80, 204, 225
Fort Graham, 80
Fort Worth, 80, 170–72
Four Mile Prairie, 5, 36–45, 54, 80, 87, 89–92, 94–99, 109, 113–15, 154, 167, 181, 196, 233, 235, 237–38, 241, 245–47, 249, 301
Fox River Colony, La Salle County, 22, 61–65, 70–72, 82–83
freemasons, 312–14. *See* masons
freethinkers, 311–14, 315–16, 330
Fredericksburg, Texas, 21, 288
Frontier Regiment, 9, 139, 166, 179, 200–209, 211, 310

Galena, Illinois, 17, 19
Galveston, 20–21, 107, 123, 130, 135, 153, 157, 160, 162, 186, 216, 236, 239, 254–55, 258, 261, 263, 265, 270, 288, 337
Gary Creek, Bosque, 213–16, 220, 224, 384
Gasmann, Hans, 17–18
Geertz, Clifford: thick description, 27, 29
German Texans, 4, 7, 21, 25, 49, 121, 149–50, 153, 162–64, 169, 171, 276–77, 284, 290–91, 311, 323, 329–33, 338–39
Gjerde, Jon, 106–7

Gjestvang, Taale Andreas, Løten, 81, 85–86, 100–101, 111–13
Gjøvdal, Aust-Agder, 36–37, 40, 84, 99, 101
Grimland family, 190–91, 245, 247–49, 299, 303, 314–15, 320, 324, 326, 328
Grimset, Johan Halvorsen 92–93
Grøgaard brothers, 52–54; in the Confederate army, 181
Grøgaard family, Christian and Thomine, 13–16, 46–54
Gulf, Colorado and Santa Fe Railroad, 300

Haabesland, Syvert Nielsen, 25, 31
Hamilton County, 103, 247, 249, 267, 275, 278, 281, 283, 287–92, 294, 208, 300, 303–6, 337
Hastvedt, Jørgen Olsen, 37–38
Health: malaria, 32, 40, 95–97, 220, 245–47; Beaver Creek and malaria 84; cholera, 68, 74, 81, 83,133–34; cleanliness, 176
Helton, J. K., 117, 127, 129, 189, 312
Henderson County, 19, 29–30
Hjelmeland, Rogaland County, 78, 82, 201
Holley, Orleans County, New York, 2, 62
Holt, Aust-Agder County, 14, 23, 24, 55, 83–4, 101, 113, 179–80, 184, 191–93, 201, 249
Houston, Sam: president Republic of Texas, 20, 79

Illinois & Michigan Canal, 62, 84
Indentured servants: in the 1600–1700s, 11, 251; in Texas, 249, 251–53, 257–64; bridging the social class

gap in Bosque, 269–83, 336; social contracts and, 256, 264–65; the cotter system and, 252–55; Poulsen and McKerral business scheme, 251–5; reinforcing Norwegianess, 307, 314, 335
Independence, Missouri, 20; wagon train, 22
Indians, *see* Native Americans
Izette, 34

Jackson, Andrew, 66, 106; Jacksonian Democracy, 66
Jansson, Eric, 87, 91–93
Jenson, Jens, 113–14, 120, 146, 148, 179
Johnson, Bernt Fredrik, 280, 283–84
Johnson, Carl, 303–6
Johnsen. John Vatne, 139
Jordan's Saline, 41, 186
Juneau, Solomon, 64

Karmøy island, Rogaland County, 82–83, 113, 129–30, 137, 140, 294
Kaufman County, 14, 41–43, 75, 167–68, 178, 180, 183–84, 186, 190–91, 193, 201, 204, 328
Kendall, New York State, 64, 70, 75, 77, 80, 83, 133, 137, 200
Kickapoo Creek, Texas, 29–30, 38, 45, 193
Knudson, Knud Salve, 36–37, 40, 71, 194–96, 235, 237, 246–57, 302, 315
Knudson, Salve, 36–37, 40, 246–47, 302, 315
Koshkonong Prairie, Wisconsin, 87, 102

La Salle County, Illinois, 66–70
La Vere, David: 239–40
Lake Erie, 68–69

Lake Michigan, 19, 64, 68–69
land office 17, 19, 29, 68, 70, 110, 117, 132
land speculation, 64, 67–68, 70, 91, 276
Landvik group, 15, 17, 23–27, 31, 34–35
Larissa College, Texas, 53, 182
Le Havre, France, 21, 26–27, 31, 36–38, 84, 105
Lee County, Iowa, 22–23, 82, 95, 121
Lillesand, Aust-Agder County, Norway, 13, 23–24, 26, 37, 46–49, 84, 101, 233
Limestone County, 119, 280, 282–84, 337
Løten, Hedmark County, 44, 55, 77, 85–90, 92, 101, 111, 113, 190, 193, 204, 208, 219, 226, 255, 259–61, 264–65, 269, 275, 280, 283, 303–36
Louisiana, S/S, 265
Lower South, 147, 149, 152–5, 157, 161, 163, 179, 180, 197
Lund, David, 138, and letters from the Civil War 173, 202–4, 245
Lund, Ole and Christopher, 266–8, 305–6
Lutheran Church: Norwegian, 307–10

Mabray, Jaspar N., 104, 110, 115, 117, 125, 127
Madison, Wisconsin, 17
Magnolia, 31,
malaria, *see* health
marriage, 74, 128, 133, 155, 247, 257, 282, 293, 333
Marshall, Harrison County, 28, 30–33, 47, 161, 179, 181–82, 190, 215,
masons, 147, 495–97. *See also* Bosque County, Norwegian freethinkers

McLennan County, 79–80, 103–6, 108–10, 115, 125, 154–60, 170, 197–98, 278, 282, 285, 287, 337
McLennan, Neil, 106–09
Meridian Valley, 220, 223–24
Meridian: administrative center, Bosque County, 104, 106, 110, 112, 114, 117, 125, 127, 142–44, 159, 161, 169, 189, 220, 222–25, 241, 246, 265, 279, 284, 289, 293–94, 298, 301–2, 317–18, 320–26, 335–36
Methodists, 55, 87
Mickelson brothers, Lars and Lucas, 259, 285
migration chains, 13, 21–23, 74, 95
mills in Bosque County, 160, 220, 284, 298–300
Milwaukee, 62, 64, 103
Mission, La Salle County, 69, 70, 137–38, 200
Missouri Synod, 490, 504; and the slavery question, 302
Mormons and Norwegians converting at Fox River, 72–73
Munger, Robert S., 283
Muskego, Wisconsin, 17, 23

Nacogdoches, 20, 27–30, 32–34, 38–41, 45–56, 78, 125, 152, 161, 190, 341
Natchitoches, 20, 28, 39, 48–49, 55, 130
Native Americans: 106–8, 116–20, 141–44; way of life, 366–68; in the border areas, 366–78; Cherokee, 29–30, 33, 50, 184; Comanche, 21, 23–29, 39, 79, 81, 238–39, 241–42; Indian raids, 79, 108, 118–19, 141–42, 206–7, 290
Neil's Creek, Bosque, 104, 110–11, 119, 122, 127, 135, 247–48, 256, 280, 283, 302, 313, 319
Nelson, Albert Aldrich, 52, 53
Nelson, Aslak, 40, 99, 235, 237–38, 247, 268, 316
Nelson, Nils Jacob, 230, 268, 305, 323–24, 326
Nes (Næs)Jernverk (Ironworks), 14, 55, 179, 193, 249
New Braunfels, Texas, 21, 164, 288
New Orleans, 20–21, 27–28, 38
New York City, 36, 44–47, 67, 70, 80, 86, 90, 200.
Nineteenth Texas infantry, 192
Norboe: John, 81; Norboe family, 75–77
Norse picnics: 17th May, 4th of July, 319–21
Norse, 58, 87, 89, 109–14, 134, 139, 160–61, 200, 206, 211, 218, 235–36, 241, 246–8, 265, 268–69, 285–86, 289, 293–96, 298–300, 304, 306, 309–11, 313, 315–17, 319–22, 326–28, 335, 338
Norway Mills, 220, 257, 278–79, 293, 298–302, 315, 320, 326
Norwegian immigrants and higher education: *See* Olsen, Peter and Helene
Norwegian and also American, 327–28
Norwegian Synod, 176, 177, 308–9, 313, 317; the slavery question, 164–67, 175–76
Norwegians in the Confederate Army, 170–75, 178, 184, 186, 188, 190, 197, 211
Nystøl, Ole: captured by the Comanches, 238–42

Olsen, Peter and Helene and higher education, 326–27
Olsen Strand, Christian, Elverum, 123, 182
Olson, Jacob, 130, 139
Olson, Lars, letters from the Civil War, 202, 204, 215
Omenson, Omen and Eline, 138, 256, 278, 298
Ommundsen, Abraham (Brown), 82, 203
Ottawa, La Salle County, 23, 67, 69, 70–71
Our Savior's Lutheran Church, 58, 234, 304, 309, 315, 317–18, 323

Pathfinder or «Veiviseren», 15, 22–24, 27, 37, 45, 85
Peerson, Cleng, 17–19, 26, 35–38, 58–60, 79–81, 103, 107–10, 115–17; his death, 140–41; and Joseph Fellows, 60–62; migration chain, 19, 129–41; Peerson monument 58, 218, 313
Peters' Colony, Texas, 77–79
Picnics and pretty girls, 319–20
Pierson, Ole, 9, 109–10, 113–14, 118, 121, 126–28, 210, 315
Pierson, Peter, 9, 210, 294, 310, 315, 316, 322
Polish Texans and Silesians; 21, 171–72, 307, 329, 332–35
Poulson, Poul, 127–29, 146–48, 250–52, 263
Prairieville, Kaufman County, 14, 42, 167, 180, 184, 187, 190, 204, 328
Pre-emption act, 103–4
Presbyterians, 53, 55, 106, 176, 313, 319; and Cumberland 78, 175, prosopography, 10

Pulteney Estates 1, 60, 62, 65–66, 70

Quakers; in Stavanger, 1, 11, 59; in Farmington, New York 59–60; in Macedon, 59; in Palmyra 59; in Rochester and the Erie Canal region, 60
Quebec and cholera epidemics: 130, 133–34, 139, 254, 262, 265
Questad, Carl and Sedsel, 88–90, 97, 101, 111–13, 120–23, 130–32, 202–9, 212–13, 215, 219, 282–83, 313

Rea, Ole Knudsen Rye (O. K.), 289, 290–93, 304–6
Rea, T.K, 291–93, 305–6
Reiersen chain, Aust-Agder and Hedmark counties, 5, 84, 271
Reiersen family, 33
Reiersen, Johan R., 13–35, 42, 93–5
Reiersen, Ole, 14, 27
Restauration, 17, 60–61
revival meetings, 49, 151, 318–19
Ringness, Jens, 88–89, 98, 101, 109–17, 120, 122–3, 141, 146–48, 174, 249, 264–65, 269, 285–87, 310, 315, 324
Rochester, New York, 36, 36, 40, 47
Rock Church, 317–18
Rogstad, Berger and Anne, 115, 146–47, 235, 268, 313
Rohne, Evan and Marianne, Romedal, Hedmark County 258–59, 264, 274–75, 280–82, 284, 294, 317, 325
Romedal, Hedmark County, 86–87, 110, 114, 122, 139, 233–35, 255–59, 266–67, 274, 278–79, 281, 284, 302, 304–7, 326, 338
Rosdail, Ovee 133–34, 137

Røraas, O., blacksmith, Kristiansand, 33
Rynning, Ole, 23, 62, 165
Rystad, John K., 316–18

Sabine River, Texas, 37–39, 41, 56, 153
San Augustine, Texas, 20, 39, 190
Scrutchfield, Lowery H., 104, 120, 125–26, 312
Second Texas infantry, 193–94
Seventh Texas infantry, 182–83
Shelby County, Missouri, 71, 72, 75, 164
Sixth Texas cavalry, 182, 184,
Skimland brothers, 139, 173–74, 208–9, 211–12, 301–2
slave culture, 11, 154, 167, 181
slavery, 33, 54, 109, 148, 149, 152–69; emancipation of slaves, Texas, June 19th, 231, 233
St. Olaf Lutheran Church, 317
Stange, Hedmark County, 8, 114, 127–9, 198, 200–1, 254–5, 257, 283
Starr, James L., dr., Nacogdoches, 27–29, 51
Stephansen, Lauritz Christian, shipowner, Arendal, 84–86, 90,100–1
Sterne, Adolphus, 32, 49–53
Stiansen, Ole, 33
Swenson family, 131, 146–48, 166, 200, 211–12, 301–02, 310, 314–15, 320, 322
Swenson, Bersvend, 146, 301
Syversen and Johnson, 15, 47, 102

Texas secession from the Union, 154, 158, 161–65, 168–9, 171, 178
Texas Rangers, 105–6, 108, 119–20, 125, 207, 292
Texas Republic, 17, 29, 45, 50, 52, 79, 106, 313
Texas-Czech 154, 333–34
Third Texas cavalry, 178–82, 184–85
Thirtieth Texas cavalry, 155, 197, 199–200
Thirty-first Texas cavalry, 197, 200–2, 204
transplanted community, 298–328
Tovdal, Aust-Agder County, 36–37, 40, 84, 101, 184, 194
Trinity River, 24, 30, 37, 75, 77–78, 80, 92, 93, 123, 153–54
Twentieth Texas cavalry, 184, 192–93
Tyler, Texas, 56–57, 190–92, 194–96, 201, 206
Tysvær, Rogaland County, 1, 59, 124, 133

Upper settlement, Bosque County, 301–3
Upper South, 147- 50, 152, 154, 162, 179

Valley Mills: gristmill, 299; sawmill, 299
Van Zandt County, 40–44, 87, 91, 98, 154, 166–69, 178–80, 187, 191, 193–94, 196–97, 240, 246, 268, 295
Victoria, 99, 100–2, 110–11, 113, 123
Vinzent, Caroline and Charles, 32, 52, 54–56

Waco, 39, 79–80, 103–4, 106, 108–9, 125–26, 140, 147, 154–59, 197–200, 202, 250, 267, 278, 280, 293, 298–304, 321, 325–27, 337
Wærenskjold, Elise, 41–42, 43–44, 53, 91, 96–98, 140; and slavery, 166–68, 190, 196–97, 233, 235, 238, 318

Wærenskjold, Wilhelm, 41–42, 44, 167–70, 190, 195, 197
wagon train, 20, 22, 155, 160–61, 220, 247, 249, 295
weddings, 55, 74, 226, 256, 268–70, 279–80, 338. *See also* marriage
Ween, Ole L.: 109, 111, 118, 120, 130–31, 141, 148, 208, 211, 215, 250

Wilson Line, 262, 280
Wooster: 154, 161–62, 168, 170–71, 179, 188, 213–14, 216
Wold, Ole, 122, 139

Yankee land, 6, 67

OTHER BOOKS IN THE TARLETON STATE UNIVERSITY
SOUTHWESTERN STUDIES IN THE HUMANITIES SERIES

◊ ◊ ◊

In the Deep Heart's Core: Reflections on Life, Letters, and Texas
Craig E. Clifford

Cannibals and Condos: Texans and Texas along the Gulf Coast
Robert L. Maril

Rough and Rowdy Ways: The Life and Hard Times of Edward Anderson
Patrick Bennett

Larry McMurtry and the Victorian Novel
Roger W. Jones

Nueva Granada: Paul Horgan and the Southwest
Robert F. Gish

Beautiful Swift Fox: Erna Fergusson and the Modern Southwest
Robert F. Gish

Texas Women Writers: A Tradition of Their Own
Sylvia A. Grider and Lou H. Rodenberger

Riding the Wind and Other Tales
James Hoggard

State of Mind: Texas Literature and Culture
Tom Pilkington

Culture in the American Southwest: The Earth, the Sky, the People
Keith L. Bryant

Donald Barthelme: The Genesis of a Cool Sound
  Helen M. Barthelme

West of the American Dream: An Encounter with Texas
  Paul Christensen

San Antonio on Parade: Six Historic Festivals
  Judith B. Sobré

Let's Hear It: Stories by Texas Women Writers
  Sylvia A. Grider and Lou H. Rodenberger

Lone Star Chapters: The Story of Texas Literary Clubs
　Betty H. Wiesepape

Pastoral Vision of Cormac McCarthy
　Georg Guillemin

Exploding the Western: Myths of Empire on the Postmodern Frontier
　Sara L. Spurgeon

Undaunted: A Norwegian Woman in Frontier Texas
　Charles H. Russell

Lost Years of William S. Burroughs: Beats in South Texas
　Robert E. Johnson

Birth of a Texas Ghost Town Thurber, 1886–1933
　Mary J. Gentry and Lindsay T. Baker

Alexandre Hogue: An American Visionary—Paintings and Works on Paper
　Susie Kalil

Fritos® Pie: Stories, Recipes, and More
　Kaleta Doolin

Faded Glory: A Century of Forgotten Texas Military Sites, Then and Now
　Thomas E. Alexander and Dan K. Utley

Marfa Flights: Aerial Views of Big Bend Country
　Paul V. Chaplo

From the Frio to Del Rio: Travel Guide to the Western Hill Country and the Lower Pecos Canyonlands
　Mary S. Black

Still Turning: A History of Aermotor Windmills
　Christopher C. Gillis

Texas Calaboose and Other Forgotten Jails
　William E. Moore

Guide to the Historic Architecture of Glen Rose, Texas: Bypassed, Forgotten, and Preserved
　T. Lindsay Baker, Photographs by Paul V. Chaplo

Wind Energy Revolution: How the 1970s Energy Crisis Fostered Renewed Interest in Electric-Generating Technology
　Christopher C. Gillis Sr.